Many Degrees of Freedom in Particle Theory

NATO ADVANCED STUDY INSTITUTES SERIES

A series of edited volumes comprising multifaceted studies of contemporary scientific issues by some of the best scientific minds in the world, assembled in cooperation with NATO Scientific Affairs Division.

Series B: Physics

RECENT VOLUMES IN THIS SERIES

Volume 23 – Photon Correlation Spectroscopy and Velocimetry
edited by H. Z. Cummins and E. R. Pike

Volume 24 – Electrons in Finite and Infinite Structures
edited by P. Phariseau and L. Scheire

Volume 25 – Chemistry and Physics of One-Dimensional Metals
edited by Heimo J. Keller

Volume 26 – New Developments in Quantum Field Theory and Statistical Mechanics–Cargèse 1976
edited by Maurice Lévy and Pronob Mitter

Volume 27 – Topics in Theoretical and Experimental Gravitation Physics
edited by V. De Sabbata and J. Weber

Volume 28 – Material Characterization Using Ion Beams
edited by J. P. Thomas and A. Cachard

Volume 29 – Electron–Phonon Interactions and Phase Transitions
edited by Tormod Riste

Volume 30 – Many Degrees of Freedom in Field Theory
edited by L. Streit

Volume 31 – Many Degrees of Freedom in Particle Theory
edited by H. Satz

Volume 32 – Electron and Ion Spectroscopy of Solids
edited by L. Fiermans, J. Vennik, and W. Dekeyser

Volume 33 – Microscopic Structure and Dynamics of Liquids
edited by J. Dupuy and A. J. Dianoux

The series is published by an international board of publishers in conjunction with NATO Scientific Affairs Division

A	**Life Sciences**	Plenum Publishing Corporation
B	**Physics**	New York and London
C	**Mathematical and Physical Sciences**	D. Reidel Publishing Company Dordrecht and Boston
D	**Behavioral and Social Sciences**	Sijthoff International Publishing Company Leiden
E	**Applied Sciences**	Noordhoff International Publishing Leiden

Many Degrees of Freedom in Particle Theory

Edited by
H. Satz
University of Bielefeld, Federal Republic of Germany

PLENUM PRESS • NEW YORK AND LONDON
Published in cooperation with NATO Scientific Affairs Division

Library of Congress Cataloging in Publication Data

International Summer Institute on Theoretical Physics, 8th, University of Bielefeld, 1976.
Many degrees of freedom in particle theory.

(NATO advanced study institutes series: Series B, Physics; v. 31)
"Published in cooperation with NATO Scientific Affairs Division."
Includes index.
1. Degree of freedom–Congresses. 2. Particles (Nuclear physics)–Congresses. I. Satz, H. II. Title. III. Series.
QC793.3.D43I57 1976 539.7'21 78-27
ISBN 978-1-4684-2816-2 ISBN 978-1-4684-2814-8 (eBook)
DOI 10.1007/978-1-4684-2814-8

Proceedings of the 1976 International Summer Institute of Theoretical Physics he at the University of Bielefeld, Federal Republic of Germany, August 23–Septembld 4, 1976, published in two volumes, of which this is the second er

Softcover reprint of the hardcover 1st edition 1978
A Division of Plenum Publishing Corporation
227 West 17th Street, New York, N.Y. 10011

Preface

Volumes 30 and 31 of this series, dealing with "Many Degrees of Freedom," contain the proceedings of the 1976 International Summer Institute of Theoretical Physics, held at the University of Bielefeld from August 23 to September 4, 1976. This Institute was the eighth in a series of summer schools devoted to particle physics and organized by universities and research institutes in the Federal Republic of Germany.

Many degrees of freedom and collective phenomena play a critical role in the description and understanding of elementary particles. The lectures in this volume were intended to show how a combination of theoretical prejudices and experimental results can lead to the crystallization of models and theories. Topics ranged from quark, parton, and bag models to dual unitarization, from cluster pictures to hadron-nucleus collisions and to astrophysical implications.

The Institute took place at the Center for Interdisciplinary Research of the University of Bielefeld. On behalf of all participants, it is a pleasure to thank the officials and the administration of the Center for their cooperation and help before and during the Institute. Special thanks go to V.C. Fulland, M. Kämper, and A. Kottenkamp for their rapid and competent preparation of the manuscripts.

The Institute was sponsored by the NATO Advanced Study Institute Programme and supported by the Bundesminister für Wissenschaft und Forschung of the Land Nordrhein-Westfalen. Last, but certainly not least, the valuable help of I. Andric, V. Enss, F. Jegerlehner, B. Petersson, and P. Stichel in organizing the institute is gratefully acknowledged.

March, 1977 H. Satz

Contents

REGGEON FIELD THEORY WITH $\alpha(0) > 1$

M. Le Bellac

CERN

CH-1211 Geneva 23, Switzerland

I will discuss in these lectures the theory of the Pomeron when the intercept $\alpha(0)$ is larger than one. In the first two parts of the lectures I will give the motivations for such an investigation, as well as a short introduction to the Reggeon calculus. In the last part I will describe what I believe to be the solution when $\alpha(0) > 1$.

I. INTRODUCTION TO THE REGGEON CALCULUS

Let me begin by explaining briefly what a Reggeon is. Of course it is first a pole in the partial wave amplitude of the crossed (t)-channel, which leads to a $s^{\alpha(t)}$ behaviour of the scattering amplitude at high energy. However this is a very general statement, and one needs more dynamical information in order to go further.

Indeed there are many dynamical models which give in a particular range of s and t, and to a first approximation, a power law of the $s^{\alpha(t)}$ type. Examples of such dynamical models are the multiperipheral model, the dual model, the first order of the topological expansion. A general characteristic of all these approaches is that besides the single Regge pole exchange, which is the first order approximation, they also predict the presence of multiple Regge exchanges which leads to Regge cuts.

The same conclusion is reached by a more abstract analysis, based on t-channel unitarity.

A most important point is that all these approaches agree on the general properties of these Regge cuts, such as the position of the n-Reggeon branch point at t=0:

$$\mu_n(0) = n(\alpha(0)-1)+1 \tag{1}$$

or $\mu_n = n\mu$

(where I have defined μ as $\alpha(0)-1$), and the negative sign of the two-Reggeon cut (more generally the alternating sign of the n-Reggeon cut). Eq. (1) implies that the high-energy behaviour of the total cross-section will be:

$$\sigma_T \sim s^{\mu} + \sum_{n=1} f_n(\log s)\, s^{n\mu} \tag{2}$$

The extra logs dependence in (2) reflects the presence of cuts. In principle at least the functions f_n can be calculated in the framework of each of the dynamical models mentioned above.

Eq. (2) suggest a very simple situation; indeed, when $\mu\to 0$, the Regge pole contribution ($\sim s^{\mu}$) clearly dominates when $s\to\infty$, and one can simply ignore the existence of cuts, at least asymptotically. Unfortunately (or maybe fortunately, since it allows to keep some theorists busy) this situation does not seem to be realized in nature: $\mu<0$ implies decreasing cross-sections, and, as is wellknown, ISR experiments showed 3 years ago that, on the contrary, total cross-sections are rising.

We are thus led to consider the case $\mu\geq 0$, where we cannot avoid the calculation of f_n's and the analysis of the sum over all multiple reggeon exchanges.

The actual computation of the f_n's depends on the specific Reggeon model under consideration. However there is a general framework that allows to calculate the f_n's for large s and small t, which incorporates all known dynamical models and is in agreement (and in fact suggested by) t-channel unitarity. This framework is Gibrov Reggeon calculus.[(1)]

In order to give an intuitive justification to Gribov's formalism, let us start from the expression for 1 Pomeron exchange (Fig.1)

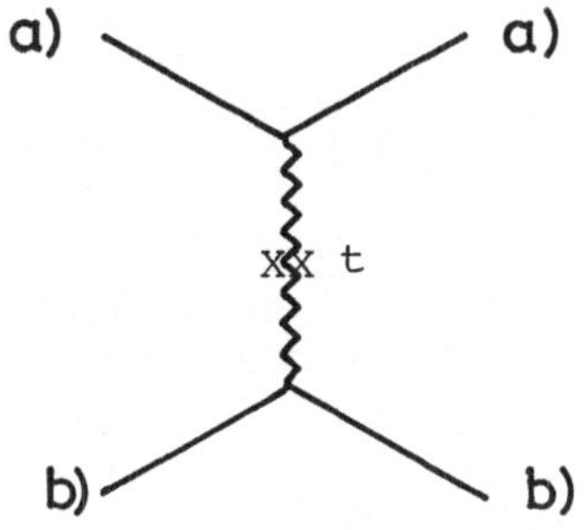

Figure 1

in the elastic scattering of particle (a) on particle (b) at high-energy. We use a linear approximation to the Regge trajectory $\alpha(t)$ close to $t=0$

$$\alpha(t) \doteqdot 1 + \mu + \alpha' t \qquad (3)$$

where α' is the Pomeron slope. Now, at high-energy, the momentum-transfer t is minus the square of a two-dimensional vector $\vec{k}$ which lies in a plane perpendicular to the collision axis: $t=-\vec{k}^2$. Finally it is useful to define the rapidity y by $y=\log s$, in order to rewrite the graph in Fig.(1) as:

$$\frac{1}{s} \operatorname{Im} T(s,t) = \beta_a(t)\, s^{\alpha(t)-1} \beta_b(t) =$$

$$\beta_a(\vec{k}^2) e^{\mu y - \alpha' \vec{k}^2 y}\, \beta_b(\vec{k}^2) \qquad (4)$$

Eq.(4) is interpreted in the following way: particle (a) acting as a Pomeron source, emits a Pomeron with an amplitude $\beta_a(\vec{k}^2)$. This Pomeron propagates with a propagator $e^{\mu y-\alpha'\vec{k}^2 y}$, and is finally absorbed by particle (b) with an amplitude $\beta_b(\vec{k}^2)$. It is then very appealing to introduce a Pomeron field $\psi(y,\vec{b})$, where $\vec{b}$ is a two-dimensional vector conjugate to $\vec{k}$: this is nothing else than the impact parameter. Notice that y plays the role of an imaginary time. In $(y,\vec{b})$ space the propagator $D(y,\vec{b})$ is obtained by a Fourier transform of the propagator in eq.(4):

$$D(y,\vec{b}) = \frac{1}{\alpha' y} e^{\mu y - \frac{\vec{b}^2}{4\alpha' y}} \theta(y) \tag{5}$$

Since this propagator is the solution of a diffusion equation (Schrödinger like equation with an imaginary time!)

$$\frac{\partial}{\partial y} D(y,\vec{b}) = \mu D(y,\vec{b}) + \alpha' \nabla_b^2 D(y,\vec{b}) \tag{6}$$

one can write at once the free part of the Lagrangian density $\mathcal{L}$ which will give rise to the propagator (4):

$$\mathcal{L}_o = \frac{1}{2} \bar{\psi} \overleftrightarrow{\partial}_y \psi - \alpha' (\vec{\nabla}_b \bar{\psi})(\vec{\nabla}_b \psi) + \mu \bar{\psi}\psi \tag{7}$$

The interaction will depend of course on the dynamical model; in general all powers of $\bar{\psi}$ and ψ are possible, i.e. the general interaction term will be $\lambda_{mn} \bar{\psi}^m \psi^n$. Fortunately there exist good arguments to be discussed later on to limit ourselves to the simplest interaction:

$$\mathcal{L}_I = -i\lambda (\bar{\psi}^2\psi + \bar{\psi}\psi^2) = \mathcal{L} - \mathcal{L}_o \tag{8}$$

Before proceeding further, let me make a couple of remarks about (7) and (8). First, since the diffusion equation (6) is of first order in time, we need to introduce two fields ψ and $\bar{\psi}$, as in ordinary non-relativistic quantum mechanics. Then one must notice the i in eq. (8), which will lead to a non hermitian hamiltonian: this i is imposed by the negative sign of the two-Pomeron cut. Finally, when we quantize the theory by imposing the canonical commutation rules:

$$[\psi(y,\vec{b}), \bar{\psi}(y,\vec{b}')] = \delta^{(2)}(\vec{b}-\vec{b}') \tag{9}$$

we have to remark that, due to the non-hermiticity of $\mathcal{L}_I$, ψ and $\bar{\psi}$ will not be in general hermitian conjugates.

The most general contribution to the high-energy scattering amplitude is represented graphically in Fig.2:

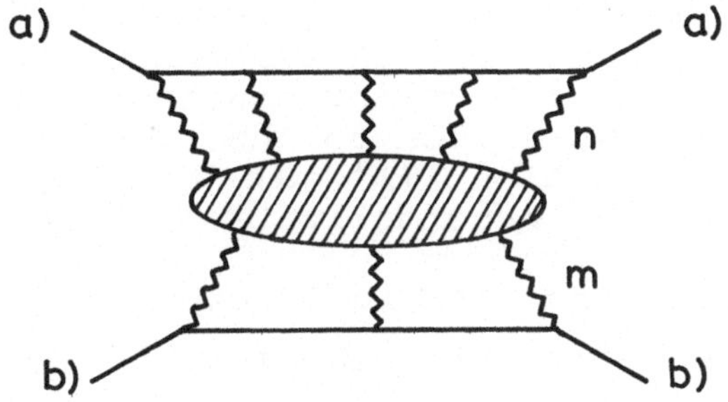

Figure 2

It is interpreted in the following way: particle (a) emits at "time" y_1 (n) Pomerons which interact among themselves, and (m) Pomerons are absorbed at "time" y_2 by particle (b); the total rapidity is $y = \log s = y_1 - y_2$. The fundamental quantities to be computed are then the Green's functions $G_{n,m}$. They can be obtained from the Lagrangian of eqs. (7),(8) by Feynman diagram techniques.

II. THE CRITICAL POMERON

As discussed before, for $\mu \geq 0$, we cannot rely on the perturbative expansion of $\mathcal{L}$ as given by eq. (2), at least if we are interested in $y \to \infty$. For y finite, the perturbative expansion is in fact an expansion in powers of $\lambda^2 y/\alpha'$, which may be convergent if y is small enough. But for $y \to \infty$, we need to sum the formal series of eq.(2), which implies finding a non-perturbative solution to the field theoretical problem. This turns out ot be possible in the asymptotic limit.

We shall first discuss briefly the $\mu=0$ case,* where the position of the pole and all branch cuts coincide at t=0; this is the typical situation of a massless theory. As is well known, the disadvantage of having to sum all terms in eq.(2) is compensated by the advantage of having lost the "mass" scale μ. This is a situation in which the renormalization group methods work. Indeed the absence of a scale allows to rescale arbitrarily all

---

* $1+\mu$ should be interpreted here as the renormalized Pomeron intercept, the bare intercept is in this case slightly larger than one.

quantities so that the asymptotic properties are given in terms of the fixed point of the theory. We can as usual define a running coupling constant λ and for $y \to \infty$ we will be always driven to the value λ^* given by the fixed point condition $\beta(\lambda^*)=0$, where $\beta(\lambda)$ is the usual function of the renormalization group.

This approach has been fully explored in the years 1973-74.[2] The asymptotic behaviour of the amplitude turns out to be:

$$\text{Im}\, T(s,t) \propto s(\log s)^{\eta}\, F(t(\log s)^{\nu}) \qquad (10)$$

where η and ν are critical exponents, which, together with λ^*, can be computed within the standard approximate methods. For example the ε-expansion gives to first order:

$$\eta=\frac{1}{6} \qquad \nu=\frac{13}{12} \qquad (11)$$

At first sight the result (10) is welcome, since it predicts rising cross-sections. However problems arise at two levels: phenomenological and theoretical.

First, from phenomenological point of view, it is very difficult to understand the rise of total cross-sections at FNAL-ISR energies. Indeed, due to the smallness of the triple-Pomeron coupling λ, a perturbative expansion in powers of $\lambda^2 y/\alpha'$ is reliable at such energies, since $\lambda^2/\alpha' \simeq 2.10^{-2}$ and $y \leq 8$, while the asymptotic solution (10) should be relevant at much higher energies. Now it has been shown that the perturbative expansion is unable to reproduce the experimental rise of the rotal cross-sections if $\mu=0$, and that one needs $\mu \doteq .1$ in order to fit the data.

The second objection is of a more theoretical nature. There is no property, as gauge invariance in QED, which allows to fix μ at a zero value independently on the other parameters of the theory. In other words, the "mass" μ is also renormalized and in order to have it zero all the parameters of the theory should be choosen in a specific, and arbitrary, correlated way. This statement, in terms or renormalized and therefore measurable quantities, implies that μ as well as λ are expressed in terms of running function . If asymptotically $\lambda \to \lambda^*$ and we want $\mu \to \mu^*=0$ then at our present energies the value of μ is not zero and should depend on the actual value of λ through the solution of the whole theory. This implies that in order to have a critical pomeron, the actual intercept should have a specific value, however only in an approximate way to be slightly larger than one. Therefore, both from a theoretical viewpoint and from the necessity of finding rising cross sections, it is clearly of interest to study the $\mu>0$ case.

III. NON-PERTURBATIVE SOLUTION FOR $\mu>0$

Since perturbation theory cannot be used when $Y\to\infty$ and $\mu>0$, we are led to look for a non perturbative solution of the Reggeon field theory. Let us note first that we shall disregard higher order Pomeron interactions such as $\lambda_{22}\bar{\psi}^2\psi^2, \lambda_3{}^{\times}\bar{\psi}^3\psi$ etc..., and keep only the 3 Pomeron coupling as in eq. (8). In the neighbourhood of $\mu=\mu_c$, this is perfectly justified: the transformations of the renormalization group drive to zero higher order Pomeron couplings $\lambda_{m,n}$ and one can start as well with only a triple Pomeron interaction. For $\mu>\mu_c$ higher order Pomeron interactions will change the quantitative behaviour of the scattering amplitude. However they will not modify the most important qualitative features, and in particular the nature of the phase transition at the critical point, which is indeed our most important result.

It is important to remark that at finite Y, there is always a region in $\vec{b}$ space in which perturbation theory is convergent. This comes from the fact that the range in impact parameter of a n-Pomeron cut is only 1/n that of the pole (exactly as in nucleon-nucleon scattering the range of 1-pion exchange is larger than that of multipion exchange, so that the nucleon-nucleon potential is determined at large distances by 1-pion exchange only).

Hence, for large impact parameters (in fact for $|\vec{b}|\gg\sqrt{4\alpha'\mu}\ y$, as seen from eq.(5)) the pomeron propagator does not blow up for $y\to\infty$, multipomeron exchange is suppressed and perturbation theory is reliable. This remark is extremely important, as it implies that there must be a region in $\vec{b}$-space in which the non-perturbative solution matches with the perturbative one. In fact, because of this, all attempts to shift the Pomeron field by a constant must necessarily fail.

Our method is based on the following observation: since the "time" evolution is given by e^{-yH}, and since we are interested in the limit $y\to\infty$, the important contribution will come from the lowest lying states of the hamiltonian H. The hamiltonian is immediately obtained from the lagrangian density (7),(8):

$$H=\int d^2b\left[-\mu\bar{\psi}\psi+\alpha'(\vec{\nabla}\bar{\psi})(\vec{\nabla}\psi)+i\lambda\bar{\psi}(\bar{\psi}+\psi)\psi\right] \tag{12}$$

As already remarked, the i in the interaction term of (12) makes H non hermitian, and $\bar{\psi}$ cannot be indentified with ψ^+. There exists however an operator L which maps $\bar{\psi}$ into ψ^+ and H into H^+:

$$L\psi^+(y)L^{-1} = -\bar{\psi}(-y) \qquad LHL^{-1} = H^+ \tag{13}$$

Physically this transformation corresponds to the exchange of the projectile and the target. It is not a real symmetry, since $H \neq H^+$, but it seems to play a role analogous to the symmetry transformation $\phi\to-\phi$ in the standard $\lambda\phi^4$ theory.

We have found convenient to put the field theory on a lattice. This is harmless at the critical point, since we are then studying an infrared phenomenon. Indeed we can exactly repeat the same discussion which allowed us to neglect the higher Pomeron interactions. If we call a the lattice spacing, H transforms into:

$$H = \sum_j \left(-\mu\bar{\psi}_j\psi_j + \frac{i\lambda}{a}\bar{\psi}_j(\bar{\psi}_j+\psi_j)\psi_j)\right)-\frac{\alpha'}{a^2}\sum_{(j\ell)}\bar{\psi}_j\psi_\ell \tag{14}$$

where $\sum_{(j\ell)}$ is a sum over nearest neighbours and ψ_j is the field operator at site j, which satisfy the commutation relations.

$$[\psi_j,\bar{\psi}_\ell] = i\delta_{j\ell}$$

The first step is to solve for H at each lattice site, by neglecting the interlattice interactions proportional to α' in eq. (14). This is equivalent to solving the original field theory in zero transverse dimension. The intersite interaction will be introduced later on, when the level structure at each site has been understood. Let us notice that the same method has been applied to the $\lambda\phi^4$ theory and is able to give the phase transition associated with spontaneous symmetry breaking. It will be instructive to show the analogies with the $\lambda\phi^4$ case, and also the basic differences: we shall show that there does exist a phenomenon analogous to a second order phase transition, associated with the appearance of a non zero order parameter above the critical point. However, due to the non-hermiticity of the hamiltonian which prevents L to be a true symmetry (see eq. (13)), the nature of the phase transition will be completely new.

Let us now come back to the single site dynamics, which can be solved thanks to its equivalence with a Schrödinger equation (3). We shall only describe the level structure when $\mu/\lambda >> 1$. Besides the vacuum χ_o, which is always eigenstate of H with zero energy, we find a second state χ_1, almost degenerate with the vacuum, and whose energy is +

$$\Delta = \frac{\mu^2}{\lambda\sqrt{2\pi}} e^{-\frac{1}{2}\frac{\mu^2}{\lambda^2}} \tag{15}$$

Then one finds pairs of closely lying levels at $\mu, 2\mu$, 3μ.... The spectrum is summarized in Fig.3
Since $\Delta << \mu$ in the limit $\mu/\lambda >> 1$, our approximation to the single site dynamics will be to keep only the two lowest lying levels χ_o and χ_1.

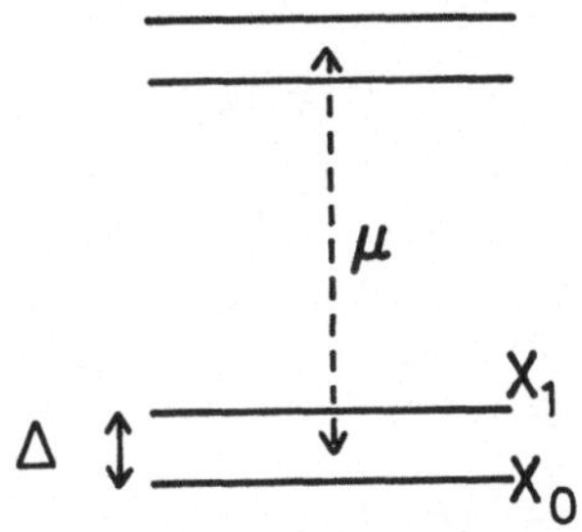

Figure 3

Now a simple approximate representation of χ_1 is as follows:

$$\chi_1 = \chi_o - e^{\frac{\mu}{i\lambda}\bar{\psi}} \chi_o \tag{16}$$

+ We set a=1 in all formulas, unless the contrary is explicitly mentioned.

where the second term in (16) is nothing but a coherent state. However one has to be careful in computing the matrix elements of say, field operators, and norms. In fact the operator L in (13) acts as a metric operator, mapping H into H^+. Since this mapping is also realized by changing λ into $-\lambda$ we can define a conjugate state $\bar{\chi}_1$ by: +

$$\bar{\chi}_1 = \chi_1^+ L = \chi_o^+ - \chi_o^+ e^{\frac{\mu}{i\lambda}} \tag{17}$$

From eq. (16) χ_1 has a negative norm with the metric L:

$$(\bar{\chi}_1, \chi_1) = -1 + O(e^{-\mu^2/\lambda^2})$$

If we wish to represent ψ and $\bar{\psi}$ in the (χ_o, χ_1) basis by 2x2 matrices, the metric operator L is simply represented by $M=\sigma_3$. We shall have

$$\chi_o = \begin{pmatrix} 1 \\ o \end{pmatrix} \quad \chi_1 = \begin{pmatrix} o \\ 1 \end{pmatrix} \quad (\bar{\chi}_i, O\chi_j) = (\chi_i, MO\chi_j) \tag{18}$$

where O is any operator ($O \equiv \mathbb{1}$ for the norm). Then ψ and $\bar{\psi}$ have the following representation:

$$\psi = \frac{\mu}{i\lambda}\begin{pmatrix} o & -1 \\ o & 1 \end{pmatrix} \qquad \bar{\psi} = \frac{\mu}{i\lambda}\begin{pmatrix} o & o \\ 1 & 1 \end{pmatrix} \tag{19}$$

as can be easily checked by using the representations (16) and (17) of χ_1 (neglecting terms $\sim e^{-\mu^2/\lambda^2}$)

We have recognized the appearance of a new scale Δ, which can be interpreted as a kind of tunnel effect, between the two regions corresponding to the fixed points of the classical limit: $(\psi=0, \bar{\psi}=\frac{\mu}{i\lambda})$ and $(\psi=\frac{\mu}{i\lambda}, \bar{\psi} = 0)$. A similar pattern occurs in the $\lambda\phi^4$ theory in zero space dimension, where the spectrum looks again like that of Fig.3. There is however a basic difference on the nature of the states and therefore on the matrix elements of the field operators. None of the two lowest lying states

+ When looking for the eigenstates of H or a single site, it is always possible to identify $\bar{\psi}$ and ψ^+.

is the vacuum; these two states are in fact symmetric and antisymmetric combinations of wave functions localized in the two wells of Fig.4.

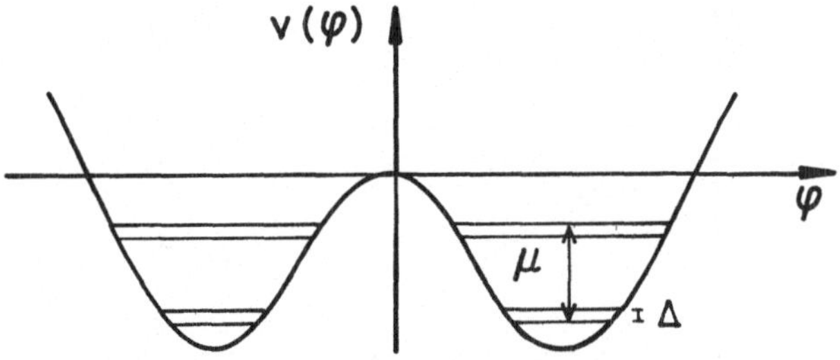

Figure 4

The energy separation Δ is given by the tunnel effect ($\Delta \alpha\, e^{-\mu^2/\lambda^2}$), and the matrix elements of the field have the usual hermiticity properties.

Suppose we now try to introduce intersite interactions in this $\lambda\phi^4$ theory (4). The gradient term, which is represented by $(\phi_{i-1}-\phi_i)^2$ will favour configurations in which all the one site wave-functions are in the same well. This is exactly opposite to the tunneling, which tends to favour the symmetric combination of the two wells. If we call the symmetric combination spin up and the antisymmetric combination spin down, the hamiltonian (restricted to the lowest starts at each site) will take the form of an Ising hamiltonian in a transverse field:

$$H = \frac{\Delta}{2} \sum_j (1-\sigma_3^j) + \frac{1}{2} J \sum_{(j\ell)} \sigma_1^j \sigma_1^\ell \qquad (20)$$

The parameter J can be of course computed from the original theory. This hamiltonian has been extensively studied in solid state physics (5), and its properties are well-known. At J fixed there is a phase transition at a critical value of Δ, $\Delta=\Delta_c \sim J$. For $\Delta>\Delta_c$ we find a disordered phase, with a single ground state. On the contrary for $\Delta<\Delta_c$ the ground state is doubly degenerate, this situation corresponding clearly to a spontaneous symmety breakdown, as expected. There is also a continuum of excited states, separated from the ground state (s) by a finite gap, except at $\Delta=\Delta_c$ (Fig.5)

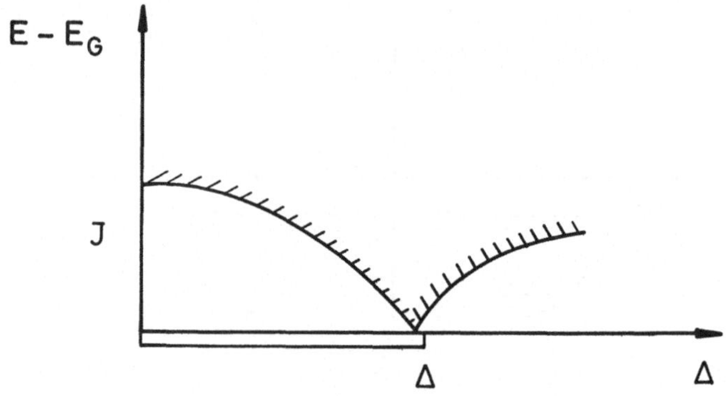

Figure 5

When $\Delta<\Delta_c$, the field operators have non zero expectation value. In other words:

$$<\sigma_1> = \begin{pmatrix} c & o \\ o & -c \end{pmatrix}$$

where c is the order parameter, which vanishes as $(\Delta_c-\Delta)^\beta$ when $\Delta\to\Delta_c$, β being a critical exponent. When the number of sites is infinite, the two ground states do not communicate with each other: the non diagonal matrix elements of field operators tend exponentially to zero.

Let us now come back to Reggeon field theory. Using the 2 x 2 representation of the field operators (19), we can rewrite H as given by (14) in the form:

$$H = \frac{1}{2}\Delta\sum_i (1-\sigma_3^i)+\frac{1}{2}\sum_{(ij)} J_{ij}\left(\frac{1+\vec{\sigma}_i\cdot\vec{n}_+}{2}\right)\left(\frac{1+\vec{\sigma}_j\cdot\vec{n}_-}{2}\right) \quad (21)$$

where $J = \sum_{(ij)} J_{ij} = \alpha'\mu^2/\lambda^2$ and

$$\vec{n}_\pm = (\pm 1,-i,-1) \quad (21')$$

The hamiltonian of eq. (21) would be the Ising one of eq. (20) if $\vec{n}_\pm = (\pm 1,0,0)$. We see that these vectors differ from those of eq. (21') by a null vector, and both satisfy

$$\vec{n}_+^2 = \vec{n}_-^2 = 1 \quad \vec{n}_+\cdot\vec{n}_- = -1$$

$\vec{n}_\pm$ of (21') are complex realization of these equalities and this is, again, a consequence of the non hermiticity of Reggeon field theory.

We can try to use for (21) the approximation methods which have been shown to work in the Ising case. For example the basic features were obtained in ref. (6) by using the mean field approximation, or equivalently variational methods. Alternatively one can use an adaptation of the block spin method (at least in 1-transverse dimension) [(7)]. Both methods are in excellent agreement, as far as the qualitative features are concerned. Of course we know that mean field theory cannot be trusted quantitatively close to the critical point, and renormalization group methods must be used for the actual computation of critical exponents.

Let us summarize the results: the spectrum looks again like that of Fig. 5. There is a phase transition and for $\Delta>\Delta_c$ we have the 1-ground state (vacuum) situation. From the relation between (Δ,J) and (λ,μ,α') we see that $\Delta>\Delta_c$ corresponds to $\mu<\mu_c$. In this region perturbation theory is applicable. For $\Delta<\Delta_c$ we have two degenerate ground states and a continuum . But contrarily to the $\lambda\phi^4$case, one of these ground states is the perturbative vacuum, Φ_o:

$$\Phi_o = \prod_i \chi_o^i \tag{22}$$

which is annihilated by the field operator ψ. The other state, ψ_o, is negative norm state. In the limit $\Delta\to 0$ one checks easily that:

$$\psi_o = \Phi_o - \prod_i \chi_1^i \tag{23}$$

The metric operator is evidently $M = \prod_i \sigma_3^i$. The basic fact is however that ψ_o and Φ_o are connected by field operators:

$$(\bar{\psi}_o, \bar{\psi}\Phi_o) = (\bar{\Phi}_o, \psi\psi_o) = \frac{-\mu}{i\lambda} c \qquad (24)$$

where c plays the role of an order parameter: it vanishes as $(\Delta-\Delta_c)^{\beta}$ when $\Delta \to \Delta_c$, and tends to one when $\Delta \to 0$.

We are now in a position to compute the asymptotic behaviour of Green functions, and then the S-matrix if we specify the coupling to external particles.

Let us call $f(\vec{b})$ and $g(\vec{B}-\vec{b})$ the impact parameter structure of the sources, namely the Fourier transforms of the external couplings $\beta_a(\vec{k}^2)$ and $\beta_b(\vec{k}^2)$ of eq. (4). As explained before, these sources act at "time" y_2 and y_1 $(Y=y_2-y_1)$ respectively. We write, in the lattice formulation, and exhibiting explicitly the a-dependence:

$$S(Y,\vec{B}) = (\bar{\Phi}_o, e^{-ia\Sigma f_j \psi_j} e^{-YH} e^{-ia\Sigma g_j \bar{\psi}_j} \Phi_o) \qquad (25)$$

Introducing a complete set of intermediate states, and letting $Y \to \infty$ at fixed $\vec{B}$ we find in the continuum limit $(a \to 0)$:

$$S \underset{\{\substack{Y\to\infty \\ \vec{B}\text{ fixed}}}{=} 1 - \left(1-e^{-\frac{c\mu}{\lambda}\int f(\vec{b})d^2b}\right)\left(1-e^{-\frac{c\mu}{\lambda}\int g(\vec{B}-\vec{b})}\right) + O(e^{-\omega_o Y}) \qquad (26)$$

In eq. (26) the 1 comes from Φ_o as an intermediate state, the second term from ψ_o, and finally the term $e^{-\omega_o Y}$ from the states of the continuum, where ω_o represents the gap between the ground state and the first excited state of the continuum. In the mean field approximation:

$$\omega_o = 2Jc = 2(J-\Delta)$$

ω_o vanishes at $J=\Delta$, since in the mean field approximation $\Delta_c=J$. One also sees that the corresponding critical exponent is one.

There are a few comments to be made on the result (26). First it holds only in the limit indicated: $\vec{B}$ fixed, $Y\to\infty$. Later on we shall show that the region of validity is an expanding disk of radius $R\sim Y$, so that the cross-section will increase like $Y^2 = (\log s)^2$.

Secondly we have assumed in (25) that the coupling to the external particles is of the eikonal type. This is by no means essential and is useful only to get compact formulae. It is clear that we can obtain from (26) by derivation with respect to f and g, the Green functions $G_{n,m}$ of the theory. We can then couple them to the external particles as we like. For example the Green function $G_{1,1}(Y,\vec{B})$ is:

$$G_{1,1}(Y,\vec{B}) \underset{\substack{Y\to\infty \\ \vec{B}\text{ fixed}}}{=} c^2\left(\frac{\mu}{\lambda}\right)^2 + O(e^{-\omega_o Y}) \tag{27}$$

Finally, since we get a cross-section behaving like $(\log s)^2$, it might look superficially that our result is similar to that of Cheng.Wu [(8)], obtained by simple eikonalization of a Pomeron with intercept larger than one without Pomeron interactions or to that of Ter Martirosyan[(9)] in which the eikonalization is done at the level of pomeron diagrams thanks to multipomeron interactions. Our approach is different. From (27) we see for instance that the (1,1) Green function satisfies by itself the Froissart bound, thanks to Pomeron interactions, while in Cheng and Wu $G_{(1,1)}$ is of course simply the bare Pomeron, which is power behaved. We also emphasize that the disk we have obtained is factorizable and grey in general, contrarily to the black disk of refs.(8) and (9). The opacity of the disk is in fact calculable, at least in principle, from the parameter of the origianl Gribov lagrangian, and it does not involve the external couplings.

We shall conclude these lectures by examining in some detail the mechanism for the vacuum instability in Reggeon field theory. This will allow us to understand the mechanism which is at the origin of the disk picture.

We have seen that on both sides of the phase transition, the perturbative vacuum Φ_0 is always one of the lowest lying states. This is not what happens in $\lambda\phi^4$ theory. There the vacuum is unstable at the critical point, due to the fact that there are other states which lie below it, and can be chosen as the new ground states of the theory.

The actual mechanism for the dynamical instability has been found in ref. (10), and we summarize the results in a simplified language.

By applying to the vacuum a local operator $\bar{\psi}_j$ at time y=0, we build a state in which the quasi-degenerate state χ_1^j is excited locally. By studying the time evolution of this state, it has been recognized that the excitation of the other level is a disturbance of Φ_0, which propagates with a finite velocity on the lattice, and leads to the disk picture. At y=0, the disturbance has filled out the complete impact parameter space, giving rise to the "second vacuum" ψ_0, and we recover at that point the expression (26) of the S-matrix. The vacuum instability has thus a dynamical character, and should not be understood as a redefinition of the vacuum, as in $\lambda\phi^4$. In other words, one does not have to change the vacuum, due to the fact that there are lower lying levels (static instability), but one has a situation in which the application of a field operator on Φ_0 destroys its unstable equilibrium, and the effect propagates in $\vec{b}$-space (dynamical instability).

One sees clearly that, at finite Y, there is a region in $\vec{b}$-space where the vacuum is not influenced. This region is a disk of radius proportional to Y. Inside this disk the expression (26) of the S-matrix is valid, and outside it $S\to 1$. Hence we obtain a cross-section rising like Y^2.

This picture looks very appealing, but there remains a number of points which should be better understood. The most important one is that the disk found in ref.(10) expands with a velocity proportional to the interlattice distance a, and not as $\sim \sqrt{\alpha'\mu}$ as expected from a classical argument. It appears that this result is due to the approximation of keeping only the two lowest states at each lattice site. "Phonon" like excitations, with a dispersion $\sim\alpha'\vec{k}^2$ can only appear if one keeps excited levels with energy $\mu, 2\mu$ etc..., while the excitations which build up the disk in ref. (10) are of the spin-wave kind, with a dispersion law $\sim a\vec{k}^2$. The "phonon" degrees of freedom should not affect the structure of the phase transition, but they may be important away from the critical point.

IV. CONCLUSION

To summarize the previous discussion, we believe that we have well understood the basic nature of the phase transition in Reggeon field theory, although there are still some open problems. We have seen that one must not make a shift of the vacuum, which would be incompatible with perturbation theory at large b and finite Y. On the contrary our mechanism for the phase transition allows matching with perturbation theory: this matching occurs around the boundary of an expanding disk. Outside the disk perturbation theory is valid, inside it the Green functions and S-matrix are given by expression like (26) and (27), which involve the "second vacuum" ψ_0. At finite Y, all Green functions are finite and satisfy the cluster decomposition property. We then believe that our results satisfy all the requirements which one would have expected a priori.

ACKNOWLEDGEMENTS

I would like to thank Daniele Amati and Marcello Ciafaloni for their collaboration in the preparation of these lecture notes.

REFERENCES

1) V.N. Gribov JETP (Sov. Phys.) 26 (1968) 414
2) A.A. Migdal, A.M. Polyakov and K.A. Ter-Martirosyan Ph. Lett. 48B (1974) 239 and JETP 67 84 (1974)
H.D.I. Abarbanel and J.B. Bronzan Ph. Rev. D9 (1974) 2397
3) V. Alessandrini, D. Amati and R. Jengo-N.Phys. B108 (1976) 425
R. Jengo N. Phys. B108 (1976) 447
J. Bronzan, J. Shapiro and R. Sugar, Santa Barbara preprint (1976)
4) For a treatement of ϕ^4 theory analogous to our approach cf. S.D. Drell, M. Weinstein and S. Yankielowicz, SLAC-PUB-1719, Phys. Rev. D in press

5) cf. for instance P. Pfeuty and R. Elliott J. Phys. C: Solid State Physics 4, 2370 (1975)
6) D. Amati, M. Ciafaloni, M. LeBellac and G.Marchesini, CERN-TH 2152 (1976). N. Phys. B in press
7) J. Cardy - SLAC-PUB 1784 (1976)
8) H. Cheng and T.T. Wu Phys. Rev. Letters 24 (1970) 1456
H. Cheng, J.K. Walker and T.T. Wu Ph. Lett. 44 B (1973) 97
9) M.S. Dubovikov and K.T. Ter Martirosyan: preprint ITEP (37) 1976
10) D. Amati, M. Ciafaloni, G. Marchesini and G. Parisi, CERN Th. 2185 (1976), Nucl. Phys. in press.

SOME REMARKS ON IMPLEMENTING HADRONS AS EXTENDED OBJECTS WITH APPLICATIONS TO DUAL MODELS*

L.C. Biedenharn**† and H. van Dam††

†Institut für Theoretische Physik der
Johann Wolfgang Goethe Universität,
Frankfurt/Main, Germany (B.R.D.)

††Physics Department, University of North
Carolina, Chapel Hill, North Carolina 27514
U.S.A.

SUMMARY

The Poincaré covariant quantal theory of a Regge trajectory (realized in Minkowski space) is reviewed and the classical (relativistic) limit for this structure is developed. It is demonstrated that this classical limit motion corresponds identically to the leading trajectory of the classical relativistic motion given by the Nambu-Goto-Nielsen action.

§1 INTRODUCTION

Dual resonance (1,2,3) models have been the object of some remarkable and brilliant theoretical inventions, which have fitted together, into a coherent whole, important concepts such as duality, Regge structure, the bootstrap hypothesis, factorization of multiperipheral reactions, and asymptotic mass distributions -- to mention but a few. The existence of specific detailed models -- obtained from such diverse considerations as quantized relativistic string models, or generalized projective invariant amplitudes -- has demonstrated the internal consistency of ideas previously thought to be contradictory (4). Current investigations (5) have indicated that spinor dual models possess local supersymmetry and have a remarkable connection to supergravity.

Yet there are serious foundational difficulties in all these models. The incorporation of spin as well as colour and flavour variables is quite ad hoc. But the real problem lies elsewhere, and to avoid misunderstanding let us focus attention entirely on the standard spinless (scalar) dual model (DRM). For this model the necessary existence of a tachyon ground state, and the internal consistency (for the quantized versions) only in a space-time of 26 dimensions is bizarre, and surreal even by modern standards. Dual models, however, are so tightly constructed that -- it is believed -- the elimination of such flaws can be achieved only internally, say through self-interactions or renormalization (6). This could very well be true.

It is our contention, however, that the difficulties of the DRM seem to stem from a different source: *the use of space-time four-vectors as the elementary constituents for model building.* To validate this remark, note that:

(a) The negative metric (time) component forces gauge constraints on the theory;

(b) The first excited state of the system is necessarily vectorial, necessarily has only d-2 physical components, which implies that this state has mass zero;

(c) Hence the ground state is necessarily a tachyon;

(d) The elimination of all negative metric states requires the Virasoro constraints;

(e) It is the consistent incorporation (quantum mechanically) of these constraint operators that forces dimension 26, (and also implies the tachyon again, i.e., intercept unity).

It is therefore an obvious question to ask if one might -- somehow -- avoid the use of four-vectors in model building. Could one not use instead a spinorial structure?

We will demonstrate, in §2, that there does exist in fact an acceptable spinorial structure, which we shall denote as 'the quantum theory of a single Regge trajectory . This structure is formulated explicitly, quantum mechanically, and Poincaré covariantly, in Minkowski space (7,8).

To illustrate the connection with the dual resonance model we will extend (in §3) some recent work of L.P. Staunton (9), and demonstrate that the classical limit of the spinorial model of §2 is precisely the classical relativistic string model moving in a mode corresponding to the leading trajectory (§4). We shall prove, in other words, that the quantal theory of a single Regge trajectory (of either integral or half-integral spin) has for its classical limit the leading trajectory of the Nambu-Goto-Nielsen classical relativistic string. The extension of this result to higher modes is sketched.

We discuss in the concluding section (§5) some implications of these results.

§2 QUANTUM THEORY OF A SINGLE REGGE TRAJECTORY

Our discussion of this topic will be rather brief, and synoptic, since most of the work is published (7,8). The emphasis will be on the concepts underlying the construction, since our purpose is to prepare the way for the specific applications in §3.

The starting point is empirical: hadrons (of given internal quantum numbers) appear to lie on approximately linear Regge trajectories having a universal slope, α'. It is reasonable to idealize this situation, much as is done in dual resonance models, and *declare a hadron to be a trajectory* with the various discrete (zero width) states of a given trajectory being excitations of an extended, internal, structure.

Accordingly a hadron belongs to an infinitely reducible, time-like, representation of the Poincaré group. This viewpoint is not new, and is familiar, for example, in the infinite component wave function approach (10,11). (The more abstract, group theoretic, approach was strongly discouraged by various theorems which were misinterpreted as ruling out discrete spectra.)

Our approach has, algebraically, much in common with the Majorana equation, and is based on a realization of the generators of the SO(3,2) group over two harmonic oscillators. It is necessary to be explicit about this structure for clarity in the applications to follow.

The ten generators of the SO(3,2) algebra are a four-vector: V_μ and an antisymmetric tensor $S_{\mu\nu}$ consisting of a rotation operator S_{ij} and a boost operator S_{i0}.

The commutation relations are:

$$[V_\mu, V_\nu] = i\, S_{\mu\nu}, \tag{1}$$

$$[V_{\mu\nu}, S_{\alpha\beta}] = i\, (g_{\mu\beta} V_\alpha - g_{\mu\alpha} V_\beta) \quad , \tag{2}$$

$$[S_{\mu\nu}, S_{\alpha\beta}] = i(g_{\mu\alpha} S_{\nu\beta} - g_{\mu\beta} S_{\nu\alpha} + g_{\nu\beta} S_{\mu\alpha} - g_{\nu\alpha} S_{\mu\beta}) . \tag{3}$$

Defining the two harmonic oscillators by the coordinates ξ_1 and ξ_2 with momenta π_1 and π_2, where $[\xi_i, \pi_j] = i\delta_{ij}$, one has the realization:

$$V_0 = (1/4)(\xi_1^2 + \xi_2^2 + \pi_1^2 + \pi_2^2), \tag{4a}$$

Four-vector:

$$V_1 = (1/2)(-\xi_1 \pi_1 + \xi_2 \pi_2), \tag{4b}$$

$$V_2 = (1/2)(\xi_1 \pi_2 + \xi_2 \pi_1), \tag{4c}$$

$$V_3 = (1/4)(\xi_1^2 + \xi_2^2 - \pi_1^2 - \pi_2^2). \tag{4d}$$

Boost:

$$S_{10} = (1/4)(\xi_1^2 - \pi_1^2 - \xi_2^2 + \pi_2^2), \tag{5a}$$

$$S_{20} = (1/2)(\pi_1 \pi_2 - \xi_1 \xi_2), \tag{5b}$$

$$S_{30} = (1/2)(\xi_1 \pi_1 + \pi_2 \xi_2) \quad . \tag{5c}$$

Rotation Operator: $S_{12}=(1/2)(\xi_1\pi_2-\xi_2\pi_1)$, (6a)

$$S_{31}=(1/4)(\xi_2^2+\pi_2^2-\xi_1^2-\pi_1^2), \quad (6b)$$

$$S_{23}=(1/2)(\xi_1\xi_2+\pi_1\pi_2) . \quad (6c)$$

To complete this structure we define the momentum operator P_μ conjugate to the space-time position (x_μ) of the 'hadron', and the space-time (orbital) Lorentz generators: $L_{\mu\nu}\equiv x_\nu P_\mu - x_\mu P_\nu$. Hence the Poincaré group generators are:

$$P_\mu; \quad M_{\mu\nu} = L_{\mu\nu} + S_{\mu\nu} . \quad (7)$$

The Majorana equation is seriously flawed, for model building, by the fact that the time-like spectrum is physically unacceptable, and more seriously, that it admits of space-like (tachyon) solutions. Our construction avoids both difficulties.

To motivate this construction, let us consider the hadronic system viewed in its rest frame. In the rest frame the invariant P•V involves only the *positive definite* operator V_0, which is *quantized*. Moreover, in the rest frame, the spin is generated by S_{ij} whose eigenvalues, s, are given by $1/2V_0$. This suggests that we relate the invariant $M^2 = P\cdot P$ to V_0, thereby achieving a quantized (and hence Poincaré invariant) mass-spin relation.

The key point in implementing this structure is to recognize that the boson operators are Poincaré covariant (carrying the irreps (1/2,0) and (0,1/2)) and that the ground state relation:

$$(i\pi_i + \xi_i)|0\rangle = 0, \text{ (for i - 1,2)} \quad , \quad (8)$$

must be *form invariant*. In other words, the internal (boson) structure -- that is, the operators (π_i, ξ_i) and the fundamental state $|0\rangle$ as well -- are to be characterized by the boost B_p which transforms the rest frame to a general frame having velocity p_μ/M. (Note that this velocity is independent of mass.)

Hence one has the more explicit relation:

$$(i\pi_i(p/M) + \xi_i(p/M))\,|0;\,\mathcal{U}_{p/M}\rangle = 0. \qquad (9)$$

The internal structure is accordingly oriented to accord with the state of motion of the hadron as a whole, a concept which is formally very similar to a gauge condition.

Once this structure is understood, it is not difficult to recognize that the constraint condition:

$$P_o|\psi;\mathcal{U}_{P/M}\rangle = +[P_i P_i + m_o^2\, V_o(P/M)]^{1/2}$$

$$(P_o - [P_i\, P_i + m_o^2\, V_o(P/M)]^{1/2})\,|\psi;\,\mathcal{U}_{P/M}\rangle = 0 \qquad (10)$$

is in fact a Poincaré covariant relation which corresponds to the trajectory constraint: $(\mathrm{mass})^2$ = linear function of spin. The spectrum is time-like and always has positive energy. Note that the spectrum splits into two distinct pieces: an integer spin Regge band (0,1,2,...) and a half-integer Regge band (1/2,3/2,...) with the same slope. *The occurrence of half-integer angular momenta is significant:* an extended indecomposable object inherently possesses spinorial properties without the necessity of an ad hoc introduction of spinorial components.

Let us remark that this construction is quite general and works for an arbitrary number of pairs of bosons. From a group theoretic viewpoint one has a 'dynamical' symmetry group realized in the rest frame (more properly a *kinematical* symmetry group).

§3 THE CLASSICAL LIMIT OF THIS EXTENDED STRUCTURE

Dirac has often emphasized that the physical interpretation of a given structure is most clearly made by using the Heisenberg picture. We shall use this approach to understand the physical significance of our extended structure, as a one-dimensional (linearly extended) spinning object in Minkowski space.

Our task is made easy by the fact that Dirac (12) has carried out this analysis already, for the spin 0 component of the present structure (13), and Staunton (9) has recently given an extension of these results to include the spin 1/2 component (14).

Staunton's work was carried out in the context of his relativistic wave equation (14) -- an elegant formulation for spin 0 and spin 1/2 which includes Dirac's spin 0 equation (13). We need only remark that his results are actually completely general, and apply to the complete Regge trajectory of the present formulation.

To validate this remark, let us note that we can use the Hamiltonian, P_o, of Eq. (10), to generate the operator equations of motion:

$$i\,\frac{d\mathcal{O}}{dt} \equiv [\mathcal{O},P_o] \quad , \tag{11}$$

where t is the c-number time conjugate to P_o. Following Dirac, however, it is more expedient to consider a Poincaré scalar operator as a generalized Hamiltonian (15). Accordingly we replace the Hamiltonian P_o by the two invariant operators:

$$\phi \equiv M^3 - P\cdot V \quad , \tag{12a}$$

$$\phi_1 \equiv P\cdot P - M^2 \quad . \tag{12b}$$

[The wave equation $\{(P^2)^{3/2}-P\cdot V\}\psi=0$ has been mentioned by Barut and Duru (16).]

We will regard ϕ_1 as a constraint: $\phi_1 \approx 0$ and ϕ as a generalized Hamiltonian:

$$i\,\frac{d\mathcal{O}}{d\tau} = [\mathcal{O},\phi] \quad , \tag{13}$$

using the procedure given by Dirac. (τ is a c-number 'time' conjugate to ϕ.)

Before proceeding we must verify that this is equivalent to using P_0 as the Hamiltonian. One sees that, in fact, we have actually adjoined negative energy solutions; this is desirable, since the motion we are seeking is analogous to the *Zitterbewegung* of the Dirac electron. To see the equivalence otherwise, one need only carry out explicitly the transformation of the boson operators in $V_0(P/M)$ to recognize that the transformed structure is precisely $M^{-1}\, P\cdot V$, evaluated in a fixed frame. In applying Dirac's methods, we are at liberty to split the Hamiltonian into ϕ and a constraint.

The operator equations of motion are thus found to be:

$$\frac{dP^\mu}{d\tau} = 0 \quad , \tag{14a}$$

$$\frac{dx^\mu}{d\tau} = V^\mu \quad , \tag{14b}$$

$$\frac{dV^\mu}{d\tau} = -S^{\mu\nu}P_\nu \quad , \tag{14c}$$

and
$$\frac{dS^{\mu\nu}}{d\tau} = V^\mu P^\nu - V^\nu P^\mu \quad . \tag{14d}$$

(We have used the commutation relations of §2 to obtain these results, noting that both P^μ and x_μ commute, by definition, with the boson structure.)

These equations may be iterated to yield, for example:

$$\frac{d^2V^\mu}{d\tau^2} + M^2\, V^\mu = P^\mu\, (M^3-\phi) \quad , \tag{15}$$

which suffices to show that the τ-dependence is harmonic.

We shall not carry through the complete integration of the equations of motion. (These results may be found in (9).) Instead we shall proceed directly to the classical limit (replacing commutators by Poisson brackets) making use of the important fact that *the SO(3,2) commutator structure*, because it is realized over two bosons, *has a well-defined classical limit of its own.*

One obtains the following results for the *classical motion:*

$$x^{\mu}_{cl.} = M\, p^{\mu}\tau + M^{-1}(A^{\mu}\sin M\tau - B^{\mu}\cos M\tau) + C^{\mu} \quad (16)$$

$$v^{\mu}_{cl.} = M\, p^{\mu} + A^{\mu}\cos M\tau + B^{\mu}\sin M\tau \quad (17)$$

$$S^{\mu\nu}_{cl.} = (A^{\mu}p^{\nu} - A^{\nu}p^{\mu})M^{-1}\sin M\tau \quad (18)$$
$$- (B^{\mu}p^{\nu} - B^{\nu}p^{\mu})M^{-1}\cos M\tau$$
$$+ D^{\mu\nu} .$$

Here the parameters A^{μ}, B^{μ}, C^{μ} and $D^{\mu\nu}$ are integration 'constants' (which, though commuting, have Poisson bracket properties). These parameters, from the analysis of (9), have the properties:

$$A^{\mu}A_{\mu} = B^{\mu}B_{\mu} = -M^{4} , \quad (19)$$

$$A^{\mu}B_{\mu} = 0 , \quad (20)$$

$$D^{\mu\nu} = M^{-2}(A^{\mu}B^{\nu} - A^{\nu}B^{\mu}) , \quad (21)$$

$$A^{\mu}p_{\mu} = B^{\mu}p_{\mu} = D^{\mu\nu}p_{\nu} = 0 . \quad (22)$$

These properties show that in a proper Lorentz frame, where $p^{\mu} \equiv (M\ 0\ 0\ 0)$, the four-vectors A^{μ}, B^{μ} become orthogonal 3-vectors which, together with $D^{\mu\nu}$, form an orthogonal triad. (The vector C^{μ}, which defines the origin for x^{μ}, is not restricted.)

To interpret these classical results it is helpful to observe that *in the rest frame* the Lorentz generators $M^{\mu\nu}$ reduce to $D^{\mu\nu}$ alone. It follows that D^{ij} is the *intrinsic spin*, which is constant in τ. Moreover, from the fact that $D^{\mu\nu}p_{\nu} = 0$, we see that in the rest frame $D^{\mu\nu}$ reduces entirely to the spin D^{ij}. This property is a defining characteristic of a classical, relativistic, rotating, extended object (17). It is quite essential to note that, *in constrast to a classical rotator* (which has degeneracy $(2J+1)^2$) our construction yields an object having only the spatial degeneracy (2J+1), which is characteristic of a one-dimensionally extended object.

It is helpful for further interpretation to introduce parameters having the usual dimensions; this corresponds to scaling ϕ to have the dimensions of energy. (We used Staunton's notation, (9), as a convenience to the reader.) Since ϕ has the dimension M^3 this implies (from Eq. (13)) that the combination $M^2\tau$ has the dimensions of time so we find: time $\equiv M^2\tau$. The rotation frequency is thus: $\omega = M^{-1}$. Then x^μ has the dimensions of length, with $M^{-2}V^\mu$ having the dimensions of velocity.

We recognize that -- in a proper frame -- the rotatory motion of x^μ, takes place in a plane perpendicular to the spin angular momentum. The essential observation now is to determine the *scaling*: the size of the rotatory motion in x^μ scales as the mass M. Thus we find:

<u>classical rotatory motion of 'hadron'</u>:

size	$\sim M$ = mass	
frequency	$\sim M^{-1}$	
spin angular momentum	$\sim M^2$	(trajectory constraint)

Although this motion has been termed *Zitterbewegung*, it should be noted that it is actually quite different in its scaling properties from the <u>Zitterbewegung</u> of the Dirac electron.

One final point. As Dirac has discussed (12), the proper definition of the spin (in a frame other than the rest frame) requires a prior definition of the coordinates entering the definition of the orbital angular momentum. For the case at hand, the required definition of the space-time coordinates is not x^μ but rather:

$$y^\mu \equiv x^\mu - M^{-1}(A^\mu \sin M\tau - B^\mu \cos M\tau). \qquad (23)$$

The orbital angular momentum, $y^\mu p^\nu - y^\nu p^\mu$, is now τ-independent, as required for a τ-independent spin. The coordinates y^μ do not show Zitterbewegung and follow a force free motion:

$$y^\mu_{cl} = M\, p^\mu\, \tau + C^\mu \ . \tag{24}$$

The *quantum* variables y^μ are, however, *non-commuting*. This fact will be of importance in the discussion of §5.

§4 CLASSICAL MOTIONS OF THE RELATIVISTIC STRING

Our discussion of the classical motion of the relativistic string will be based on the Nambu-Goto-Nielsen action functional, as discussed in standard references (1,2,3). In constrast to the usual treatments, we will find it essential to work in the gauge where the equations of motion are non-linear. (We denote this guage by NL; the orthonormal gauge being denoted by ON.)

The action functional is based on the area of a two-dimensional surface (τ,σ) in space-time; this is a direct generalization of the mechanics of a mass point based on a length functional of proper time. Thus one has:

$$L = -(2\pi\alpha')^{-1} \int d\tau \int d\sigma\ (-\det g)^{1/2} \tag{25a}$$

$$\det g \equiv \begin{vmatrix} (\dot{x},\dot{x}) & (\dot{x},x') \\ (\dot{x},x') & (x',x') \end{vmatrix} , \tag{25b}$$

$$\text{with}\quad \frac{\partial x^\mu}{\partial \tau} \equiv \dot{x}^\mu \ , \quad \frac{\partial x^\mu}{\partial \sigma} \equiv x'^\mu \ , \tag{25c}$$

and (Lorentz) inner products denoted as $x^\mu x_\mu = (x,x)$, metric $(1,-1,-1,-1)$.

In the standard way one finds for the momentum density:

$$P^{\mu} = \frac{\delta L}{\delta \dot{x}^{\mu}} = (2\pi\alpha')^{-1}(-\det g)^{-1/2}(\dot{x}^{\mu}(x',x')-x'^{\mu}(\dot{x},x')) \tag{26a}$$

$$\Pi^{\mu} = \frac{\delta L}{\partial x'^{\mu}} = (2\pi\alpha')^{-1}(-\det g)^{-1/2}(x'^{\mu}(\dot{x},\dot{x})-\dot{x}^{\mu}(\dot{x},x'))\,. \tag{26b}$$

The variational principle implies that the equations of motion are:

$$\frac{\partial}{\partial \tau} P^{\mu} + \frac{\partial}{\partial \delta} \Pi^{\mu} = 0 \quad , \tag{27a}$$

with boundary condition: $\Pi^{\mu}=0$ at $\sigma=0,\pi$. (27b)

The equations defining P^{μ} and Π^{μ} imply the primary constraints:

$$(2\pi\alpha')(P,P) + (2\pi\alpha')^{-1}(x',x') = 0, \tag{28a}$$

$$(x',P) = 0 \quad , \tag{28b}$$

and the two additional constraints:

$$2\pi\alpha' \ (\pi,\pi) + (2\pi\alpha')^{-1}(\dot{x},\dot{x}) = 0, \tag{28c}$$

$$(\dot{x},\pi) = 0 \quad . \tag{28d}$$

The equations of motion are non-linear, and few would choose to solve them, since an easily solvable linear version can be found. We will show why one must resist this temptation.

One solution of these non-linear equations is known (1). This solution is:

$$x^{\mu}(\tau,\sigma) = (\tau, A\rho \cos \omega\tau, A\rho \sin \omega\tau, 0), \qquad (29)$$

where for convenience we write: $\frac{\pi}{2}\rho = (\sigma - \pi/2)$.

From this one obtains the momentum densities:

$$P^{\mu}(\tau,\sigma) = \left(\frac{A}{\pi^2\alpha'}\right)(1-\rho^2)^{-1/2}(1, -\rho \sin\omega\tau, \rho \cos \omega\tau, 0) \qquad (31)$$

$$\pi^{\mu}(\tau,\sigma) = (-)(2\pi\alpha')^{-1}(1-\rho^2)^{1/2}(0, \cos\omega\tau, \sin\omega\tau, 0). \qquad (32)$$

To satisfy the boundary conditions one finds that: $\omega A = 1$.

It is readily verified that this is indeed a solution of the equations of motion and of all constraint conditions.

Let us interpret this solution by calculating the total linear and angular momenta:

$$p^{\mu}(\tau) \equiv \int_0^{\pi} d\sigma\, P^{\mu} = (A/2\alpha')(1,0,0,0), \qquad (33)$$

$$j^{i} \equiv \int_0^{\pi} d\sigma\, e_{ijk}\, x^j P^k = (A^2/4\alpha')(0,0,1). \qquad (34)$$

This solution has the following features:

(a) It is realized entirely in Minkowski space, in a proper Lorentz frame and is a rotation in the x-y plane with frequency ω, with v=c at the end points. Note that τ is the proper time and σ is the radius vector in the x-y plane. *There are no unphysical variables.*

(b) The frame of reference is the *rest frame*, with p_i being zero. It follows that J^1 is the spin.

(c) The energy (mass=M) and angular momentum (J) are continuously variable and obey the linear Regge trajectory constraint:

$$\alpha' M^2 = J. \qquad (35)$$

(d) Noting that the energy (Mass=M) varies with A, we can obtain the following scaling properties:

Radius of rotation ~ Mass = M

Frequency of rotation = $\omega \sim M^{-1}$

Angular momentum ~ M^2.

These results suffice for our main purpose, but it is helpful (in extending our model) to consider now a more general solution to the nonlinear equations of motion. It is quite remarkable that -- as one can easily verify directly -- *a general solution results if we simply replace* ρ *by* $f(\rho)$ *in the special solution given for* $x^\mu(\tau,\sigma)$. The functional dependence is essentially *arbitrary* except for: (a) boundary conditions: $f^2=1$ at $\rho = \pm 1$; (b) conditions of differentiability; and (c) the limitation $f^2 \leqslant 1$, (to avoid $v > c$ and ambiguity in the square roots).

Using this functional freedom one may take as 'simple modes' (one cannot add modes) the family: $f(\sigma) = \cos n\sigma$. The results (18) reproduce the earlier solution (for n=1) and yield the more general modes:

$$M = n \cdot (A/2\alpha'); \tag{35a}$$

$$J = n \cdot (A^2/4\alpha'); \tag{35b}$$

which together imply:

$$(\alpha'/n)M^2 = J \quad . \tag{35c}$$

In other words these solutions yield, as desired, *a family of Regge trajectories*.

Note that the effective slope of these -- continuously variable -- $(\text{Mass})^2$-spin trajectories is: α'/n.

Let us mention now our reason for working in the NL gauge. The NL gauge is that gauge in which the motion is realized in Minkowski space, in a proper Lorentz frame, with the momentum and angular momentum densities both being *additive*. This gauge is accordingly the physical gauge for Hamiltonian dynamics based on the point form.

By contrast the ON gauge is adapted to dynamics in the front form (null plane) and the additivity (which would contradict the trajectory constraint) is lost.

§5 COMPARISON OF §3 AND §4; FURTHER DISCUSSION

It is probably clear already from the scaling properties, given in §3 and §4, that the fundamental features -- at the classical level -- of the quantal Regge trajectory (§3) and of the Nambu-Goto-Nielsen relativistic string (rotating in the leading mode) are essentially identical. In both systems (in a proper frame) some sort of linearly extended object is rotating in a plane, with the end points moving at light velocity. The angular momentum in both structures is perpendicular to the rotation plane and is time independent.

The center of the rotation in both systems is fixed, and time independent. This uniquely determines the positional coordinates to be y^μ, not x^μ. Trivial as this may seem the implication is quite important: the quantal version of the y coordinates are non-commuting and pose a logical difficulty for any straightforward quantization of the string. (By constrast, the model of §3 is already quantum mechanical.)

Let us recall next that the quantal Regge trajectory possessed both integer and half-integer spins. This clearly implies that the quantized relativistic string should already possess half-integer modes, without the necessity of introducing spinorial properties 'by hand'.

Let us note that it is quite easy to achieve the other pure modes of the relativistic string (the modes with n>1 in §4). We simply adjoin additional oscillators $\xi_i \rightarrow \xi_i^{(n)}$, using:

$$V_o \rightarrow \sum_{n=1} n\, V_o\, (n) \quad .$$

For the pure modes (only $\xi_i^{(n)}$, n=fixed, entering), the construction goes through precisely as in §3, changing only the trajectory slope: $\alpha' \rightarrow \alpha'/n$. (It should be noted that when a *general* mode (of this quantal structure) is excited, because of degeneracy, the trajectories are actually parallel in the resulting lattice of $(\text{mass})^2$ - spin points.)

It is not profitable to discuss these results in terms of the quantized (DRM) string models, not only because the quantization is only consistent in an unphysical number of dimensions, but also because the time-like dimension in the oscillators causes non-trivial angular momentum changes. Thus, for example, one cannot ascribe the leading trajectory to the lowest harmonic oscillator mode, or even to the lowest rotational mode.

One is thus forced for precision to make all comparisons at the classical level to the relativistic string. At his level we have demonstrated the existence of an acceptable alternative structure having no difficulties of principle whatsoever. *This construction indicates that the three basic difficulties of the scalar dual model:*

(i) existence of a tachyon ground state
(ii) 26 space-time dimensions for the quantized model
(iii) restriction to integer spin,

are not inherent consequences of the quantization of the classical relativistic string.

ACKNOWLEDGEMENTS

We would like to thank Professor L.P. Staunton for the favour of a preprint of his paper (6), and for several discussions on his work.

FOOTNOTES AND REFERENCES

* Work supported in part by the National Science Foundation

** Alexander von Humboldt Foundation Senior U.S. Scientist Award 1976; on leave from Duke University, Durham, North Carolina 27706, U.S.A..

1. Rebbi, C. (1974), Phys. Reports, 12C, 1-73.
2. Mandelstam, S. (1974) Phys. Reports, 13C, 259-353.

3. Frampton, P.H. (1974) "Dual Resonance Models" (W.A. Benjamin, Inc., Reading, Mass.)
4. Mandelstam, S. (1971) Brandeis University Summer Institute, Vol. 1, 165 (M.I.T. Press, Cambridge, Mass.)
5. Gliozzi, F., Scherk, J., and Olive, D. (1976) PTENS preprint, Sept. Cf. also the lecture by deVecchia at this conference.
6. Work by Volkov, Sjelest and Sjeldakhin on this possibility was pointed out to us at the conference by Dr. K. Litwin.
7. Biedenharn, L.C., and van Dam, H. (1974) Phys. Rev. D9, 471-486.
8. van Dam, H., and Biedenharn, L.C. (1976) Phys. Rev. D14, p. 405-417.
9. Staunton, L.P. (1976), Phys. Rev. D13, 3269.
10. Nambu, Y. (1976) Proc. International Conf. on Particles and Fields, (Interscience, N.Y.).
11. Barut, A.O. (1971), "Dynamical Groups and Generalized Symmetries" (U. of Canterbury Press, Chch. N.Z.).
12. Dirac, P.A.M. (1972) Proc. R. Soc. Lond. A328, 1-7.
13. Dirac, P.A.M. (1971) Proc. R. Soc. Lond. A322, 435-445.
14. Staunton, L.P. (1974) Phys. Rev. D10, 1760-1767.
15. Dirac, P.A.M. (1964) "Lectures on Quantum Mechanics" (Yeshiva Univ. Press, N.Y.).
16. Barut, A.O. and Duru, I.H. (1973) Lettere al N.C. 8, 768-771.
17. Hanson, A.J. and Regge, T. (1974) Ann. of Phys. (N.Y.) 87, 498-566.
18. The calculations are not entirely trivial (singularities enter). The physical requirements that P^o be positive, and additive, resolves these ambiguities.

HIGH DENSITY MATTER IN THE UNIVERSE

V. Canuto

Institute for Space Studies, NASA
New York, New York 10025

J. Lodenquai

Department of Physics, UWI
Kingston, Jamaica

I. INTRODUCTION

We can identify three main areas in the universe where matter at high densities occurs. These are (1) neutron stars, (2) high energy p-p collisions, and (3) the early stages of the universe itself.

The densities involved in neutron stars span a wide spectrum, ranging from about 10^6g.cm^{-3}, at the surface, to about 10^{16}g.cm^{-3} at the core. In the case of high energy p-p collisions we meet densities that are even higher, typically in excess of 10^{17}g.cm^{-3}, while in the early stages of the universe the matter density approaches infinity.

In this article we shall review the results of the description of the properties of matter at high densities, including the equation of state, with specific references to the three areas mentioned above. We shall see that up to a density $\rho \sim 5.10^{14}$g.cm^{-3}, the properties of matter are well understood, but beyond this point uncertainties about these properties increase rapidly with increasing density. However, in the asymptotic region of high density it is expected that the equation of state should take the simple form

$$p = c_s^2 \varepsilon \qquad (1)$$

where p is the pressure, $\varepsilon = \rho c^2$ the energy density and c_s the speed of sound in matter (in units of c) assumed to be constant. However, there is wide disagreement in the predicted values of c_s in the superhigh density region. Some theories predict $c_s \to 1$, whereas others predict $c_s \to 1/\sqrt{3}$ or even lower values.

The available data on neutron stars, particularly the moment of inertia, will be seen to be insufficient to pin down a specific value of c_s at superhigh densities. A possible answer however might be found in the analysis of the results of p-p collisions at high energies where $c_s \to 1$ seems to be favored, thus providing a way to continue the well-established equation of state into the superhigh density region.

In Part II of this article, we shall review the predicted properties of matter in the various density regions appropriate to neutron stars. We shall then test the theoretical predictions based on derived equations of state with observable parameters of an actual neutron star. In Part III, the Landau hydrodynamical model for multiparticle production, which depends upon c_s through the equation of state (1), is used to study the multiplicity and transverse momentum distribution in high energy p-p collisions. The value of c_s that appears in the predictions of the model can be varied to bring theory and experiment into agreement, thereby providing (hopefully) the correct value for c_s. In Part IV we review the early stages of the universe and the problem of galaxy formation based on eq. (1).

II. DENSITY RANGE UP TO 10^{16}g.cm^{-3}

1. Review of Neutron Stars and Pulsars

The real impetus for the study of matter at high densities came from the fortuitous discovery of pulsars in the late sixties[(1)] and their subsequent identifications with rotating, magnetized neutron stars.[(2)] As far back as 1932, the year the neutron was discovered, Landau argued that a stable gravitating system composed essentially of degenerate neutrons could theoretically exist.[(3)] In 1934 Baade and Zwicky[(4)] proposed a method whereby such a "neutron star" could originate from the remnant of a supernova explosion, which is the end-point of evolution of the more massive stars.

A neutron star is expected to have very unusual properties by terrestrial standards. A typical neutron star has a mass comparable to the solar mass $M_\odot \sim 2.10^{33}$g., but a radius of only about 10 km. These two parameters imply an enormous average density $\sim 10^{15}$g.cm^{-3}, so that the matter density at the cores of some neutron stars can easily exceed nuclear matter density ($\sim 2.8 \times 10^{14}$g.cm^{-3}) by an order of magnitude. If the equation of state of matter at such high densities is known, then it is straightforward to integrate the equations of hydrostatic equilibrium to obtain the mass M and radius R of the star in terms of the assumed central density ρ_c. However, since the potential energy of an object on the surface of a neutron star is an appreciable fraction (typically one-tenth) of its rest mass energy, the equations of hydrostatic equilibrium must be the general relativistic analog of the classical equations.[5] In particular, the moment of inertia can be computed, since it is a function of M and R only.

The first attempt at a detailed model of neutron stars was made by Oppenheimer and Volkoff in 1939.[6] These authors used the equation of state appropriate for a cold Fermi gas and their calculation verified Landau's prediction and yielded in addition an upper limit to the mass a stable neutron star can have. This limit is the so-called Oppenheimer-Volkoff mass limit, M_{OV}. This limit is present in all subsequent model-calculations of neutron stars but differs from the originally predicted value of $\sim 0.72 M_\odot$, depending on the particular equation of state used.

The assumption of a cold star is quite reasonable since the typical Fermi energy of the neutron is $\sim$150 Mev, which is much greater than the expected ambient temperature of a real neutron star.[7] However, the assumption that the neutrons can be treated as a Fermi gas is clearly unrealistic and subsequent calculations are all based on attempts at a more realistic equation of state.

2. Dynamics[8,9]

Interest in neutron stars died down because of the general belief that there was little chance they could be discovered. However, when the first pulsar was discovered in 1967 and was later identified by Gold as a rotating, magnetized neutron star, interest in neutron stars rapidly picked up. The present consensus of opinion is that a pulsar is a highly magnetized neutron star that is rotating about an axis non-coincident with its magnetic axis and radiating electromagnetic radiation along two oppositely directed narrow cones (Fig. 1). The youngest and most observed of the pulsars is NP0532 located in the Crab nebula, which is in turn known to be a remnant of a supernova explosion that occurred in 1054. This pulsar has a present rotation period $P_0 = 0.033$sec., which is increasing very slowly, $\dot{P}_0 = 4.63 \times 10^{-13}$sec./sec. (The subscript 0 indicates that the parameter to which it is affixed is evaluated today.) There is a limit to the minimum period a pulsar can have if it is to remain stable. This limit is obtained by noting that the centrifugal acceleration at the surface of the star must be less than or equal to the gravitational acceleration there, i.e.

$$\Omega^2 R \lesssim GM/R^2 \qquad (2)$$

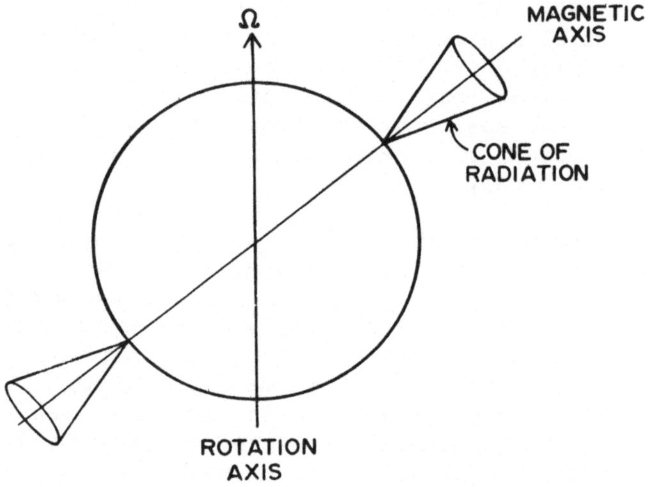

Fig. 1. A neutron star as a rotating magnetized dipole.

where $\Omega = 2\pi/P$ is the angular frequency of rotation. This implies

$$\Omega_{max} \simeq (GM/R^3)^{1/2} \tag{3}$$

If we choose $M \approx M_\odot$ and $R \approx 10$km., we get $\Omega_{max} \approx 10^4 sec^{-1}$. We therefore see that the frequency of the Crab pulsar ($\sim 190 sec^{-1}$, the highest known) is well within this limit. Eq. (2) may also be expressed in another way. Since $\rho \approx M/V \approx M/(\frac{4}{3}\pi R^3)$ and $P = 2\pi/\Omega$, eq. (2) implies

$$\rho \gtrsim 3\pi/GP^2 \tag{4}$$

If $P = 0.033$sec., Crab pulsar, then eq. (4) implies that the average density must be greater than $1.3 \times 10^{11} g.cm^{-3}$.

Let us now look at the energetics of pulsars. It is observed that all pulsars are gradually slowing down, i.e. $\dot{P}_0 > 0$, and ranges typically from 10^{-15}sec/sec to 10^{-13}sec/sec, among the known pulsars. (There are anomalous instances where $\dot{P}_0$ is found to be <u>negative</u> for a brief moment for some pulsars but we shall not discuss the reasons for this effect here.) The above observatons then imply that a given pulsar is gradually converting rotational energy to some other form. If we let I be the moment of inertia of a pulsar, then from energy conservation

$$\frac{d}{dt}(\tfrac{1}{2}I\Omega^2) = \dot{E}_r \tag{5}$$

i.e.

$$I = (\dot{E}_r/\Omega\dot{\Omega}) \tag{6}$$

where $\dot{E}_r$ is the rate of loss of rotational energy. Now it is generally accepted that the luminosity of the Crab nebula (.3 x 10^{38} ergs/sec at 2kpc) is sustained by just such a process involving the pulsar NP0532. The luminosity can actually be measured, and depends on the accepted value of the distance D from us; this distance is quite uncertain and can have any value between 1.2 and 2.5kpc.(10) (1kpc = 3.09 x 10^{21}cm). If we use the presently observed values of Ω and $\dot{\Omega}$ (= $-2\pi\dot{P}/P^2$) we deduce from eq. (6) that

$$I(\text{Crab}) = \begin{cases} 0.62\times10^{44}\,\text{g.cm}^2 & D=1.2\text{kpc.} \\ 1.72\times10^{44}\,\text{g.cm}^2 & D=2.0\text{kpc.} \\ 2.69\times10^{44}\,\text{g.cm}^2 & D=2.5\text{kpc.} \end{cases} \tag{7}$$

It should be noted that no details of the mechanism of the conversion of rotational energy to nebular luminosity need be assumed.

The above possible values of the moment of inertia of the Crab pulsar can be tested against those computed from specific equations of state of the neutron star matter, a topic which we now face.

Fig. 2 is a schematic summary of what a cross sectional view of a typical neutron star ($M \approx \tfrac{1}{2}M_{\odot}$) is expected to look like. First, there is a magnetosphere a few meters thick composed mainly of iron. Below this is an outer crust about 1km thick composed of nuclei and free (i.e., unbound) electrons. This is followed by an inner crust, about 2km thick, composed of nuclei and free electrons and neutrons. The type of nuclei expected within a given layer of the crust depends on the density in that layer. Below the inner crust the nuclei dissolve, resulting in a liquid mixture of n, p, e^- and some hyperons. Towards the core where the density exceeds about 10^{15} $g.cm^{-3}$, the neutrons and hyperons might solidify, but there is some amount of disagreement on that point. We shall now summarize the reasons for this picture of a neutron star in terms of the properties of matter in the various density ranges of the neutron star.

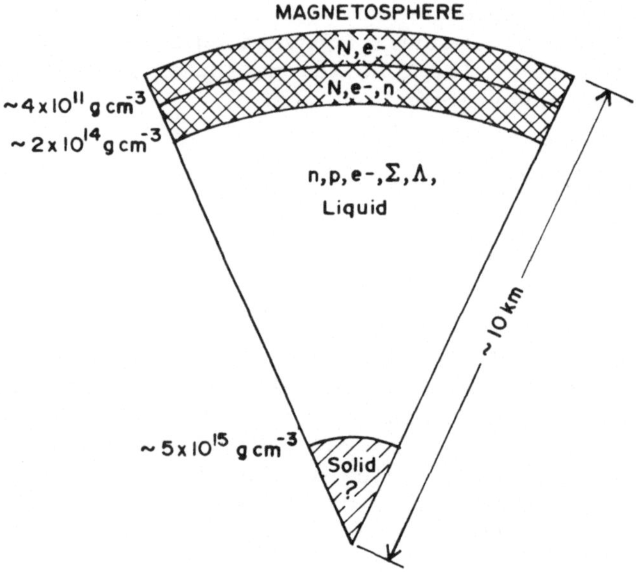

Fig. 2. A cross section of a neutron star.

3. The Various Density Regimes

(i) $\rho < 10^4 g.cm^{-3}$. This range of density may be found in the magnetosphere. Normally, there would be nothing new about this density regime. Matter here would be composed mainly of partially ionized iron, which is the most stable type of nuclei in this region. The equation of state in this density regime has been calculated by Feynman, Metropolis and Teller in 1949,(11) using the Thomas-Fermi method. However, in the magnetosphere of an acutal pulsar, the magnetic field can be enormously strong, typically $\approx 10^{12}$gauss(12) and the properties of matter can be drastically affected by it.(13,14) However, the actual effects on neutron star models is negligible, since the atmosphere is only a small fraction (both in size and mass) of the star itself.

(ii) $10^4 \leq \rho \leq 2 \times 10^{14} g.cm^{-3}$.(15) At a density $\rho \approx 10^4 g.cm^{-3}$, the electrons become essentially free and degenerate, rapidly becoming relativistic at about $10^7 g.cm^{-3}$. One of the main features characterizing this density region is the occurrence of inverse beta decay when the Fermi energy of the electrons reaches a value where the reaction

$$e^- + p \to n + \nu \tag{8}$$

becomes energetically favorable, as opposed to the more familiar beta decay

$$n \to p + e^- + \bar{\nu}$$

The inverse beta decay reaction occurs when the total electron Fermi energy exceeds the neutron-proton mass difference of 1.29 Mev. An immediate consequence of this reaction is that matter becomes increasingly neutron rich with increasing density.

Another important feature of this density region is the onset of crystallization of the nuclei present. As the electron Fermi energy ε_f increases with density, the electron kinetic energy begins to exceed the Coulomb interaction energies between the electrons themselves and between electrons and nuclei, since ε_f goes like $\rho^{2/3}$ (non-relativistically) while the Coulomb energy $\sim e/r$ goes like $\rho^{1/3}$. The upshot is that the perturbation exerted by the nuclei on the electrons diminishes with increasing density and the electrons begin to act as a uniform background gas whose screening properties become less effective. In fact, when $\varepsilon_f \gtrsim 1$ Mev, we have

$$\frac{r_s}{r_i} \sim \frac{6}{Z} > 1 \tag{9}$$

where r_s is the screening length, r_i the interionic spacing and Z the atomic number of the nucleus.[16] Because of the greatly reduced screening of the electrons, the interaction energy between ions will be considerably greater than normal. For example, for iron at a density of 10^8g.cm^{-3}, the Coulomb interaction between neighboring nuclei is $\sim$1 Mev, which is about 10^5 greater than that for normal iron. This enormous repulsion among the nuclei causes them to arrange themselves in a body-centered cubic lattice with a melting temperature estimated to be[17]

$$T_m \sim 1.6 \times 10^7 (\rho/10^6)^{1/3} \; (Z/8)^{5/3} \; {}^{o}K \tag{10}$$

This temperature is $\sim 4 \times 10^8$ ^{o}K for $\rho \sim 5 \times 10^7$g.cm^{-3}, $Z \sim 26$ and is much less than the expected surface temperature of pulsars (except immediately after formation).[7]

From the astronomical point of view, the existence of a solid crust is a far-reaching conclusion since one of the presently accepted explanations for the anomalous occurrences of negative $\dot{P}_0$ mentioned previously is related to a starquake phenomenon when the crust cracks in order to readjust itself to a decreasing ellipticity as the pulsar slows down. This decrease in ellipticity reduces the moment of inertia, causing a slight increase in angular frequency in order to conserve angular momentum. From the point of view of the equation of state, however, it is of little importance whether the crust exists, but it is very important in determining the particular types of nuclei that are present.

As the density increases with depth through the crust, the nuclei become increasingly neutron-rich due to the inverse beta decay mentioned previously. At a density of about $4 \times 10^{11} g.cm^{-3}$, nuclei are so neucron-rich that neutrons begin to appear in the energy continuum. This point is the so-called "neutron drip point" (NDP) and marks the boundary between the outer and inner crusts. At this point the predominant nucleus present has an atomic weight $A \approx 100$ and atomic number $Z \approx 35$. The concentration of free neutrons increases with density until at $\rho \sim 2 \times 10^{14} g.cm^{-3}$ the nuclei dissolve into a liquid of n, p, e^-, and this point marks the end of the inner crust. A summary of the above results is tabulated in Table 1. The equation of state of matter can be found from the thermodynamic definition of pressure, $p = -(dE/dV)$, where the total energy E is given by

$$E = n_e \varepsilon_e + n_N \varepsilon_N (A,Z) + n_N \varepsilon_L(Z) \tag{11}$$

n_k denotes the concentration of the k-th species. ε_e, the energy per electron, is given by the analytic expression

$$\varepsilon_e = \frac{m_e^4 c^5}{8\pi^2 \hbar^3} \{(2t^2+1)t(t^2+1)^{\frac{1}{2}} - \ln|t + (t^2+1)^{\frac{1}{2}}|\} \tag{12}$$

where $t \equiv p_f/m_e c$. Similarly ε_N is the energy per nucleus and ε_L the lattice energy per nucleus. ε_L depends slightly on the particular lattice and takes the form[18]

$$\varepsilon_L = -(Z^2e^2/r_c) \times \begin{matrix} 0.89593 & \text{bcc} \\ 0.89588 & \text{fcc} \\ 0.88006 & \text{hcp} \end{matrix} \tag{13}$$

where r_c is defined by the relation

$$\frac{4}{3}\pi n_N r_c^3 = 1 \tag{14}$$

The evaluation of the equilibrium concentrations n_e and n_N at a given density may be done by minimizing E subject to the conservation laws of baryon number and charge.

There is a short history of disagreement concerning the properties of the matter in the inner crust region. The main point of disagreement was whether an extrapolated version of the usual semi-

TABLE 1

DENSITY	SIGNIFICANT EVENT	MATTER COMPOSITION	STRUCTURE
$<10^4 g.cm^{-3}$	NOTHING SPECIAL (UNLESS STRONG MAGNETIC FIELD PRESENT)	...Mg,Ca,$\underline{\underline{Fe}}$,Co,...	
$\sim 10^4 g.cm^{-3}$	IONIZATION	e^-, NUCLEI	
$\sim 10^7 g.cm^{-3}$	ONSET OF CRYSTALLIZATION. INVERSE β-DECAY $e^- + p \to n + \nu$	e^-, NUCLEI	IONS FORM BCC LATTICE IN SEA OF ELECTRONS
$\sim 10^{11} g.cm^{-3}$	NEUTRON DRIP POINT	e^-, NUCLEI, n.	IONIC LATTICE IN SEA OF ELECTRONS & NEUTRONS
$\sim 2 \cdot 10^{14} g.cm^{-3}$	TRANSITION TO LIQUID	e^-,n,p,Y	LIQUID

empirical mass formula could be used for ε_N in view of the high densities prevailing. The problem is fully reviewed in Ref. (19) where the presently accepted solutions are given along with a "best choice" equation of state.

(iii) $2\times10^{14} \lesssim \rho \lesssim 10^{15} g.cm^{-3}$. After the nuclei have dissolved into a fluid, the system is left in a state composed predominantly of neutrons, so that, to a good approximation, the equation of state applicable in this region is probably that for a pure neutron fluid. This equation of state has been derived by several authors, for example Bethe and Johnson,[20] who employed a variational many-body technique with a new phenomenological N-N potential. However, in a more realistic treatment of the problem, one has to take account of the fact that the appearance of hyperons becomes energetically favorable, as first pointed out by Cameron[21] and Ambartsumyan and Saakyan.[22] Salpeter[23] concluded on general grounds that the Σ^- hyperon should appear first, a conclusion confirmed by all subsequent calculations. Before going into details we should like to point out that all calculations of the hyperonic liquid published so far have reached the same conclusion, that is, the equation of state is not severely altered from the one corresponding to a pure neutron fluid. However, one must accept this conclusion cautiously because all the computations share the same defect: the hyperonic two-body potential is assumed to be almost indentical to the corresponding N-N case. The exception is Ref. (22), where the hyperon interactions are neglected and is even therefore more unrealistic. The results of this calculation are shown in Fig. 3, where we see the presence of Σ^- and Λ hyperons as well as μ^- mesons.

The formulation of the problem in this regime is as follows. Consider a system composed of k different species, characterized by a number concentration n_k. We want to find n_k subject to the conservation laws of baryon number and charge. The problem is solved in principle by evaluating a set of simultaneous equations among the chemical potentials, obtained by minimizing the Gibbs free energy. In treating the problem realistically, two difficulties are met before the chemical potentials can actually be computed:

1. The two-body N-N potential is required. Several forms of the N-N potential have been proposed, among which one of the favorites is the one constructed by Reid.[24]

2. The situation is however quite different when we come to the hyperon-hyperon (Y-Y) or hyperon-nucleon (Y-N) interactions. The experimental data are still insufficient to derive reliable interactions involving hyperons. The only data involving hyperon interactions are those derived from low-energy Λ-N and Σ-p scatterings and from the production of double hypernuclei, i.e. nuclei containing two hyperons. There is no information on Σ-Σ interaction so

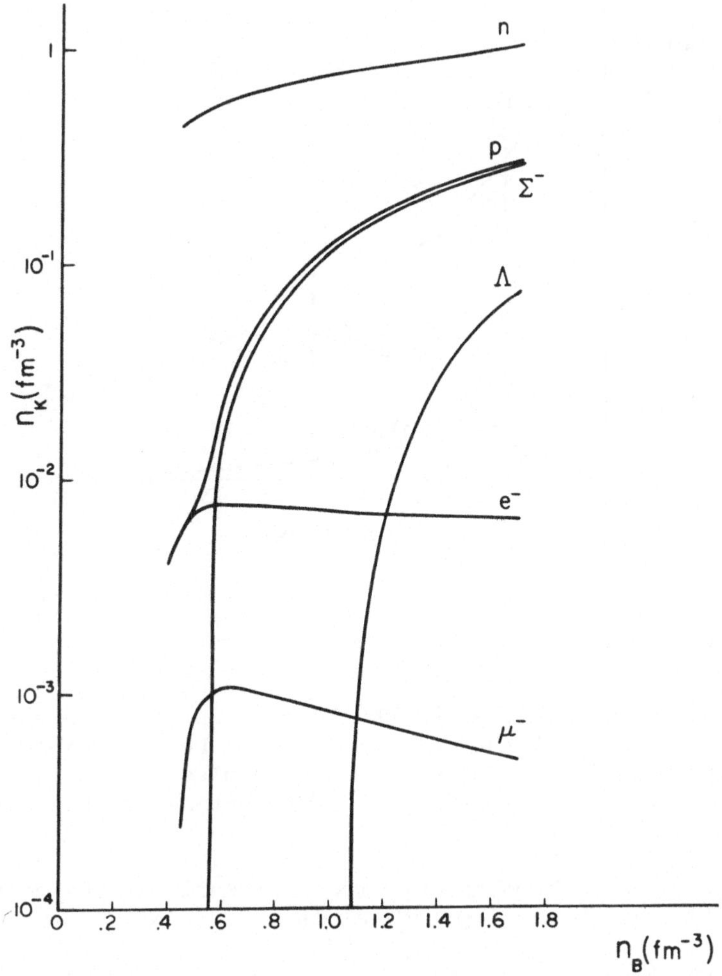

Fig. 3. Free hyperons concentration vs. baryonic number.

far. The simplest possible choice for the hyperonic potentials consists in assuming that they are the same as in the N-N case, if one excludes the one-pion contribution. However, this choice is not entirely satisfactory as can be seen from Fig. 4 where the cross-section calculated from the Reid potential, minus the one-pion contribution, curve A, is compared with the experimental cross-sections for Λ-N scattering.

To improve on this, a theoretically derived hyperonic potential based on meson exchange must be employed. This derivation

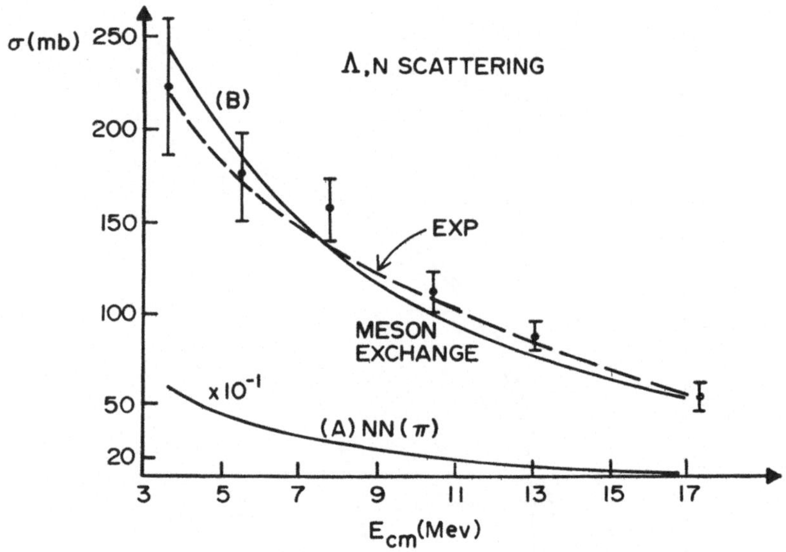

Fig. 4. ΛN scattering cross section vs. E_{cm}. The curve (A) is obtained from using NN potential minus the one pion contribution.

is possible in principle although the coupling constants entering into the final form of the potential cannot be determined with the same accuracy as in the N-N case. The results of the cross-section computations from such a derived potential is shown in Fig. 4 curve B.

Assuming we do know the hyperonic potentials, the general expression for the chemical potential is

$$\mu = (p^2c^2 + m^2c^4)^{1/2} + U(p) \tag{15}$$

where U(p) is the one-body potential felt by a particle with Fermi momentum p. From Brueckner's many-body theory[25], the one-body potential felt by particle i is given by

$$\begin{aligned}(2\pi)^2 U(p_i) = & \int_0^{p_i} p_j^2\, dp_j \,|K(p_i/p_j) - \text{exch.}| \\ & + \sum_{\ell \neq i}^{n_k} \int_0^{p_\ell} p_j^2\, dp_j\, K(p_i/p_j)\end{aligned} \tag{16}$$

The first integral refers to the interaction of i with other identical species, hence the exchange term. The second integral refers to the interaction of i with non-identical particles. The upper limits of these integrals are the Fermi momenta of the indicated particles and are obviously not known since they are related to the concentrations. The K-matrix is proportional to the matrix element of the two-body potential taken between the perturbed and unperturbed wave functions. The perturbed wave function is in turn the solution of an integral equation whose Green's function is again given in terms of U(p). This self-consistency is the most important feature of the Brueckner theory. In practice, the above method is extremely complicated to use, especially when we deal with, say, 8 or 10 different species, and none of the calculations of the hyperonic liquid published so far has treated the self-consistency feature in a satisfactory way.

One of the early attempts to take an hyperonic interaction into account is that of Langer and Rosen[26], who used the Levinger-Simmons non-local potential [27] for any two hadrons. The authors considered the following particles: n, p, e^-, μ^-, Σ^-, Σ^o, Σ^+, Δ^- and Δ^o. The treatment was based on a Hartree-Fock definition of U(p) but no self-consistency requirement nor a many-body theory as described above was employed. Their results are shown in Fig. 5, where we see that the first hyperon to appear is indeed Σ^-.

A more recent attempt at dealing with the hyperonic liquid is that of Moszkowski[28] whose work is reviewed in detail in Ref. (29). A many-body treatment of the type described previously was employed but the problem of full consistency was avoided by choosing two specific forms of the K-matrix between any two hadrons.

The difficulty of choosing the Y-Y potentials was also faced by Moszkowski, who adopted two methods of approach. The first method involved the use of the usual N-N potential for all the hadrons. We have mentioned previously that this potential does not seem to agree with the known results of Λ-N scattering. We shall call this choice 1. The second method involved the alteration of the Reid N-N potential to incorporate the different nature of the mechanism of exchange of mesons when we deal with hyperons. This choice we shall refer to as choice 2.

The four possible permutations of the above two choices with two specific forms of the K-matrix are labelled 1(a), 1(b), 2(a) and 2(b), and Moszkowski's results for these are shown in Fig. 6. The combinations 1(a) and 1(b) give rather strange results, for instance negative energy or negative pressure which is presumably associated with the unrealistic form used for the hyperonic potentials.

In view of the difficulties that plague the hyperonic equation of state published so far, we feel it is safe to treat the region 2×10^{14} to 10^{15} g.cm^{-3} as a pure neutron fluid whose equation of state is reliably known.

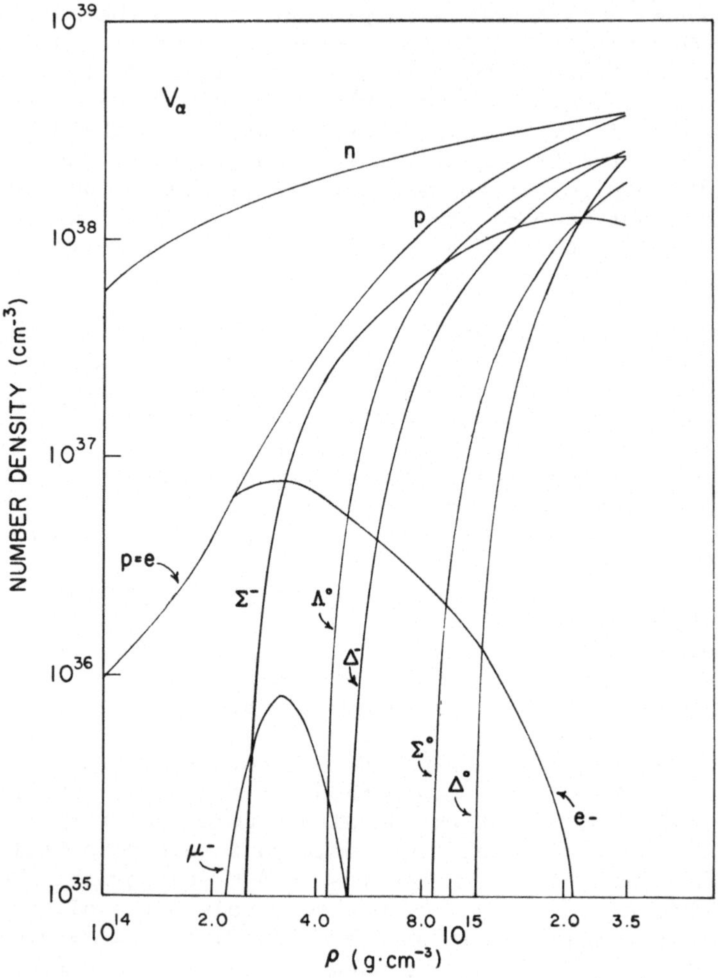

Fig. 5. Hyperons concentrations vs. matter density using the Levinger-Simmons two-body potential.

Before we leave the liquid region, we would like to mention one more interesting phenomenon that most likely occurs among the neutrons in the inner crust and liquid interior, and this is the phenomenon of superfluidity. This possibility was first pointed out by Ginzburg and Kirzhnits[30] and the argument goes as follows. According to the BCS theory of superconductivity, the gap width Δ is given by

$$\Delta = kT_c \sim (p_f\hbar/Ma)e^{-(N(o)g)^{-1}} \tag{17}$$

where T_c is the superfluid transition temperature, a the width of the effective region of interaction ($\sim$1 fm in our case), M the

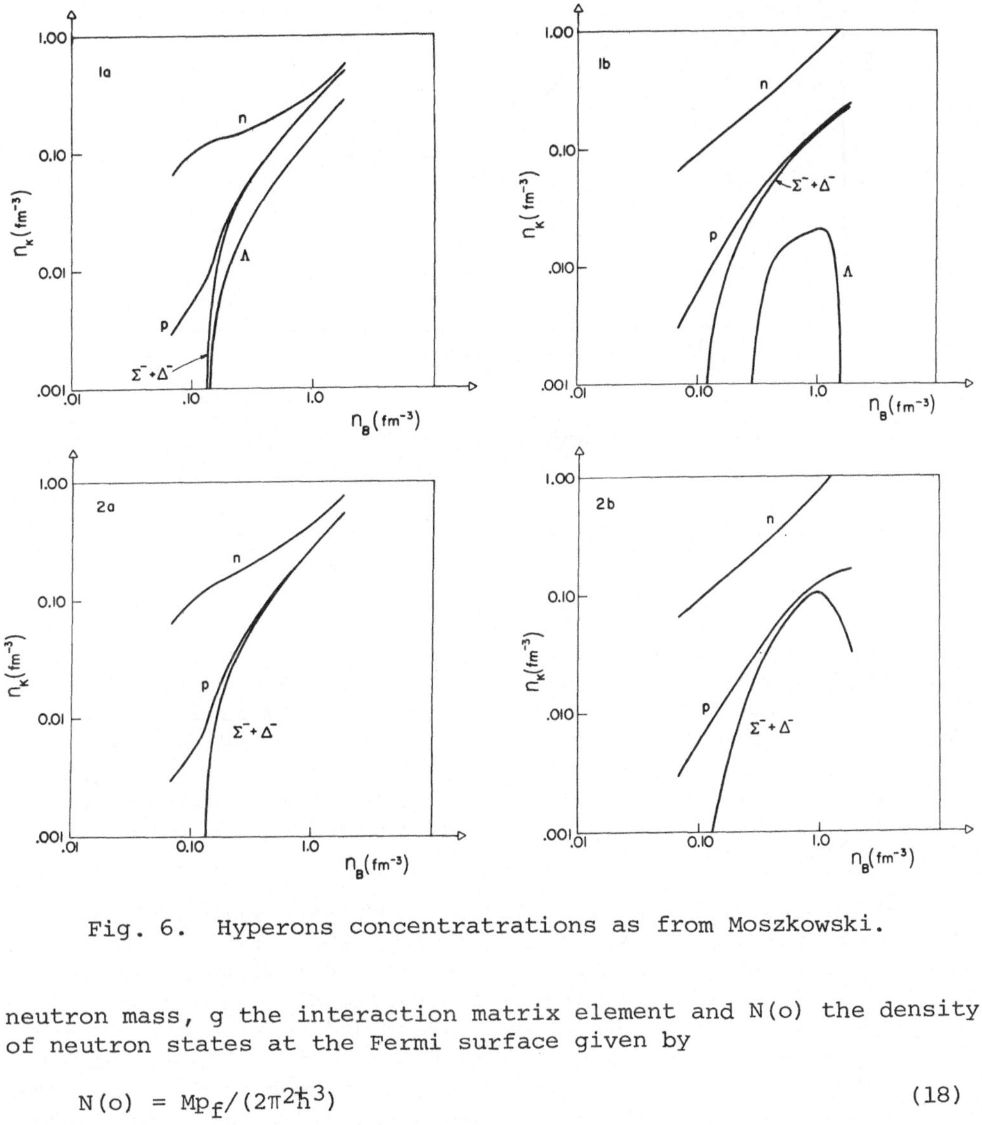

Fig. 6. Hyperons concentratrations as from Moszkowski.

neutron mass, g the interaction matrix element and N(o) the density of neutron states at the Fermi surface given by

$$N(o) = Mp_f/(2\pi^2\hbar^3) \tag{18}$$

For densities of the order 10^{14}g.cm^{-3}, N(o) $\sim 10^{42}$ (erg$\cdot$cm^3)$^{-1}$. If we now assume that the $1S_o$ interaction is described by a potential well of depth of 15 Mev and width 2.5 fm, then

$$g \sim \int V(r)d^3\vec{r} \sim 0.5\text{x}10^{-42}\text{erg}\cdot\text{cm}^{3} \tag{19}$$

We thus obtain $\Delta \sim 10$ Mev, i.e., a transition temperature $T_c \sim 10^{10}$ °K implying superfluidity in this range.

A detailed study of the gap Δ was made by Hoffberg et al.[31] with the result that

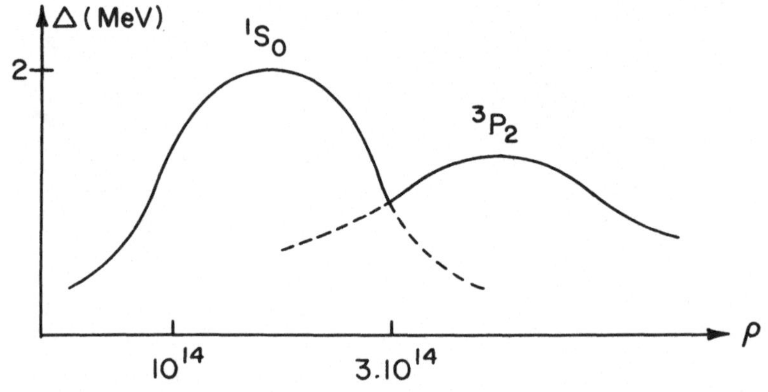

Fig. 7. The gap Δ vs. matter density.

$$\Delta = kT_c \sim 2\, \varepsilon_f \exp\left(-\frac{\pi}{2}\cot\delta\right) \tag{20}$$

where δ is the phase shift. This expression is valid for either ^1S_o or $^3P_2$ pairing, whichever is dominant. Fig. 7 shows curves of transition temperature versus density for both ^1S_o and $^3P_2$ pairings as obtained from Ref. (31). The conclusion is that within the density regions where the curves peak, we most likely have superfluidity of the neutrons. The same conclusion is reached when the less dense proton fluid is examined as shown by the most careful recent calculations of Chao et al.(32) However, the electron binding is not expected to be strong enough to produce electronic superconductivity.

The presence of superfluid layers in pulsars will again not seriously affect the overall equation of state, but it will have important consequences on the dynamics(33) and cooling rates of pulsars.(7) The dynamical effect of superfluidity should be especially important immediately after a starquake in a pulsar. Because of the negligible viscosity of superfluids, there will be differential rotation of those portions of the pulsar separated by the superfluid layer. The important link between these regions is the superstrong magnetic field that permeates the pulsar.

(iv) $10^{15} \lesssim \rho \lesssim 10^{16}$g.cm^{-3}. As we approach the core of our typical neutron star, we reach densities in the range 10^{15} to 10^{16} g.cm^{-3} and the outstanding question here is "Do the neutrons solidify in this density regime?" If we assume the N-N potential to be a hard-core with a range a, then it is almost inevitable that solidification will occur at a baryon density $n_B \approx a^{-3}$. For example if a=1 fm, then n_B at solidification is $\approx 10^{39}$cm^{-3}, i.e., $\rho \approx 1.7 \times 10^{15}$

$g.cm^{-3}$. At these densities each particle feels an infinitely strong repulsion all around, i.e., it is caged, and such a localization leads to a crystalline structure.

However, realistic N-N potentials are not hard-cores, so it is not evident whether or not solidification will actually occur in the present density regime. In the following, we shall review some of the main attempts at answering this important question.

The early attempts. It seems that the first paper which discusses the topic of solidification is that of Cazzola et al.[34], which appeared in 1966. The authors did not actually show that a solid would exist, but postulated that the N-N potential is repulsive enough to produce localization and thereby proceeded to solve the Dirac equation for a particle in a one-body square well potential responsible for localization. Their resulting equation of state does not join smoothly to present-day results of the neutron liquid computation and the model is of historical interest only.

In 1970, Banerjee et al.[35] performed a computation of the equation of state at high densities by localizing the neutrons in a bcc lattice and performed classical lattice-dynamics computations. The N-N potential between neighboring neutrons was chosen to be an average of the Reid $^1S_0$ potential and a solid structure was found at a density of 8×10^{14} $g.cm^{-3}$. Evidently the application of classical lattice-dynamics is not justified in this case, since at these densities the quantum mechanical zero-point vibration of a neutron about an assumed lattice site is a significant fraction of the mean distance between neighboring lattice sites. In fact at $\rho \sim 5 \times 10^{15} g.cm^{-3}$, this fraction is ~ 0.3 and correlation between particles must be taken into account. Such a solid, with a significant zero-point to interparticle separation ratio, is termed a "quantum solid." Quantum solids are by no means rare and can be produced under conditions of low temperature and high pressure in the laboratory, for example from the noble gases such as He, Ne, and A.

Anderson and Palmer (AP)[36] attempted to exploit a well-known law among quantum solids, viz. deBoer's law of corresponding states.[37] This law is applicable when the two-body potential V(r) for different molecular species can be obtained simply by scaling. For example, the two-body potentials for the noble gases mentioned previously are well represented by the Lennard-Jones potential

$$V(r) = \varepsilon \left[(\frac{r}{\sigma})^{12} - (\frac{r}{\sigma})^{6} \right] \equiv \varepsilon f(\frac{r}{\sigma}) \qquad (21)$$

where ε and σ depend on the particular species. The Lennard-Jones potential is quite close in form (if scaled down by a factor of

$\sim 10^5$) to that expected for an averaged N-N potential, so AP included the neutrons in the family of noble gases as far as the applicability of the law of corresponding states is concerned.

A summary of deBoer's law is as follows: if the two-body potentials between various molecular species can be obtained by simple scaling, and if the pressure p, volume V and temperature T are expressed in terms of reduced parameters

$$p^* = p\sigma^3/\varepsilon,\ V^* = V/N\sigma^3,\ T^* = kT/\varepsilon,$$

then the reduced equation of states is a universal relation, i.e., $p^* = f(V^*, T^*)$. This reduced equation of state is found to depend on a quantum parameter Λ^* defined by

$$\Lambda^* = h(m\varepsilon\sigma^2)^{-\frac{1}{2}}$$

and a graph of p_s^* versus Λ^* is a smooth curve, where p_s^* is the reduced solidification pressure. What AP did essentially was to approximate the N-N potential by a core-shifted Lennard-Jones form i.e. of the form $\varepsilon f[(r+a)/\sigma]$, where a is the core-shift distance. AP then determined ε and σ from empirical data on nuclear matter and hence Λ^*. The solidification pressure for the neutron solid was then read off from the existing p_s^* versus Λ^* curve for the noble gases. The value obtained by AP was a solidification pressure of around 10^{28} atmospheres, which corresponds to a density of around 5×10^{14}g.cm^{-3}. However, the above work is too severely linked to the possibility of representing the N-N potential by a Lennard-Jones form which avoids the treatment of the angular momentum dependence of the real potential. In fact there is evidence(38) that the Yukawa nucleon core leads to quite different solidification behavior compared to the harder Lennard-Jones core in the context of deBoer's law. Despite several tricks devised by many people, the idea of Anderson and Palmer seems to have exhausted its fruitfulness.

The early microscopic attempts. The calculations described above were exploratory in nature, whereas the subsequent computations are microscopic in nature and the best available nucleon-nucleon data and many-body techniques were employed. The techniques employed in the recent works may be divided into two groups, to be discussed in more detail later: (1) variational, and (2) t-matrix methods.

One general method of approach is to assume that the solid exists and compute the energy per particle versus density. Next, assume the liquid exists and compute its energy per particle versus density curve, using the same many-body technique and potentials. Compare the solid and liquid curves. If at some point the solid curve falls below the liquid curve, the solid phase will be the stable phase in this region, whereas if the solid curve remains higher than the liquid curve always, the liquid will be the stable phase always.

Another procedure to adopt if the liquid curve cannot be calculated with the same technique as the solid, is to (mathematically) subject the solid to a small shearing stress. The shear modulus C_{44} of the lattice is then plotted versus density and the transition from negative to positive values of the shear modulus marks the transition from liquid or gel to a crystalline solid, if a solid state does exist.

Canuto and Chitre[39] employed the t-matrix method in their investigation of the neutron solidification problem. This method has been developed and applied by Guyer[40] in his treatment of solid He^3, which is another quantum solid.

The t-matrix method offers the important advantage of being able to deal with state-dependent interactions, which is one of the prominent features of the N-N potential. In this method, the Hamiltonian for a system of N particles

$$H \equiv \sum_{i}^{N} T_i + \tfrac{1}{2} \sum_{i \neq j}^{N} V_{ij} \tag{22}$$

is separated into two parts, i.e.

$$H = H_o + H_1$$

where

$$H_0 = \sum_{i}^{N} T_i + \tfrac{1}{2} \sum_{i \neq j}^{N} W_{ij}$$

$$H_1 = \tfrac{1}{2} \sum_{i \neq j}^{N} (V_{ij} - W_{ij}) \tag{23}$$

The Hamiltonian H_0 is supposed to be exactly solvable, leading to localized orbitals. Expanding W_{ij} about the lattice site and retaining up to quadratic terms, the eigenfunctions of H_0 turn out to be made up of harmonic oscillator wave functions of unknown frequency ω. For example, the ground state eigenfunction $\Phi_0(1,2,\ldots.N)$ satisfies the relation

$$H_0\Phi_0(1,2,\ldots.N) = E_0\Phi_0(1,2,\ldots.N) \tag{24}$$

where $\Phi_0(1,2,\ldots.N)$ is a product of N single-particle wave functions of gaussian form

$$\Phi_0(1,2,\ldots.N) = \prod_i^N \phi(i)$$

$$\phi(i) = (\alpha^{3/2}/\pi^{3/4})e^{-\alpha^2(\vec{r}_i - \vec{R}_i)^2/2} \tag{25}$$

$\vec{r}_i$ is the position of particle i which is associated with the lattice site situated at $\vec{R}_i$. The parameter $\alpha^{-1} = (\hbar/m\omega)^{1/2}$ represents the spread of the wave function about $\vec{R}_i$. An unperturbed wave function such as (25), which is localized at a lattice site and hence contains correlation, is supposed to exhibit the essential features of the actual system. The Hamiltonian H_1 is used as a perturbation, and the Rayleigh-Schrödinger expansion of the energy shift gives rise to the following expression for the total energy per particle[41]

$$E/N = \frac{3}{4}\hbar\omega + \frac{1}{2N}\sum_{i\neq j}^{N} \frac{\int\psi_{ij}V_{ij}\phi(i)\phi(j)d^3\vec{r}_i d^3\vec{r}_j}{\int\psi_{ij}\phi(i)\phi(j)d^3\vec{r}_i d^3\vec{r}_j} \tag{26}$$

ψ_{ij} is the correlated two-body wave function that satisfies the Beth-Goldstone equation

$$\left[T_i + T_j + U(i) + U(j) + V_{ij}\right]\psi_{ij} = \varepsilon_{ij}\psi_{ij} \tag{27}$$

Here U(i) is the self-consistent one-body potential which we expand in a Taylor series up to quadratic terms

$$U(i) = U(0) + \tfrac{1}{2}M\omega^2 (\vec{r}_i - \vec{R}_i)^2 \tag{28}$$

If we use relative and center-of-mass coordinates $\vec{r}$ and $\vec{R}$ defined by

$$\vec{r} = \vec{r}_i - \vec{r}_j \ , \quad \vec{R} = (\vec{r}_i + \vec{r}_j)/2$$

then

$$U(i) + U(j) \equiv U(ij) = U(\vec{r}) + U(\vec{R}) \tag{29}$$

where

$$U(\vec{r}) = 2U(0) + \tfrac{1}{4}M\omega^2(\vec{r} - \vec{\Delta})^2$$

$$U(\vec{R}) = M\omega^2 \left[\vec{R} - \tfrac{1}{2}(\vec{R}_i + \vec{R}_j)\right]^2 \tag{30}$$

The Bethe-Goldstone equation separates into terms involving $\vec{r}$ and $\vec{R}$. The more important part is that involving $\vec{r}$ and is given by

$$\left[\frac{-\hbar^2}{M}\nabla_r^2 + \tfrac{1}{4}M\omega^2(\vec{r}-\vec{\Delta})^2 + V(\vec{r})\right]\psi(\vec{r}) = (\varepsilon_{ij} - 2U(0) - \frac{3}{2}\hbar\omega)\psi(\vec{r}) \equiv \tilde{\varepsilon}_{ij}\psi(\vec{r}) \tag{31}$$

where $\vec{\Delta} = \vec{R}_i - \vec{R}_j$. Eq. (31) was then solved at a given density for various values of α by constructing a partial wave expansion of $\psi(\vec{r})$ belonging to definite quantum states S and M_S. E/N was then computed for a given lattice and spin configuration. The value of α that produced the same value for two independent expressions for U(0) was taken to be the correct value for the lattice. The E/N so computed is then considered the correct value at that density.

The most unsatisfactory feature of eq. (31) is the term $\vec{r}\cdot\vec{\Delta} = r\Delta\cos\theta$, which couples even and odd partial waves. All previous works that dealt with such an equation invariably averaged over angles thus avoiding the angular momentum problem. To judge the t-matrix method and the handling of eq. (31), the most appropriate test is solid He^3, and some results of this test are presented in Fig. 8. The two theoretical curves to compare for the moment are those of Guyer and Canuto et al. Guyer applied the t-matrix method but averaged the $\vec{r}\cdot\vec{\Delta}$ term over angles. Canuto et al.[42] expanded ψ into partial waves up to $\ell = 25$ and solved the resulting set of coupled differential equations, using the same

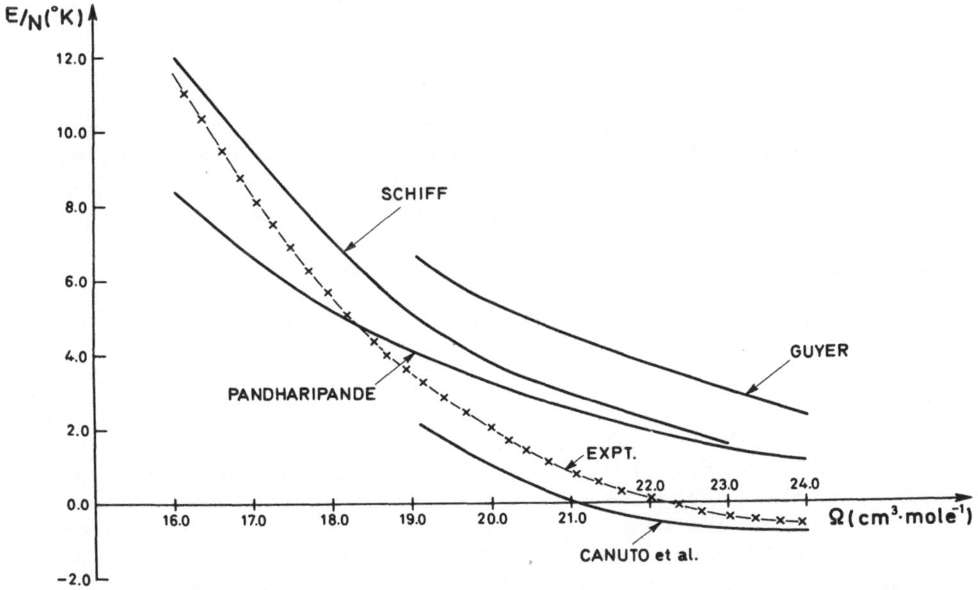

Fig. 8. Ground state energy for solid He^3 vs. molar volume.

two-body potential as Guyer, viz. the (state-independent) Lennard-Jones potential, eq. (21). The improvement in the previous result is significant and should be even more so in the neutron case where the potential is state-dependent.

In the case of the neutron solid, partial waves up to $\ell = 6$ were taken, with the Reid potential for $V(\vec{r})$. The coupling of even and odd ℓ waves by the $\vec{r}\cdot\vec{\Delta}$ term brought in unphysical potentials and the choices made to deal with this problem are discussed in (39). It is found in (39) that the fcc configuration with antiparallel spins gives the lowest energy for the solid in the relevant density range. However, to determine the density where the transition from liquid to solid actually took place, Canuto and Chitre adopted the procedure of computing the shear modulus C_{44} of their lattice as a function of density. This approach was adopted because the liquid phase cannot be readily computed by the t-matrix method, which is designed from the beginning to deal with solids. Briefly, the idea involved in computing C_{44} is as follows. If a unit cell of the lattice is subjected to an infinitesimal deformation δ, its energy can be represented by[43]

$$E(\delta) = E(0) + \Delta E \tag{32}$$

where $E(0)$ is the energy of the undeformed lattice and ΔE, the deformation energy, is given by

$$\Delta E = \tfrac{1}{2}C_{11}(e^2_{xx} + e^2_{yy} + e^2_{zz}) + C_{12}(e_{yy}e_{zz} + e_{zz}e_{xx} + e_{xx}e_{yy}) + \tfrac{1}{2}C_{44}(e^2_{yz} + e^2_{zx} + e^2_{xy}) \tag{33}$$

where e_{xy} etc. are the components of the strain tensor. The crystal structure is deemed to be stable if $\Delta E > 0$, i.e. if there is a gain of energy while undergoing a small deformation. Eq. (33) is positive-definite provided the following conditions are satisfied

(a) $C_{11} + 2C_{12} \equiv 3/\kappa > 0$ (κ = compressibility)

(b) $C_{44} > 0$

(d) $C_{11} - C_{12} > 0$

If (a) is violated, the lattice has no cohesion and it is generally unstable. If (b) is violated, there is an elastic resistance to shearing and the lattice melts, while if (c) is violated, there may yet be an elastic resistance to shearing stresses, but the material exists in a "gel" form. Physically, C_{44} represents the shear modulus, so the condition for the solid to exist is $C_{44} > 0$.

In the computations of Canuto and Chitre for the fcc configuration, the shear modulus C_{44} becomes positive at a density $\rho \sim 1.6 \times 10^{15} g.cm^{-3}$ and this density is taken to be the solidification density for the neutrons.

Pandharipande[44] employed the variational technique in his treatment of the solidification problem. The normalized variational wave function is assumed to be

$$\Psi_v = A\{ \prod_{m<n} f_{mn}(|\vec{r}_{ij}|) \prod_m \phi_m(\vec{r}_i)\} \tag{34}$$

where $\phi_m(\vec{r}_i)$ are gaussians centered at lattice points R_m, i.e.

$$\phi_m(\vec{r}_i) = (\alpha^{3/2}/\pi^{3/4})\, e^{-\frac{\alpha^2}{2}(\vec{r}_i - \vec{R}_m)^2} \tag{35}$$

Long-range correlations are included in $\phi_m(\vec{r}_i)$, while $f_{mn}(|\vec{r}_{ij}|)$ represent the short-range correlations. Both α and f are to be determined variationally by minimizing the energy expectation value

$$E = (\Psi_v, H\Psi_v) / (\Psi_v, \Psi_v) \tag{36}$$

Van Kampen[45] has given a cluster expansion of E in which each term contains a finite number of bodies and in which the convergence depends on the range of f. In order to assure convergence, Pandharipande applied suitable constraints on f such that only lowest order (up to two-body) clusterings are significant. These conditions are

$$\begin{aligned} f(r > d) &= 1 \\ \nabla f(r = d) &= 0 \end{aligned} \tag{37}$$

where d is the range of f and satisfies

$$r_{rms} = (3/2)^{\frac{1}{2}} \alpha^{-1} = 2r_o - d$$

and r_o is given by

$$\frac{4}{3} \pi n r_o^3 \equiv 1$$

Pandharipande calls this the lowest order constrained variational (LOCV) method.

Although the restriction of the correlation function to render many-body contributions of higher orders negligible is an advantage of the variational method, its inability to handle state-dependent potentials in a straightforward manner is a distinct disadvantage.

When the van Kampen expansion for the ground-state energy E is truncated at the second order, we get

$$E = \frac{3}{4} N\hbar\omega + \tfrac{1}{2} \sum_{ij} C_2(ij) \tag{38}$$

where C_2 contains 3 parts. Two of these do not involve the potential and almost cancel each other. The third part is required to be minimized and gives rise to a differential equation for Ψ

$$\left[-\frac{\hbar^2}{M} \nabla_r^2 + \tfrac{1}{4} M\omega^2 (\vec{r} - \vec{\Delta})^2 + V(\vec{r}) \right] \psi(\vec{r}) = \varepsilon\psi(\vec{r})$$

which is precisely equation (31) in the formalism of Canuto and Chitre, but is usually changed into an equation for f by the use of eq. (34).

Pandharipande applied the LOCV method to the test case, solid He^3 and obtained the curve shown in Fig. 8. His curve is not as steep as the experimental curve and those of Guyer and Canuto et al. and gets worse with increasing density. This is surprising in view of the fact that the chief merit of the LOCV is its ability to handle the high density regime through the constraints imposed on the correlation function.

Pandharipande calculated the neutron solid energies using for the potentials the central part of Reid $^1D_2$ for all even-ℓ states except $\ell = 0$, where he used the Reid $^1S_0$. For odd-ℓ states he used the central part of Reid $^3P_2 - {}^3F_2$. The LOCV can also be adapted to treat the liquid case as well (46) and when the liquid and solid energies were compared, Pandharipande found that the E/N for the liquid was always lower than that for the solid over the entire relevant density range. His conclusion therefore is that the neutrons do not solidify.

Nosanow and Parish[47] also evaluated both the solid and liquid energies of the neutron system. These authors formulated their problem within the framework of the variational method, but used Monte-Carlo technique to compute the energies, and included effects of exchange to lowest order in a cluster expansion. The results of their computations are shown in Fig. 9 where the liquid and solid energies per particle are plotted against density. The potentials used were the Reid $^1S_0$ for even-ℓ states and the central part of the Reid $^3P_2 - {}^3F_2$ for all odd-ℓ states. A liquid-solid transition does occur at rather low density $\approx 4.4 \times 10^{14} g.cm^{-3}$. Nosanow and Parish do not give any indication of the sensitivity of the results to the choice of potentials however.

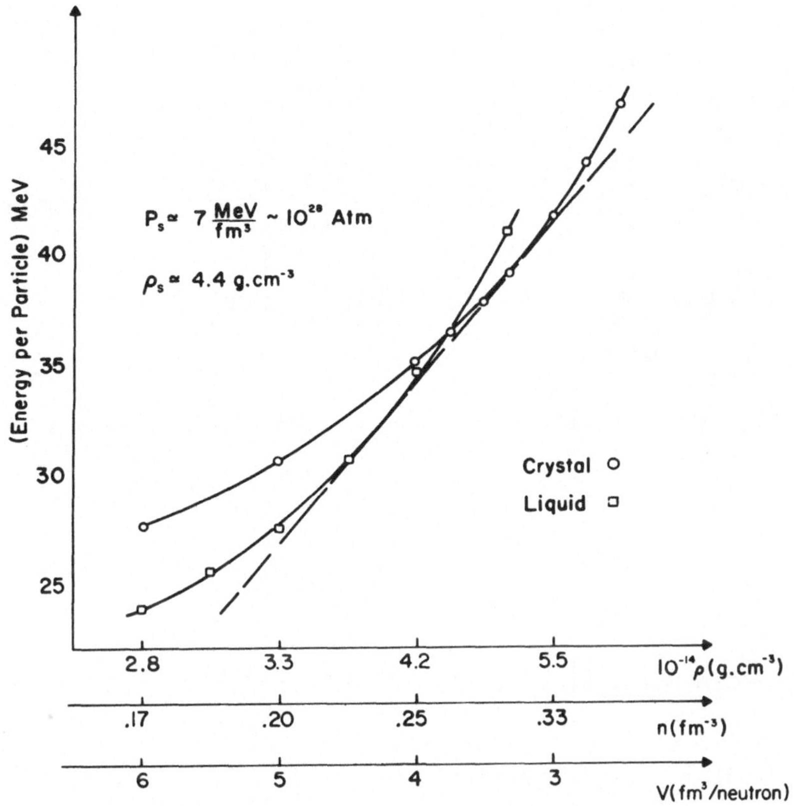

Fig. 9. Liquid-solid transition as from the work of Nosanow and Parish.

The neutron solidification problem has been investigated by several other authors including Coldwell[(48)] and Schiff,[(49)] both getting solidification at $\sim 10^{15}$ and 3×10^{15} g.cm^{-3} respectively. A summary of the results of investigations carried out on the solidification problem up to 1974 is shown schematically in Fig. 10.

Recent results. We have mentioned that the most unsatisfactory feature of the t-matrix approach to the solidification problem is the $\vec{r}\cdot\vec{\Delta}$ term in eq. (31) which couples even and odd ℓ-waves, and therefore brings in unphysical potentials. The origin of this term lies in the self-consistent one-body potentials of eq. (30). This potential is clearly not symmetric, i.e., $U(\vec{r}) \neq U(-\vec{r})$. Recently, Canuto et al.[(50)] have obtained a symmetric one-body potential, denoted $U_S(\vec{r})$, constructed in such a way that the ground-state eigenfunction of the Bethe-Goldstone equation (31) with $V(\vec{r}) = 0$, be a symmetrized linear combination of gaussians, i.e.

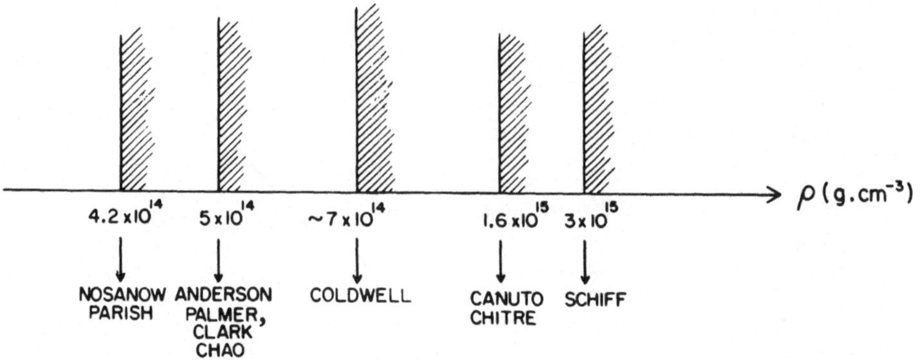

Fig. 10. The solidification densities as from several authors.

$$\left[T_i + T_j + U_s(i,j)\right] \quad \Phi^{\pm}(i,j) = \varepsilon_o^{\pm}\, \Phi^{\pm}\,(i,j) \tag{39}$$

where

$$\Phi^{\pm}(i,j) = \frac{1}{\sqrt{2}}\left[\phi_i(i)\phi_j(j) \pm \phi_j(i)\phi_i(j)\right] \tag{40}$$

and

$$\phi_i(j) = \frac{\alpha^{3/2}}{\pi^{3/4}}\, e^{-\frac{\alpha^2}{2}(\vec{r}_j - \vec{R}_i)^2}$$

Let $\varepsilon_o^{\pm}$ be chosen equal to $3\hbar\omega + 2U(0)$, which is the same eigenvalue of eq. (31) with $V(\vec{r}) = 0$ (and therefore equivalent to the neglect of exchange energy) and write $U_s(ij) \equiv U_s(\vec{R}) + U_s(\vec{r})$. We then obtain

$$U_s^{\pm}(\vec{r}) = 2U(0) + \tfrac{1}{4}M\omega^2 \left[r^2 + \Delta^2 - 2\vec{r}\cdot\vec{\Delta}\left\{\begin{matrix}\tanh(\frac{\alpha^2}{2}\vec{r}\cdot\vec{\Delta}) & S=0 \\ \coth(\frac{\alpha^2}{2}\vec{r}\cdot\vec{\Delta}) & S=1\end{matrix}\right.\right] \tag{41}$$

We notice now that $U_s(\vec{r})$ is symmetric with respect to the transformation $\vec{r}\to-\vec{r}$. However, when $\frac{1}{2}\alpha^2\vec{r}\cdot\vec{\Delta} >> 1$, U_s reduces to the non-symmetric $U(\vec{r})$.

When this symmetric one-body potential was used in the treatment of the solid He^3 problem, the results for the energy per particle versus molar volume were almost identical to those of (42). (This is not surprising in view of the fact that the Lennard-Jones potential is state-independent and $U_s(\vec{r})$ is quite similar to $U(\vec{r})$ in this case.) Canuto and Lodenquai (51) then incorporated this symmetric potential in the neutron solidification problem. First, the repulsive part of the $^1S_0$ Bethe-Johnson potential was used as a test case. When a similar repulsive potential was used in the formalism with the non-symmetric $U(\vec{r})$, there was convincing evidence for solidification.[52] The present method did indicate solidification but the spread of the gaussian wave function was rather large and so cast some doubt on the validity of the claim for solidification. When the full Bethe-Johnson potential was then applied, there was no evidence for solidification. This latest result seems to point out the importance of overcoming the problem of the unphysical waves introduced by the non-symmetric potential $U(\vec{r})$. Around the same time, Takemori and Guyer[53] also applied their t-matrix formalism to the neutron solidification problem and also came up with similar negative results.

(v) Overall equation of state. Although the lack of unanimity concerning the neutron solidification problem is regrettable, it fortunately does not seriously affect the overall equation of state of the neutron star. For example, in Fig. 11 we present various possible equations of state in the range 5×10^{14} to 10^{16}g.cm^{-3}. The lowest curve H represents the equation of state for a free Fermi gas which was originally used by Oppenheimer and Volkoff, while the uppermost curve I is the stiffest equation of state found so far. (Both H and I represent unrealistic equations of state.) The intermediate curves are all based on attempts at more realistic treatments. It should be noticed that there is really very little scatter among these curves.

When the above equations of state are used in the relativistic equations of hydrostatic equilibrium to compute the mass versus central density, the curves in Fig. 12 are obtained. It should be noticed that all equations of state give rise to maxima which range from about 1.4 to 1.9 solar masses among the realistic calculations. The corresponding moments of inertia are shown in Fig. 13, where I is plotted against M. When we compare these values of I with the possible range of values of that for the Crab pulsar, eq. (7), we see that all the equations of state give compatible results. If we reverse the argument and try to deduce the distance D of the pulsar, it seems that D = 1.2 kpc. can be eliminated since it gives too low a value for the moment of inertia.

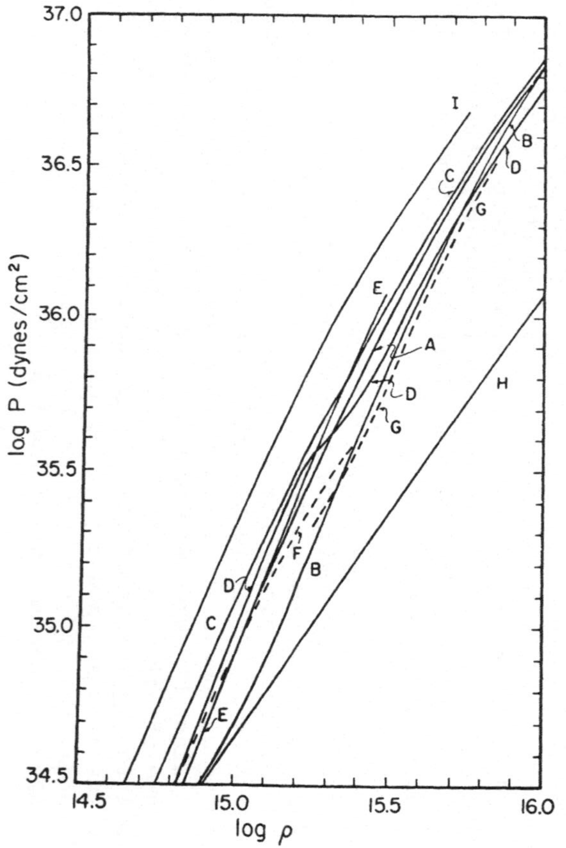

Fig. 11. The equation of state p vs. ρ as from different authors. For the details on the curves see Canuto, V., 1975. E. Fermi, Varenna Summer School

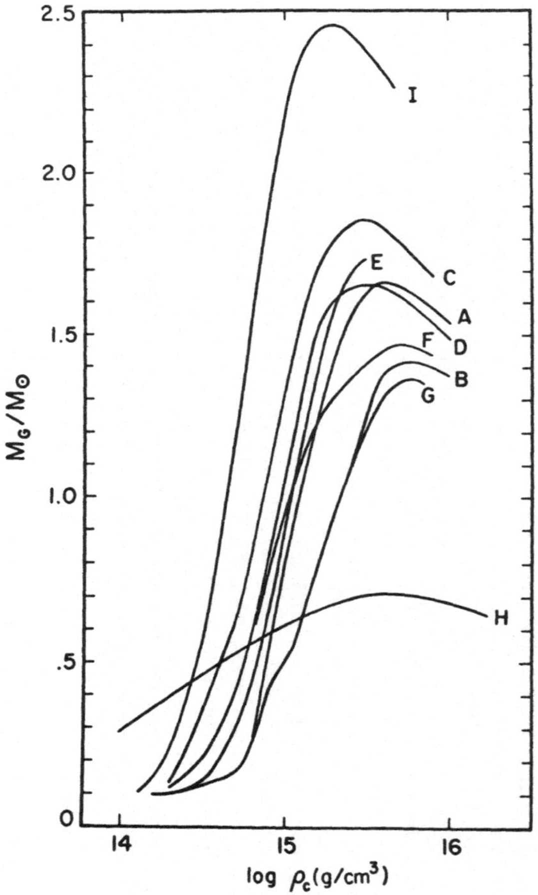

Fig. 12. The maximum mass of a neutron star vs. central density as from the equation of state of Fig. 11.

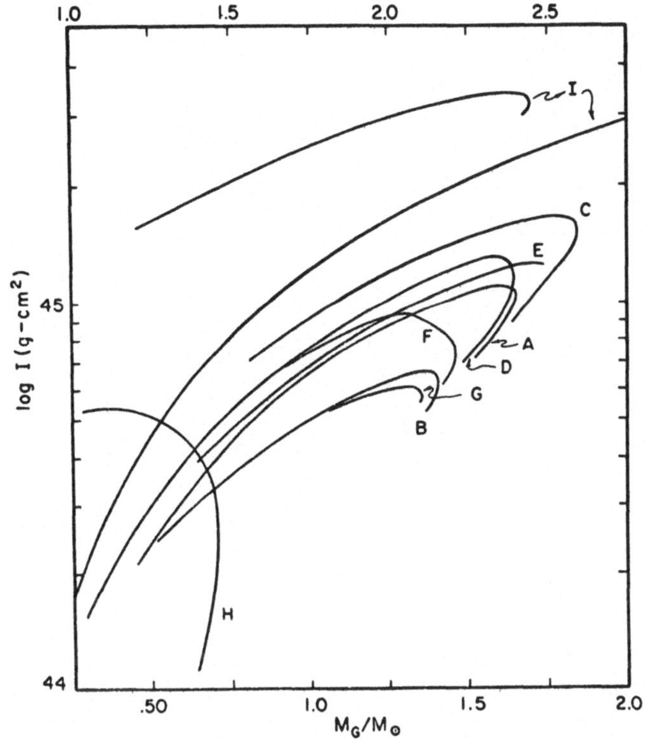

Fig. 13. The moment of inertia vs. M_G corresponding to Fig. 11.

III. DENSITY RANGE GREATER THAN 10^{16} $g.cm^{-3}$

Once we exceed a density of around $10^{16} g.cm^{-3}$, the behavior of matter can no longer be described by the two conventional tools: (1) non-relativistic many-body theories, and (2) the concept of a potential. Even before we reach a density where the concept of the potential breaks down, we already face another problem. Different potentials that are equally reliable on the basis of their phase-shift predictions, give rise to different ground-state energies. This is a manifestation of the fact that the region around $10^{15} g.cm^{-3}$ is extremely sensitive to the hard-core region which comes into effect at such a high density.

The above considerations call for a fully relativistic treatment of a many-particle system, but the works published so far in this area are still exploratory in nature. It is interesting to find out the asymptotic limit of the p versus ε curve. In Fig. 14 the solid curve represents a typical realistic result up to $\sim 10^{16}$ $g.cm^{-3}$. The question arises: as ε increases, would the continuation of the curve approach the limiting lines $p = \varepsilon$, $p = 1/3\,\varepsilon$ (appropriate for a non-interacting system) or some other value such as p = constant?

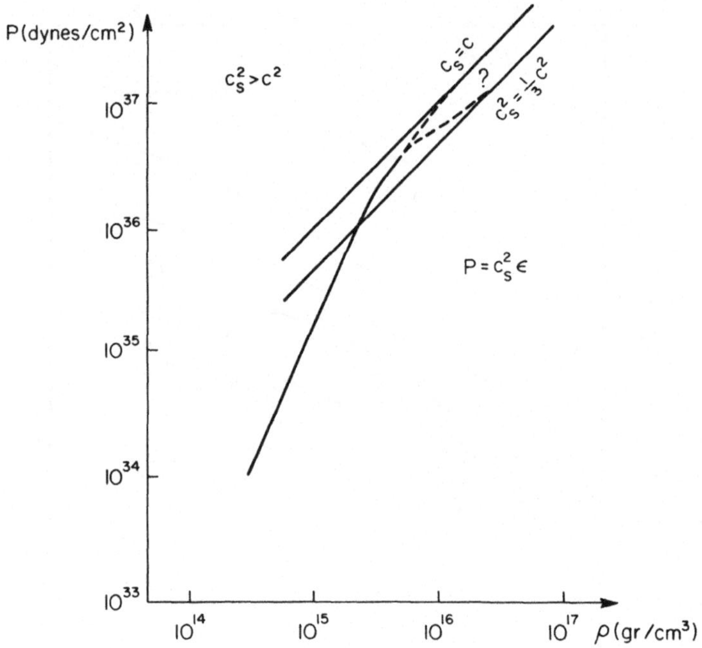

Fig. 14. Possible behavior of p vs. ρ at superhigh densities

TABLE 2

$$P = c_s^2 \varepsilon$$

	c_s^2
1) Highly relativistic free fermions	1/3
2) All "respectable" equations of state non-relativistic treatment	>1/3
3) Leung and Wang model (54)	∿.12
4) Zeldovich model (55)	1
5) Mean field theories (σ,ω exchange) (56)	1
6) Green functions technique (57)	incomplete
7) Hagedorn (58)	0 $p \sim \ln \varepsilon$
8) Spin 2 model (59)	<0

Many attempts to deal with this asymptotic density regime have been made so far. A tabulated summary of these results is given in Table 2, along with references. The corresponding asymptotic values of the sound speed squared, c_s^2 of eq. (1), are also listed. The disagreement in the values of c_s^2 is glaring and therefore unsatisfactory. In order to obtain an unambiguous value for c_s^2 in the asymptotic limit, Canuto and Lodenquai[60] decided to adopt a semi-empirical approach. It should be evident from our discussion of the predicted values of the moments of inertia of a neutron star that an approach based on the comparison of moments of inertia would be quite insensitive to the equations of state in the asymptotic limit. However, in the case of high energy proton-proton collisions, the matter density can exceed 10^{17} g.cm^{-3} in the center of mass (c.m.) system as we shall see. Landau has developed a relativistic hydrodynamical model to predict the multiplicity and distribution of the resulting particles, and these predictions depend quite critically on the assumed equation of state. It was the intention of Canuto and Lodenquai to use the Landau model to predict the multiplicity and transverse momentum distribution, leaving c_s^2 in the equation of state as a free parameter to be chosen so as to bring theory and experiment into agreement.

(i) Summary of the Landau Model[61]

When two ultrarelativistic protons collide, each will be Lorentz-contracted into flattened disk in the c.m. frame. The volume of each disk is given by ($\hbar = c = 1$)

$$V_F \sim \frac{4\pi}{3} \frac{1}{m_\pi^3} (M/E) \tag{42}$$

where the radius of a proton in its rest frame is taken to be of the order of the pion Compton wavelength m_π^{-1}; m_π and M are the rest-masses of the pion and proton respectively and E the c.m. energy of each proton. The factor (M/E) is the Lorentz contraction factor. When the protons collide, a dense, flat blob of hadronic matter results, with density in the c.m. frame given by

$$\rho \sim E/V_F \sim 1.5 \times 10^{14} E_L \ (\text{g.cm}^{-3}) \tag{43}$$

where E_L is the laboratory kinetic energy of the projectile proton measured in Gev. and is related to the c.m. energy E through the relation

$$E = \sqrt{2ME_L} \tag{44}$$

For example, if we take $E_L \sim 10^3$ Gev., then $\rho \sim 10^{17}$g.cm^{-3}. Immediately after the collision of the two protons, the constituents of the resulting hadronic blob will be strongly interacting and will experience mutual transformation. Subsequent to the collision, this blob will expand in essentially one dimension in two oppositely directed jets as a relativistic fluid. When the mean distance between the constituents of this hadronic liquid exceeds $\sim m_\pi^{-1}$, the range of the strong interaction, the fluid system breaks up into the resulting product particles.

The first detailed model that described the evolution of the blob was proposed by Landau in 1953. During the expansion of the blob, the run of the thermodynamic and hydrodynamic variables such as the temperature T and fluid four-velocity u_μ versus space and time may be derived from the solution of the relativistic Navier-Stokes equation

$$T_{\mu\nu,\nu} = 0 \tag{45}$$

where the energy-momentum tensor $T_{\mu\nu}$ can be written as

$$T_{\mu\nu} = (p+\varepsilon)u_\mu u_\nu + p\delta_{\mu\nu} + \tau_{\mu\nu} \equiv T^{(o)}_{\mu\nu} + \tau_{\mu\nu} \tag{46}$$

In order for $T_{\mu\nu}$ to be used in eq. (45), an equation of state $p = p(\varepsilon)$ is required. Originally, Landau used the relation appropriate for a non-interacting gas, $p = 1/3\,\varepsilon$, and neglected the viscous effects, i.e., $\tau_{\mu\nu} = 0$. In this case the total entropy S remains constant throughout the expansion. Since the multiplicity N_s is proportional to the entropy, one gets

$$N_s \sim sV \sim \frac{\varepsilon}{T}V \tag{47}$$

where s is the specific entropy. Now $s \sim \varepsilon^{3/4}$ (cf. black-body radiation). If we take $V = V_F$, then $\varepsilon = E/V_F$ and so

$$N_s \sim E_L^{\frac{1}{4}} \tag{48}$$

From dimensional arguments, eq. (48) can be written

$$N_s = K(E_L/2M)^{\frac{1}{4}} \tag{49}$$

When this result was fitted to the experimental points known in the 1950's, K turned out to have a value of about 2. When Landau's result is compared with the present experimental points, the agreement is not too impressive, Fig. 15.

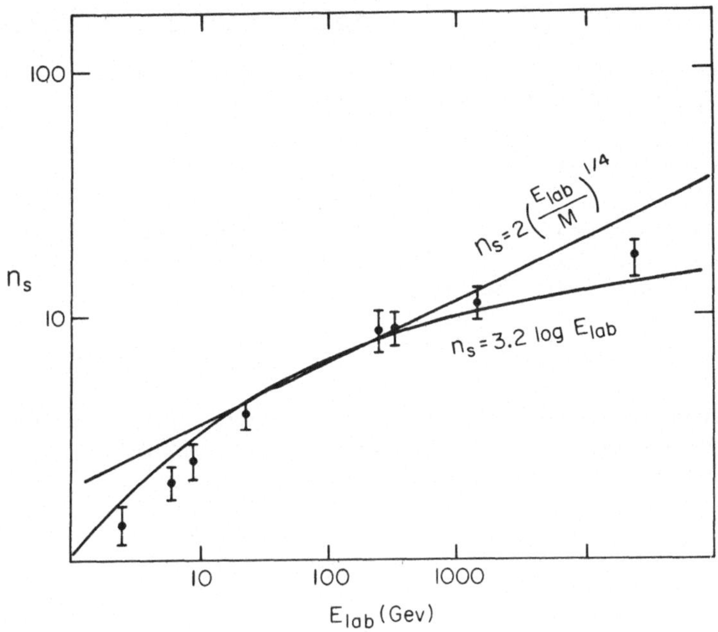

Fig. 15. The multiplicity vs. $E_{Lab.}$ (see Table 4).

Landau neglected the dissipative term $\tau_{\mu\nu}$ in the energy-momentum tensor because when he made an estimate of the Reynolds number Re of the hadronic fluid, he found it to be much greater than unity. Since Re is inversely proportional to the viscosity, then Re >> 1 is the condition for the viscosity to be negligible. However, Iso, Mori and Namiki,[62] using a specific Lagrangian and employing a perturbative approach to the Kubo formulas for the transport coefficients, found that both the viscosity and the thermal conductivity were proportional to T^3 and therefore substantial. Feinberg[63] then showed that the T^3 dependence followed more generally from dimensional arguments if the coupling constant entering in the given Lagrangian is dimensionless and if T is high enough such that the particle masses can be neglected.

The above analyses implied that in the initial stage of the Landau model, thermodynamic equilibrium cannot be established and therefore Landau's results were invalid. It is only after the fluid has expanded to a certain volume V_q, where the temperature has fallen sufficiently so that the viscosity η may be neglected, that the Landau model may be applied. It is evident that as the fluid expands from V_F to V_q, the entropy increases and hence the multiplicity, which is proportional to the entropy at V_q, will be greater than that in the Landau case of negligible viscosity.

(ii) Landau Model with General Equation of State

If the general equation of state $p = c_s^2 \varepsilon$ is used, it is shown in (60) that for dimensionless coupling constant,

$$\eta \sim T^{1/c_s^2} \tag{50}$$

so c_s becomes the only free parameter in such a model. In this general case, the first law of thermodynamics gives $\varepsilon \sim T^{(1+c_s^2)/c_s^2}$, and so the multiplicity becomes

$$N_s \sim sV_q \sim \frac{\varepsilon}{T} V_q \sim E^{1/(1+c_s^2)} V_q^{c_s^2/(1+c_s^2)} \tag{51}$$

This relation is quite general, and the only problem is in specifying V_q. The evaluation of the transverse momentum is done using the solution of eq. (45) and is given by

$$\langle p_T \rangle \simeq N_s^{-1} E^{(2c_s^2 - 1 - c_s^4)/2c_s^2} \, *$$

$$* \int d\lambda \exp\left(\frac{1-c_s^4}{2c_s^2} (L^*-\lambda^2)^{\frac{1}{2}}\right); \quad L^* = \ln(E/2M)$$

and to leading power in E we have

$$\langle p_T \rangle \simeq E^{1-c_s^2} N_s^{-1} \tag{52}$$

which again is of quite general validity. With these equations, we can investigate the predictions of the model in two important cases.

(a) Inviscid flow, $\eta = 0$. If we accept that $\eta = 0$ always, then the entropy S remains constant in the expansion, so we can choose $V_q = V_F$. In this case

$$N_s \simeq E^{(1-c_s^2)/(1+c_s^2)}$$

$$\langle p_T \rangle \simeq E^{c_s^2(1-c_s^2)/(1+c_s^2)} \tag{53}$$

These results have been discussed by Suhonen et al.[64] and by Chaichian et al.[65]

(b) Viscous flow, $\eta \neq 0$. In this more realistic case, the solution to eq. (45) for the temperature profile $T = T(x,t)$ with finite η is not known. However, we can examine the Reynolds number Re ($\sim\eta^{-1}$) and decide whether it is $\gtrsim 1$. It can be shown that[63]

$$Re = \frac{\varepsilon L}{\eta} \tag{54}$$

where L is a typical length of the system. Let us write $\eta \sim T^k$, where k is to be specified. Then eq. (54) can be written

$$Re = L(E/V_q)^{[1+(1-k)c_s^2]/(1+c_s^2)} \tag{55}$$

which is still quite general. Since we are dealing with one-dimensional flow, the volume V_q which is Lorentz-contracted, can be written

$$V_q \simeq L_0^2 L \tag{56}$$

where $L_0 \sim m_\pi^{-1}$. Eq. (55) can then be rewritten for the one-dimensional case as

$$Re = (L/L_1)^{kc_s^2/(1+c_s^2)}$$

$$L_1 \equiv E^{-\left[1+(1-k)c_s^2\right]/kc_s^2} \tag{57}$$

Since the exponent of Re is positive, it follows that Re is large, i.e. η is small only for $L > L_1$, L_1 defining the minimum distance at which one can start applying the concept of constant entropy.

If we now substitute the above definition of L_1 into eqs. (51) and (52), we get

$$N_s \sim E^{(k-1)/k}$$

$$\langle p_T \rangle \simeq E^{(1-kc_s^2)/k} \tag{58}$$

If we choose $k = c_s^{-2}$, eq. (50), we get

$$N_s \simeq E^{1-c_s^2} \tag{59}$$

$$\langle p_T \rangle \simeq E^0 = \text{const.} \tag{60}$$

TABLE 3

INVISCID FLOW

(A) $\eta = 0$ (LANDAU)

$L \sim E^{-1}$

$N_s \sim S \sim E^{(1-c_s^2)/(1+c_s^2)}$

$\langle p_T \rangle \sim E^{c_s^2 (1-c_s^2)(1+c_s^2)}$

VISCOUS FLOW

(B) $\eta \sim T^3$ (FEINBERG)

$L_1 \sim E^{-(1-2c_s^2)/3c_s^2}$

$Re = (\frac{L}{L_1})^{3c_s^2/(1+c_s^2)}$ $N_s \sim E^{2/3}$

$\langle p_T \rangle \sim E^{(1-3c_s^2)/3}$

(C) $\eta \sim T^{1/c_s^2}$ (CANUTO, LODENQUAI)

$L_1 \sim E^{-c_s^2}$

$Re = (\frac{L}{L_1})^{1/(1+c_s^2)}$ $N_s \sim E^{1-c_s^2}$

$\langle p_T \rangle \sim E^{o}$

TABLE 4

c_s^2	η	N_s	$\langle p_T \rangle$
1/3	0	$E^{1/2}$	$E^{1/6}$
1/3	$\sim T^3$	$E^{2/3}$	const
1	0	const	const
1	$\sim T^3$	$E^{2/3}$	$E^{-2/3}$
1	$\sim T$	const	const

On the other hand, if we choose k = 3, corresponding to $c_s^2 = 1/3$ in eq. (50) we get

$$N_s \sim E^{2/3} \tag{61}$$

$$\langle p_T \rangle \sim E^{(1-3c_s^2)/3} \tag{62}$$

In Table 3 we summarize the results for N_s and $\langle p_T \rangle$ for the viscid and inviscid one-dimensional case. In each case we list the distance at which the system becomes inviscid, the corresponding multiplicity and the average transverse momentum. In Table 4 we summarize the predictions of the model for the top two competing values of c_s^2 along with the possible corresponding temperature dependence of the viscosity. The viscosity can be either 0 (Landau), vary like T^3 (62,63) or vary like T^{1/c_s^2}.(60) (Since the formulas given in Table 4 are modulus lnE, whenever const appears, we should understand lnE.) The experimental result for $\langle p_T \rangle$ is shown in Fig. 16. From Fig. 16 we can see the well-known fact that in p-p collisions the average transverse momentum $\langle p_T \rangle$ is constant (or perhaps increases logarithmically) with energy over a wide range, from about 10 to 10^5 Gev. This constancy excludes the combination

$$c_s^2 = 1, \; \eta \sim T^3 \tag{63}$$

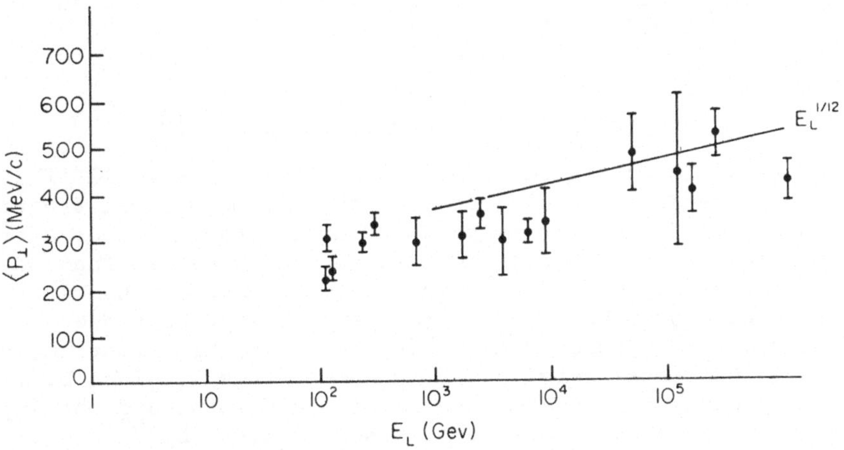

Fig. 16. The transverse momentum vs. E_L. See Ref. (60).

The choice

$$c_s^2 = 1/3, \; \eta \simeq 0$$

yields $\langle p_T \rangle \sim E_L^{1/12}$. Such a behavior does not contradict the experimental data, as shown in Fig. 16. In fact, the present data do not rule out any of the combinations in Table 4, except that of eq. (63). The same conclusion is reached when the particle multiplicities are analyzed. A possibility of discriminating among the remaining combinations comes from comparing the previous results with those of the multiperipheral model, i.e., a lnE behavior instead of a power law. Within the hydrodynamic model, such a behavior can be achieved if we limit ourselves to the combinations

$$c_s^2 = 1; \quad \eta \simeq 0, \; T \tag{64}$$

If we believe this chain of arguments, we are led to a unique value for c_s^2.

In order to choose between the two alternatives $\eta \simeq 0$, T, we can analyze e^+e^- collisions in terms of a hydrodynamic model as well. Because the particles involved in this case are point particles, the Lorentz contraction is irrelevant, and the resulting hadronic fluid subsequently expands with spherical symmetry in the c.m. frame. There is therefore no average transverse momentum to deal with, so the analysis must be limited to the multiplicity. It is shown in (60) that the choice

$$c_s^2 = 1; \quad \eta \sim T \tag{65}$$

of eq. (64) is consistent with e^+e^- multiplicity. In conclusion, we propose the choice eq. (65) along with eq. (50) for a superdense hadronic fluid.

(iii) Turbulence in the Hadronic Fluid

Before we leave the topic of p-p collision, we would like to mention an interesting consequence of the hydrodynamical treatment of the hadronic fluid that is produced as a consequence of ultra-relativistic p-p collisions. We have mentioned before that as the hadronic fluid expands to the volume V_q, the Reynolds number increases and approaches unity. As the expansion continues beyond this volume, the Reynolds number can become so large that the condition for turbulence sets in. At a temperature T, given by $T \sim m_\pi$, the fluid breaks up into the individual components, and the Reynolds number just before this point is reached is given by[60]

$$Re \simeq \frac{5\pi^2}{2} (E/m_\pi) \tag{66}$$

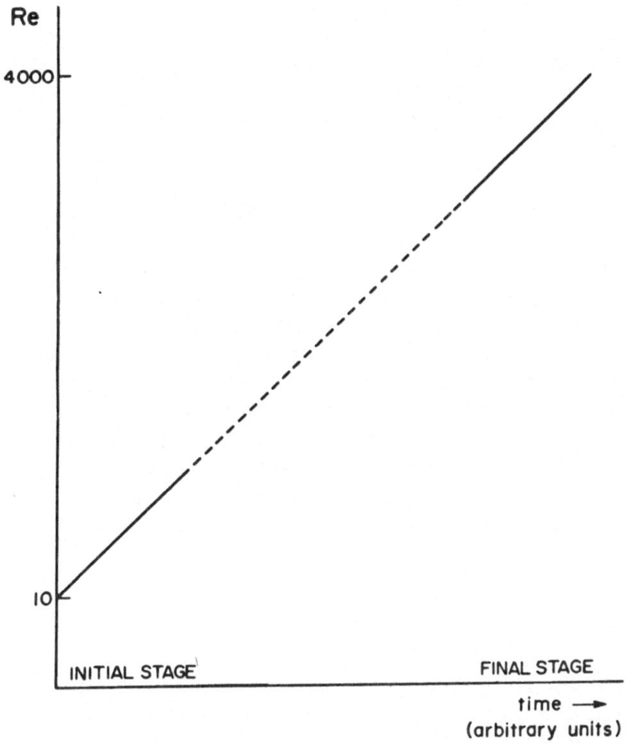

Fig. 17. The Reynolds number vs. expansion time for a p-p collision (Ref. 60).

This relation is quite general, provided the coupling constant of the corresponding Lagrangian for the hadronic field is dimensionless. At a typical laboratory kinetic energy of 10^3 Gev, we get a value for Re of around 4000, a value sufficiently high for turbulence to set in, Fig. 17. Although not much is known about turbulence in a one-dimensional relativistic fluid, we can make the following observations if we limit ourselves to familiar three dimensional turbulence occurring under laboratory conditions. The point at which turbulence sets in depends critically on the geometrical boundaries, and sometimes Reynolds numbers as high as 10^4 are required to produce turbulence. In the case of p-p collisions, the hadronic fluid is produced in a rather violent manner, so one would expect turbulence to occur at a somewhat lower Re. Even so, any turbulence that results is probably not fully developed.

A possible experimental consequence of the turbulence in the expanding hadronic fluid is the formation of clusters of pions, which are here identified with those eddies of the turbulent fluid that detach themselves in the final stage of the expansion. Such clustering of pions resulting from p-p collisions have indeed been identified.[66]

IV. GALAXY FORMATION

One of the major problems in cosmology is the existence of galaxies, a departure from a uniform distribution of matter that takes place on scales of the order of a few Mpcs. Several mechanisms have been proposed but no agreed-upon explanation is yet in sight.[67] The very first model was proposed by Jeans back in the twenties. Jeans studied the rate of growth of an otherwise arbitrary fluctuation in the initial cosmic plasma. If the initial fluctuation $\delta(0)$ is of the order of $N^{-\frac{1}{2}}$, where N is the number of particles involved (for a galaxy $N \sim 10^{68}$), Jeans showed that if the medium is static, the time evolution of $\delta(0)$ is governed by the following law

$$\delta(t) \approx \delta(0)\, e^{\alpha t} \;, \quad (\alpha > 0) \tag{67}$$

Since today $\delta(t_0) \approx 10^6$, the fast exponential growth is capable of securing at least a $\delta(t) \approx 1$ at some time within the age of the universe, $\sim 15 \times 10^9$ years. Regrettably, Jeans' analysis does not include the most important cosmological fact presently known: the expansion of the universe, which historically speaking was established only after Jeans' analysis.

The first fundamental analysis taking the expansion into account is due to Lifshitz. The result was that the exponential factor is degraded to a power law, i.e., (67) is changed to

$$\delta(t) = \delta(0)\, t^{\omega} \tag{68}$$

with $\omega \sim 1$. With such a slow growth in time, one can easily show that the initial $\delta(0) \sim 10^{-34}$ has no time to reach unity within the age of the universe.

If Jeans' idea has to be pursued any further, the only possibility of offsetting the slow polynomial behavior is by trying to exploit the freedom in choosing the initial $\delta(0)$. From statistical mechanics we know that a fluctuation can be written in general as

$$\delta^2(0) = \frac{kT}{V} K \tag{69}$$

where the compressibility K is defined as $(n = N/V)$

$$K = \frac{1}{n}\left(\frac{dp}{dn}\right)^{-1} \tag{70}$$

The use of the perfect gas law, $p = nkT$, immediately gives

$$\delta(0) = N^{-\frac{1}{2}} \approx 10^{-34} \tag{71}$$

for $N \sim 10^{68}$, typical of a galaxy. If, however, one could have $dp/dn = 0$ at some n, say n^*, then from the theory of critical phenomena we learn that in the vicinity of the $K \to \infty$ point, $N^{-\frac{1}{2}}$ is changed to $N^{-1/6}$ a considerable increase, that can offset the slow polynomial behavior t^{ω} to the point of making the model workable again. One possibility of achieving $dp/dn = 0$ consists in considering that at early enough times, the energy density $\varepsilon(=E/V)$ might have been changed by gravitational effects to take the form[67]

$$\varepsilon = \varepsilon_N - \hbar c\left(\frac{\hbar G}{c^3}\right) n^2 - \hbar c\left(\frac{\hbar G}{c^3}\right)^2 n^{8/3} \tag{72}$$

where ε_N is the contribution from nuclear forces, as discussed in the proceeding sections. The negative sign is due to the attractive nature of gravity. The coefficients are the only possible combinations of $\hbar$, c and G and finally the n^2 and $n^{8/3}$ direct and exchange interaction respectively. As discussed in the previous sections, the two most probable alternative forms of the energy density at superhigh densities are

$$\text{a)}\quad \varepsilon_N = \hbar c\, n^{4/3} \;; \qquad \text{b)}\quad \varepsilon_N = g_\omega^2 n^2 \tag{73}$$

Upon computing the pressure $(= n(d\varepsilon/dn)-\varepsilon)$, it is easy to find that $dp/dn = 0$ occurs at

$$\text{a)}\quad n_* = \ell_p^{-3} \approx 10^{99}\ \text{cm}^{-3} \;; \qquad \text{b)}\quad n_* = (\hbar/mc)^3 \ell_p^{-6} \approx 10^{60}\ \ell_p^{-3} \tag{74}$$

Here ℓ_p is the so-called Planck length $(\hbar G c^{-3})^{\frac{1}{2}}$. It is therefore possible to enhance $\delta(0)$, if one considers that the initial fluctuation occurred when $n \approx 10^{99}\ \text{cm}^{-3}$, or $\rho \approx 10^{75}\ \text{g}\cdot\text{cm}^{-3}$, $t \approx 10^{-40}$ sec, $T \approx 10^{30}$ °K. There is a snag in this approach, however. At such early times (even granting that the physics we have done is legitimate!), the number of nucleons within the horizon, $R_H \sim ct$, was too small to form a galaxy. In fact at any time t during the cosmic expansion, the mass within the horizon (= ct) is easily computed to be

$$M_H(t) = (ct)^3 \rho(t) \approx M_\odot 10^5\, t\ (\text{sec})$$

In order to have $M(t) \approx 10^{13}\ M_{\odot}$, i.e. a galaxy, we must wait until $t \approx 10^{13}$ sec $\approx 10^{6}$ yrs. We then conclude that if a phase transition is to be used, it must not occur too early in the history of the universe.

Recently one of the authors (V.C.), in collaboration with B. Datta and G. Kallman has shown that a system of nucleons interacting via scalar mesons (σ), vector mesons (ω) and spin two meson ($f°$) exhibits a phase-transition at baryonic densities higher than 10 baryons per fm^3, i.e., around 10^{16}-10^{17} $g \cdot cm^{-3}$. The resulting equation of state $p = p(n)$ is shown in Figure 18, where it is seen that the presence of f^o (giving attraction) is responsible for the bending of the curve. Due to the uncertainty of the coupling constant, we have studied two limiting cases. The true curve must lie somewhere in the middle. Also plotted is the case without f^o which behaves like (73b) at high n's. The point where $dp/dn = 0$ occurs, is characterized by a density

$$10^{16} \leq \rho_* \leq 10^{17} \quad (g \cdot cm^{-3}) \tag{75}$$

or a time and temperature given by

$$t \approx 10^{-11}\ \text{sec}\ , \qquad T \approx 10^{16}\ {}^oK \tag{76}$$

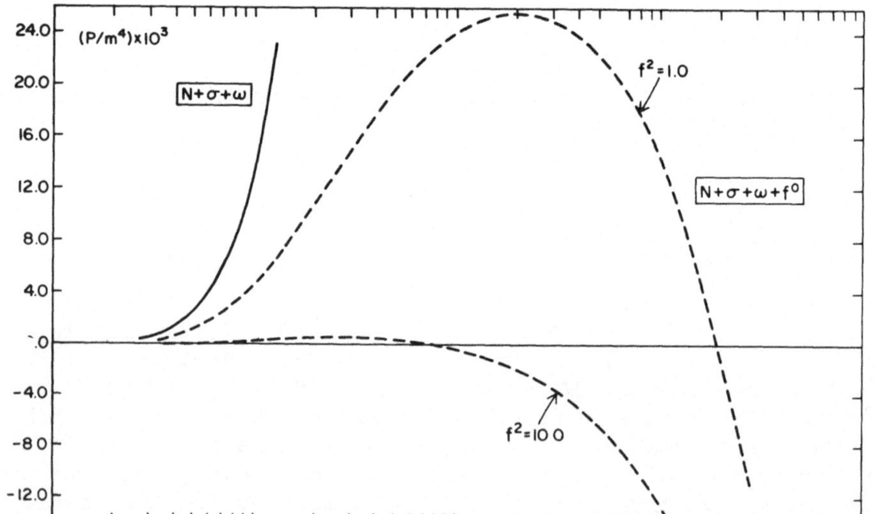

Fig. 18. The equation of state p vs. n for a gas of nucleons interacting with and without the spin 2, f^o meson.

At this point in time, the mass within the horizon is still much smaller than that of a galaxy. In fact, at $t \approx 10^{-11}$ sec, we have

$$M(t) \approx 10^{27} \text{ g.} \tag{77}$$

just about the mass of the Earth. In spite of this, we believe that this mechanism stands a better chance than one based upon Eq. (72).

The main feature is that the phase transition occurs when the quantum-gravitational effects occurring at $t \approx 10^{-40}$ sec have already passed. Even though the mass within the horizon is still too small to be considered a galaxy, we can think of generating not galaxies but black holes, which in some models are considered as seed-galaxies. Even though black-holes with masses less than 10^{14}g. will have evaporated by now, we can still use (77) to form a great number of more massive black holes, which will in the subsequent evolution attract enough matter to become galaxies.

We cannot claim to have solved the problem of galaxy formation. However, we think we have made good headway by presenting a reasonable model whereby primordial black holes can be generated.

CONCLUSIONS

We believe we can conclude this review with a positive note. It is true we don't have a full description of the behavior of matter from 10^6 g.cm^{-3} all the way up. It is also true, however, that only a few years ago we knew much less, actually a great deal less, than today. The discovery of neutron stars has been the *leitmotif* for such an upsurging of interest in the subject of dense matter. We have tried to indicate how not only neutron stars, but also p-p scattering data and early cosmological data can be used meaningfully to complete a scenario that a few years ago was not only poorly understood but also uncorrelated.

It is important to notice how the observational data have played a decisive role in discriminating among different models. It is our opinion that up to 10^{15} - 10^{16} g.cm^{-3} we have indeed understood what is going on. Future data on p-p scattering and on black-hole physics should allow us to make a step ahead of a few powers of ten in densities.

REFERENCES

1. Hewish, A., Bell, S.J., Pilkington, J.D.H., Scott, P.F., and Collins, R.A., Nature 217, 709 (1968).
2. Gold, T., Nature 218, 731 (1968).
3. Landau, L., Physik. Zeit. Sowjetunion 1, 285 (1932).
4. Baade, W., and Zwicky, F., Proc. Nat. Acad. Sci. U.S.A. 20, 259 (1934).
5. Harrison, B.K., Thorne, K.S., Wakano, M., and Wheeler, J.A., "Gravitation Theory and Gravitational Collapse" (University of Chicago Press, 1965).
6. Oppenheimer, J.R., and Volkoff, G.M., Phys. Rev. 55, 374 (1939).
7. Tsuruta, S., Canuto, V., Lodenquai, J., and Ruderman, M., Ap.J. 176, 739 (1972).
8. Ostriker, J.P., and Gunn, J.E., Ap.J. 157, 1395 (1969).
9. Ruderman, M., Ann. Rev. Astron. and Ap. 10, 427 (1972).
10. Börner, G., "Springer Tracts Mod. Phys." 69, 1 (1973).
11. Feynman, R.P., Metropolis, N., and Teller, E., Phys. Rev., 75, 1561 (1949).
12. Ginzburg, V.L., Soviet Physics Doklady 9, 329 (1964).
13. Cohen, R., Lodenquai, J., and Ruderman, M., Phys. Rev. Lett. 25, 467 (1970).
14. Ruderman, M.A., Phys. Rev. Lett. 27, 1306 (1971).
15. Baym, G., Pethick, C., and Sutherland, P., Ap.J. 170, 299 (1971).
16. Ruderman, M.A., "Superdense Matter in Stars," Talk at Colloque du C.N.R.S.: Physique fondamentale et astrophysique, (1969).
17. Mestel, L., and Ruderman, M., Mon. Not. R. Astron. Soc. 136, 27, (1967).
18. Coldwell-Horsfall, R.A., and Maradudin, A.A., J. Math. Phys. 1, 395 (1960).
19. Canuto, V., Ann. Rev. Astron. and Ap. 12, 167 (1974).
20. Bethe, H.A., and Johnson, M.B., Nucl. Phys. A. 230, 1 (1974).
21. Cameron, A.G.W., Ap.J. 130, 884 (1959).
22. Ambartsumyan, V.A., and Saakyan, G.S., Sov. Astron. 4, 187 (1960).
23. Salpeter, E.E., Ann. Phys. 11, 393 (1960).
24. Reid, R.V., Ann. Phys. 50, 411 (1968).

25. Brueckner, K.A., Coon, S.A., and Dabrowsky, J., Phys. Rev. 168, 1184 (1968).

26. Langer, W.D., and Rosen, L., Ap. Space Sci. 6, 217, (1970).

27. Levinger, J.S., Simmons, L.M., Phys. Rev. 124, 916 (1961).

28. Moszkowski, S., Phys. Rev. 9, 1613 (1974).

29. Canuto, V., Ann. Rev. Astron. and Ap. 13, 335 (1975).

30. Ginzburg, V.L., and Kirzhnits, D.A., JETP 20, 1346 (1965).

31. Hoffberg, M., Glassgold, A., Richardson, R., and Ruderman, M., Phys. Rev. Lett. 24, 775 (1970).

32. Chao, N.-C., Clark, J.W., and Yang, C.-H., Nucl. Phys. A 179, 320 (1972).

33. Pines, D., Shaham, J. and Ruderman, M., IAU Symposium (1972).

34. Cazzola, P., Lucaroni, L., and Scarinci, C., Nuovo Cimento B 43, 250 (1966).

35. Banerjee, B., Chitre, S.M., and Garde, V.K. Phys. Rev. Lett. 25, 1125 (1970).

36. Anderson, P.W., and Palmer, R.G., Nature Phys. Sci. 231, 145 (1971).

37. deBoer, J., Physica 14, 139 (1948).

38. Cochran, S., and Chester, G.V., Phys. Rev. (1975).

39. Canuto, V., and Chitre, S.M., Phys. Rev. D 9, 1587 (1974).

40. Guyer, R.A., Solid State Phys. 23, 413, ed. F. Seitz, D. Turnbull and H. Ehrenreich (Academic Press, N.Y., 1969).

41. Guyer, R.A., and Zane, L.I., Phys. Rev. 188, 445 (1969).

42. Canuto, V., Lodenquai, J., and Chitre, S.M., Phys. Rev. A 8, 949, (1973).

43. Born, M., Proc. Camb. Phil. Soc. 36, 160 (1940).

44. Pandharipande, V.R., Nucl. Phys. A 217, 1 (1973).

45. van Kampen, N.G., Physica 27, 783 (1961).

46. Pandharipande, V.R., Nucl. Phys. A 178, 123 (1971).

47. Nosanow, L.H., and Parish, L.J., In "Proc. Texas Symp. Relativistic Ap., 6th Ann. N.Y. Acad. Sci." 224, 226 (1973).

48. Coldwell, R.L., Phys. Rev. D 5, 1273 (1972).

49. Schiff, D., Nature Phys. Sci. 243, 130 (1973).

50. Canuto, V., Lodenquai, J., Parish, L.J., and Chitre, S.M., J. of Low Temp. Phys. 17, 179 (1974).

51. Canuto, V., and Lodenquai, J., Phys, Rev. D 12, 2033 (1975).

52. Canuto, V., Lodenquai, J., and Chitre, S.M., Nucl. Phys. A 233, 521 (1974).

53. Takemori, M.T., and Guyer, R.A., Phys. Rev. D 1, 2696 (1975).

54. Leung, Y.C., and Wang, C.G., Ap.J. 170, 499 (1971).

55. Zel'dovich, Ya. B., JETP 14, 1143 (1962).

56. Walecka, J.D., Ann. Phys. 83, 491 (1974).

57. Bowers, R.L., Campbell, J.A., and Zimmerman, R.L., Phys. Rev. D 8, 1089 (1973), and ref. therein.

58. Hagedorn, R., Astron. and Ap. 5, 184 (1970).

59. Canuto, V., Datta, B., and Kalman, G. (to be published).

60. Canuto, V., and Lodenquai, J., Phys. Rev. D 11, 233 (1975).

61. Landau, L.D., Izv. Nauk. SSSR 17, 31 (1953). English trans. in Collected Papers of L.D. Landau, ed. D. ter Haar. (Gordon and Breach, N.Y., 1965).

62. Iso, C., Mori, K., and Namiki, M., Prog. Theor. Phys. 22, 403, (1959).

63. Feinberg, E.L. in "Proc. of the P.N. Lebedev Phys. Inst." ed. D.V. Skobel'tsyn (Consultants Bureau, N.Y., 1967) vol. 29, 151.

64. Suhonen, E., Enkenberg, J., Lassila, K.E., and Sohlo, S., Phys. Rev. Lett. 31, 1567 (1973).

65. Chaichian, M., Satz, H., and Suhonen, E., Phys. Lett. 50B, 362 (1974).

66. Jacob, M., (private communication).

67. Adams, P.J., and Canuto, V., Phys. Rev. 12, 3793 (1975).

DUAL UNITARISATION - A NEW APPROACH TO HADRON REACTIONS*

Chan Hong-Mo

Theory Division, Rutherford Laboratory

Chilton, Didcot, Oxon OX11 OQX, England

and

Tsou Sheung Tsun

Department of Theoretical Physics

Oxford University

ABSTRACT

The new approach of dual unitarisation (or topological expansion) to hadron reactions is reviewed with particular emhpasis on its aspect as a practical calculational method and the consistent manner in which it has been applied to a wide range of hadronic phenomena.

INTRODUCTION

'Dual Unitarisation' is the name of a scheme developed by the Oxford-Rutherford collaboration over the last few years, intended for the practical calculation of hadronic reactions in general. It has close relations,

* Based on lectures given at the Ecole Polytechnique, Paris in April 1976, and at the summer institute at Bielefeld, Germany, in August 1976.

though with different emphases, to a scheme conceived simultaneously by Veneziano which he calls the 'topological expansion', and also to the subsequent work by Chew, Schmid and many others. Our purpose here is to describe the program and to summarize the results so far obtained in this general area. We stress in particular the aspect of dual unitarisation as a practical calculational method and the consistent manner in which it has been applied to a wide range of hadronic phenomena.

The idea is as follows.[1-7] Hadronic reactions show some marked regularities such as exchange degeneracy, ideal mixing of states (pure quark states), OZI rule for resonance decays etc. Violations of such are in general small. This suggests that one should start with a system preserving these regularities, then calculate their violations as corrections.

Now the dual model preserves these regularities at the tree diagram level. Indeed, that was how the dual model was first conceived. However, the hadron coupling being large, a simple perturbation expansion has no meaning. Tree diagrams are not expected to be an acceptable first approximation. Some unitarity corrections (loop diagrams) must be included right from the beginning.

This leads then to the question: what loop diagrams preserve exchange degeneracy, the purity of quark states and the OZI rule, etc.? An immediate answer is: all planar loop diagrams. For example, the loop diagram in Fig. 1a has exactly the same quark content as the tree diagram in Fig. 1b. Algebraically, the closed quark loop in Fig. 1a is represented by a trace of λ-matrices[8], which gives just a numerical factor to the amplitude. Therefore if the tree diagram preserves exchange degeneracy, purity of quark states, and the OZI rule, so does the planar loop diagram.

On the other hand, non-planar loops in general do not preserve exchange degeneracy, (e.g. Fig. 2a carries only zero isospin in the s-channel, and therefore must break the exchange degeneracy between ω with $I = 0$ and ρ with $I = 1$), nor the purity of quark states (e.g. Fig. 2b mixes ϕ and ω), nor yet the OZI rule (e.g. Fig. 2c allows the OZI-forbidden $\phi \to \rho\pi$ decay).

We propose therefore to start, as a first approximation, with amplitudes which have already all planar

Fig. 1. Planar (a) loop diagram, (b) tree diagram.

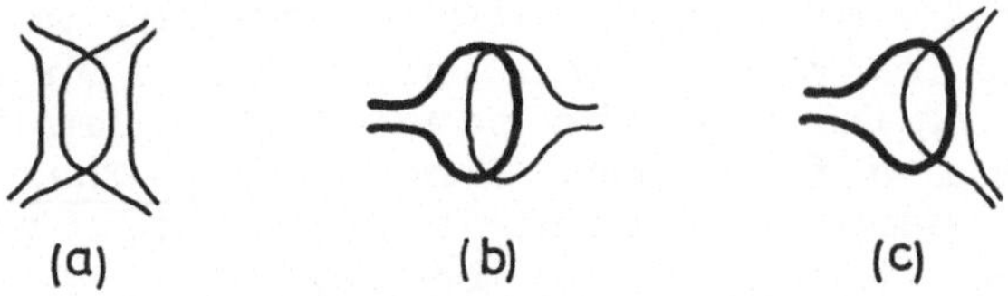

Fig. 2. Non-planar loop diagrams.

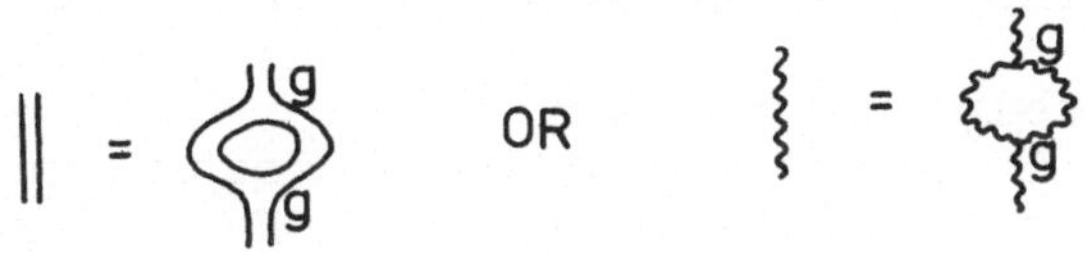

Fig. 3. Planar propagator bootstrap.

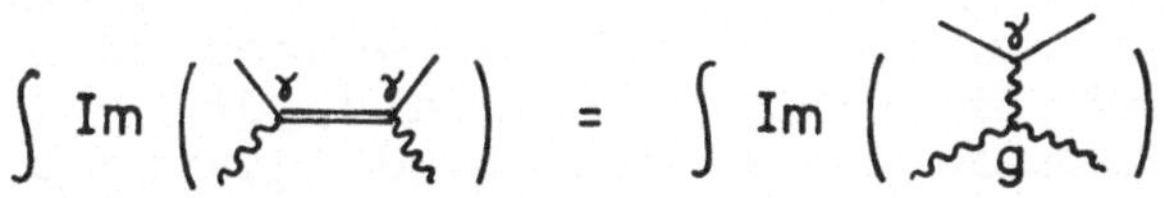

Fig. 4. Finite energy sum rules relating various couplings.

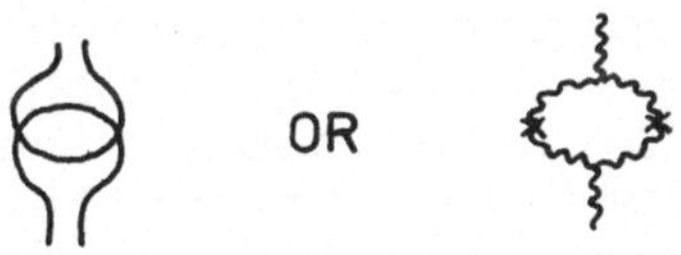

Fig. 5. Crossed loop (or cylinder) insertions.

Fig. 6. Diagram containing the Pomeron-Pomeron cut.

loops included, but to consider non-planar diagrams as higher order 'corrections'.

How can one, however, calculate these initial planar amplitudes? Since one does not have an explicit theory, even the tree diagram is unknown, let alone the sum of all planar loop diagrams. What one seeks is in effect a non-perturbative solution to an unspecified theory. However, in spite of this basic ignorance, we believe we can still extract a lot of information on such planar amplitudes by exploiting the consistency requirement intrinsic in their definition. Since our planar amplitude is supposed already to include all possible planar loops, it follows that it should not be further renormalized by planar insertions. Thus, for example, the reggeon propagator, should be required to satisfy the consistency equation represented symbolically in Fig. 3.

Unfortunately, as in many other instances, such bootstrap equations are not easy to formulate precisely. Indeed, all existing formulations of Fig. 3 resort some approximations and/or assumptions about the form of the amplitudes. As long as this is so, the true extent of their basic theoretical significance must remain uncertain. There is, however, little doubt of their predictive power in practice. Thus, for example, Fig. 3 represents an inhomogeneous equation in the triple-reggeon coupling g, which in principle determines g in absolute values. Indeed such an equation has been solved, giving a value of g in fair agreement with experiment.[7)]

The solution of planar consistency fixes the scale of strong interactions in general. First, at the planar level, analyticity in the form of finite energy sum rules implies many homogeneous relations among Regge couplings. An example of these is symbollically represented in Fig. 4. Being homogeneous, they give only the relative sizes of couplings, but not their absolute strength. However given a value of g, from the reggeon bootstrap, the scale is now fixed. In simple cases one can even imagine inverting the sum rules of Fig. 4 and solve for the couplings γ.

Second, and even more importantly, the planar bootstrap sets the scale for the nonplanar 'corrections' also. Take for example the loop diagram of Fig. 5, which, as we shall see later, breaks exchange degeneracy and the OZI rule, and in addition generates the Pomeron. Its magnitude depends again on this same constant g. Therefore, knowing g, one should be in a position to calculate the Pomeron and to check the consistency of our method, by demonstrating that violations to exchange degeneracy and the OZI rule are indeed small as required.

Now, non-planar 'corrections' are small in some circumstances, (e.g. in OZI violations), but large in others (e.g. in the Pomeron). The convergence of our method will thus depend on the quantity calculated and on the kinematic region. However in the problems so far examined, it seems that convergency is quaranteed, at least for finite s and t.

In this connection, there is an important observation first made by Veneziano.[5-9] The planar loop diagram of Fig. 3 is proportional to g^2N, where the factor N arises from the sum over quark flavours in the internal quark loop, for the symmetry group SU(N). The crossed loop of Fig. 5, on the other hand, is proportional only to g^2, since all quark lines there are fixed by the external reggeons. Now, from the reggeon bootstrap of Fig. 3, one has essentially $g^2N \sim 1$. Hence, it would appear that the crossed reggeon loop is $1/N$ down from the planar loop. In general, it can be seen that diagrams with more complicated topologies give higher powers in $1/N$. For example, the diagram of Fig. 6, which represents among other things a Pomeron-Pomeron cut, is of order g^6, or equivalently $1/N^3$, compared with the planar diagram. Notice however that heavy quarks are hard to excite and do not contribute significantly to loop diagrams. Hence, even if many more quarks exist, for our purpose the effective N is only ~ 2.5. The preceeding observation by itself therefore does not quarantee convergence, especially since diagrams with different topologies can have very different dependence on s and t. Moreover there are in general many more diagrams with higher topologies. Nonetheless, it is a great help in the classification of diagrams.

Hopefully, then, we have here a general method for hadronic reactions. The method involves no drastically new ideas - only conventional concepts, sharpened and reinforced. At the planar level, exchange degeneracy is exact, quark states are pure, OZI rule is inviolate, there is no Pomeron and no diffraction.

The introduction of non-planar corrections which are all in principle calculable leads to the breaking of exchange degeneracy, the mixing of quark states, the violation of OZI rule, the generation of the Pomeron and diffraction effects and hence to absorption, Regge cuts and eventually all of Reggeon calculus. The program is still in progress - its formulation is far from perfect and there are many consequences not yet explored or consolidated. We shall describe it here in its present state, including even some preliminary results and open questions.

The following is a list of the main predictions so far obtained which have all been checked against experiments:

I. Planar bootstrap
(i) triple-reggeon vertex $g \sim 1.3 \times$ expt.
(ii) + FESR gives other couplings γ.

II. Non-planar propagator renormalization
(i) leading vacuum trajectory with C = + (P):
$\alpha_P(o) \sim 1$, $\alpha'_P(o) \sim .3$, $\gamma_P(o)/\gamma_R(o) = \text{const.} \sim 1$.
(ii) leading vacuum trajectory with C = - (ω):
$\alpha_\omega(o) \sim .4$, large coupling to baryons
(iii) mixing of quark states (i.e. deviations from ideal mixing in SU(3)), e.g. $\lambda\bar{\lambda}$ component in Pomeron at $t = o$ gives $\sigma_T(\phi p) \sim 8$ mb.
(iv) breaking of exchange degeneracy:
$\alpha_\rho(o) - \alpha_{A_2}(o) \sim \alpha_{K^*}(o) - \alpha_{K^{**}}(o) \sim .1$
(v) Deviations from exchange degeneracy and ideal mixing increase for large $t < o$, e.g. Pomeron tends to SU(3) singlet.
(vi) Deviations from exchange degeneracy and ideal mixing small in resonance region ($t > o$); e.g.:
$m^2_{A_2} - m^2_f \sim .1$ GeV^2
(v) Exchange degeneracy and purity of quark states asymptotically exact for $t > o$.

III. Nonplanar vertex renormalization:
(i) Violations of OZI rule for resonance decay in general small: $|A(\phi \to \rho\pi)/A(\phi \to K\bar{K})|^2 \sim 1o^{-2}$
(ii) Violations of OZI strongly mass dependent:
$|A(\psi \to \rho\pi)/A(\phi \to \rho\pi)|^2 \sim 1o^{-6}$
(iii) Violations of OZI not simply perturbative:
$|A(\psi' \to \psi\varepsilon)/A(\psi' \to \rho\pi)|^2 \sim 1o^5$ <u>not</u> 1

I. PLANAR BOOTSTRAP

As indicated in the introduction, this appears to be the hardest part of the program to formulate. Apart from the theoretical question of its fundamental significance, even at the practical level, one has to face a couple of difficult tasks. Firstly, the symbolic equation depicted in Fig. 3 has yet to be given a meaning; without an explicit theory, we do not know how to write down a reggeon loop. Secondly, we argued that Fig. 3 was a necessary condition imposed on the planar reggeon by virtue of its definition, but we do not at all know whether it is sufficient. On the surface at least, it would seen that there should be consistency conditions on the planar vertex also, as represented symbolically in Fig. 7, and there may be even others. At present, we have only partial answers to the first point - the second problem has hardly even been considered.

To begin with, for simplicity, let us assume that internal (flavour) symmetry is exact, i.e. all N quarks have the same mass. Consider then the unitarity condition represented symbolically in Fig. 8. We are required to calculate the part of the overlap function on the right which is entirely planar, having no twists or crosses anywhere. Now, in the dual model any particles which are not separated by a cross can resonate. A good representation of the overlap function is therefore that depicted in Fig. 9 where the intermediate states form resonant clusters.[6)]

Such a picture is well-attested by recent phenomenological analyses of multiparticle data.[10)]

Here, however, a difficulty already arises. Two resonances which are not separated by a cross can also resonate together; how can one then distinguish between the terms in Fig. 9 having different numbers of resonances produced? Clearly, it is profitable to represent a multiparticle system by a resonance only when the resonance is narrow. When the widths of resonances become comparable to their separation, they strongly interfere; duality then tells us that it would be better instead to represent the sum of these resonances by the exchange of a Regge pole. Therefore one practical way of distinguishing the diagrams in Fig. 9 would be to insist that all cluster masses s_i be less than a certain cut-off value $\bar{s}$, and all inter-cluster energies s_{ij} be greater than $\bar{s}$. One would then ensure that all resonant clusters

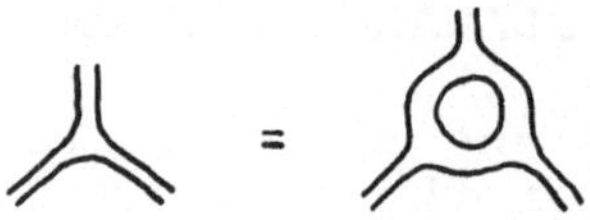

Fig. 7. Planar vertex bootstrap.

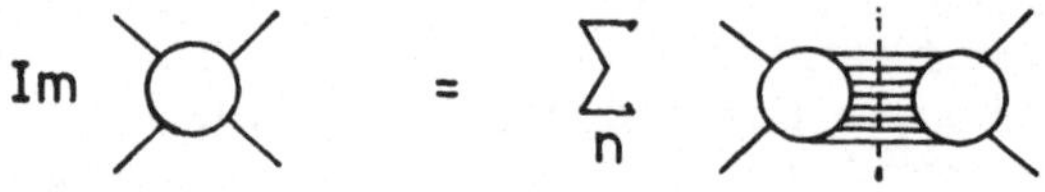

Fig. 8. Unitary condition.

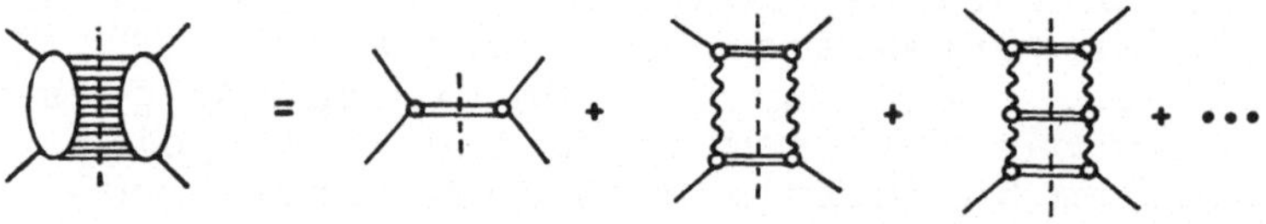

Fig. 9. Cluster approximation of overlap function.

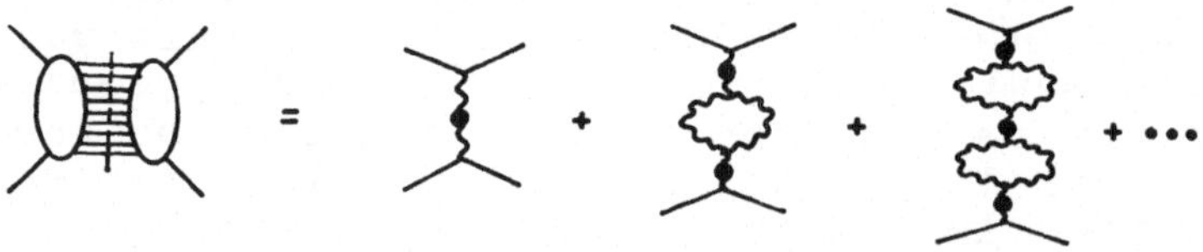

Fig. 1o. Overlap function in terms of reggeon loops.

Fig. 11. Integral equation for the propagator bootstrap.

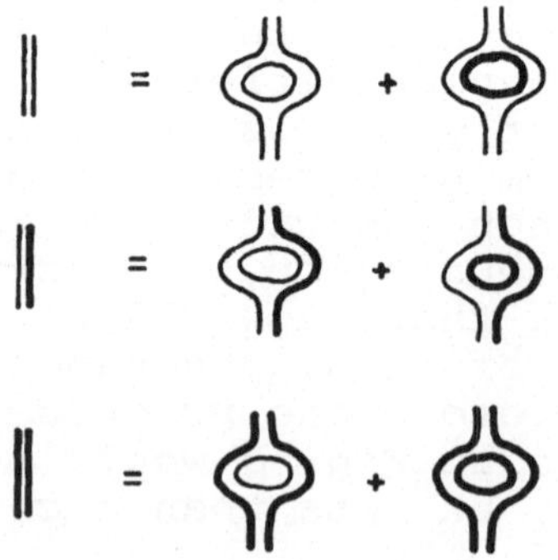

Fig. 12. Propagator bootstrap in broken SU(3).

are reasonably narrow, and that the Regge formula is employed only above a certain threshold. A sensible value to choose for $\bar{s}$ is around 6 GeV^2, where empirically resonances begin strongly to overlap.[6]

Now different diagrams in Fig. 9 cover, by definition, different sectors of phase space and do not therefore interfere. However, there are regions in phase space where the decay products of different clusters on the same diagram overlap. Our treatment here[6] neglects the possible interference between clusters in these regions and can lead to errors of the order (resonance width/ resonance separation), which are hopefully not important when resonances are narrow. This is, however, an important point to which we shall return later.

In each diagram of Fig. 9, one still has to sum over the mass of each resonance. This sum can be performed by semi-local duality using sum rules of the type illustrated in Fig. 4. Repeating this for every rung of the ladder, we obtain the representation of Fig. 10[6], where we have introduced the notation of a dotted-reggeon line to remind ourselves that the invariant masses s_i and s_{ij} are by definition subject to the constraints:

$$s_i < \bar{s} \quad ; \quad s_{i,i+1} > \bar{s} \tag{1}$$

Such reggeon-loop diagrams are given in terms of the trajectory function $\alpha(t)$ and the triple-reggeon vertex $V(t;t_1,t'_1)$. As an example, we write down below the expression for the one-loop diagram:[6]

$$V_1 = g^2 N \int_{s_{th}}^{\bar{s}} ds_1 ds_2 \int dt_1 dt'_1 \frac{\theta(-\Delta)}{|\Delta|^{1/2}} s_1^{\alpha(t)} s_2^{\alpha(t)}$$

$$(s/s_1 s_2)^{\alpha_1(t_1)+\alpha_{1'}(t'_1)} V^2(t;t_1,t'_1)\,\xi^*(t_1)\,\xi(t'_1)\,\theta(s-\bar{s}) \tag{2}$$

Δ is the usual Jacobian determinant whose zeros give the boundary of the physical region, and $\xi(t)$ is the signature factor of the trajectory exchanged, which for our planar reggeons gives[3][6]

$$\xi^*(t_1)\ \xi(t'_1) = \exp i\pi\left[\alpha_1(t_1) - \alpha_{1'}(t'_1)\right] \tag{3}$$

Of course, we should in principle also sum over the various possible reggeons in each channel. However, to begin with let us assume that the leading trajectories dominate, namely the mesons belonging to the vector-tensor nonets.

Fig. 1o has the general form of a multiperipheral series, and can be summed as usual to generate a Regge pole. By bootstrap consistency now, one requires that the pole generated be identified again with the leading vector-tensor trajectories. One obtains then a consistency condition, which can be written as an integral equation for the reggeon propagator P; symbolically,

$$P = \dot{P} + UP \qquad (4)$$

where the kernel U is very similar in structure to the one loop integral, (2). This is illustrated in Fig. 11, the iteration of which gives back the series of Fig. 1o. Notice that for $s>\bar{s}$ the inhomogeneous term $\dot{P}$ in Fig. 11 is by definition zero, so that one has an equation the form of Fig. 3 as required.

The first bootstrap equation of this type was given in ref. 7)*. Since then there have appeared many variations on the same theme[13)-18)]. These differ essentially only in the means of avoiding the error mentioned above in the interference region between the decay products of neighbouring clusters - the so-called 'double-counting error' - which is handled with varying degrees of sophistication and realism. The details of their differences are not important for our general discussion.

Now as noted in the introduction the equation (4) is inhomogeneous in the coupling g. Given thus the trajectory function α and the triple-reggeon vertex V, it can be solved for g. Such a solution is obtained in ref. 7) by supplying $\alpha(o) = .5$, and assuming an exponential form for V in t_1, t'_1 with the exponents taken from experiment. This gives

$$g \sim 7.0 \qquad (5)$$

---

* Bootstrap equations in the same general spirit but without clustering of intermediate particles have also been considered in ref. 1,2,11,12.

in GeV units, as compared with the value $g \sim 6.3$ estimated phenomenologically from inclusive data. One also obtains the dependence of V on t, which however cannot be checked directly against experiment. Notice that as it stands, the equation (4) is insufficient to determine $V(t,t_1,t'_1)$ as a function of all its arguments. This may be possible, however, once vertex bootstrap conditions such as that in Fig. 7 is taken into account.

Because of the somewhat artificial kinematic constraints (1) introduced to avoid 'double counting', the full bootstrap equation is rather complicated whose solution has to rely heavily on numerical methods[7]. For exploratory purposes it is therefore often more convenient first to average over the transverse motion, leaving only the rapidity as variable. In this 'one-dimensional' approximation, one loses the interesting information on the t-dependence, but the equation becomes almost diagonal under a Mellin transform, leading to a simple equation in j which can be explicitly solved. In particular, in the formulation of Kwiecinski and Sakai[15], which is closest in spirit to the original 3-dimensional version of ref. 7), one obtains the solution[15]:

$$\begin{aligned}(\alpha - \bar{\alpha}_1 - \bar{\alpha}_{1'} + 1)\,\frac{\bar{X}}{2} &= 1.9\\ (\bar{g}^2 N)\left(\frac{\bar{X}}{2}\right)^2 &= 1.24\end{aligned} \qquad (6)$$

$\bar{g}$ and $\bar{\alpha}_i$ represent respectively the triple-reggeon vertex and the trajectory functions, averaged appropriately over the loop integtal. $\bar{X}$ is the rapidity equivalent of the cut-off parameter $\bar{s}$ introduced above. Notice that for cluster sizes $\bar{x} \sim 2$ as favoured by phenomenologists, $\bar{g}^2N$ is indeed of order 1 as suggested in the introduction.

The simple system considered above can be generalised in several directions. First, although the equation (4) was obtained from s-channel unitarity, and hence originally valid only for $t < o$, it can be expanded to $t > o$ by analytic continuation, at least in principle, to calculate for example the coupling g or the trajectory α in this region. However, no work in this direction has yet been done.

Secondly, one can relax the initial simplifying assumption that all N quarks have identical masses. For example, consider broken SU(3) symmetry with two distinct quark masses $n = p$, and λ. The bootstrap equation (4) of

Fig. 3 now becomes those of Fig. 12. From our previous discussion, one knows that there exists a completely symmetric solution with all intercepts α (i.e. quark masses) and all couplings g identical. One now raises the question whether there are also other solutions with broken symmetry. The question can be asked at two levels: (i) assuming symmetry is broken infinitesimally, what is the breaking pattern? or (ii) since the equation is non-linear, may there in addition be spontaneously broken solutions with finite, calculable breaking? The first question has been studied in the one-dimensional approximation resulting as expected in the Gell-Mann-Okuba formula[19,13,20]. To the second, there are some interesting suggestions[20] but as yet no definite answers. Notice, however, that even when SU(3) is broken, quark states remain pure at the planar level since there is no diagram in Fig. 12 which mixes them.

Thirdly, we may generalize our considerations to trajectories other than the vector-tensor nonet, e.g. to pions, charmed mesons, baryons and exotics. In the pion bootstrap, one meets with the usual difficulties and little progress has been made. On the other hand for charmed mesons, baryons and exotics, bootstrap equations are readily obtained by replacing the λ-quark in Fig. 12 with the charmed quark c, or the diquark qq. Moreover, because of the heavy masses of c and qq, the second term in all the equations of Fig. 12 can be neglected, giving equations almost identical to Fig. 3, which are easily solved in the one dimensional approximation. For example, in the formulation of ref. 15), the solution (6) remains valid in each case with the appropriate trajectories and coupling constants. Thus given $\alpha_B \sim -.1$, one predicts that $\alpha_E \sim -.7$, as compared with the phenomenological estimate $\alpha_E \sim -.5 \pm .3$. There are other consequences which will be of use later.

To conclude this section, we mention some attempts at calculating particle-reggeon couplings (say γ) in terms of the bootstrapped triple-reggeon vertex g. As indicated in the introduction, by saturating finite-energy sum rules with a small number of resonance states, as illustrated in Fig. 4, one obtains homogeneous relations between the couplings γ and g. Further independent relations are derived from sum rules with different moments. For simple systems saturated by a few resonances, there may be enough conditions to solve for γ in terms of g. Some such calculations[21)22)] were done for meson-

meson (both particles and reggeons) scattering, which though crude, give quite sensible answers. In particular, using the so-called Schwarz-type sum rules with 'wrong' moments, one is able to make estimates even for the residues of the wrong-signatured fixed poles. As we shall see, they turn out to be very useful for certain calculations of non-planar corrections.

II. PROPAGATOR RENORMALIZATION

1. Vacuum Trajectories

In contrast to planar insertions which are required to leave the reggeon propagator invariant, non-planar insertions are expected in general to renormalize the Regge trajectories. To begin with, consider the simplest non-planar insertion to the reggeon propagator, namely that shown in Fig. 5. This is formally of order g^2 or equivalently $1/N$. Notice, however, that when it is iterated, one obtains Fig. 13, which because of the closed quark loop, is of order $g^2(g^2N) \sim g^2$ also. Thus, even formally, one has to consider an infinite iteration of this insertion.

In the calculation of the overlap function such insertions occur as follows[6]. In Fig. 9 and 1o we have limited ourselves by choice only to those diagrams without crosses or twists anywhere. However, in the full overlap function between planar production amplitudes, there are many more diagrams in which any two neighbouring clusters can be separated by a crossed reggeon loop.

Now such a diagram may of course be regarded as chains of consecutive uncrossed loops separated by crosses. Summing then over all these chains with the help of Fig. 11, one obtains the series shown in Fig. 14, where all reggeons are undotted and all loops are crossed.

Again, these reggeon-loop diagrams are given in terms of $\alpha(t)$ and the triple-reggeon vertex $V(t;t_1,t'_1)$. For example, the one-loop diagram reads:[6]

$$C_1 = g^2N \int_{s_{th}}^{\infty} ds_1 ds_2 \int dt_1 dt'_1 \, \frac{\theta(-\Delta)}{|\Delta|^{1/2}} \, s_1^{\alpha(t)} s_2^{\alpha(t)}$$

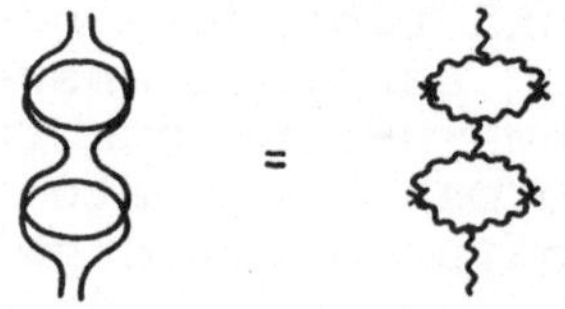

Fig. 13.   Iteration of the crossed loop.

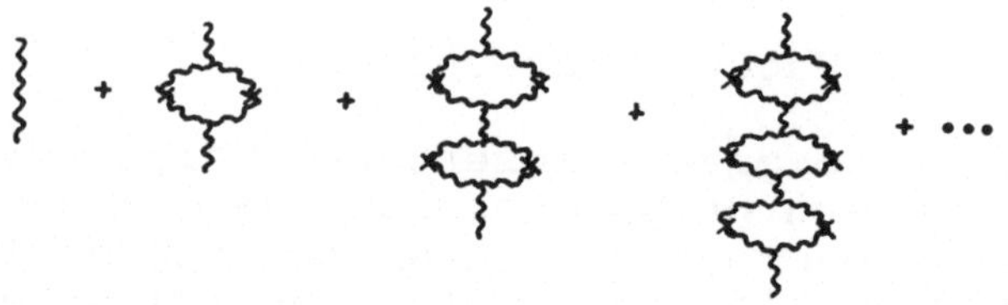

Fig. 14.   Sum of all iterations of the crossed loop.

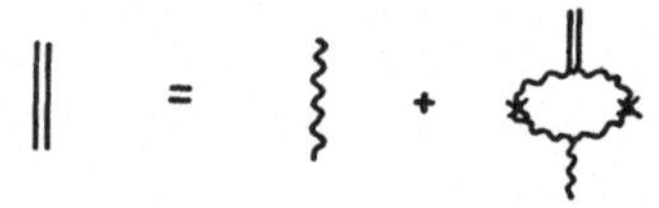

Fig. 15.   Integral equation for the renormalized vacuum
trajectory.

$$P = \begin{pmatrix} \| & 0 \\ 0 & \| \end{pmatrix} \qquad C = \begin{pmatrix} & \\ & \end{pmatrix}$$

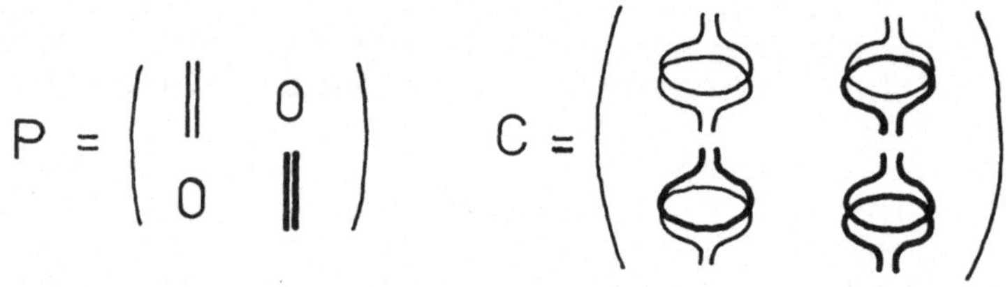

Fig. 16.   Crossed loop insertions in broken SU(3).

(A)   (B)   (C)   (D)

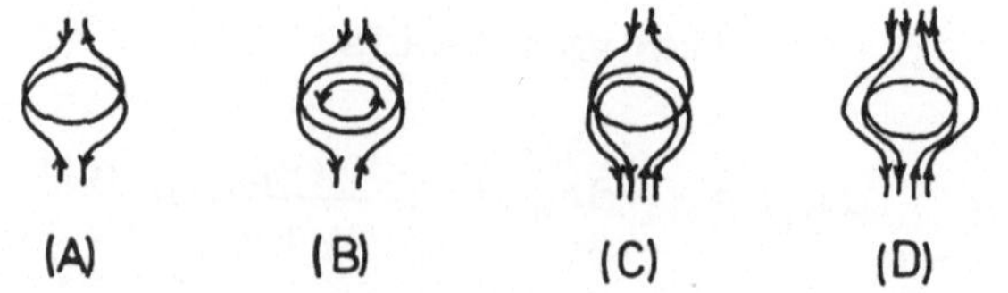

Fig. 17.   Non-planar baryon loop insertions.

$$(s/s_1s_2)^{\alpha_1(t)+\alpha_{1'}(t'_1)} V^2(t,t_1,t'_2) \qquad (7)$$

It is very similar in structure to V_1 in (2), except for the integration limit and the signature factors $\xi^*\xi$ of the exchanged reggeons, which is replaced here by 1 because of the crosses. Also, the θ-function in (2) is missing, because, in the dual model, two resonances separated by a cross do not resonate, whatever their rapidity separation. We shall see that these differences which follow directly from duality, are essential for understanding the properties of the 'Pomeron'.[6)7)]

As is obvious from the quark diagram in Fig. 5, the insertion can transmit only zero quantum numbers. Also, it contributes with opposite signs to $\sigma = \pm$ (or equivalently $C = \pm$) trajectories, [23,24,25)] which can be seen as follows.[23] Because of the direction of quark lines entering and leaving the insertion, only diagrams with an odd number of crossed loops contribute to BB scattering, and only those with an even number to $\bar{B}B$. Forming then the even- and odd-signatured combinations, one has

$$\begin{aligned} 2\sigma_+ &= \sigma(\bar{B}B) + \sigma(BB) = 1 + C + C^2 + C^3 + \ldots \\ 2\sigma_- &= \sigma(\bar{B}B) - \sigma(BB) = 1 - C + C^2 - C^3 + \ldots \end{aligned} \qquad (8)$$

Therefore the insertion will renormalize opposite signatures in opposite directions. (Incidentally, this explains, for example, why $m_f < m_{A_2}$ but $m_\rho < m_\omega$).

Under the insertion of Fig. 5 then, trajectories with non-zero quantum numbers such as ρ and K^* are not affected. Zero quantum number trajectories however will have their propagators renormalized according to Fig. 11, or equivalently by the integral equations:[7)23)24)25)]

$$P'_\pm = P \pm CP'_\pm \qquad (9)$$

where C is similar in structure to C_1 in (7). This is illustrated in Fig. 15. Since all the unknown quantities in (7) are supposedly already fixed in the planar bootstrap, the equation (9) can be solved directly.

To begin with, consider only t = o, exact SU(3) symmetry, and that all exchanges are dominated by the leading vector-tensor nonet. The crossed loop C is then the square of an amplitude and positive definite. It will therefore generate a leading vacuum trajectory with C = +. Indeed, a solution of (9), with bare reggeon intercept α =.5, and triple-reggeon coupling g as determined in the planar bootstrap (5), gives[7)]

$$\alpha_+(o) \approx 1. \pm .1 \; ; \quad \alpha'_+(o) \approx .3 \pm .2. \qquad (1o)$$

where the errors represent uncertainties in the approximate calculations and in the parametrisation of the triple-reggeon vertex. The properties (1o) are both attributes of the phenomenological 'Pomeron', and since it is also generated by the shadowing of multiparticle intermediate states as expected of diffraction phenomena, we shall identify it as such.[7)]

The result $\alpha_P \sim 1.$ is qualitatively easy to understand in the one-dimensional approximation. Diagonalising the equation (9) under a simple Mellin transform, one has[15)]

$$P = \frac{1}{j - \alpha_M} \; ; \quad C = \frac{g^2 N}{j - 2\bar{\alpha}_M + 1}$$
$$P' = P\left[1 - CP\right]^{-1} = \left[j - \alpha_M - \frac{g^2 N}{j - 2\bar{\alpha}_M + 1}\right]^{-1} \qquad (11)$$

The pole in P' is displaced by a positive amount from $j = \alpha_M$. Indeed, substituting the result in (6) from the planar bootstrap, one obtains an intercept similar to (1o). The fact that $\alpha'_P < \alpha'_M$ is a consequence of the difference in phase between the integrands of U and C, in (2) and (7) respectively which cannot be seen in the one-dimensional approximation. The oscillating signature factors $\xi^* \xi$ in U which are absent in C, lead to cancellations in the integral and hence to a steeper dependence on t, which when iterated give faster shrinkage to the bare reggeon.[3)6)7)]

Notice that in contrast to common usage, the 'Pomeron' here is not a new singularity, but is the same trajectory as the f, which is simply renormalized upwards by the non-planar insertions.* This, however, is not in contradiction to experiment for the energy dependence of σ_T[26] nor for tests of semi-local duality which deals only with the planar part of the amplitude, namely the bare f-trajectory.

The couplings of this 'Pomeron' are proportional to those of the bare reggeons with a calculable proportionality constant[6] $\beta(t)$ which is independent of the system it couples to and which according to ref. 7), takes the value:

$$\beta(o) = \gamma_P(o)/\gamma_R(o) \sim 1 \tag{12}$$

in GeV units at t = o. The consequences of this prediction have been tested against experiment quite stringently, especially with some rather detailed Mueller-Regge vertices extracted from inclusive experiment[27].

The generation of diffraction as a shadow of multiparticle processes entails of course also a model for the production mechanism. In our case, as seen in Fig. 14 and 9, the mechanism is the multiperipheral production of resonant clusters, of the type generally favoured by phenomenologists.[10] However, in contrast to usual cluster models, our production amplitudes are almost entirely determined through the stringent conditions imposed by duality and the planar bootstrap. In particular, the cluster mass spectrum and cluster density are both governed by the Regge intercept and the triple-reggeon coupling. They constitute therefore a specific model which may be applied to the phenomenological analysis of multiparticle data. Some work in this direction has already been done, giving reasonable answers to production cross sections of, for example, strange particles [28][29], to single particle distributions[30], as well as to two-particle correlation effects.[31][32][33]

---

* This important point was actually missed in the original calculation of ref. 7). It was known to Huan Lee, but its importance was first emphasized by Chew and Rosenzweig[24] and Schmid and Sorensen.[25]

The calculation described above for the 'Pomeron' can be repeated for the trajectory with C = -, namely the renormalised ω. One finds then that the resulting pole tends to become complex with a very low intercept:

$$\mathrm{Re}\,\alpha_{-}(o) \sim 0 \qquad (13)$$

As we shall see, however, this should not be identified with the ω-trajectory observed empirically, which acquires some special properties from mixing with ϕ and from baryon exchange contributions.

One can generalise the preceeding considerations in several directions. First, we may relax the assumption of exact SU(3) symmetry. The equation (9) then remains formally valid except that both the propagators P(P') and the insertion C are now matrices, as depicted in Fig. 16. The important point to observe here is that the off-diagonal terms in C will lead to mixing of the original planar trajectories.[23,24,25] Thus for example, f and ω which have remained pure $n\bar{n}$ and $p\bar{p}$ states at the planar level in spite of SU(3) breaking, will acquire each a $\lambda\bar{\lambda}$ component. Similarly, f' and ϕ which were pure $\lambda\bar{\lambda}$ before, will now have a mixture of the ordinary, non-strange quarks, leading in particular to violations of the OZI rule in their decay. Moreover, since the couplings have already been determined in the reggeon bootstrap, such effects are all in principle calculable. Thus, the renormalized trajectories in (9) are given by the zeros of the determinant det $|P(1-CP)^{-1}|$. For each signature, $\sigma = \pm$, there are two poles corresponding to f,f' and ω,ϕ respectively. The coupling strength and mixing angle of each are obtainable from its residue matrix. For example for the renormalized f pole or 'Pomeron', the residue matrix can be written in the form:

$$\mathrm{Res} = \rho\begin{pmatrix}\cos^2\theta & \cos\theta\,\sin\theta \\ \sin\theta\,\cos\theta & \sin^2\theta\end{pmatrix}$$

where θ is its mixing angle into the $\lambda\bar{\lambda}$ state. This allows the 'Pomeron' to couple to ϕ for example and gives non-vanishing ϕp diffractive cross sections. The mixing angles for the other trajectories are similarly calculated. Notice that, as they are defined in (14) the mixing angles as well as ρ are functions of t. The problem is easily investigated in the one-dimensional approximation. For example, for the 'Pomeron',[15)]

$$\tan\theta = \frac{(\alpha_P - \alpha_M)(\alpha_P - 2\bar{\alpha}_M + 1)}{(\alpha_P - \alpha_\phi)(\alpha_P - 2\bar{\alpha}_{K^*} + 1)} \tag{15}$$

Substituting conventional values of α_M, α_{K^*} and α_ϕ, and using (6), one obtains an estimate of $\sigma_T(\phi P) \sim 8$ mb, as compared with empirical values of $\sim 1o$ mb. The mixing between the odd signatured states ω and ϕ at $t = o$ is even stronger.[24] However, its effect is harder both to detect experimentally, and to estimate because of complications from baryon exchanges as we shall see later.

Secondly, we may include the exchanges of trajectories other than those of the vector-tensor nonet. At first sight, pion exchange may be expected to be large because of the proximity of the π-pole to the physical region. However, in our loop integrals, the masses of the resonant clusters are sizeable on the average, while the energies separating them are not big. Thus $t_{min} \sim -s_1 s_2/s$ remains finite, and the effect of the pion pole is minimised.* Indeed, when the triple-reggeon vertex for pion exchange is extracted from inclusive experiment and substituted into loop integrals, one obtains an estimate for the pion contribution of only 1o% of the total.[34]

In contrast the effect of baryon exchanges are more intriguing. Phenomenologically, baryon exchanges are known to be important for describing production processes.[35] However, to incorporate baryons in a dual scheme, one must also include $qq\bar{q}\bar{q}$ exotic states, and consider the bootstrap equations for them, as mentioned in section I. Consider now the non-planar insertions of leading order in 1/N[36-39]. These are listed in Fig. 17[39], where loops with exotic exchanges have been neglected. We note first that whereas the meson loop (A) contributes with opposite signs to $\sigma = \pm$ trajectories, all the baryon loops (B), (C) and (D) contribute with the same sign.[38,39] This can be seen by repeating the previous analyses for Fig. 5. The insertion (D) can carry octet quantum numbers and will therefore renormalise the octet components of the $qq\bar{q}\bar{q}$ multiplet, but this need not concern us here.

---

* The same effect is well-known in usual multiperipheral models. Experimentally, also, $\langle t \rangle$ is known to be 1 GeV^2.

On the other hand, all the insertions in Fig. 17 including (D) will renormalize the vacuum trajectories. The renormalized propagator is given by:

$$P'_{\pm} = P + C_{\pm}P'_{\pm} \tag{16}$$

where

$$C_{\pm} = \begin{pmatrix} \pm(A) + (B) & \sqrt{2}(C) \\ \sqrt{2}(C) & 2(D) \end{pmatrix} \tag{17}$$

The factors 2 and $\sqrt{2}$ in (17) arise from the fact that there are two ways of joining the quark lines when multiplying (C) and (D) together.

For the moment, ignore the mixing term (C), and also the contribution (B) from baryon-pair production which is effective only at high energy. The first diagonal term in (14) is then the same as (9) which has already been discussed, giving renormalized trajectories with intercepts (1o) and (13). In the second diagonal term (D) renormalizes both signatures upwards. Since both the intercept, and coupling to baryons of the bare exotic trajectory have already been determined by bootstrap, the actual amount of renormalization can be calculated. In one-dimensional approximation, this gives:[39)]

$$\alpha_{\pm}(qq\bar{q}\bar{q}) \sim .1 \tag{18}$$

Consider now the effect of the off-diagonal term (C), which will mix the old trajectories $\alpha_{\pm}(q\bar{q})$ with these new promoted exotics $\alpha_{\pm}(qq\bar{q}\bar{q})$. For the $\sigma = +$ trajectories, the mixing is small since the two trajectories (1o) and (18) are far apart. (Nonetheless the effect may still be observable.) In the case of $\sigma = -$, however, the proximity of (13) to (18) implies that the states will mix strongly and repel each other. As a result, the leading vacuum trajectory with $\sigma = -$, which we identify as the ω, will acquire a respectable intercept as well as an enhanced coupling to baryons via its exotic components both of which are phenomenologically describable features. Indeed, a calculation of (C) taking proper account of threshold barriers, with couplings and threshold parameters taken from antiproton production experiments, give:

$$\alpha_{\omega} = .37, \quad \sin\theta = .76 \tag{19}$$

as compared with the value α_ω(expt) = .41 obtained from K^o regeneration experiment. These estimates are subject to some doubt since for example SU(3) breaking effects have not been taken into account. However this picture of the ω at $t \sim o$ being a complicated mixture of n, p, λ and diquark states has to be taken seriously and is presently under investigation.

Above $s \sim 1oo$ GeV, both the 'Pomeron' and the ω receive a further boost from effects of baryon-pair production as represented by insertion (B), leading to new estimates of all effective intercepts:

$$\alpha_+(o) = 1.o4, \quad \alpha_-(o) = .5o \tag{2o}$$

and a rising total cross section as expected.

We may also generalize our considerations to the renormalization of other trajectories with zero quantum numbers. A particularly interesting possibility here is the splitting of η from exchange degeneracy with π, thus resolving one of the greatest puzzles in the dual model. At the planar level, it is natural to have π-η degeneracy as is the case for ρ-ω, or A_2-f. Then, in the same way that ω and f were renormalized by the insertion of Fig. 5, one may expect η also to be so affected. There are two points to note.

(i) For η the dominant exchanges in the loop of Fig. 5 are expected to be $\pi\rho$, for example, which is asymmetric (in contrast to f) and therefore not positive definite. There is thus a possibility of renormalizing the even-signatured η downwards from the bare (π) trajectory, as required by experiment.

(ii) The η pole at $t = m^2_\eta$ is close to the region $t \sim o$, where the effect of the insertion of Fig. 5 was shown to be big, as measured by $\alpha_P - \alpha_M \sim .5$ in (1o). It is quite possible therefore to obtain a large mass splitting between π and η. However, since we know of no sensible way of bootstrapping the pseudoscalar as we did for the vector-tensor nonet, neither the sign nor the size of the renormalization can strictly as yet be predicted.

Finally, the generalization of our discussion to $t \neq o$ will lead to very far-reaching results, which we are reserving for a later section.

2. Non-Vacuum Trajectories

In section 1, we have considered only the simplest non-planar insertions of the lowest order in g^2 to the reggeon propagator. We would like of course to extend the investigation to higher order, so as to check the consistency and convergence of our method. This is however not so easy for two obvious reasons: (i) higher order effects are in general smaller and therefore harder to detect experimentally, (ii) there are numerous diagrams which are also more complicated to evaluate. For exploration, therefore, it is sensible to select a problem with clear experimental implications, and yet requires the calculation of only a few diagrams.

The breaking of exchange degeneracy in non vacuum octet trajectories, such as ρ-A_2 and K^*-K^{**}, is a good example, as suggested in ref. 4o).

(i) The extensive data now available up to 2oo GeV for the reactions:[41)]

$$\pi^- p \rightarrow \pi^o n \ : \ \rho \tag{21}$$

$$\pi^- p \rightarrow \eta n \ : \ A_2 \tag{22}$$

show unambiguously that:

$$\alpha_\rho(o) - \alpha_{A_2}(o) \sim .1 \tag{23}$$

Also, the old problem of violations in line reversal invariance, in such pairs of reactions as,

$$K^- p \rightarrow \bar{K}^o n \quad ; \quad K^+ n \rightarrow K^o p \tag{24}$$

$$K^- p \rightarrow \pi^- \Sigma^+ \quad ; \quad \pi^+ p \rightarrow K^+ \Sigma^+ \tag{25}$$

shows that exchange degeneracy is broken.[42)] If the breaking is ascribed mainly to the trajectories themselves, one requires $\alpha_\rho > \alpha_{A_2}$, $\alpha_{K^*} > \alpha_{K^{**}}$, with a difference in intercept similar to (23). For example, an analysis of (25) gives an estimate:[43)]

$$\alpha_{K^*}(o) - \alpha_{K^{**}}(o) \sim .o6 \tag{26}$$

(ii) The splitting of such trajectories can arise only from insertions which (a) carry octet quantum number, and (b) distinguish between even and odd signatures.

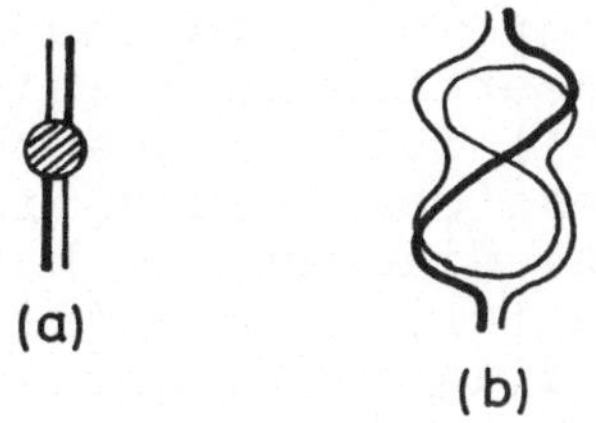

Fig. 18. Diagrams breaking exchange degeneracy of octet trajectories.

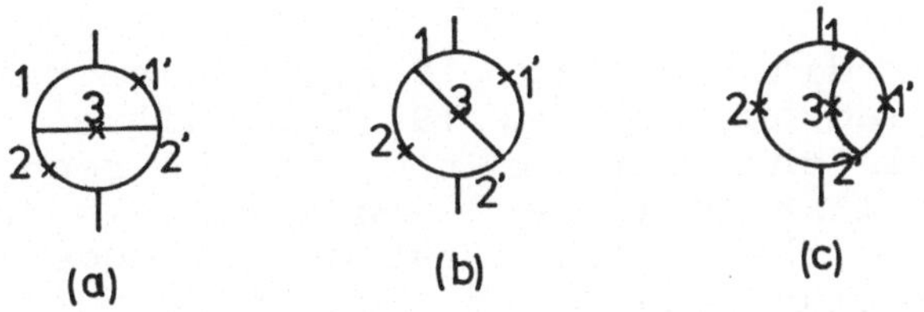

Fig. 19. Reggeon diagrams corresponding to Fig. 18(b).

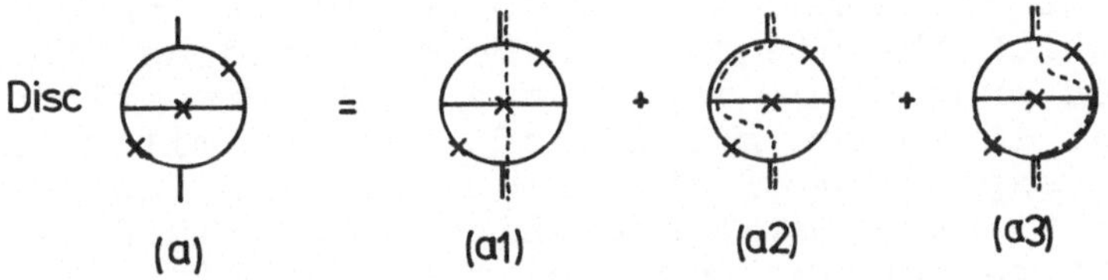

Fig. 2o. Elementary discontinuities of Fig. 19(a).

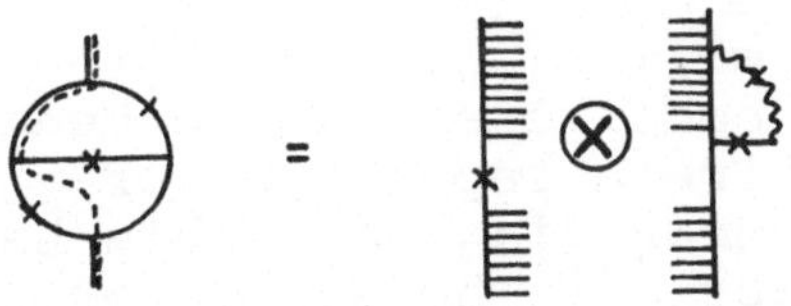

Fig. 21. s-channel content of Fig. 2o, (a2).

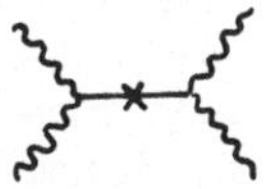

Fig. 22. Central vertex of Fig. 19(a).

In terms of quark diagrams, this means that the quark lines entering the insertion must (a') continue through the insertion, and (b') interchange positions in transit, as in Fig. 18(a). These conditions limit considerably the number of relevant diagrams to be calculated. Indeed, up to order g^4, there is only the diagram of Fig. 18(b) - the next relevant diagram is already of order g^8.*

Apart from the physical interest in the effect itself, the problem one faces in its evaluation are common to non-planar corrections of higher order in general. We shall therefore consider it here in some detail.

The quark diagram, Fig. 18(b), can be realised by the reggeon diagrams of Fig. 19. They differ from one another only in the configuration of the reggeon line 3 which is time-like in (a) and space-like in (b) and (c). They enter into the renormalization of Regge trajectories through their imaginary parts, in the calculation of the unitary overlap function. These imaginary parts, in turn, can be considered as sums of several elementary discontinuities, or cuttings of the reggeon diagrams, which e.g. for diagram (a) are listed in Fig. 2o. Different cuttings of the same diagram correspond to the different ways in which the diagram can be obtained as the overlap of various production amplitudes. An asymmetric cutting means that it is the product of two dissimilar terms in the Regge expansion of the production amplitudes themselves. For instance, the cutting (a2) arises as the overlap of two production amplitudes in Fig. 21, one of which already contains a non-planar loop insertion. Such cuttings are familiar to experts of Reggeon calculus; we have not met them here before only because, in the simpler cases so far dealt with, they did not arise.

What is not familiar, however, are the planar vertices normally not considered in the Reggeon calculus. They therefore require a repetition of the original Gribov analysis of Feynman diagrams. This has been done by Kwiecinski and Konishi[44] who give, for our special cases, the relation between various cuttings of the same diagrams. In particular, for the diagram (a), and for $\alpha_M \approx .5$, the

---

* This is most easily seen using Veneziano's topological counting rule.[5)9)] Condition (a) fixes the number of boundaries to be 1, so only the number of handles can change. Fig. 18(b) has the topology of a torus, or a a sphere with one handle, hence order g^4. The next diagram must already have two handles, i.e. order g^8.

total discontinuity is shown to be essentially minus the symmetric discontinuity (a1), as in the more familiar case of the diagram for the Pomeron-Pomeron cut. On the other hand, the total discontinuities of both diagrams (b) and (c) are expected to be small for $\alpha_M \approx .5$ because of cancellations in the various cuttings.

It remains then to evaluate these elementary discontinuities. Unfortunately, inadequate knowledge of regge vertices limits one at present to rather crude estimates. For example, in the same one-dimensional approximation of ref. 15) as in (11), where one replaces all quantities integrated over the loop by appropriate average values, we obtain for the discontinuity (a1) and its complex conjugate:

$$I = \frac{2\, g^4 \varepsilon}{\{\alpha_M - 2\bar{\alpha}_M + 1\}^2} \qquad (27)$$

where $g^2\varepsilon$ represents the central four-reggeon vertex depicted in Fig. 22.

Now the quantity ε is related to the fixed pole residue in the reggeon-reggeon amplitude, and can be roughly estimated using FESR in a way analogous to ref. 21), (see section I), giving values $.25 \leq \varepsilon \leq .5$. Taking the value $\varepsilon = .4$ suggested by OZI violating in resonance decays to be reported in section III, we obtain then for the splitting in intercepts

$$\alpha_+(o) - \alpha_-(o) \approx - 2I \approx - .o8 \qquad (28)$$

which is compatible with (23) and (26) both in sign and magnitude.

Unfortunately, there is as yet no consensus among authors on these estimates. In several recent papers[45-48] in which some (but not all) of the diagrams and/or cuttings in Fig. 19 are also considered, the authors, using different arguments arrive at quite different conclusions. We believe however, that at least the sign and order of magnitude in (28) are indeed correct.

The preliminary success of this exploratory investigation, plus the technical development in ref. 44) makes it hopeful that one will soon be able to handle quantitatively higher order effects in some generality, including such old problems as Regge cuts and absorption.

In this area, the general connection of the present approach to the Reggeon Calculus has already been established by Ciafaloni, Marchesini and Veneziano.[9)]

3. t- Dependence

The equation (9), for example, for the renormalised propagator is in principle valid for a range of $t < o$, as well as for $t > o$ by analytic continuation. We have restricted ourselves so far to $t = o$, only because we have better knowledge here of the reggeon vertices which is needed for actual calculations. We now attempt to generalize our considerations to $t \neq o$, which as we shall see, leads to very far-reaching results.

Consider first again the simplest insertion of Fig. 5 with vector-tensor nonet exchange in the exact SU(3) limit. As quoted in (1o), the slope of the 'Pomeron', α'_P, was found to be much smaller than the input α'_M, due to the difference in signature factors for the crossed and uncrossed loops.[3)6)7)] This means that the amount of trajectory renormalization, $\Delta\alpha = \alpha_P - \alpha_M$, is a decreasing function of t around $t = o$. It seems natural to expect that this behaviour will continue for some distance away from $t = o$. Indeed, using somewhat simplified kinematics, it can readily be demonstrated that $\Delta\alpha$ decreases approximately as an exponential[49)5o)], so that by the resonance region, violations of exchange degeneracy are already quite small. One can do even better, simply by extending the calculation reported in ref. 1o), with exactly the same parameters and exponential parametrisation. One then obtains the renormalised trajectory shown in Fig. 23.[51)] The analytic continuation to positive t is trivial since the integral can be done analytically and there is no freedom whatsoever in the calculation. The result strongly supports the proposed identification of the f to the Pomeron trajectory. Because of the exponential approximation to the vertices which for example, neglects t-channel threshold, Fig. 24, cannot be trusted for $|t| \geq 1$ GeV^2. Nonetheless we note:

(i) Exchange degeneracy is broken in the resonance region by roughly the right amount. From Fig. 23, one obtains for example

$$m^2_{A_2} - m^2_f \sim .15 \text{ GeV}^2$$

as compared with the experimental value of $\sim .1$ GeV^2.

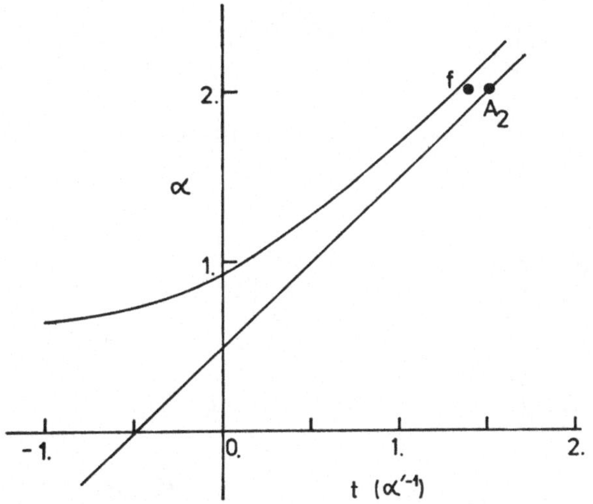

Fig. 23. The renormalized f-trajectory, as calculated in ref. 51).

(ii) The 'Pomeron' trajectory continues to flatten for $t < o$ becoming essentially flat for $t \sim -1$ GeV^2. There are some experimental indications that this may indeed be the case.

Next, we relax the assumption of exact SU(3) symmetry. As in Fig. 16 there will now be mixing between the originally pure quark states. The amount of mixing is determined by the strength of the insertions, which as we have seen, is a decreasing function of t. We expect therefore:[49)51)]

(i) Quark states will become increasingly pure for large positive t, (minimal mixing).

(ii) Mixing between quark states will increase for large negative t, with the 'Pomeron' becoming eventually an SU(3) singlet (maximal mixing).

Both of these assertions have important experimental consequences.

First, in the resonance region, the mixing is expected to be already quite small. Thus ϕ is almost a pure $\lambda\bar{\lambda}$ state, and f' even purer still. This means that we expect the OZI rule to be approximately valid in the resonance region in spite of unitarity corrections. Later in section III, we shall return to quantitative calcula-

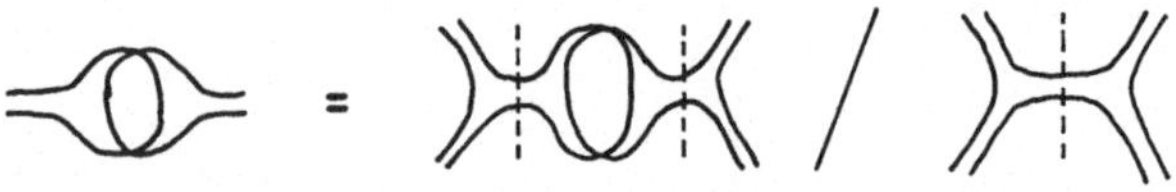

Fig. 24. Crossed loop insertions obtained by factorisation.

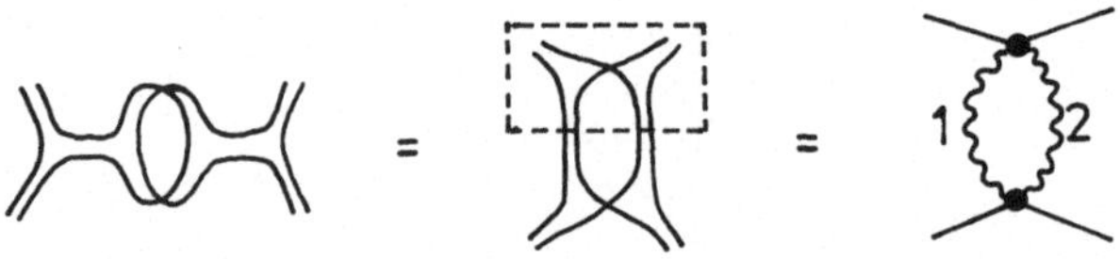

Fig. 25. Reggeon-reggeon cut.

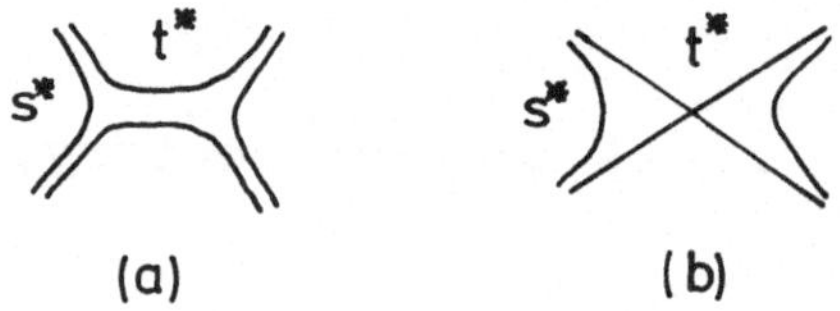

Fig. 26. (a) s^*, t^* term, and (b) s^*, u^* term.

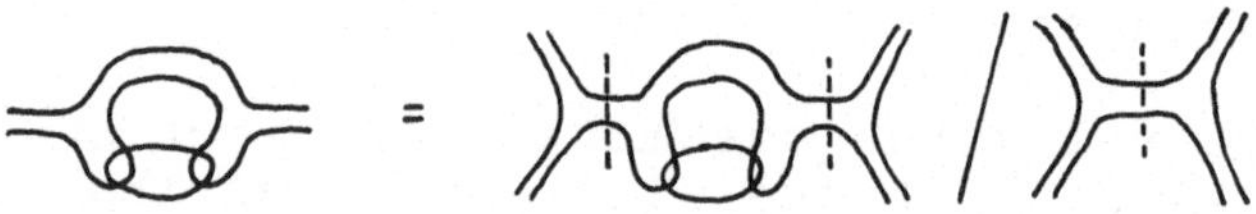

Fig. 27. Example of non-vanishing insertion.

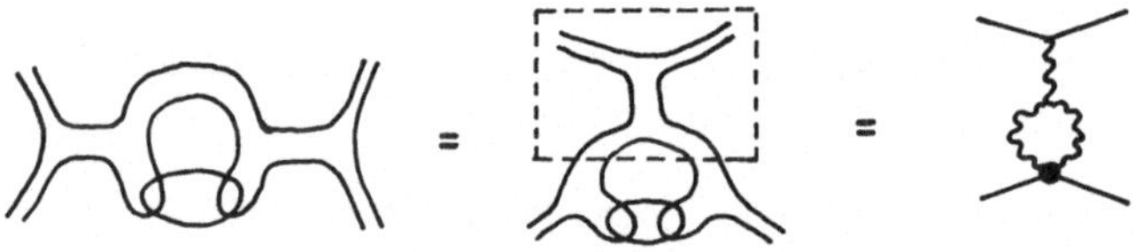

Fig. 28. Pole dominated non-planar amplitude.

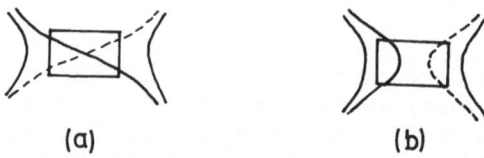

Fig. 29. General form of diagram (a) breaking exchange degeneracy, and (b) mixing $\bar{q}q$ states.

tions of OZI violations. We only note here that experimentally,

$$[\sigma_T(\phi p)/\sigma_T(\rho p)]^2 \sim \frac{1}{4}, \text{ but } \Gamma(f' \to \pi\pi)/\Gamma(f' \to K\bar{K}) \sim 1/100$$

representing a decrease in the mixing angle between the f-Pomeron and f' trajectories, from a value $\tan\theta \sim .5$ at $t = 0$ to $\tan\theta \sim .1$ at the f' mass. This is in qualitative agreement with our prediction.

Second, the Pomeron tends to an SU(3) singlet for large negative t.[52] Unfortunately, this does not imply directly that $d\sigma/dt$ is the same for, for example πp and Kp elastic scattering. We believe that for $t \leq -1$ GeV^2, multi-Pomeron cuts will dominate, which involve always the Pomeron coupling at small t. Therefore in order to check this prediction against experiment, one has first to unraffle the cuts to obtain the single Pomeron term. At present this can be done only in a model-dependent manner. For example, one may identify the single-Pomeron contribution with the eikonal extracted from the experimental cross sections.[52] One finds indeed then that the eikonals for πp, Kp, ϕp and even ψp tend to approach one another.

On the other hand one predicts also that the Pomeron has an octet component at $t \sim 0$, which quickly decreases as t becomes more negative. Consider then the following interesting example, suggested by Roberts[52]

$$\Sigma^- p \to \Sigma^{*-} p \qquad (30)$$

where Σ^{*-} can be any resonance belonging to the Δ-decuplets. The reaction can proceed only via octet exchange. Nonetheless, according to this scheme, it is still dominated at high energies by the 'Pomeron', though with a reduced octet coupling. Moreover, one predicts that $d\sigma/dt$ for (30) is a much sharper function of t than ordinary diffractive reactions exchanging the Pomeron singlet. Phenomenological tests for these predictions should soon be available from hyperon beam experiment.

An extension of the calculation in Fig. 23 for broken SU(3) is under way,[51] the preliminary results from which are qualitatively as expected. When properly refined, it should give the actual values of the mixing angles between ω-ϕ and f-f', as well as the trajectory renormalization $\Delta\alpha$ for $-1 \leq t \leq 1$ GeV^2. One should then be able to evaluate quantitatively the effects discussed above.

One can further imagine a similar calculation for other trajectories with zero quantum numbers, such as η and η' in the manner discussed at the end of section II.1. The qualitative features are expected to be similar. Even before such calculations are available, however, interesting phenomenology can already be done, by estimating the mixing angles through fitting $\Delta\alpha$ to the known masses of the resonances.[53)54)55)] The special feature of this approach is that mixing angles are now allowed to depend strongly on t, so that, for example the resonances η and η' are related effectively not by the mixing angle, but by two angles, one at the η mass and one at η', with $\theta_\eta > \theta_{\eta'}$. Indeed, adopting this, one is able to reconcile the data for η,η' decay with those on their production.[53)]

The most interesting feature in the preceeding discussion is the rapid decrease of renormalization effects with increasing t, which reconciles the apparent discrepancy between the sizes of non-planar corrections in respectively the resonance and exchange regions of Regge trajectories. We have thus partly achieved one of the main objectives set out in the introduction. The fact that this property results directly from the Regge phase suggests that it is more general than merely for the insertion of Fig. 5 around t = o. It is expected to apply also to Fig. 18(b), for example,[4o)] which was responsible for the splitting between the ρ and A_2 trajectories. We wish now to examine more carefully under what conditions this prediction is expected to be valid.

First, we inquire whether this initial decrease in $\Delta\alpha$ will continue for larger values of t > o. This question cannot easily be answered by analytic continuation of equation (9) as we have done so far, since one would need to know the large t behaviour of the Regge vertices, which is clearly inaccessible at present. Fortunately, there is another approach available which yields the result directly.[56)57)] The insertion of Fig. 5, for example, for t > o can be obtained by factorisation in the direct channel, as represented symbolically in Fig. 24. Now for large positive t, the numerator, namely Fig. 25, is dominated asymtotically by a reggeon-reggeon cut, while the denominator has the usual pole behaviour, t^α. Therefore, since the cut has a lower effective intercept $\alpha_c = \alpha_1 + \alpha_2 - 1$ than the pole, we expect the insertion to decrease asymptotically like a power for large t > o.[+]

One may wish in certain circumstances to consider the insertion in states of definite t-channel angular momentum ℓ. In our case, one expects the power decrease to remain valid only so long as $\ell \leq \sqrt{t}$. It may not be true, for example, on the leading trajectory with $\ell \sim t$, since this depends on the behaviour of the cut amplitude at large scattering angles which is largely unknown. However, since the leading trajectories in the direct channel are hard to excite, such an exception may not be physically interesting.

That the amplitude of Fig. 25 has asymptotically a cut behaviour can be seen as follows. First, the diagram can be considered as the overlap of two amplitudes, each representing the production of two resonant clusters via a reggeon exchange. Examine now the section of the diagram enclosed by the dashed line in Fig. 25 which is reproduced in Fig. 26(b). This, in dual-resonance model language, is called an (s^*, u^*) diagram, which has a peak at small u^*. At fixed t^*, therefore, it is like viewing a backward peak in the forward direction - the amplitude will decrease rapidly (exponentially, for example) as s^* increases. In terms of Fig. 25, now, s^* represents the mass of a resonant cluster. Because of its rapid cut-off, high mass resonances are not excited when the total energy t is increased. Instead, all the energy goes into exciting the exchanged reggeons 1 and 2, hence the cut-dominated asymptotic behaviour.

The rapid decrease of Fig. 26(b) with s^* can also be understood in terms of the direct channel resonances. Whereas for the (s^*,t^*) term of Fig. 26(a) all resonances contribute positively at $t^* = o$ to give ordinary Regge asymptotic behaviour, they alternate in sign for the (s^*,u^*) term in (b) interfering thus destructively and cutting off sharply at large s^* values. This change in sign, of course, is in turn closely related to the signature factor $\exp(i\pi\alpha)$ for the reggeon in the direct

---

+ A power behaviour has also been conjectured by Chew and Rosenzweig[49] by direct analytic continuation of (9). However, for lack of information on the large t behaviour of Regge vertices, they obtained a different power from here which is independent of α and probably incorrect.

channel of (a) which is absent in (b). It is exactly this same factor which gives the difference in phase between the crossed and uncrossed loops invoked in Section II.1 to explain the reduced Pomeron slope at t = o.[3)6)7)]

Similar considerations can be applied to insertions other than Fig. 5. For example, Fig. 18(b), which splits the exchange degeneracy of octet trajectories, is seen to have asymptotically a triple-reggeon cut behaviour, and will vanish also for large t > o.[4o)]

However not all nonplanar insertions will decrease asymptotically. For example, the insertion in Fig. 27 will not.[58)59)] The numerator here, as illustrated in Fig. 28, may be regarded as the overlap of two cluster-production amplitudes, the dependence of which on the top cluster mass is governed by Fig. 26(a), whose direct channel resonances are dual to ordinary reggeon exchange. Therefore when the total energy t is increased, high mass resonances are easily excited at the top vertex, leading to a pole-dominated asymptotic behaviour for the whole amplitude. The insertion in Fig. 27 will thus approach a constant asymptotically.[58)59)]

In general, a convenient criterion for judging whether a nonplanar quark diagram has pole behaviour or not, is to 'squeeze' the diagram all along its length. If somewhere one can make two quark lines touch, as in the top half of Fig. 28, it has pole asymptotic behaviour, but not otherwise. Insertions obtained by factorising pole-dominated diagrams will approach a constant, but all others will vanish, asymptotically.

One sees therefore that there are many non-planar insertions which can yield non-vanishing renormalization effects, even as $t \to +\infty$. Amplitudes are therefore <u>not</u> 'asymptotically planar' as conjectured in ref. 49) and 5o) based on the study of the simple insertion of Fig. 5.

However, all nonplanar insertions which break exchange degeneracy and/or mix quark states must have cut behaviour and vanish asymptotically. This can be seen as follows. Any diagram, however complicated, which breaks exchange degeneracy between odd and even signatures, must contain a twist as shown in Fig. 29(a), while any diagram which mixes quark states must have the general form of Fig. 29(b). By the criterion given above, one easily sees that both must vanish asymptotically. Hence, we deduce the following important theorem:[58)59)] Regge trajectories for $t > o$ become asymptotically pure quark states and exactly exchange degenerate. Clearly, this can have very far-reaching phenomenological consequences. Some more quantitative studies of these may prove extremely valuable.

III. VERTEX RENORMALISATION

In the same way that reggeon propagators are renormalised by non-planar insertions, reggeon vertices will also be so affected. The most dramatic manifestation of these corrections occur in the decay of resonances which are below the threshold of normal planar modes, such as $J/\psi \to D\bar{D}$. They will then have to decay via non-planar vertices, into those channels forbidden by the Okubo-Zweig-Iizuka rule, such as $J/\psi \to \rho\pi$. Indeed, recent intensive studies of J/ψ and related decays have led to quite detailed information about OZI violations, which serves ideally as a testing ground for dual unitarisation. It is on these therefore that we shall now direct our attention.

We list first of all, some outstanding features of OZI forbidden decays, observed experimentally:[60)]

(i) In spite of being strong unitarity corrections, OZI violations are in general small. For example, from the allowed decay $\phi \to K\bar{K}$ and the forbidden decay $\phi \to \rho\pi$, one estimates that the decay amplitude B is suppressed*:

$$|B(\phi \to \rho\pi)/B(\phi \to K\bar{K})|^2 \sim 10^{-2} \quad (31)$$

(ii) The suppression of forbidden decay amplitudes is apparently strongly mass dependent. For example,*

$$|B(f' \to \pi\pi)/B(f' \to K\bar{K})|^2 \sim 10^{-3} \quad (32)$$

as compared with (31) at the ϕ-mass. Also, even more dramatically,*

$$B|(\psi \to \rho\pi)/B(\phi \to \rho\pi)|^2 \sim 10^{-6} \quad (33)$$

---

* Such estimates of the decay amplitudes depend on the method for handling phase space effects. Our assertions, however, should remain qualitatively valid. The specific numbers given here assume a fixed effective radius of interaction.

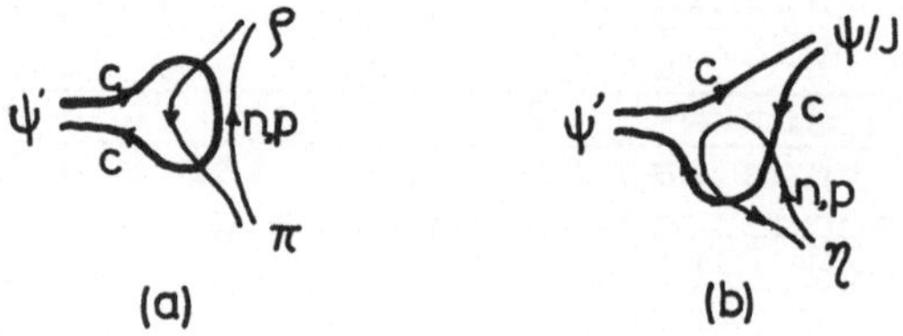

Fig. 3o. Two types of OZI-violating decays.

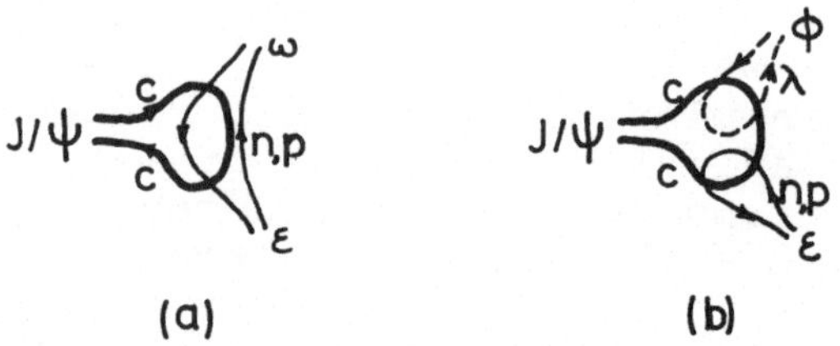

Fig. 31. J/ψ → (a) ωε, and (b) φε.

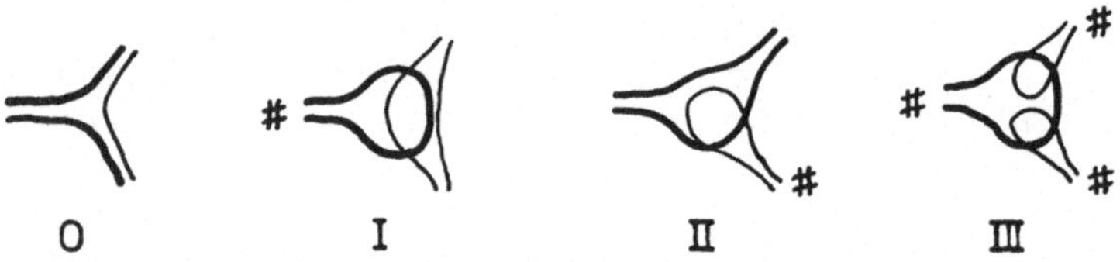

Fig. 32. Four types of decay vertices.

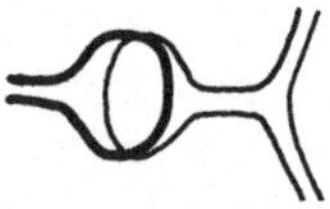

Fig. 33. Interpretation of OZI-violating vertex as due to mixing of states.

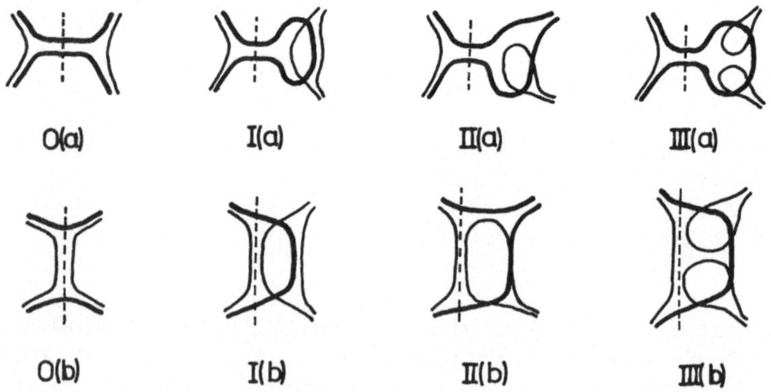

Fig. 34. Vertices obtained by factorisation.

(iii) Apparently there exist different types of OZI violations. For example, both $\psi' \to \rho\pi$ and $\psi' \to \psi\eta$ are forbidden and must proceed via non-planar amplitudes such at those in Fig. 3o. Yet*

$$|B(\psi' \to \psi\eta)/B(\psi' \to \rho\pi)|^2 \sim 10^5 \qquad (34)$$

(iv) OZI violations are apparently non perturbative, i.e. not simply due to a small coupling constant. For example, the decay $\psi \to \phi\varepsilon$ is doubly OZI forbidden, in the sense that its quark diagram, as shown in Fig. 31, is doubly disconnected. Therefore, if OZI violations are simply perturbative, one expects from (i) that $\psi \to \phi\varepsilon$ is suppressed by another factor of 10^{-2} compared with the singly forbidden mode $\psi \to \omega\varepsilon$. Experimentally, however[61]

$$\Gamma(\psi \to \phi\varepsilon)/\Gamma(\psi \to \omega\varepsilon) \sim 1/5 \text{ not } 1/100 \qquad (35)$$

(v) in OZI violating decays, products with zero quantum numbers are preferred. For example, the width for $\psi \to \gamma\eta$ is sizeable, but the very similar mode $\psi \to \gamma\pi$ has never been observed. Estimates from vector dominance give:

$$\Gamma(\psi \to \gamma\pi)/\Gamma(\psi \to \gamma\eta) \lesssim 10^{-3} \qquad (36)$$

(vi) Internal symmetry seems oddly broken in some OZI forbidden decays. For example, from the experimental fact that $\psi \to K\bar{K}$ is not seen one deduces that ψ is an SU(3) singlet. If so, one predicts $\Gamma(\psi \to \rho\pi)/\Gamma(\psi \to K^*K) \sim 1$, including already phase space effects. Experimentally, however,[62]

$$\Gamma(\psi \to \rho\pi)/\Gamma(\psi \to K^*K) \sim 2\text{-}3 \qquad (37)$$

a value more appropriate for ψ being an SU(3) octet.

(vii) The momentum spectra of decay products in forbidden multi particle modes differ widely. For example, in the modes $\psi \to \omega\pi^+\pi^-$ and $\psi \to \phi\pi^+\pi^-$ which are kinematically similar, the $\pi^+\pi^-$ mass spectra are completely different. Whereas the spectrum is sharply cut-off for large $\pi\pi$ masses in $\psi \to \phi\pi\pi$, the opposite is true in $\psi \to \omega\pi\pi$.[61]

Clearly, these features contain a great deal of dynamical information, which no hadronic theory can afford to ignore. Since they represent just those unitarity corrections which we set out to calculate, to explain them is a direct challenge. Indeed, although we have as yet no accurate techniques for calculating such effects, we shall show below how they can all, at least, be semi-quantitatively understood. We follow here the treatment of ref. 57). Related work can be found in ref. 54) and 63).

First, we classify our three-point vertices according to their topology as in Fig. 32, where in each case we retain only the diagram of the lowest order.[57) 58)] Types I and II are actually equivalent topologically, but we distinguish them here for convenience, according to whether the decaying particle or a decay product is disconnected. Channels marked with ⧣ admit only zero quantum numbers, e.g. ω, η, ε etc. Typical examples of decay modes, which proceed via the four types of vertices are: (o) $\psi \to D\bar{D}$ (below threshold), (I) $\psi \to \rho\pi$, (II) $\psi' \to \psi\eta$, (III) $\psi \to \phi\varepsilon$. Some modes belong to several types: for example, $\psi \to \omega\varepsilon$ can proceed via both types I and III.

Next we shall see how these vertices can in practice be calculated. At first sight, it might appear that this problem can be reduced to the one of propagator renormalisation already considered in section II. Indeed, by duality, the diagram (I) in Fig. 32 can be distorted into Fig. 33. The latter may be interpreted, say for $\psi \to \rho\pi$, as the ψ state first acquiring via propagator renormalisation a $p\bar{p}$ and $n\bar{n}$ component, which then decays into $\rho\pi$ through ordinary planar vertices. Such a picture is exact if one sums in Fig. 33 over a complete set of intermediate states. However, in all calculations so far of propagator renormalisation, only the mixing between leading trajectories at fixed t is considered.[23) 24) 25) 51)]. Nor is it easy to extend such calculations to trajectories other than the parent. This is clearly inadequate for studying OZI violations in resonance decays, which involves the mixing of all states at fixed j. In particular, for ψ decay, the main contributions to the sum of Fig. 33 will come from j = 1 states with mass $t \sim m_\psi^2$, namely very low-lying daughters of ω. The mixing of ψ with the leading ω-trajectory[54)] is here practically irrelevant.* In view of this, it is easier to attempt calculating vertex renormalisations directly.[57) 58)]

---

* This is not to say that phenomenological mixing angles cannot be profitably employed (see, e.g., Refs. 54 and 64).

The vertices of Fig. 32 can be obtained by factorisation from the more familiar scattering amplitudes of Fig. 34, as indicated. We are particularly interested in the vertices where the decaying resonance has a large mass. We seek therefore the leading asymptotic behaviour of these amplitudes. By the criterion given at the end of section II.3, we ascertain readily that O, II, III are all pole-dominated, but that I is governed asymptotically by a reggeon-reggeon cut, as illustrated in Fig. 35. We remind the reader that pole or cut behaviour, corresponds to whether high mass resonances in the intermediate state can, or cannot be excited.

The reggeon diagrams in Fig. 35(b) are all in principle calculable with the methods developed, although as yet rather inaccurately due to our poor knowledge of Regge vertices. For our present purpose of a qualitative understanding, it suffices to work with the one-dimensional approximation of ref. 15) which has proved successful in previous explorations. As in (11) we write in j-representation the following formulae:

$$\text{O:}\ \frac{\gamma^2}{j - \alpha_1} \tag{38}$$

$$\text{I:}\ \frac{\gamma^2\varepsilon^2 g^2}{j - \bar{\alpha}_1 - \bar{\alpha}_2 + 1} \tag{39}$$

$$\text{II:}\ \frac{\gamma^2\ \varepsilon^2 g^4}{(j - \alpha_2)(j - \bar{\alpha}_1 - \bar{\alpha}_2 + 1)} \tag{4o}$$

$$\text{III:}\ \frac{\gamma^2\varepsilon^2 g^4}{(j - \alpha_2)(j - \bar{\alpha}_1 - \bar{\alpha}_2 + 1)^2} \tag{41}$$

where as before, γ represents the reggeon coupling to particles, g the triple-reggeon vertex, and $\gamma\varepsilon g$ the crossed four-point vertex depicted in Fig. 26(b), (see also Fig. 2). All quantities ε, g and $\bar{\alpha}_i$ are understood to be appropriate averages over the loop integrals.

In s-representation, the corresponding formulae to leading Regge approximation read:

$$\text{O:}\ \gamma^2 s^{\alpha_1} \tag{42}$$

$$\text{I:}\ \gamma^2\varepsilon^2 g^2 s^{\bar{\alpha}_1 + \bar{\alpha}_2 - 1} \tag{43}$$

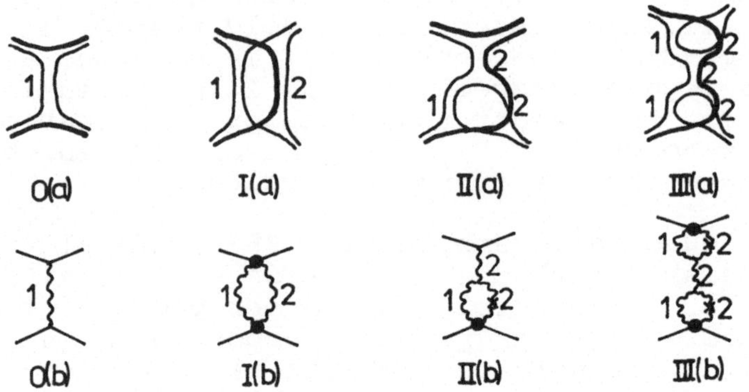

Fig. 35. Asymptotic behaviour of diagrams in Fig. 34.

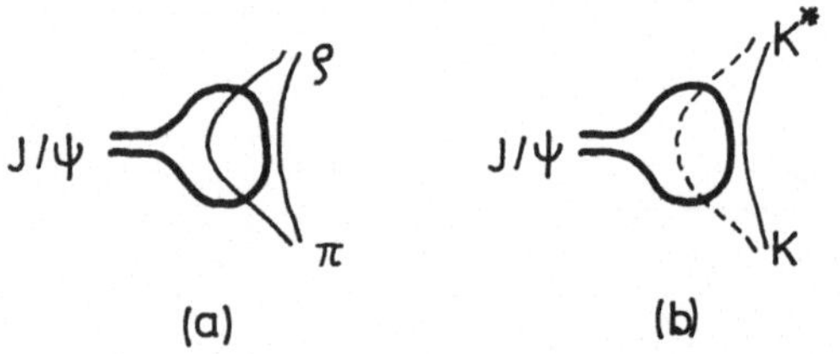

Fig. 36. J/ψ → (a) ρπ and (b) K^*K.

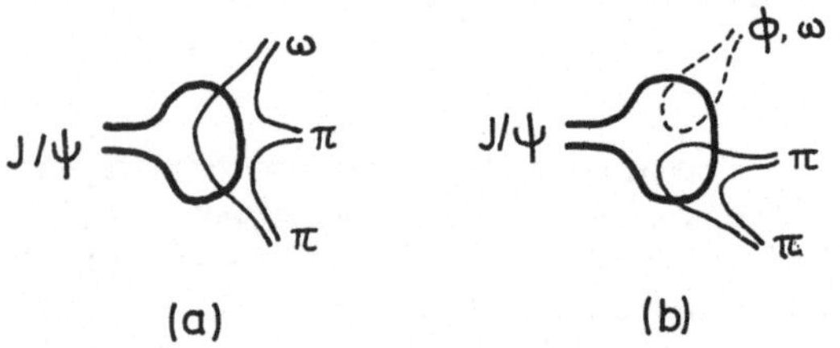

Fig. 37. (a) J/ψ → ωππ
(b) J/ψ → φππ or ωππ.

$$\text{II:}\quad \frac{\gamma^2\varepsilon^2 g^2}{(\alpha_1 - \bar{\alpha}_1 - \bar{\alpha}_2 + 1)}\, s^{\alpha_2} \tag{44}$$

$$\text{III:}\quad \frac{\gamma^2\varepsilon^2 g^4}{(\alpha_1 - \bar{\alpha}_1 - \bar{\alpha}_2 + 1)^2}\, s^{\alpha_2} \tag{45}$$

Next we have to project out the contribution of a fixed partial wave ℓ. This however depends on the angular dependence of the amplitudes which has been lost in the one-dimensional approximation. For exploration, let us assume simply an exponential form exp bt, with b constant and the same for all amplitudes. This is equivalent to assuming a fixed effective radius of interaction. Hence factorising off the redundant vertex in Fig. 34, which is essentially just $\gamma s^{1/2\alpha_1}$, we have then the following formulae for the widths of two-body modes: $a \to b + c$,

$$\text{O:}\quad F\,\gamma^2 s_a^{2\alpha_1-\alpha_1}\, \frac{(bq)^{2\ell+1}}{(2\ell+1)!!s_a} \tag{46}$$

$$\text{I:}\quad F\,\gamma^2\varepsilon^4 g^4\, s_a^{2(\bar{\alpha}_1+\bar{\alpha}_2-1)-\alpha_1}\, \frac{(bq)^{2\ell+1}}{(2\ell+1)!!s_a} \tag{47}$$

$$\text{II:}\quad F\,\frac{\gamma^2\varepsilon^4 g^4}{(\alpha_2-\bar{\alpha}_1-\bar{\alpha}_2+1)^2}\, s_a^{2\alpha_2-\alpha_1}\, \frac{(bq)^{2\ell+1}}{(2\ell+1)!!s_a} \tag{48}$$

$$\text{III:}\quad F\,\frac{\gamma^2\varepsilon^4 g^8}{(\alpha_2-\bar{\alpha}_1-\bar{\alpha}_2+1)^4}\, s_a^{2\alpha_2-\alpha_1}\, \frac{(bq)^{2\ell+1}}{(2\ell+1)!!s_a} \tag{49}$$

s_a is the squared mass of the decaying resonance, q the final c.m. momentum, and F is a statistical weight factor depending on the quantum numbers and spins of the particles involved.

As they stand, these formulae are asymptotic. They are strictly valid therefore only for resonances way above the thresholds of intermediate states. However, it is ψ, ψ' and other resonances close to thresholds which at present supply the bulk of empirical information. For such cases, we shall simply assume that our formulae still hold approximately when extrapolated, as is the case for other dual amplitudes.

Clearly, the crude assumptions made in deriving (46)-(49) can be accepted at best only as working hypotheses which must be justified by further theoretical studies. Nevertheless, they suffice to explain semi-quantitatively all the special features (i)-(vii), enumerated above, of OZI violations.[57)58)]

Proceeding then in the same order as before, we note:

(i) All the OZI violating amplitudes I, II and III are suppressed by factors of the form $\varepsilon^a g^b$, where as noted before, $g^2 \sim 1/N$ with effective $N \sim 2.5$ and $.25 \lesssim \varepsilon \lesssim .5$. Hence forbidden modes are indeed expected to be small, in spite of being strong unitarity corrections. In order to obtain (31), we need $\varepsilon \sim .4$ which is within the predicted range. It is this value which has been used in section II.2 to estimate the size of ρ-A_2 splitting.

(ii) Type I decays to which both $f' \to \pi\pi$ and $\psi \to \rho\pi$ belong, are further suppressed because of their cut-dominated asymptotic behaviour. Due to the low effective intercept $\alpha_c = \bar{\alpha}_1 + \bar{\alpha}_2 - 1$, the amplitudes are strongly dependent on the decaying particle mass, leading to quite drastic effects at large masses $\sim m_\psi^2$. The actual amount of suppression depends on the intercepts of trajectories involved. To obtain (33), one needs:

$$\alpha_\psi(o) = -1.1 \qquad (5o)$$

Using (15) and (6), appropriately generalised to include the charmed quark c, we deduce from (5o):[57)]

$$\sigma_T(\psi p) \sim 3.5 \text{ mb} \qquad (51)$$

which is quite compatible with the recent experimental value of 2.7 ± 1 mb.[65)]

(iii) Modes of type II and III, being pole dominated, are much less strongly suppressed at large masses than those of type I. Indeed, factors like (34) are readily obtained for the value of α_ψ given in (5o) above.

(iv) Type III amplitudes, being doubly disconnected, are indeed smaller in coupling that those of type I as seen in (47) and (49). However, because of (ii), they become comparable in magnitude around m_ψ^2. Thus, $\psi \to \omega\varepsilon$ is not much suppressed compared with $\psi \to \phi\varepsilon$. Indeed assuming pure type III decay one predicts for (35) a value of $\sim 1/3$.

(v) Since the pole-dominated amplitudes II and III admit only zero quantum numbers in those channels marked with ≠ in Fig. 32, neutral decay products are favoured. One naturally obtains strong suppressions of the order of (36) for the so-called 'octet-photon' modes such as $\psi \to \gamma\pi$.

(vi) For type I amplitudes, symmetry breaking occurs in the exponent of the decaying resonance mass. For example, for the decays $\psi \to \rho\pi$ and $\psi \to K^*K$, the exponents of s_a in (47) are respectively $2(\bar{\alpha}_M + \bar{\alpha}_D - 1) - \alpha_M$ and $2(\bar{\alpha}_{K^*} + \bar{\alpha}_D - 1) - \alpha_M$ as can be seen in Fig. 36. Hence, even a small splitting in the intercept $\alpha_M - \alpha_{K^*} \sim .2$ is enough to explain the observed deviation (37) from the symmetry prediction.

(vii) These considerations can all be generalised to multiparticle modes.[66] For example, for the decays $\psi \to \omega\pi\pi$ and $\psi \to \phi\pi\pi$, the relevant quark diagrams are shown in Fig. 37. One notices there that in (b), the two pions are attached to the charmed quark loop in much the same way as $\rho\pi$ is attached in Fig. 36(a). Therefore, the same argument which applies in the strong suppression of the $\psi \to \rho\pi$ amplitude for large $\rho\pi$ (i.e. ψ) mass, will also apply in Fig. 37(b) to suppress the spectrum for large $\pi\pi$ masses in $\psi \to \phi\,\pi\pi$ decay. On the other hand $\psi \to \omega\pi\pi$ can proceed also by Fig. 37(a), which though suppressed in overall normalisation by the same mechanism, has no reason to favour low $\pi\pi$ masses more than any planar multipion amplitudes. Similar considerations for $\psi \to \omega K\bar{K}$ and $\psi \to \phi K\bar{K}$ decays, suggest in contrast that the situation is there reversed with the former favouring low $K\bar{K}$ masses but not the latter. All these expectations are qualitatively well borne out by experiment.

One sees therefore that there is no difficulty with dual unitarisation to explain the observed special features of OZI-violating decays, which at first sight seem so astounding. Indeed, one can be even more ambitious, taking the formulae (46)-(49) seriously to calculate a whole list of partial widths for ψ and ψ', which are in fair semi-quantitative agreement with experiment.[58)54]

In principle, the same method can also be used to investigate OZI forbidden vertices in particle production, obtaining reasonable estimates of, for example, $\pi^- p \to \phi n$ cross sections. Here, however, our method is not in general useful for ψ production, which, because of the large ψ mass, occurs at present energies at t values $\gtrsim$ 5 GeV^2, where the small t approximations so far adopted are inapplicable.

CONCLUDING REMARKS

As it stands, dual unitarisation is clearly no substitute for a genuine theory of strong interactions - this is evident from the lack of precision in its formulation. Nonetheless, starting from a few basic assumptions, simply by exploiting duality and unitarity to the full, one is able to correlate a wide range of hadronic phenomena, much beyond what has so far been attempted. Yet this seems only the beginning - one can easily envisage great improvements both in the precision and in the generality of its application. Besides the more obvious extensions in the purely hadronic sector to higher order corrections and higher-point functions than those already considered, one can imagine grafting a current to the scheme so as to incorporate lepton- and photon-induced reactions also.

One particularly interesting point to note is the strikingly significant role played by quark diagrams. In dual unitarisation, one deals with hadrons, not with their constituents. Yet, though introduced originally only as a short-hand notation for quantum numbers, quark diagrams seem to have taken on a life of their own. One cannot help feeling that the quark lines we use here merely to denote the topology are in fact trajectories traced by genuine quarks swimming round inside the hadron. If so, then by combining some parton ideas with dual unitarisation, one may be able to bridge the existing gap between our understanding of the hidden interior world of constituents and the observable world of hadrons outside.

ACKNOWLEDGEMENT

We wish to thank our colleagues K. Konishi, J. Kwiecinski, J.E. Paton and R.G. Roberts, with whom most of this work has been done, for the numerous pleasant discussions this past year which have contributed greatly to our understanding of the subject.

REFERENCES

1. Huan Lee, Phys. Rev. Letters 3o (1973) 719.
2. G. Veneziano, Phys. Letters 43B (1973) 413.
3. Chan Hong-Mo and J.E. Paton, Phys. Letters 46B (1973) 228.
4. G. Veneziano, Nucl. Phys. 52B (1974) 22o.
5. G. Veneziano, Nucl. Phys. B74 (1974) 365.
6. Chan Hong-Mo, J.E. Paton and Tsou Sheung Tsun, Nucl. Phys. B86 (1975) 479.
7. Chan Hong-Mo, J.E. Paton, Tsou Sheung Tsun and Ng Sing-Wai, Nucl. Phys. B92 (1975) 13.
8. J.E. Paton and Chan Hong-Mo, Nucl. Phys. B1o, (1969) 516.
9. M. Ciafaloni, G. Marchesini and G. Veneziano, Nucl. Phys. B98 (1975) 472, 493.
1o. See e.g. A. Biatas, IVth International Colloquium on multiparticle reactions (1973), p. 93.
11. C. Rosenzweig and G. Veneziano, Phys. Letters 52B, (1974) 335.
12. M. Schaap and G. Veneziano, Lett. Nuovo. Cim. 12 (1975) 2o4.
13. N. Papadopoulos, C. Schmid, C. Sorensen and D. Webber, Nucl. Phys. B1o1, (1975) 189.
14. M. Bishari and G. Veneziano, Phys. Lett. 58B (1975) 445.
15. J. Kwiecinski and N. Sakai, Nucl. Phys. B1o6 (1976) 44.
16. J.R. Freeman and Y. Zarmi, to be published in Nucl. Phys. (1976).
17. S. Feinberg and D. Horn, Tel Aviv University preprint TAUP-526-76.
18. J. Finkelstein and J. Koplik, Columbia University preprint CO-2271 (1975), to be published in Phys.Rev.
19. J. Diaz de Deus and J. Uschersohn, Rutherford Laboratory preprint RL-75-o42 (1975), to be published in Physica Scripta.
2o. Ken-ichi Konishi, Rutherford Laboratory preprint RL-76-o36 (1976).
21. P. Aurenche, Nuovo Cim. 33A, (1976) 64.
22. R.G. Roberts (1976) unpublished, quoted in ref. 4o).
23. N. Sakai, Nucl. Phys. B99, (1975) 167.
24. G.F. Chew and C. Rosenzweig, Phys. Letters 58B (1975) 93; Phys. Rev. D12 (1975) 39o7.
25. C. Schmid and C. Sorensen, Nucl. Phys. B96, (1975) 2o9.
26. P. Stevens, G.F. Chew and C. Rosenzweig, Nucl. Phys. B11o (1975), 355.
27. T. Inami and R.G. Roberts, Nucl. Phys. B93 (1975)497.
28. J.E. Dodd, Oxford University preprint, 54/76 (1976).

29. D. Ponting, private communication (1976).
3o. C. Cronström, J. Dias de Deus and J. Uschersohn, University of Helsinki preprint, No. 1o-76 (1976).
31. A. Gula, Lett. Nuovo Cim. 13 (1975) 432.
32. A. Gula, Rutherford Laboratory preprint RL-75-o8o (1975) to be published in Acta Physica Polonica.
33. P. Aurenche and F. Bopp, Ecole Polytechnique preprint (1976).
34. J.E. Dodd, Nucl. Phys. B1o7 (1976) 179.
35. See e.g. Chan Hong-Mo, J. Loskiewicz and W. Allison, Nuovo Cim 57A (1968) 93.
36. Y. Eylon and H. Harari, Nucl. Phys. B8o (1974) 349.
37. A. Kalinowski, Warsaw University preprint IFT/23/75 (1975) to be published in Acta Physica Polonica.
38. B.R. Webber, Phys. Lett. 62B (1976) 449.
39. Chan Hong-Mo and Tsou Sheung Tsun, Rutherford Laboratory preprint RL-76-o54 (1976), to be published in Nucl. Phys. B.
4o. Chan Hong-Mo, Ken-ichi Konishi, J. Kwiecinski and R.G. Roberts, Phys. Letters 63B (1976) 441.
M. Bishari, Helsinki preprint (1976).
41. A.V. Barnes et al., Phys. Rev. Lett. 37 (1976) 76; O.I. Dahl et al., Phys. Rev. Lett. 37 (1976) 8o; V.N. Bolstov et al. Nucl. Phys. B73 (1974) 365, 387; P. Sonderegger et al., Phys. Letters 2o (1966) 75; O. Guisan et al., Phys. Letters 18 (1965) 2oo.
42. A. Berglund et al., Phys. Letters 57B (1975) 1oo.
43. R.G. Roberts, unpublished (1976) quoted in ref. 4o).
44. Ken-ichi Konishi and J. Kwiecinski, Rutherford Laboratory preprint RL-76-o68 (1976).
45. J. Uschersohn, Helsinki University preprint (1976).
46. Y. Eylon, University of California Berkeley preprint (1976).
47. M. Fukugita, T. Inami, N. Sakai and S. Yasaki, Tokyo University preprint (1976).
48. G.F. Chew and C. Rosenzweig, private communication (1976).
49. G.F. Chew and C. Rosenzweig, Nucl. Phys. B1o4 (1976) 29o.
5o. M. Bishari, Phys. Letters 55B (1975) 4oo.
51. Tsou Sheung Tsun, Phys. Lett. 65B (1976) 81.
52. R.G. Roberts, Rutherford Laboratory preprint, RL-76-o46 (1976), to be published in Nucl. Phys. B.
53. T. Inami, K. Kawarabayashi and S. Kitakado, Phys. Letters 61B (1976) 6o, and Tokyo University preprint UT-Komaba 76-5 (1976).
54. C. Rosenzweig, Phys. Rev. D13 (1976) 3o8o.
55. G.F. Chew and C. Rosenzweig, preprint (1976).
56. G. Veneziano, Nucl. Phys. B1o8 (1976) 285.

57. Chan Hong-Mo, J. Kwiecinski and R.G. Roberts, Phys. Letters 6oB (1976) 367.
58. Chan Hong-Mo, Ken-ichi Konishi, J. Kwiecinski and R.G. Roberts, Phys. Letters 6oB (1976) 467.
59. Chan Hong-Mo, Ken-ichi Konishi, J. Kwiecinski and R.G. Roberts, Phys. Lett. 64B (1976) 3o1.
6o. See e.g. G.S. Abrams, Stanford Conference on Leptons and Photons (1975).
61. F. Vannucci, Seminar at Rutherford Laboratory (oct. 1976) SLAC-PUB (in preparation).
62. See e.g. review talk by V. Martyn at the XVIII International Conference on High Energy Physics, Tbilisi (1976).
63. C. Schmid, C. Sorensen and D. Webber, ETH preprint (1976).
64. D.P. Roy, Phys. Lett. 62B (1976) 315.
65. R.L. Anderson et al.. The latest value quoted by Camvini at Tbilisi is $\sigma_T(\psi p) = 3.48 \pm .8$ mb.
66. Chan Hong-Mo, J. Kwiecinski and Tsou Sheung Tsun, Lett. Nuovo Cim. 16 216 (1976).

COLLECTIVE EFFECTS IN HIGH ENERGY INTERACTIONS WITH NUCLEI*

G. Eilam

Department of Physics, Technion - Israel Institute of Technology, Haifa, Israel

ABSTRACT

Multiparticle production in high energy particle-nucleus and nucleus-nucleus collisions is discussed in the framework of the Collective Tube Model. Comparison between experiments and theory is presented, and agreement between experiment and theory is demonstrated. Important implications for high energy particle physics research are emphasized.

I. INTRODUCTION

When a high energy particle with laboratory momentum p_{lab} collides with a target nucleon at rest, the center of mass energy squared is approximately given by

$$s \simeq 2mp_{lab},$$

where m is the nucleon mass. When the target particle is a nucleus of atomic weight A, the center of mass energy for coherent production is A times larger and it is given by

$$s_A \simeq 2mAp_{lab} \simeq As.$$

What is the center of mass energy squared available for particle production in incoherent reactions in:

a) particle-nucleus collisions at high energies?

b) nucleus-nucleus collisions at high energies?

The various models that have been suggested so far for high energy particle-nucleus collisions and for high energy nucleus-nucleus collisions can be roughly divided into two categories according to their answer to this question:

The first category includes models like Intranuclear Cascade Models[1)], Leading Particle Cascade Models[2)], Energy Flux Cascade Models[3)], Multiperipheral Regge Type Models[4)] and various types of Statistical and Hydrodynamical Models[5)].

The second category includes all models that assume that a particle-nucleus collision is a single step process where a few nucleons in the nucleus interact collectively with the incident particle. In such models the effective center of mass energy squared available for the production of particles is approximately given by

$$s_{eff} \simeq 2m_{eff}p_{lab},$$

where m_{eff} is the mass of the system that interacts collectively with the incident particle and is of the order of a few times the mass of a single nucleon.

The possibility that in inelastic particle-nucleus collisions at very high energies few nucleons react collectively with the incident particle has been independently suggested by a few authors[6)]. First experimental results that strongly indicate such collective behavior were obtained at JINR, Dubna by Baldin and coworkers[7)]: In their investigations the authors bombarded nuclear targets with high energy deuterons (several GeV/c per nucleon), and they observed produced particles at forward angles with energies that significantly exceed the energy per nucleon in the incident deuteron. In a related series of experiments[8)] they bombarded nuclear targets with high energy protons and observed particles at backward angles with energies that significantly exceed the maximum energy allowed if the reaction took place on a single nucleon. These results supported the initial observations because if one considers them in the antilab system where the incident proton is at rest, they look like experiments where high energy nuclei collide with protons at rest and produce particles in the forward direction with energies that are a few times larger than the average energy per nucleon in the incident nucleus. Similar results were obtained at the Berkeley Bevalac in heavy ion induced reactions at considerably lower energy per nucleon in the incident ion[9)] but the authors claimed that they could explain most of the effect only by taking into account the internal motion in the colliding nuclei without postulating collective interactions. It should be stressed that if Baldin's results will be confirmed at much higher energies, they will provide a major difficulty for models of the first category.

The present review will be organized in the following way: In chapter II the Collective Tube Model (CTM) will be formulated for high energy particle-nucleus inelastic reactions. In chapter III it will be used to predict cross sections, multiplicity distributions, rapidity distributions and momentum distributions for particles inclusively produced in high energy particle-nucleus collisions. In chapter IV a detailed comparison between the CTM predictions and experimental results on particle production in high energy particle nucleus inelastic reactions will be presented.

II. THE COLLECTIVE TUBE MODEL

High Energy Particle-Nucleus Collisions

The Collective Tube Model (CTM)[10] for high energy particle-nucleus collision is based on two assumptions. Assumption (1) is:

The Tube Assumption: In a high energy particle-nucleus collision the struck nucleon and all the nucleons that lie behind it within a tube of cross section σ recoil collectively, i.e. a high energy particle nucleus collision is actually a particle-tube collision and all the nucleons outside this tube are spectators. If there are i nucleons inside the tube then its effective mass is im and the c.m. energy squared for the particle-tube collision is approximately given by

$$s_i \simeq 2imp_{lab}. \qquad (1)$$

The collective recoil of the tube can be considered as a "Nuclear Mössbauer Effect". Unfortunately there is no reliable hadrodynamic theory at high energies which can be used to derive such a Mössbauer Effect in analogy to its derivation in low energy solid state physics. However, it is not unreasonable to assume that when the shock velocity inside a dense tube is smaller than the incident projectile velocity, then the whole tube recoils collectively.

The tube cross section σ, is in principle, a free parameter in the model since the effective cross section of the recoiling nucleon within nuclear matter can be different from the free nucleon cross section. (Except perhaps in the simple case of a deuteron target where one may expect that $\sigma \simeq \sigma_t^{pp}$; σ_t^{pp} is the total nucleon cross section). The Collective Tube Model thus assumes that the collective interaction takes place between the incident particle and a long tube of nuclear matter (which is Lorentz contracted in the rest frame of the projectile), while the other collective models assume that the interaction takes place with a multinucleon cluster or a coherent fluctuation of nuclear matter that occupies only a small volume.

Assumption (1) can explain the Cumulative Nuclear Effect that has been observed by Baldin et. al[7] and can be further tested by looking for the production of massive particles in particle-nucleus collisions at incident energies well below the threshold for their production in particle-nucleon collisions. However, in order to increase the predictive power of the model one must introduce additional assumptions regarding the particle-tube interaction. The Collective Tube Model in its minimal form is based on the following additional assumption (Assumption (2)):

The Universality Assumption: Quantities that in high energy particle-particle collisions are found to be independent of the quantum numbers of the colliding particles are assumed also to be independent of the quantum numbers of the tube in high energy particle-tube collisions. For such quantities we assume that in the center of mass systems the particle-tube collision looks like a particle-nucleon collision at the same center of mass energy. Such quantities, for instance, include the average multiplicity, the dispersion, the multiplicity distribution and various inclusive cross sections for production of particles with small transverse momenta. However universality is badly violated in many high energy exclusive reactions, and there are even a few examples of universality violation in high energy inclusive reactions (e.g. π^0 production with large transverse momenta by high energy protons and pions[11], ψ production by high energy protons and pions[12]). For such processes the universality assumption has to be replaced by a specific assumption regarding the dependence of the particle-tube cross section on the quantum numbers of the tube. As a first guess one may try to generalize models for high energy particle-particle collisions to high energy particle-tube collisions. Such attemps have been made already with the Statistical Model of Fermi[13] with the Hydrodynamical Model of Landau[5], with Parton Models[14] and with Regge Exchange Models. In this paper we will not discuss these attempts but rather try to exhaust and to examine the validity of the Collective Tube Model in its minimal form.

The CTM in its minimal form can be used to calculate particle-nucleon collision provided one knows the probability P(i,A) that the incident particle encounters a tube of i nucleons in particle-nucleus collisions. Detailed calculations based on low energy nuclear models show that P(i,A) is not very sensitive to the specific choice of a nuclear model. Moreover, in most of our calculations we have found that final results are not sensitive to

P(i,A). Therefore, in this review we will present results which are based only on two simple and "opposite" nuclear models:

An Independent Particle Model: The simplest independent particle model of the nucleus assumes that the nuclear wave function of a nucleus A is a product wave function of A identical nucleon wave functions properly normalized to give the correct nuclear density. In such a model the probability P(i,A,b) to find i nucleons within a tube of cross section σ at transverse coordinate b from the center of a nucleus A is given by

$$P(i,A,b) = \binom{A}{i} \left(\frac{\sigma T}{A}\right)^{i} \left(1 - \frac{\sigma T}{A}\right)^{A-i}, \quad (2)$$

where T(b) is the nuclear thickness at transverse coordinate b:

$$T(b) = \int_{-\infty}^{\infty} \rho(b,z)dz. \quad (3)$$

ρ is the nuclear density function normalized so that $\int \rho(r)d^3r=A$. (In the independent particle model the nucleon numbers of different tubes in the same nucleus are independent). From Eq. (2) one obtains that P(i,A) is given by

$$P(i,A) = \frac{\int d^2bP(i,A,b)}{\int d^2b\sum_{i=1}^{A} P(i,A,b)} \underset{A>>1}{\cong} \frac{\int d^2bP(i,A,b)}{\sigma_{in}^{pA}} \quad (4)$$

where

$$\sigma_{in}^{pA} = \int d^2b\sum_{i=1}^{A} P(i,A,b) = \int d^2b[1-(1-\frac{\sigma T}{A})^A] \underset{A>>1}{\cong} \int d^2b(1-e^{-\sigma T}) \quad (5)$$

P(i,A) as given by expressions (4) and (5) represents the probability to interact with a tube of i nucleons in inelastic particle-nucleus collisions.

Nuclear Droplet Model: If the target nucleus is described by a nuclear droplet of density $\rho(r)$ then

$$P(i,b) = \delta_{i,n(b)} \quad (6)$$

where

$$n(b) = \sigma T(b) \quad (7)$$

Furthermore, if the nucleus is chosen to be represented by a spherical droplet of constant nuclear density ρ and with a radius $r_A = R_1 A^{1/3}$, where R_1 is the nucleus radius constant, then

$$n(b) = \begin{cases} \frac{3}{2}\frac{\sigma}{\pi R_1^2}\left[A^{2/3} - b^2/R_1^2\right]^{1/2}; & b<r_A \quad (8A) \\ 0 & ;\ b \geqslant r_A \quad (8B) \end{cases}$$

As for the Independent Particle Model also for the Nuclear Droplet Model we will use only the probability distributions for the cases where nucleons are either opaque or transparent to the incident particles.

III. CTM PREDICTIONS FOR HIGH ENERGY PARTICLE-NUCLEUS COLLISIONS

Multiparticle Production

Many features of multiparticle production in high energy particle collisions are found to be practically independent of the quantum numbers of the colliding particle, e.g.

the average charge multiplicity,[15)]

$$\langle n\rangle \equiv \sum_n n\sigma_n / \sum_n \sigma_n, \tag{9}$$

the multiplicity dispersion,

$$D \equiv \langle n^2\rangle - \langle n\rangle^2, \tag{10}$$

and the KNO scaling function

$$\psi \equiv \langle n\rangle\sigma_n / \sum_n \sigma_n , \tag{11}$$

where σ_n is the cross section for producing n charged particles and the summation over n is over all inelastic channels that have charged particles.

Let us now apply the CTM to calculate these quantities for high energy particle nucleus collisions:

Average Multiplicity and Multiplicity Ratio: According to the CTM the probability to produce n charged particles in particle-tube collision is given by

$$\frac{\sigma_n^{pi}(s)}{\sigma_{in}^{pi}(s)} = \frac{\sigma_n^{pp}(is)}{\sigma_{in}^{pp}(is)} \tag{12}$$

where i is the number of nucleons in the tube. Consequently

$$\langle n(s)\rangle_{pA} = \frac{1}{\sigma_{in}^{pA}}\int d^2b \sum_{i=1}^{A} P(i,A,b) \sum_n n \frac{\sigma_n^{pp}(is)}{\sigma_{in}^{pp}(is)}$$

$$= \sum_{i=1}^{A} P(i,A) \ \langle n(is)\rangle_{pp}. \tag{13}$$

The average multiplicity in particle-particle collision can be well described by a power law[15]

$$\langle n(s)\rangle_{pp} = \langle n(s_o)\rangle_{pp} (s/s_o)^{\alpha}, \tag{14}$$

therefore the average multiplicity in pA collision is given by

$$\langle n(s)\rangle_{pA} = \langle n(s)\rangle_{pp} \sum_{i=1}^{A} P(i,A) \ i^{\alpha} \equiv \langle n(s)\rangle_{pp} \ \langle i^{\alpha}\rangle_A, \tag{15}$$

and the multiplicity ratio is given by

$$R_A = \frac{\langle n(s)\rangle_{pA}}{\langle n(s)\rangle_{pp}} = \langle i^{\alpha}\rangle_A \tag{16}$$

In the case that the nucleus is described by a nuclear droplet with a constant density then the average over P(i,A) can be performed analytically.

The average multiplicity and the multiplicity ratio are then given by:

$$\langle n(s)\rangle_{pA} = R_A \ \langle n(s)\rangle_{pp}, \tag{17}$$

where

$$R_A \underset{A>>1}{=} \left(\frac{3\sigma}{2\pi r_o}\right)^{\alpha} \frac{A^{\alpha/3}}{1+\alpha/2} \tag{18}$$

r_o is related to the nuclear radius r_A through $r_A = r_o A^{1/3}$; Eq.(17) represents an opaque nucleus. We do not discuss here photon and leptons interacting with nuclei; such a discussion will be presented elsewhere[16].

Multiplicity Dispersion and KNO Scaling: From Eqs. (11) and (12) one can write

$$\psi_A \equiv \langle n\rangle_{pA} \frac{\sigma_n^{pA}}{\sigma_{in}^{pA}} = \langle n(s)\rangle_{pA} \sum_{i=1}^{A} P(i,A) \frac{1}{\langle n(is)\rangle_{pp}} \psi_p\left(\frac{n}{\langle n(is)\rangle_{pp}}\right) \tag{19}$$

and by using (16) and (19) one obtains

$$\psi_A = R_A \sum_{i=1}^{A} P(i,A)\, i^{-\alpha} \psi_p(i^{-\alpha} R_A z);\ z \equiv \frac{n}{\langle n(s)\rangle_{pA}} . \tag{20}$$

R_A is energy independent, thus ψ_A obeys KNO scaling. Since ψ_A scales, the scaled dispersion $D_A/\langle n\rangle_{pA}$ is energy independent and it can be shown to satisfy the inequality

$$1 \leq \frac{[D/\langle n\rangle]_{pA}}{[D/\langle n\rangle]_{pp}} \leq \frac{\langle i\rangle A^{1/2}}{R_A^2} \tag{21}$$

Note that this inequality is independent of ψ. For a nucleus that is represented by a nuclear droplet one can show that

$$\frac{\langle n^2\rangle_{pA}}{\langle n\rangle^2_{pA}} = \gamma \frac{(1+\alpha/2)^2}{1+\alpha} , \tag{22}$$

where

$$\gamma = \int dz z^2 \psi_p(z) \tag{23}$$

If one uses the best fitted KNO scaling function ψ to πp data, one obtains the value $\gamma=1.2519$ and with the world average value $\alpha=.285$[15] one arrives at the prediction that

$$D_{\pi A} = .52\langle n\rangle_{\pi A}. \tag{24}$$

Multiplicity as Function of Tube Disintegration: In some bubble chamber experiments[17] the difference between the number of positive and negative charged particles among shower particles has been measured. From this difference it has been concluded that the shower particles contain a significant number of fast protons that result from the disintegration of the target nucleus. In the CTM these fast protons result from the disintegration of the recoiling tube. One may expect that if a nuclear tube contains N_p protons, it also contains on the average $\frac{A-Z}{Z} N_p$ neutrons. Therefore one may assume that on the average the N_p protons come from the dissociation of a tube of $i = \frac{A}{Z} N_p$ nucleons. The N_p fast protons thus register a particle-tube collision where the c.m. energy is approximately given by $\frac{A}{Z} N_p$ s. For such collisions the average multiplicity is given by

$$<n(s)>_{N_p} = <n(s)>_{pp} \left(\frac{A}{Z} N_p \right)^{\alpha} , \tag{25}$$

and the multiplicity ratio is given by

$$R(N_p) \equiv \frac{<n(s)>_{Np}}{<n(s)>_{pp}} \cong \left(\frac{A}{Z} N_p \right)^{\alpha} . \tag{26}$$

More precise expression for the average multiplicity and the multiplicity ratio as function of N_p can be found in reference (10a).

Inclusive Production of Particles

According to the CTM the inclusive cross-section for the reaction $p+i \rightarrow c$ + anything, where (without loss of generality) the projectile is called a proton and i is the number of nucleons in a tube, is given by

$$\frac{E}{\sigma_{in}^{pi}} \frac{d^3\sigma_c^{pi}}{dp^3} (s, E+p_{\parallel}, p_T) = \frac{E}{\sigma_{in}^{pp}} \frac{d^3\sigma_c^{pp}}{dp^3} (is, i^{1/2}(E+p_{\parallel}), p_T). \tag{27}$$

$E, p_{\parallel}, p_T$ are the energy, longitudinal and transverse momentum, respectively, of the measured particle c, and pp denotes proton-nucleon collision. For proton-nucleus collision, $p+A \rightarrow c$ + anything, we average Eq. (27) over the probabilities to find i nucleons at impact parameter b in a tube of cross section σ along a beam direction, and integrate the contributions from all impact parameters, i.e.

$$E \frac{d^3\sigma_c^{pA}}{dp^3} (s, E+p_{\parallel}, p_T) = d^2b \sum_{i=1}^{A} P(i,A;b) \frac{E}{\sigma_{in}^{pp}} \frac{d^3\sigma_c^{pp}}{dp^3} (is, i^{1/2}(E+p_{\parallel}), p_T), \tag{28}$$

where pA denotes proton-nucleus collision. Using Eq.(4) we obtain the sum rule

$$\frac{E}{\sigma_{in}^{pA}} \frac{d^3\sigma_c^{pA}}{dp^3}(s,E+p_{\parallel},p_T) = \sum_{i=1}^{A} P(i,A) \frac{E}{\sigma_{in}^{pp}} \frac{d^3\sigma_c^{pp}}{dp^3}(is,i^{1/2}(E+p_{\parallel}),p_T). \quad (29)$$

Integration of relation (29) over p_T yields the rapidity distribution:

$$\frac{1}{\sigma_{in}^{pA}} \frac{d\sigma^{pA}}{dy}(s,y) = \sum_{i=1}^{A} P(i,A) \frac{1}{\sigma_{in}^{pp}} \frac{d\sigma^{pp}}{dy}(is, \frac{1}{2}\ln i + y), \quad (30)$$

where

$$y \equiv \frac{1}{2} \ln (\frac{E+p_{\parallel}}{E-p_{\parallel}}) \equiv \ln (\frac{E+p_{\parallel}}{p_T}) \quad (31)$$

is the rapidity of particle c in the laboratory system.

Similarily, for the integrated cross section one obtains

$$\sigma_c^{pA}(s) = \sigma_{in}^{pA} \sum_{i=1}^{A} P(i,A) \frac{1}{\sigma_{in}^{pp}} \sigma_c^{pp}(is), \quad (32)$$

Note that in formulae (27) to (32) σ_{in}^{pp} stands for $\sigma_{in}^{pp}(is)$.

Let us now consider special cases of formulae (28) - (31) that are of particular interest:

Pseudo Rapidity Distributions: In many experiments on multiparticle production only angles are measured and rapidity distributions can not be obtained from such measurements unless additional assumptions are introduced regarding the dependence of the inclusive cross section on p_T. In such experiments usually a new variable η, called pseudo rapidity, is defined through

$$\eta \equiv - \ln(\mathrm{tg}\,\frac{\theta}{2}) = \ln (\frac{p+p_{\parallel}}{p_T}), \quad (33)$$

where p is the magnitude of the three-momentum of particle c.

From Eqs. (31) and (33) it is easy to see that

$$\frac{d^2\sigma}{d\eta dp_T^2} = \frac{p}{E} \frac{d^2\sigma}{dy dp_T^2} = \pi\frac{p}{E} E \frac{d^3\sigma}{dp^3} \quad (34)$$

Since the relation between η and y involves p_T (η= y only for $m_c^2 \ll p_T^2$), both for pp collisions and for pA collisions one needs an assumption on the p_T dependence of the inclusive cross section in order to be able to predict pseudo rapidity distributions.

Production of Particles with Large p_T: For large c.m. angles experimental data indicate[18] that

$E \frac{d^3\sigma_c^{pp}}{dp^3}$ is independent of $p_{\parallel}$. It then follows from Eq.(29) that

$$\frac{E}{\sigma_{in}^{pA}} \frac{d^3\sigma_c^{pA}}{dp^3}(s,p_T) = \sum_{i=1}^{A} P(i,A) \frac{E}{\sigma_{in}^{pp}} \frac{d^3\sigma_c^{pp}}{dp^3}(is,p_T). \tag{35}$$

Production of Particles with Small p_T: For $p_T^2+m_c^2 << p$ where m_c is the mass of particle c, Eq. (29) reduces to

$$\frac{E}{\sigma_{in}^{pA}} \frac{d^3\sigma_c^{pA}}{dp^3}(s,x\ ,p_T) = \sum_{i=1}^{A} P(i,A) \frac{E}{\sigma_{in}^{pp}} \frac{d^3\sigma_c^{pp}}{dp^3}(is,x\ ,p_T); \tag{36}$$

where $x = 2p_{\parallel}/\sqrt{s}$. In case that $E \frac{d^3\sigma^{pp}}{dp^3}(s,x\ ,p_T)$ for large s and small p_T, is a function of x and p_T only, then

$$\frac{Ed^3\sigma_c^{pA}/dp^3}{Ed^3\sigma_c^{pp}/dp^3} \underset{\substack{s\ \text{large} \\ p_T\ \text{small}}}{\cong} \frac{\sigma_{in}^{pA}}{\sigma_{in}^{pp}} \tag{37}$$

Approximate Scaling Laws: Comparison between the predictions of the CTM and experimental results on particle production in high energy hadron-nucleus collisions requires both knowledge of low energy nuclear properties to calculate P(i,A) and experimental data on hadron-nucleon collisions at c.m. energies squared s_i up to As, which in most cases of interest lie far beyond the energy range of present accelerators. In order to avoid ad-hoc parametrizations of cross sections at energies where no data are available, we approximate the averaging over P(i,A) in the following way: We have found that for $A \gtrsim 10$ the averaging over P(i,A) in the relevant expressions can be well approximated by $A^{1/3}$ the average nucleon number of a tube. The CTM predictions for incident hadrons then reduce to the following simple expressions:

Average Multiplicities:

$$\langle n(s)\rangle_{pA} \cong \langle n(A^{1/3}s)\rangle_{pp} \tag{38}$$

Multiplicity Ratios: If $<n(s)>_p \sim s^\alpha$ then

$$R_A \cong \frac{<n(A^{1/3}s)>_{pp}}{<n(s)>_{pp}} \sim A^{\alpha/3} \qquad (39)$$

KNO Scaling Functions:

$$\psi_A(z) \cong \psi_p(z) \qquad (40)$$

Particle Dispersions:

$$D_A(s) \cong D_p(A^{1/3}s) \qquad (41)$$

Dispersion Ratios: If $D_p(s) \sim c <n(s)>_{pp}$ then

$$\frac{D_A(s)}{<n(s)>_{pA}} \cong \frac{D_p(A^{1/3}s)}{<n(A^{1/3}s)>_{pp}} \sim c \qquad (42)$$

Inclusive Cross Sections:

$$\frac{\sigma_{in}^{pp}}{\sigma_{in}^{pA}} \sigma_c^{pA}(s) \cong \sigma_c^{pp}(A^{1/3}s) \qquad (43)$$

Momentum Distributions:

$$\frac{\sigma_{in}^{pp}}{\sigma_{in}^{pA}} E \frac{d^3\sigma_c^{pA}}{dp^3}(s, E+p_{\parallel}, p_T) \cong E \frac{d^3\sigma_c^{pp}}{dp^3}(A^{1/3}s, A^{1/6}(E+p_{\parallel}), p_T) \qquad (44)$$

Rapidity Distributions:

$$\frac{1}{\sigma_{in}^{pA}} \frac{d\sigma_c^{pA}}{dy}(s,y) \cong \frac{1}{\sigma_{in}^{pp}} \frac{d\sigma_c^{pp}}{dy}(A^{1/3}s, \frac{1}{6}\ln A+y) \qquad (45)$$

Large p_T Behavior:

$$\frac{\sigma_{in}^{pp}}{\sigma_{in}^{pA}} E \frac{d^3\sigma_c^{pA}}{dp^3}(s,p_T) \cong E \frac{d^3\sigma_c^{pp}}{dp^3}(A^{1/3}s, p_T) \qquad (46)$$

Small p_T Behavior:

$$\frac{\sigma_{in}^{pp}}{\sigma_{in}^{pA}} E \frac{d^3\sigma_c^{pA}}{dp^3}(x_{\parallel}, p_T) = E \frac{d^3\sigma_c^{pp}}{dp^3}(x_{\parallel}, p_T). \tag{47}$$

IV. COMPARISON WITH EXPERIMENT

In this chapter we present detailed comparison between the predictions of the CTM and experimental results on high energy particle-nucleus inelastic collisions. Since most of the available experimental data are on hadron induced reactions we will present comparisons between the CTM predictions and experimental results mostly for hadron induced reactions. More comparisons can be found in references (10). We hope that in a short time more data will be available on both high energy photon and lepton inelastic collisions with atomic nuclei so that it will be possible to test the CTM predictions against such data.

Multiplicity Distributions: Detailed information on multiplicity distributions in high energy particle-nucleus collisions is now available from emulsion experiments[19], from buble-chamber experiments and from counter experiments[20]. They practically cover the whole periodic table and a wide energy range. They also include information on the dependence of the multiplicity on the nuclear fragmentation. They exhibit the following main features:

1) The multiplicity ratio R_A, is energy independent above 50 GeV. It slowly increases with A, by about a factor 2 from hydrogen to uranium.

2) The multiplicity ratio depends only on the number of fastly recoiling protons from the target nucleus. For a fixed number of fastly recoiling protons it does not depend on A or on the incident energy.

3) The multiplicity distribution obeys KNO scaling with a scaling function that is practically A independent.

All these features are indeed reproduced by the CTM as can be seen from Eqs. (16), (26) and (20). However a prcise test of the CTM (and of other theoretical models for multiparticle production in particle-nucleus collisions) is still difficult because of the following reason - the CTM has well defined predictions only for the particles that are produced during the particle-tube collisions. However, most experiments do not

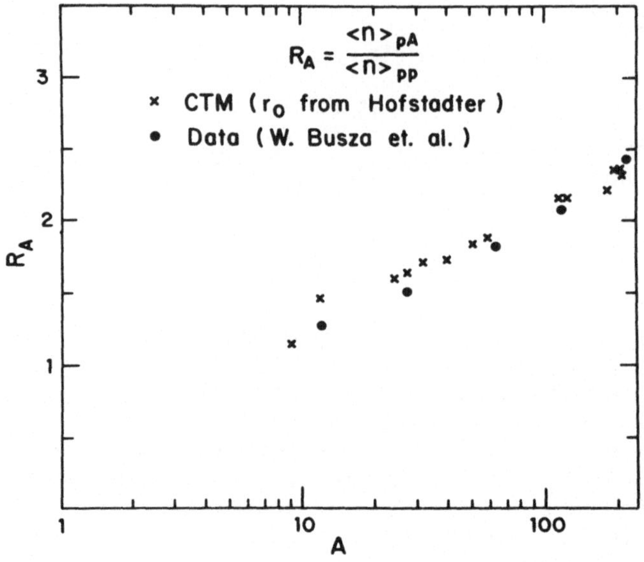

Fig. 1. Comparison between experimental results[20] and the CTM prediction Eq. (48) for R_A as function of A for high energy π-A collisions.

distinguish between such particles and nuclear fragments that result from the decay of the excited nucleus. Thus, when comparing the CTM predictions with experimental results one faces the problem of separating nuclear fragments (most protons) from produced particles. For some time it was believed that this separation is automatically achieved by considering only shower particles, i.e. it was believed that shower particles provide a sample of produced particles clean of nuclear fragments. However, at least two recent buble chamber experiments have indicated that this is not the case and that shower particles contain a significant number of fast protons stemming from the dissociation of the target nucleus[17]. (The difference between the number of positively charged and negatively charged shower particles together with charge conservation was used to determine the number of fast nuclear fragments among shower particles). Since most experimental data on charge multiplicity in high energy particle-nucleus collisions are on total shower multiplicity therefore Eq. (16) should be modified to include the contribution of the fast protons stemming from the nucleus fragmentation.

In the CTM the fast protons from the target among the shower particles result from the dissociation of the recoiling tube.

If there are i nucleons in such a tube then on the average $\frac{Z}{A}i$ of them are protons. Consequently, the multiplicity ratio for shower particles is given by

$$R_A = \frac{\langle n_s \rangle_A}{\langle n_s \rangle_p} \cong \langle i^{\alpha} \rangle_A + \frac{Z}{A} \frac{\langle i-1 \rangle_A}{\langle n_s \rangle_p} \tag{48}$$

where the subscript s stands for shower particles. In Fig. 1 Eq. (48) is compared with recent MIT data on π-A interactions at $50 \leq p_{lab} \leq 200$ GeV/c[)]. The CTM predictions were calculated for a uniform sphere with a radius $r_A = r_o A^{1/3}$ taken from Hofstadter's measurements [21]. Agreement between theory and experiment is very good. The theoretical predictions are not sensitive to the specific choice of a nuclear model. In Fig. 2a we compare Eq. (20) for ψ_A (solid line) with experimental data on π-N_e collisions[18]. The dashed line is the best fit for ψ_p from π-p collisions. Fig. 2a indicates that ψ_A as predicted by the CTM well reproduces the π-N_e data (and also the p-emulsion data that is not shown here) and

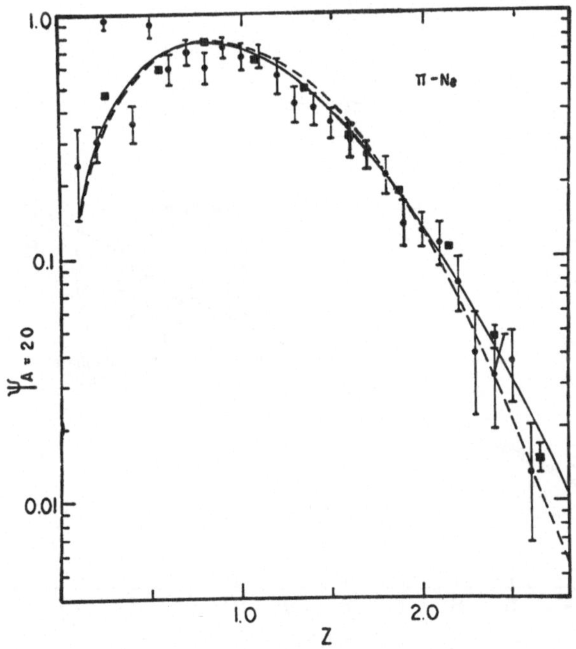

Fig. 2a. ψ_A versus $z = n/\langle n \rangle_A$ for π-Ne. Data from Ref. (17) (circles: 200 GeV; squares: 10.5 GeV). The solid line is the CTM prediction; the dashed line is the best fit for particle-particle collisions.

that ψ_A is indeed not much different from ψ_p except at large z values. Fig. 2b compares experimental data on $\frac{D_A/<n>_A}{D/<n>}$ as a function of A, with the CTM prediction as given by Eq. (20) and the upper and lower bounds set by inequality (21).

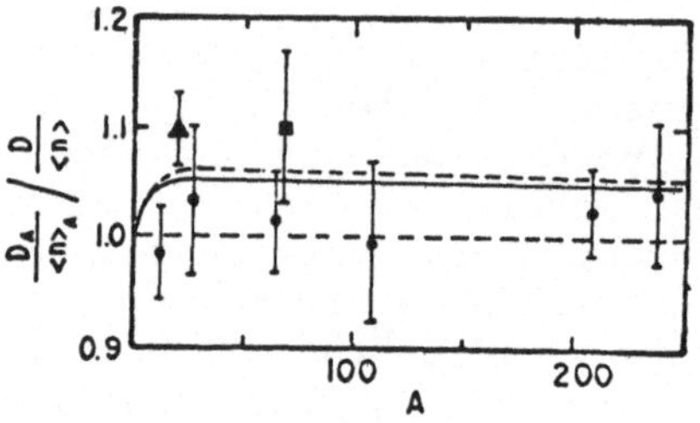

Fig. 2b. $(D_A/<n>_A)/(D/<n>)$ versus A for π- and p- nucleus. The solid line is the prediction of the CTM and the dashed lines are the predicted bounds.

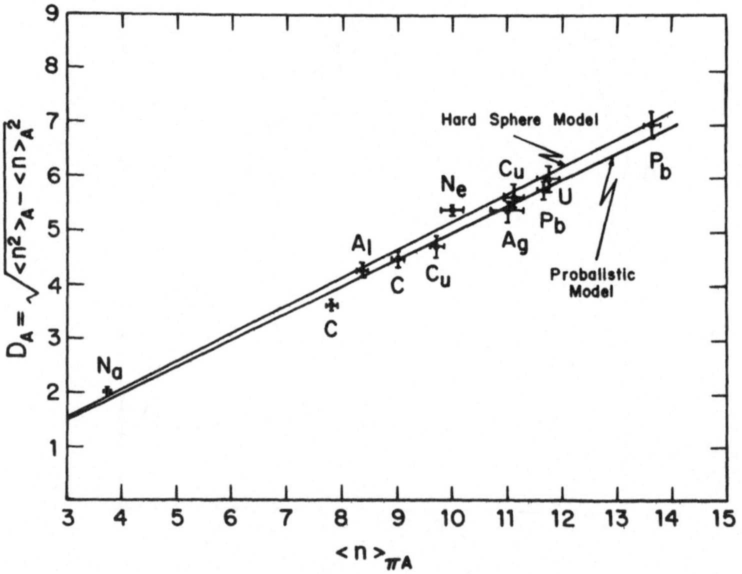

Fig. 3. Comparison between experimental results [22] and the CTM predictions Eq. (24) for $D_{\pi A}$ versus $<n>_{\pi A}$ in π-A collisions.

Further comparison between the CTM prediction (24) and experimental data on D_A as function of A in π-A collisions[22], is presented in Fig. 3. Very good agreement between experiment and theory is demonstrated.

Fig. 4 presents a comparison between the multiplicity ratio (26) as function of the number of fastly recoiling protons and experimental data from π-N_e[17] collisions and p-emulsion collisions [23]. For p-emulsion collisions we assumed that the number of fastly recoiling protons is proportional to the number of heavy tracks in the emulsion. Good agreement between the CTM predictions and the experimental data is demonstrated.

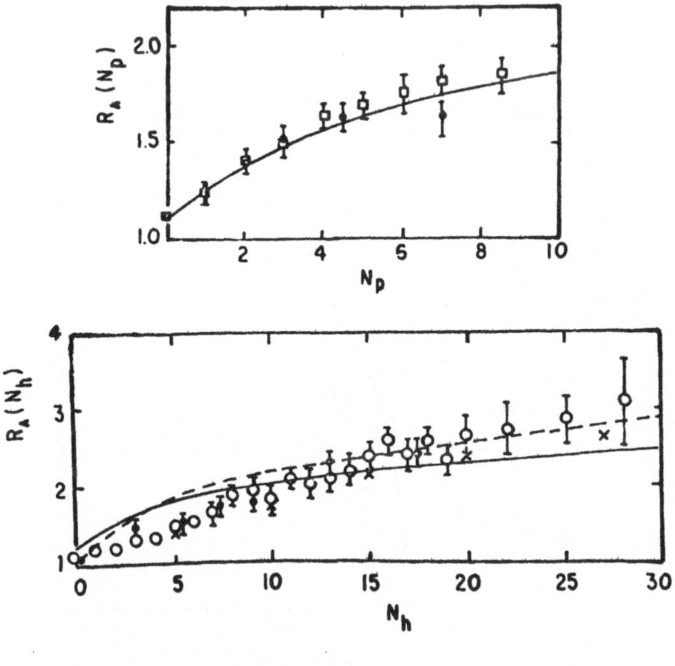

Fig. 4. (a) R_A versus the number of final protons for π-Ne. Data from Ref. (17). Circles, 200 GeV; squares, 10.5 GeV. (b) R_A versus the number of heavy tracks for p-emulsion. Solid circles (300 GeV) and open circles (200 GeV): The solid lines are the predictions of Eq. (14), and the dashed line is the prediction of Eq. (15) (N_p = $1.2N_h$) of Ref. (10a).

<u>Inclusive Cross Sections</u>: In the CTM the collective recoil of the target tube leads to rescaling of the effective center of mass energy $s_{eff} = A^{1/3}s$. Consequently the CTM prediction

$$\sigma_c^{pA}(p_{lab}) \cong \frac{\sigma_{in}^{pa}}{\sigma_{in}^{pp}} \sigma_c^{pp}(A^{1/3}p_{lab}), \tag{49}$$

is certainly different from predictions that result when an A^{α} dependence, without energy rescaling, is assumed. An A^{α} dependence leads to:

$$\sigma_c^{pA}(p_{lab}) = A^{\alpha} \sigma_c^{pp}(p_{lab}), \tag{50}$$

where α is usually taken as 1 or 2/3. (2/3 is supposed to

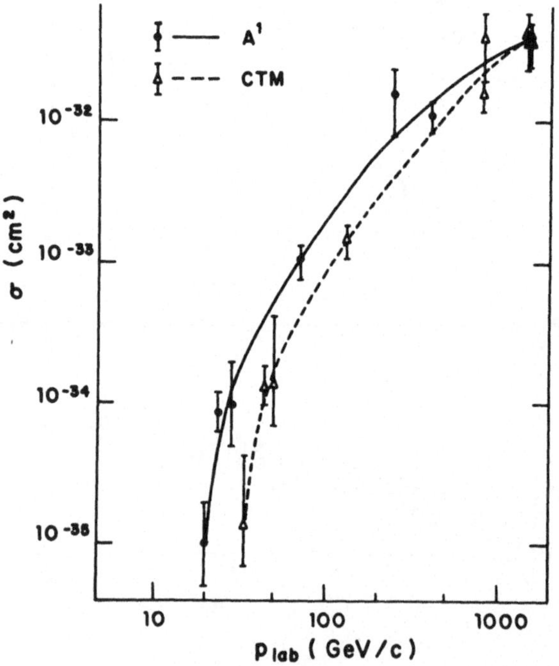

Fig. 5. $\sigma(pp \to (J/\psi \to \mu^+\mu^-) + X)$ as function of p_{lab} extracted from measurements of J/ψ productions from various targets[25]. The circles were obtained by assuming an A^1 dependence of the nuclear cross section. The triangles are the predictions of the CTM as summarized in Eq. (32) with $A^{1/3}p_{lab}$ replaced by $\bar{p}_{lab}(A^{1/3})$. The lines were drawn to guide the eye.

represent the A dependence of σ_{in}^{pA}). The main difference is due to energy rescaling which is present in Eq. (49) but is absent from Eq. (50). The energy rescaling leads to a dramatic A-dependence of inclusive particle-nucleus cross sections that depend strongly on the c.m. energy. Typical examples such as J/ψ production, W production and inclusive production of particles at large transverse momenta in high energy particle-nucleus collisions will be discussed below.

Cumulative Enhancement of J/ψ Production. The cross section for inclusive production of J/ψ in hadron induced reactions shows a dramatic increase with energy from BNL to ISR energies, resembling a threshold phenomenon[24]. Experimental data have been obtained mainly from nuclear targets[25]. Cross sections for nucleons have been extracted by dividing the nuclear cross section by either A or $A^{2/3}$ where A is the atomic number of the target nucleus. Here we show that such a procedure may lead to an overestimate of J/ψ production in pp collisions and to an underestimate of J/ψ production off heavy nuclei, mainly at BNL energies. Verification of our predictions for p+A→J/ψ + X has practical consequences for production of new particles.

The analysis of Albini et al.[15] of multi-particle production in high energy collisions, as well as our analysis of inclusive production of particles with large transverse momenta in high energy hadron-nucleus collisions, indicate that the available energy E_{av}^{ab} for the collision a+b,

$$E_{av}^{ab} \equiv \sqrt{s_{ab}} - (m_a+m_b), \tag{51}$$

is a better universal parameter for the CTM than the center-of-mass energy s_{ab}. Therefore, in comparing the predictions of the CTM with experimental results, at present accelerator energies, one should rather use Eq. (27), (28) and (32) where ip_{lab} is replaced by $\bar{p}_{lab}(i)$. $\bar{p}_{lab}(i)$ is obtained from the simple equation

$$E_{av}^{pi}(p_{lab}) = E_{av}^{pp}(\bar{p}_{lab}(i)) \tag{52}$$

i.e.

$$\bar{p}_{lab}(i) = ip_{lab} + (i-1)(im-\sqrt{m^2(i^2+1)+2imp_{lab}}) \underset{p_{lab}\to\infty}{\sim} ip_{lab} \tag{53}$$

Similar relations hold for i replaced by $A^{1/3}$. In Figs. 6 and 7 the CTM predictions (dashed lines), respectively for $\sigma(pp\to J/\psi+X)$ and $\sigma(pU\to J/\psi+X)$, as calculated from Eq. (49) with $\bar{p}_{lab}(A^{1/3})$ and from measurements of J/ψ production from various nuclei[25] are compared with the predictions resulting from

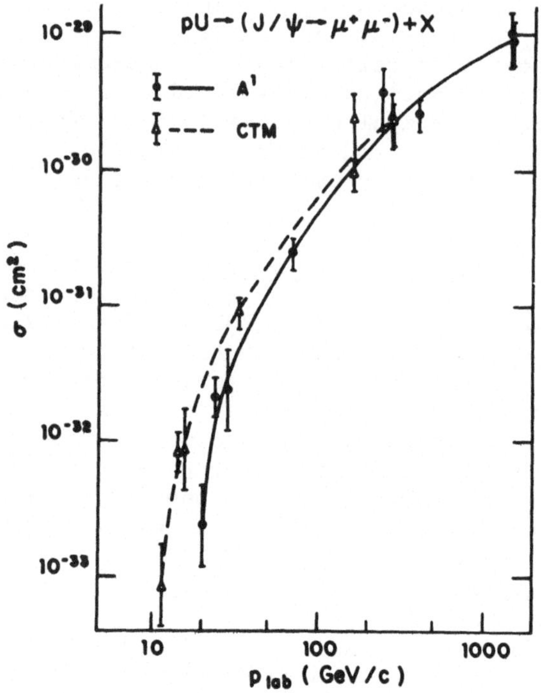

Fig. 6. $\sigma(pU\to(J/\psi\to\mu^+\mu^-)+X)$ as function of p_{1ab} deduced from measurements of J/ψ production from various nuclear targets[25]. The circles were obtained by assuming an A^1 dependence of the nuclear cross section. The triangles are the predictions of the CTM as summarized in Eq. (32) with $A^{1/3}p_{1ab}$ replaced by $\bar{p}_{1ab}(A^{1/3})$. The lines were drawn to guide the eye.

a linear A dependence of the nuclear cross section as in Eq. (50) with α=1 (solid lines)[10c]. Figs. 5 and 6 demonstrate that if J/ψ production from nuclei is correctly described by the CTM then the common procedure for obtaining nucleon cross sections from nuclear cross sections by dividing the nuclear cross section by either A or $A^{2/3}$ leads to a large overestimate of $\sigma(pp\to J/\psi+X)$ and to a large underestimate of $\sigma(pU\to J/\psi+X)$ at BNL energies. In Fig. 7 the CTM prediction for the A dependence of $\sigma(nA\to J/\psi+X)$ at p_{1ab} = 300 GeV/c are compared with recent experimental results from FNAL[26]. The CTM prediction was calculated from Eq. (49) with $\sigma(pp\to J/\psi+X)$ as given by the dashed line in Fig. 5. Complete agreement within statistical errors between experiment and theory is demonstrated, note in particular that the CTM predictions for A ⩾ 9 in Fig. 7 fall approximately on a line A^α with $\alpha=1.00\pm.10$ in good agreement with the value $\alpha=.93\pm.04$ found in the FNAL experiment[26].

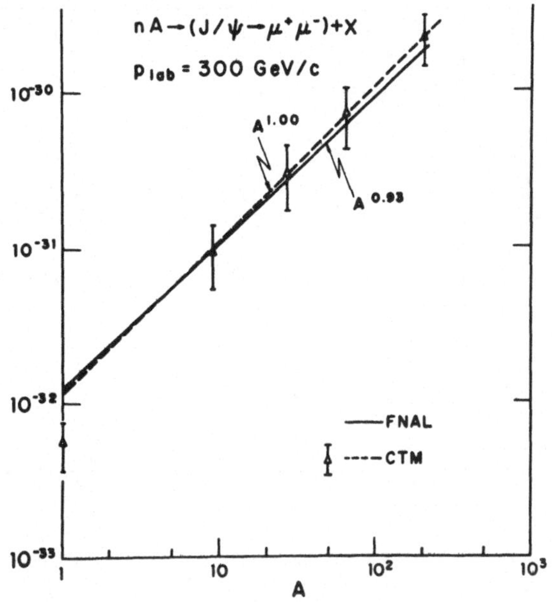

Fig. 7. $\sigma(nA \to (J/\psi \to \mu^+\mu^-) + X)$ as function of A for p_{lab}=300 GeV/c. The triangles represent the predictions of the CTM. Error bars reflect the experimental errors in Ref. (25). The dashed line is the best fitted A^α line to the CTM predictions for $A \geqslant 9$. The solid line is the best fitted A^α line to the experimental points of Ref. (26) ($A \geqslant 9$).

Cumulative Enhancement of W Production. In order to evaluate the inclusive cross section for W production in high energy particle -nucleus collisions one needs as an input $\sigma(pp \to WX)$. We have used as an input the predictions of E.A. Paschos and L.L. Wang[27] for $\sigma(pp \to WX)$ which are based on the Drell-Yan[28] mechanism. In Fig. 8 we compare our predictions for $\sigma(pU \to W^- x)$ as given by Eq. (49) (dashed line) with the predictions resulting from Eq.(50) with the choice α=1 (full line) and with the predictions resulting from Eq. (50) when A is replaced by $\sigma_{in}^{pA} = 1.4\sigma_{in}^{pp}A^{.69}$(dotted line). Note the large enhancement of the CTM predictions (dashed line) over the usual predictions especially for low values of s/m_W^2. Note also that if we use the estimate of T.K. Gaiser et al.[29] for $\sigma(pp \to W\ X)$, which are about an order of magnitude larger than the predictions based on a Drell-Yan mechanism, we obtain even larger cross sections for $\sigma(pU \to WX)$.

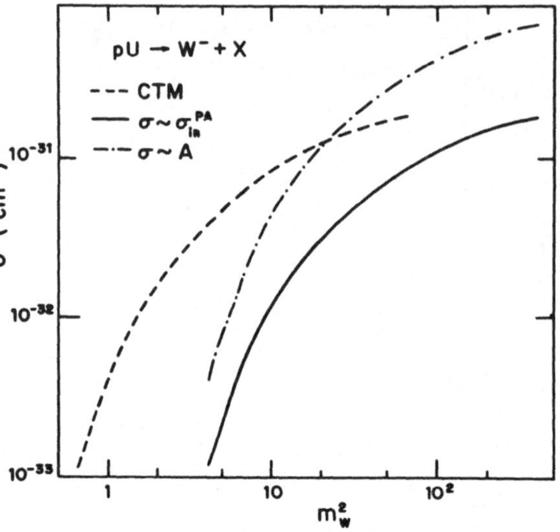

Fig. 8. $\sigma(pU\to W^- X)$ as function of s/m_W^2 . The dashed line is the CTM prediction as given by Eq. (49). The solid line is the prediction resulting from Eq. (50) with the choice $\alpha=1$. The dotted line is the prediction resulting from Eq. (50) when A^α is replaced by $\sigma_{in}^{pA}=1.4\ \sigma_{in}^{pp}A^{.69}$.

<u>Rapidity Distributions</u>. Detailed predictions require knowledge both of $\frac{1}{\sigma_{in}}\frac{d\sigma}{dy}$ over a wide range of p_{lab}, and of P(i,A).

For illustration purposes we use the following input data: We parametrize $\frac{1}{\sigma_{in}}\frac{d\sigma}{dy}$ for the process $p+p\to\pi_c+X$ where π_c denotes a charged pion, such that: (1) its rise over the ISR range[30] is correctly reproduced, (2) its measured values in pp collisions at $p_{lab}=205$ GeV/c[31] are correctly reproduced (3) its integral reproduces the measured charged multiplicity $\langle n\rangle_{pp}$ over the whole ISR range. We use the following parametrization:

$$\frac{1}{\sigma_{in}^{pp}}\frac{d\sigma^{pp}}{dy} = B\left\{1-\exp[-(|y-yo|/w)^P]\right\} \tag{54}$$

where w=3.2, P=4,y_o is the half length of the rapidity interval, and B was normalized to give[15] $\langle n\rangle_{pp}=1.16\ s^{0.285}$. Examples of the above parameterization are shown in Fig. 9. For simplicity

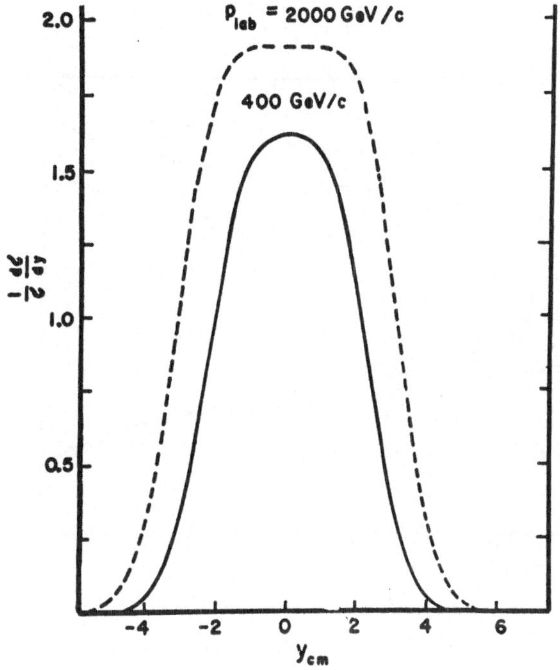

Fig. 9. Examples of the parametrization used for $\frac{1}{\sigma}\frac{d\sigma}{dy}$ versus y_{cm} in $p+p\to\pi_c+X$; π_c denotes a charged pion.

reasons we represent the nucleus by a uniform sphere with a radius $r_oA^{1/3}$; and again the parameter $\beta=\sigma/2\pi r_o^2$ is taken to be 1.

Predictions based on Eq. (30) are presented in Figs. 10 and 11, respectively, for p_{lab} fixed as function of A, and for A fixed as function of p_{lab}. Predictions based on Ref. (10a) for A fixed and p_{lab} fixed as function of N_p are presented in Fig. 12.

To predict pseudo-rapidity distributions we have assumed that $E\frac{d^3\sigma^{pp}}{dp^3} \propto \frac{d\sigma^{pp}}{dy}(s,y)g(p_T)$ and only for illustration purposes, we assumed

$$g(p_T) = \exp(-7.3p_T + 1.1.p_T^2) \tag{55}$$

$\frac{1}{\sigma_{in}^{pA}}\frac{d\sigma^{pA}}{d\eta}$ is compared with $\frac{1}{\sigma_{in}^{pA}}\frac{d\sigma^{pA}}{dy}$ for A=70 and p_{lab}=200 GeV/c in Fig. 13.

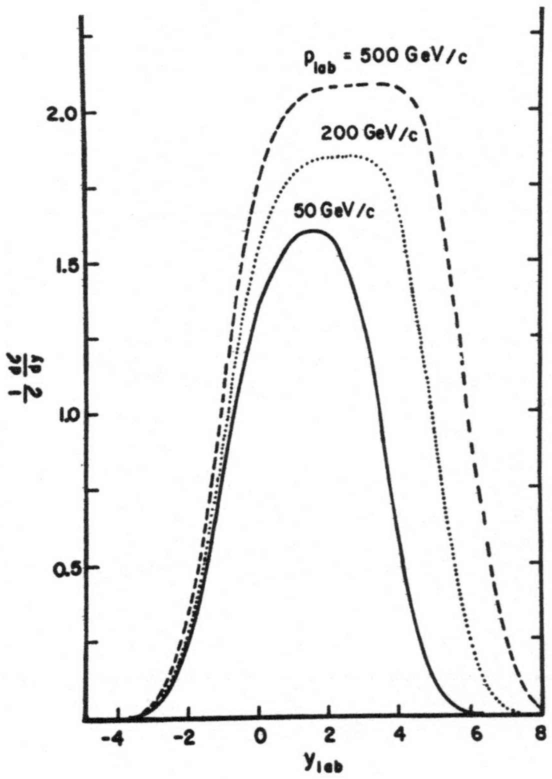

Fig. 10. Predictions from Eq. (30) for $\frac{1}{\sigma}\frac{d\sigma}{dy}$ versus y_{lab} in $p+A\rightarrow\pi_c+X$, p_{lab} = 200 GeV for different A values.

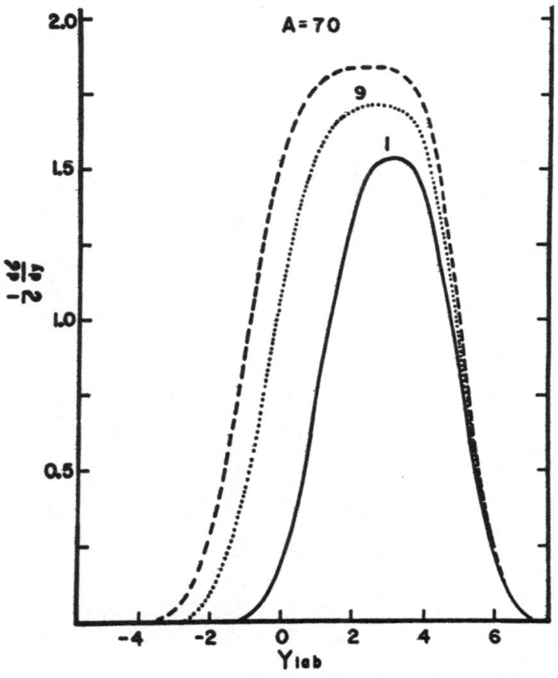

Fig. 11. Predictions from Eq. (30) for $\frac{1}{\sigma}\frac{d\sigma}{dy}$ versus y_{lab} in $p+A\rightarrow\pi_c+X$, A=70 for different p_{lab} values.

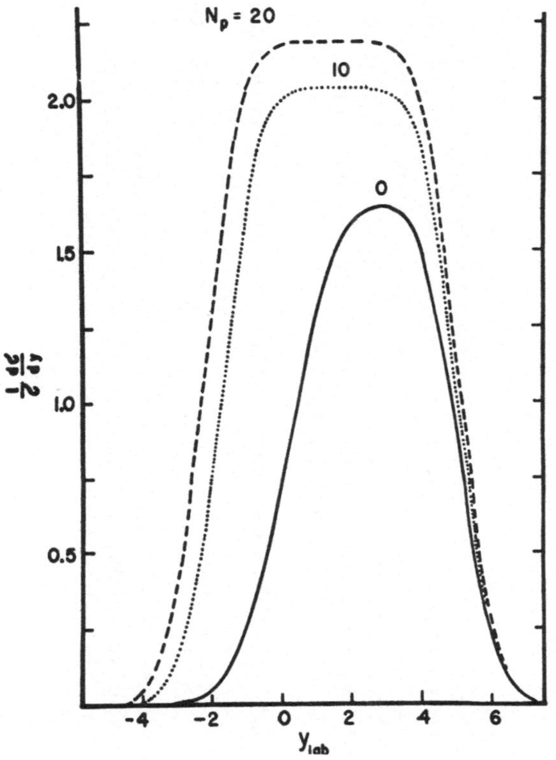

Fig. 12. Predictions for $\frac{1}{\sigma}\frac{d\sigma}{dy}$ versus y_{lab} in $p+A\to\pi_c+X$, $A=70$ and $p_{lab}=200$ GeV for different N_p (the number of knocked out protons) values.

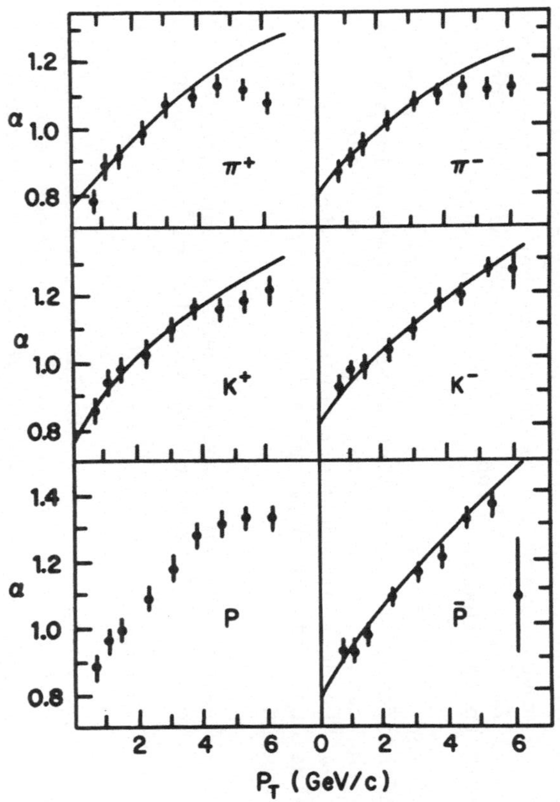

Fig. 13. Comparison between the CTM approximate prediction (59) for the power α and experimental results of Ref. (32).

From Figs. 10 - 12 one may draw the following conclusions regarding the CTM predictions for rapidity distributions: (1) At fixed p_{lab} when A increases the beam fragmentation region changes only slightly, the central region of the rapidity distribution rises and moves toward the expanding target fragmentation region. (2) At fixed A as p_{lab} increases the distribution changes, similarly to the change of $\frac{1}{\sigma^{pp}} \frac{d\sigma^{pp}}{dy}$ as a function of p_{lab}. (3) For fixed A and fixed p_{lab} as N_p increases, the central region rises and shifts to the expanding target fragmentation region. (For a fixed N_p the center of mass energy squared is at least $2mN_p\, p_{lab}$). (4) Pseudo-rapidity distributions are similar to rapidity distributions; differences are mainly in the left side of the central region.

The above conclusions are not sensitive both to the parametrization chosen for $\frac{1}{\sigma^{pp}_{in}} \frac{d\sigma^{pp}}{dy}$ and to the specific nuclear model.

Predictions based on Eq. (45) are similar to predictions based on Eq. (30) although there are some numerical differences between them. Preliminary experimental data on pseudo-rapidity distributions in particle-nucleus collisions with $50 \leqslant p_{lab} \leqslant 200$ GeV/c were given by the MIT group[20]. Indeed our conclusions (1) and (2) are strongly supported by these data. Unfortunately the comparison made by these authors between their experimental measurements and our theoretical predictions is misleading: It is not possible to compare directly the data with our predictions presented in Figs. 10 and 11 since:

(a) In Ref. (20) no distinction is made among the various charged particles which are measured. Thus, fast protons resulting from the dissociation of the target nucleus are not separated from the produced particles. Extra assumptions are needed regarding the number of these fast protons, their distribution in rapidity, and their dependence on the atomic number of the target nucleus.

(b) Pseudo-rapidity distributions rather than rapidity distributions were measured in Ref. (20). Additional assumptions regarding p_T dependence in p+p c+X are therefore needed in order to relate the two.

(c) Because of the use of lucite hodoscopes in Ref. (20), only particles with $v \geqslant .85c$ were counted there. In the theoretical calculations an explicit assumption on the p_T dependence of

$E \frac{d^3\sigma}{dp^3}$ has to be introduced in order to calculate the effect of this velocity selection on the predictions, before comparing them with the experimental dat of Ref. (20).

For detailed demonstration of the importance of points (a)-(c) we refer the reader to Ref. (10d).

Conclusions: The multiplicity $\langle n\rangle_{pA}$ of shower particles measured in high energy particle-nucleus collisions is a slowly rising function of A. From the MIT experiment[20)] one may conclude that the excess of $\langle n\rangle_{pA}$ over $\langle n\rangle_{pp}$ come from the target fragmentation region, but it is also due to a rise over the whole y region as predicted by the CTM. Other models for particle-nucleus interaction expect the excess in $\langle n\rangle_{pA}$ to come mainly from the target fragmentation region; these include the Multiperipheral Model (MPM)[4)] and the Energy Flux Model (EFM)[3)]. However, the detailed shape of $\frac{1}{\sigma}\frac{d\sigma^{pA}}{dy}$ in the CTM, as described below, differs significantly from the shape predicted by the MPM and the EFM. The main difference is that the CTM predicts a rise in $\frac{1}{\sigma}\frac{d\sigma^{pA}}{dy}$ for a fixed p_{lab} (the incident laboratory momentum) as A increases, over the whole y region and an expansion in the target fragmentation region (see Fig. 11). The MPM and EFM predict a certain $\bar{y}$, such that $\frac{1}{\sigma}\frac{d\sigma}{dy}$ rises mainly for $y<\bar{y}$, thus causing a bulge; in the MPM $\bar{y}$ depends on A but not on p_{lab} and in the EFM $\bar{y}$ depends on p_{lab} but not on A. We do not know of any quantitative (and parameter free) predictions of the MPM and the EFM that can be tested in detail aginst the MIT experimental results. Therefore the MIT data cannot be used to draw any final conclusions regarding these models.

Large p_T Phenomena

Introduction: A recent experiment at FNAL[32)] on production of hadrons at large transverse momentum p_T with 200, 300 and 400 GeV protons incident on nuclear targets, showed a strong dependence of the invariant cross-section for the inclusive production of

$\pi^{\pm}$, $K^{\pm}$, $\bar{p}$, p on the atomic number A of the target nucleus. While for low p_T the dependence is close to $A^{0.7}$ at high p_T the power rises, reaching numbers larger than 1.[33]. Such a strong A dependence is hard to understand unless collective effects are present. Indeed such behavior is natural in the CTM.

Both at ISR and FNAL it has been observed that at high energy and fixed transverse momentum the inclusive proton-proton cross section for particle production increases with energy. The larger the p_T the faster the increase. This increase is reflected onto hadron-nucleus collisions by the energy rescaling effect of the CTM, and shows itself in a steepened A dependence observed in these processes. In particular, if we parametrize the pp data as

$$\sigma_c^{pp}(E_{av}^{pp}, p_T) \sim (E_{av}^{pp})^{\beta_c(p_T)} \qquad (56)$$

Then the CTM approximate prediction is

$$\sigma_c^{pA}(E_{av}^{pA}, p_T) \sim \frac{\sigma_{in}^{pA}}{\sigma_{in}^{pp}} (E_{av}^{pA})^{\beta_c(p_T)}. \qquad (57)$$

Since $\sigma_{in}^{pA}/\sigma_{in}^{pp} \sim A^{2/3}$ and $E_{av}^{pA} \sim A^{1/3}E_{av}^{pp}$ consequently the CTM predicts that approximately

$$\frac{\sigma_c^{pA}(p_T)}{\sigma_c^{pp}(p_T)} \sim A^{\alpha_c(p_T)}, \qquad (58)$$

where

$$\alpha_c(p_T) = \frac{2}{3} + \frac{1}{3} \beta_c(p_T). \qquad (59)$$

The effective A dependence $\alpha_c(p_T)$ is therefore determined by two contributions: 2/3 is of a geometrical origin while $1/3\ \beta_c$ is due to the energy rescaling effect of the CTM. The CTM approximate prediction (59) for the power α is compared in Fig. 13 with the experimental data of reference (32).

A better test of the CTM predictions for large p_T phenomena is provided by comparing directly Eq. (46) with experimental results, without ad hoc parametrizations of any experimental data. (We do not compare the CTM "precise prediction" (29) with experimental data since a direct comparison of Eq. (29) with experiment requires both knowledge of low energy nuclear properties to calculate P(i,A), and data for inclusive pp cross-sections at momenta up to Ap_{lab}, which in the cases considered here lie far beyond the energy range of present accelerators.)

In Figs. 14 - 18 we compare the approximate sum rule Eq. (46) with the data as given in Ref. (32). We have plotted

$$\frac{\sigma_{in}^{pp}}{\sigma_{in}^{pA}} E \frac{d^3\sigma^{pA}}{dp^3}$$ from Ref. (32) for W and Ti for s = 19.4, 23.8 GeV

respectively, as a function of p_T; defining $s_{eff} = A^{1/3}s$, we find that its values are almost equal (s_{eff} = 46.2 and 45.2 GeV respectively). Data for $E \frac{d^3\sigma^{pp}}{dp^3}$ were plotted at s = 44.6 GeV.

According to Eq. (46)

$$\frac{\sigma_{in}^{pp}}{\sigma_{in}^{pA}} E \frac{d^3\sigma^{pA}}{dp^3}$$ for W, Ti, p at s = 19.4, 23.8, 44.6 GeV respectively,

should be approximately equal. Good agreement between experiment and theory is found for $c = \pi^{\pm}, K^{\pm}, \bar{p}$, and $p_T \leqslant 4$ GeV/c as shown in Figs. 14 - 18. Deviations from our predictions that are found at $p_T > 4$ GeV/c can be partly due to the replacement of the sum over $P(i,A)$ with the average $A^{1/3}$ and the use of s rather than E_{av} as the universal parameter of the CTM. Indeed the use of E_{av} as the universal parameter of the CTM improves the agreement between theory and experiment but does not remove completely the discrepancy at large p_T values, $p_T > 4$ GeV/c. We tend to believe that this discrepancy is due to the break-down of "universality" at such large p_T. More realistic modesl than the minimal model of particle-tube interactions should be utilized in order to correctly predict the behavior in this (low probability) kinematical region[14].

A most sensitive test for the necessity of energy-rescaling is provided by the high-p_T particle ratios[32]. The trivial target geometry dependence (be it $A^{2/3}$ or A^1) is cancelled in these ratios, and in the absence of energy rescaling, particle ratios from all nuclear targets at a given lab. energy and p_T should coincide. On the other hand if energy-rescaling is present, then, particle-ratios from different nuclear targets at fixed p_T when plotted as a function of the available energy

$$E_{av} = A^{1/3}E_{lab}$$

should follow a single universal line.

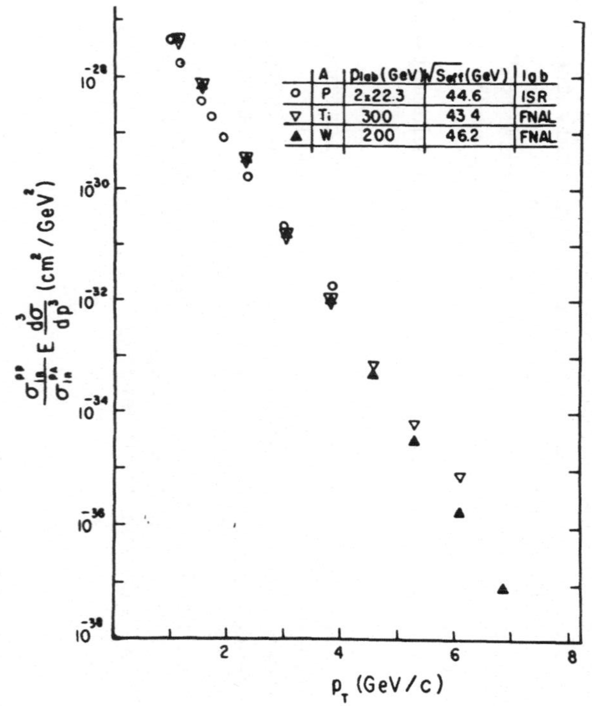

Fig. 14. Comparison between data on inclusive production of π^+ in pA collisions for A=184, 48, 1 at s=19.4, 23.8, 44.6 GeV respectively, from Refs. (32) and (18). According to our approximate scaling law all these data should lie on the same line, since they all have the same $s_{eff}=A^{1/3}s$.

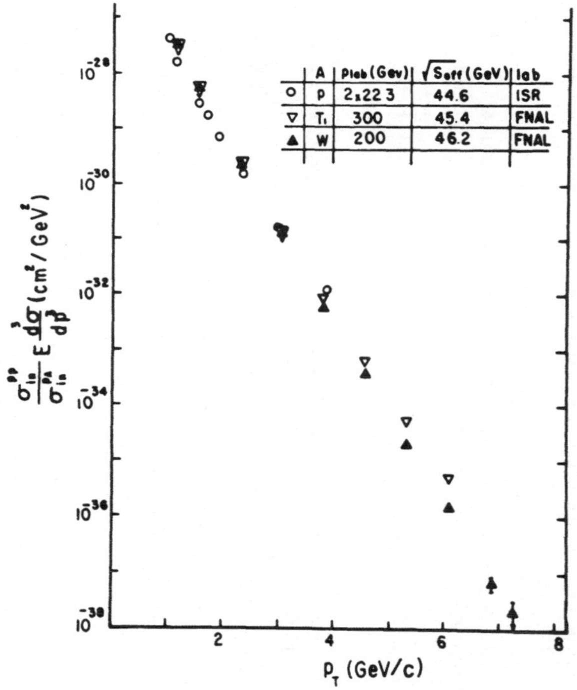

Fig. 15. The same as in Fig. (14), for π^-.

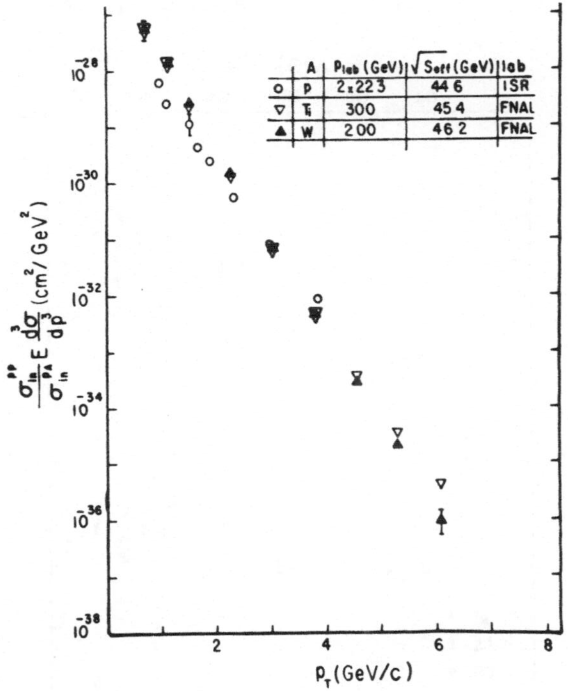

Fig. 16. The same as in Fig. (14), for K^+.

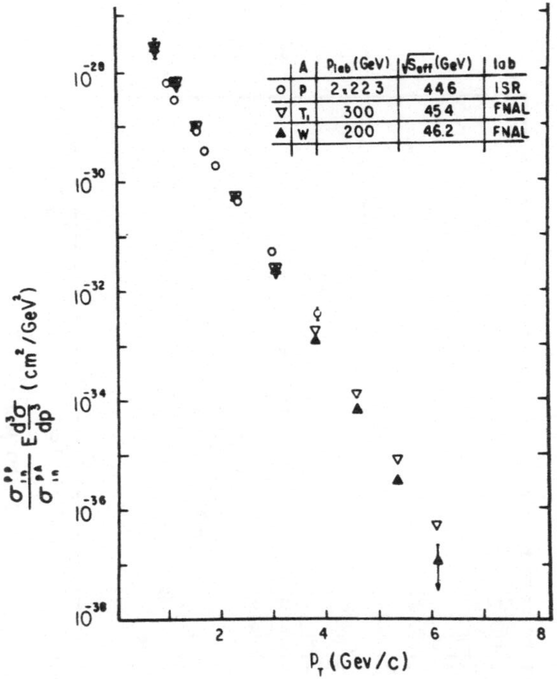

Fig. 17. The same as in Fig. (14), for K^-.

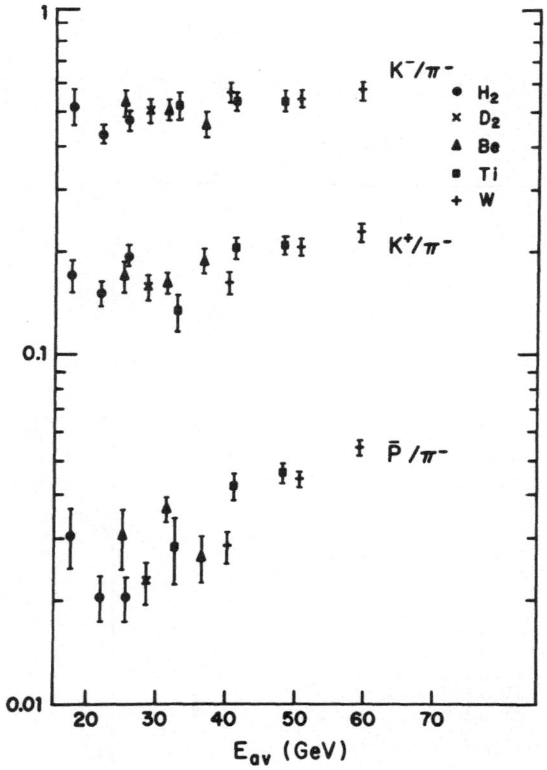

Fig. 18. The same as in Fig. (14), for p$^-$.

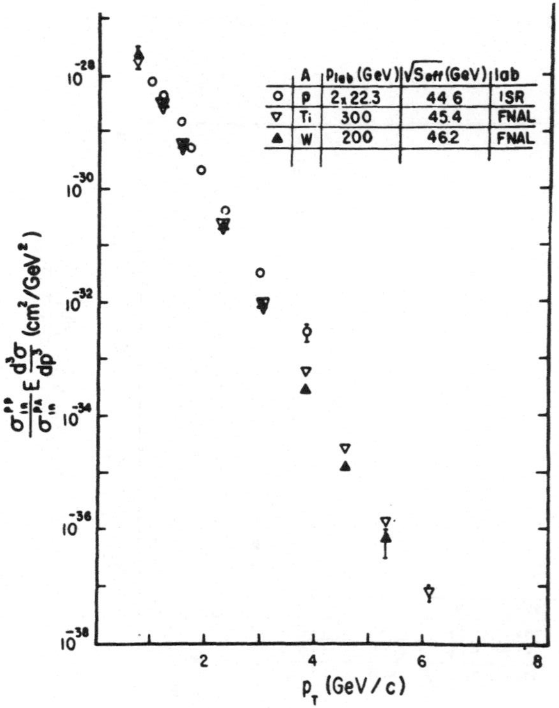

Fig. 19. Particle prediction ratios[32]) at p_T=4.58 for proton nucleus collisions as function of the effective available energy.

Fig. 19 presents particle production ratios for different nuclei as function of the available energy as calculated from the CTM. Note that the data tend to align along universal lines. It is hard to see how multistep models can expalin at the same time both the almost A independent K^-/π^- ratio and the strongly A dependent $\bar{p}/\pi^-$ ratio.

Note that pA→p+X data cannot be explained in the framework of the CTM without further assumptions regarding the disintegration of the tube.

Conclusions: Nuclear targets are commonly used in high energy experiments in order to obtain larger cross sections. However, there is accumulating evidence that generally nuclear cross sections divided by A or $A^{2/3}$ do not yield correct nucleon cross sections. The CTM provides an alternative procedure for extracting nucleon cross sections from nuclear cross sections, which is in good agreement with various experiments. It was demonstrated here for hadron, J/ψ and W production. Its full verification for J/ψ production requires more experiments on both the E dependence of $\sigma(pp\to J/\psi+X)$ and the A dependence of $\sigma(pA\to J/\psi+X)$. Such verification is highly desirable since the CTM suggests a way to obtain information on future energy domains of particle physics from experiments on nuclear targets with present accelerators. In particular the CTM suggests that the use of nuclear targets considerably lowers the kinematical thresholds for producing new massive particles which as the intermediate vector boson and it leads to enhancement of nuclear cross sections, much larger than commonly believed.

We have not presented here, for lack of time, CTM predictions for deuterium targets[10f)], for nucleus-nucleus collisions[10e)], for deep inelastic interactions of photons with nuclei[16)], for anti-shadowing[16)] and for parton models as applied to the tube[15)].

Acknowledgments: I would like to thank Y. Afek, G. Berlad, A. Dar, S. Fredriksson and Y. Zarmi for many helpful discussions, and H. Satz and the members of the Bielefeld group for their very warm hospitality.

* Research supported in part by the Israel Commission for Basic Research.

References and Footnotes

1) A. Dar and J. Vary, Phys. Rev. D6, 2412 (1972).
2) A. Dar and J. Vary, Phys. Rev. D6, 2412 (1972).
P.M. Fishbane and J.S. Trefil, Phys. Rev. D9, 168 (1974).
E.M. Friedlander, Nuovo Cimento Lett. 9, 349 (1974).
A. Bialas and W. Czyz, Phys. Lett. 51B, 179 (1974).
B. Andersson and I. Otterlund, Nucl. Phys. B88, 349 (1975).
3) K. Gottfried, Phsy. Rev. Lett. 32, 957 (1974).
4) J. Koplik and A.H. Mueller, Phys. Rev. D12, 3638 (1975); A.H. Mueller's talk in these proceedings.
L. Bertocchi, Proceedings of the VI Int. Conf. on High Energy Physics and Nuclear Structure, Santa-Fe, June 1975 and references therein.
5) B. Andersson, talk presented at the Topical Meeting on Multiparticle Production from Nuclei at Very High Energies, ICTP Trieste, June 1976.
6) F.C. Roesler and C.B.A. McCusker, Nuovo Cimento 10, 127 (1953).
W.Heitler and C.H. Terreaux, Proc. Phys. Soc. London A66, 929 (1953).
S.Z. Belenkij and L.D. Landau, Nuovo Cimento 3, 15 (1956).
D.S. Narayan and K.V.L. Sarma, Prog. of Theor. Phys. 31, 93 (1964).
L.D. Landau in "Collected Papers of L.D. Landau, Pergamon Press London 1965 (Editor D. Ter Harr).
A.M. Baldin, Proceedings of the VI Int. Conf. on High Energy Physics.
A. Dar, MIT Preprint 1972 (unpublished).
K. Gottfried, in "High Energy Physics and Nuclear Structure" North Holland Publ. Com. 1974 (Editor G. Tibell).
A. Dar, Proceeding of the ICTP Topical Meeting on High Energy Reactions Involving Nuclei, Trieste 1974 (Editor L. Bertocchi).
A.Z. Patashinskii, JETP Lett. 19, 338 (1974).
G. Berlad, S. Dar and G. Eilam, Phys. Rev. D13, 161 (1976).
F. Takagi, Nuovo Cimento Lett. 14, 559 (1975).
S. Fredriksson, Nucl. Phys. B111, 167 (1976).
Meng-Ta-Chung, talk presented at the Topical Meeting on Multiparticle Production from Nuclei at Very High Energies, ICTP Trieste June 1976.
7) A.M. Baldin, Proceedings of the VI Int. Conf. on High Energy Physics and Nucl. Structure, Santa Fe, June 1975 and refs. therein.
8) A.M. Baldin et al., Yad. Fiz. 20, 1201 (1974) (Engl. Trans. 20, 629 (1975).
9) H.H. Heckman in "High Energy Physics and Nuclear Structure" North Holland Comp. 1974 (Editor G. Tibell).
J. Papp et al., Phys. Rev. Lett. 30, 601 (1975).
J. Papp Ph.D. Thesis NC Berkeley 1975 (unpublished).

10) a. G. Berlad, A. Dar and G. Eilam, Phys. Rev. D13, 161 (1976).
Y. Afek, G. Berlad and A. Dar:
b. PH-76-12 to be published in Phys. Rev. D.
c. Phys. Rev. Lett. 37, 947 (1976).
d. PH-76-48 to be published in Nucl. Phys. B.
e. PH-76-78 to be published in Nucl. Phys. B.
f. A. Dar and Tran Thanh Van, Phys. Lett 65B, 455 (1976).
g. G. Fredriksson, see Ref. (6).
11) F. Donaldson et. al., Phys. Rev. Lett. 36, 1110 (1976).
12) G.J. Blanar et. al., Phys. Rev. Lett. 35, 346 (1975).
13) Meng Ta Chung, see Ref. (6).
14) G. Eilam and Y. Zarmi, University of Bielefeld, Preprint BI-TP 76/28.
15) E. Albini et al., Nuov. Cimento 32A, 102 (1976) and references therein.
16) Y. Afek, G. Berlad and G. Eilam, to be published.
17) J.R. Elliot et al., Phys. Rev. Lett. 34, 607 (1975).
D.J. Miller and R. Nowak. Nuovo Cimento Lett. 13, 39 (1975).
18) B. Alper et al. Nucl. Phys. B87, 19 (1975).
B. Alper et al. Nucl. Phys. B10, 237 (1975).
19) I. Otterlund, talk presented at the Topical Meeting on Multiparticle Production from Nuclei at Very High Energies, ICTP Trieste, June 1976.
20) W. Busza et. al. FNAL preprint, June 1976; C. Halliwel, talk presented at the Topical Meeting on Multiparticle Production from Nuclei at Very High Energies, ICTP Trieste, June 1976.
21) R. Hofstadter, Ann. Rev. Nucl. Sci., 7, 231 (1954).
22) W. Busza et al. Phys. Rev. Lett. 34, 836 (1975).
23) J. Babecki et al., Phys. Lett. 47B, 268 (1973). $\bar{}$
24) Such a threshold can result from production of $\psi D\bar{D}$ where D is a charmed meson; see for instance J. Ellis in "Electromagnetic Interactions and Field Theory", Ed. P. Urban, Spring 1975, p. 143.
25) J.J. Aubert et al., Phys. Rev. Lett. 33, 1404 (1974);
S.C.C. Ting, in Proceedings of the International Conf. on High Energy Physics, Palermo, Italy 1975 (to be published).
J.J. Aubert in Proceedings of the XIth Rencontre de Moriond Flaine, France 1976 (to be published).
B. Knapp et al., Phys. Rev. Lett. 34, 1044 (1975).
G.J. Blanar et al., Phys. Rev. Lett. 35, 346 (1975).
F. Büsser et al., Phys. Lett. 56B, 482 (1975).
Y.M. Antipov et al., Phys. Lett. 60B, 309 (1976).
K.J. Andersson et al., Phys. Lett. 36, 237 (1976).
E. Nagy et al., Phys. Lett. 60B, 96 (1975).
F. Binon et al., Phys. Lett. to be published.
H.D. Snyder et al., Phys. Rev. Lett. 36, 1415 (1976).
26) M. Binkley et al. Phys. Rev. Lett. 37, 571 (1976).
27) R.B. Palmer, E.A. Paschos, N.P. Samios and L.L. Wang, Phys. Rev. 14D, 118 (1976).
28) S.D. Drell and T.M. Yan, Phys. Rev. Lett. 24, 181 (1970.

29) T.K. Gaisser, F. Halzen and E.A. Paschos, BNL - 21489.

30) R. Stroynowski, Proceedings of the VI Int. Colloquium on Multiparticle Reactions, Oxford, July (1975).

31) J. Whitmore, Phys. Rep. 10, 273 (1974).

32) J.W. Cronin et al., Phys. Rev. D11, 3105 (1975); D. Antraesian et al., Paper 753/A4-17 submitted to the Tbilisi Conf., July 1976.

33) It is not clear that a power of A describes the invariant inclusive cross-section for all energies and p_T up to A=1. The form $A^{\alpha(p_T)}$ is used in Ref. (32) as an additional assumption; for its effect see G.R. Farrar, Phys. Lett. 56B, 185 (1975).

UNIFIED INTERACTIONS OF LEPTONS AND QUARKS

H. Fritzsch

CERN

1211 Geneva 23, Switzerland

Science is organized common sense where many a beautiful theory was killed by an ugly fact.

Thomas Huxley

1. INTRODUCTION

Since about 1970 a rather definite theoretical picture of the world of particle physics has emerged. This picture can be characterized by the following main aspects.

a) The rôle played by quarks in hadron physics has become rather transparent: In building up the particle spectrum, they act as constituents such that states of the type qqq describe the baryons, and states of the type $\bar{q}q$ the mesons.

The hadronic electromagnetic and weak currents act as if they were simply bilinears in quark fields which propagate near the light cone essentially like free fields ("pointlike").

b) The theoretical picture of the weak interactions has become rather definite. At present a consistent description of the known weak interactions can be given in terms of a gauge theory based on the group $SU_2 \times U_1$; the simplest version of such a theory is the one advocated by Salam, Ward and Weinberg. Which of the various $SU_2 \times U_1$ schemes, if any, is realized either exactly or at least approximately in nature, remains to be seen.

c) Nonabelian gauge theories may have something to do with nature: They are utilized by theoreticians both for the description of the weak and electromagnetic interactions as well as for the strong interactions.

Although a definite pattern for the theoretical description of particle physics has emerged, many questions have yet to be answered. In the field of the strong interactions the problem of quark confinement is, no doubt, the most challenging one. In weak interaction physics there are so many questions waiting for an answer that I can list only the most important ones: How many leptons and quarks exist? Why are the electric charges quantized? What is the physical origin of the lepton and quark masses? What is the physical origin of the Cabibbo angle and possibly existing other weak mixing angles? Are lepton and baryon number conserved? What is the mechanism which breaks parity? What is the pattern of the weak currents for possibly existing new quarks and leptons?...

It is likely that at least some of these questions cannot be answered within the conventional weak interaction physics itself, i.e. within the set of interactions whose strengths are given by the Fermi constant and that the weak interactions are only a subset of a larger set of interactions. Perhaps even the strong interactions can or must be included in such a larger scheme, which then could be called a unified theory of all interactions (Somebody really ambitious should, of course,try to include the gravitational interactions in such a scheme as well).

In the first two parts of these lectures I shall discuss the general aspects of the strong, electromagnetic and weak interactions, needed in order to understand the following parts, dealing with new ideas which have emerged recently in connection with possibly existing new types of quarks and leptons. Especially I shall concentrate on vectorlike theories of the weak interactions and their importance for the construction of a unified theory of all interactions. Some examples of such theories based on the exceptional gauge groups will be discussed in the last part.

These lectures should not be regarded as a review but rather as an overview of the present theoretical landscape in lepton-quark physics. I have tried to give a relatively broad list of references, in which the interested reader can find the details of the various subjects; it is, however, far from being a complete list.

PART I: STRONG INTERACTIONS AS CHROMODYNAMICS

2. WHY DOES ONE NEED COLOR?

The idea that quarks come in three "colors" has emerged slowly during the last twelve years. It has been realized very soon after the proposition of the quark model that a reasonable description of the baryon spectrum can only be achieved if quarks obey parastatistics of rank three[1)], or, alternatively, if there are three colors of each quark (see e.g. the Han-Nambu version of the quark model[2)]). Both possibilities have been shown to be essentially equivalent. More recently many other virtues of the color idea have been discussed:[3)] One obtains the correct magnitude and sign for the decay $\pi_0 \to 2\gamma$ and the correct cross section for the annihilation of electron - positron pairs into hadrons. One can postulate the confinement condition such that all physical particles and observables (currents etc.) are required to be singlets under the color group SU_3. (This leads automatically to the correct spectrum and eliminates quarks (q), diquarks (qq) etc. as physical particles.) The existence of the internal symmetry group SU_3^{color} allows oneself to introduce a gauge theory in color space, with the color charges acting as exactly conserved quantities[4, 5)]. Such a theory is asymptotically free[6)], and may show "infrared slavery" such that all states but color singlets are permanently bound by the color forces. If this hypotheses is correct, the gauge theory of colored quarks and gluons (quantum chromodynamics, QCD) is indeed a serious candidate for the theory of the strong interactions.

Before I discuss the details of QCD, let me repeat in a simplified version the original arguments which led to the introduction of color. For simplicity we consider the spectrum of nonstrange baryons. The lowest lying states are the isodoublet proton/neutron and the I=3/2 and spin 3/2 baryons, altogether 2×2 + 4×4 = 20 states (see Fig. 1)

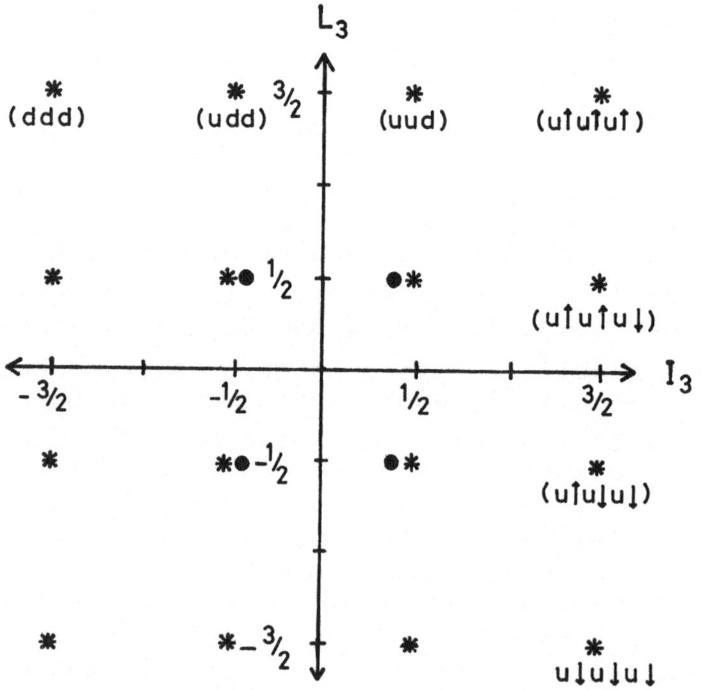

Fig. 1: The 20 nonstrange ground state baryons.
I_3: 3rd component of isospin
L_3: 3rd component of angular momentum.
O : nucleon
* : N^* resonance
(in parentheses the various quark wave functions).

These 20 states are precisely the states of quark composition $|qqq\rangle$ one can form out of two quark flavors u,d (charges 2/3,-1/3,) which are symmetric under flavor and spin (altogether $\frac{4\cdot5\cdot6}{1\cdot2\cdot3} = 20$ states). Note that in particular there exists a state with the wave function $|u\uparrow u\uparrow u\uparrow\rangle$. Since this state is symmetric under spin and flavor, one has to require the space part of the wave function to be antisymmetric in order for the state to obey Fermi - Dirac statistics; something odd for a ground state in quantum mechanics. No problem exists, however, if we introduce three "colors" of each quark flavor (say red, green, and blue)

$$\begin{pmatrix} u \\ d \end{pmatrix} \longrightarrow \begin{pmatrix} u_r & u_g & u_b \\ d_r & d_g & d_b \end{pmatrix}$$

and antisymmetrize the wave function with respect to color:

$$|q\ q\ q\rangle \rightarrow \frac{1}{\sqrt{6}}\ \varepsilon_{rgb}\ \ |q_r\ q_g\ q_b\rangle. \qquad 2.1$$

This configuration is not only the one desired for the observed spectrum, but it is in addition the most simple possibility to construct a singlet under the unitary group in color space SU_3. Thus a consistent way to reproduce the observed hadron spectrum emerges, if we require all hadrons and all physical observables (currents, energy - momentum tensor, etc.) to be color singlets (confinement hypotheses). States which are not wanted (since not seen) like quarks q (color triplets) or diquarks (color triplets or sextets) are eliminated by construction.

Although the confinement hypotheses is thus far in accordance with observation, it reaches much further than required by experiment. What is required by phenomenology is simply that the color nonsinglet states are much heavier than the color singlets ("partial confinement"). A color threshold of several GeV is still consistent with the experiments. To set it to infinity, as done by imposing absolute color confinement, is certainly the simplest and most attractive possibility, but perhaps not the one chosen by nature. Only the experiments can decide!

If one introduces the color quantum number, the nature of the strong interactions becomes rather well specified. Since we want color singlets to be distinguished from color nonsinglets in energy, the interaction itself must distinguish color, i.e. the effective potential between quarks must distinguish the various color quantum numbers. This requirement eliminates, for example, the old vector gluon theory, in which a vector gluon field was coupled to the baryon current. Such a theory would not distinguish between flavor and color, i.e. color singlets and nonsinglets would be degenerate.

The simplest realistic possibility to introduce an effective potential between the quarks is to couple it to the various color octet charges:

$$\text{Potential} = \chi_1 \cdot \chi_2 \cdot V(r) \tag{2.2}$$

(χ_i: color SU_3 matrices).

In this case one can easily envisage a situation in which the color singlets are the lowest energy eigenstates. Let us consider, for example, a quark - antiquark system which can form either a color singlet or a color octet: $\underset{\sim}{\bar{3}} \times \underset{\sim}{3} = \underset{\sim}{1} + \underset{\sim}{8}$. The total color spin of the system $\vec{C}$ is equal to the sum of the individual color spins: $\vec{C} = \vec{C}_1 + \vec{C}_2$, and one has

$$\vec{C}_1 \vec{C}_2 = \frac{1}{2} (\vec{C}^2 - \vec{C}_1^2 - \vec{C}_2^2) . \tag{2.3}$$

Table 1. The eigenvalues of the Casimir operator C^2 for the various representations.

representation	$\underset{\sim}{1}$	$\underset{\sim}{3}$	$\underset{\sim}{6}$	$\underset{\sim}{8}$	$\underset{\sim}{10}$
C^2	0	4/3	10/3	3	6

For a quark - antiquark system one finds

$$\chi_1 \cdot \chi_2 \sim \vec{C}_1 \cdot \vec{C}_2 = 1/2\ (\vec{C}^2 - 8/3) . \tag{2.4}$$

Thus, if V(r) is an attractive potential (e.g. a Coulomb potential), the color singlet state ($C^2=0$) is the lowest energy eigenstate. A similar conclusion is true for a three quark system representation a baryon.

In such a crude nonrelativistic approach the color threshold depends on the form of the potential. One can easily envisage a situation in which it is at infinite energy, e.g., if V(r) represents a harmonic oscillator potential or a linear potential.

3. CHROMODYNAMICS

The previous analysis leads us in a natural way to the hypotheses that the forces among the quarks are generated by the exchange of color octet vector gluons coupled to the quarks in a gauge invariant manner. The Lagrangian of this theory (quantum chromodynamics, QCD) is given by

$$\mathcal{L} = -1/4 \sum_{A=1}^{8} G^{A}_{\mu\nu} G^{\mu\nu}_{A} + \bar{q}\left[i\gamma^{\mu}\left(\partial_{\mu} - ig\,\frac{\chi^{A}}{2}\,B_{A\mu}\right) - \mathcal{M}\right] q$$

$$(q = u,d,s,\ldots;\ G^{A}_{\mu\nu} = \partial_{\nu} B^{A}_{\mu} - \partial_{\mu} B^{A}_{\nu} - g\, f^{ABC} B_{\nu B}\, B_{\mu C} \qquad (3.1)$$

$$B^{A}_{\mu}\text{: gluon field)}$$

$$\left[\frac{\chi_A}{2}, \frac{\chi_B}{2}\right] = i f_{ABC}\, \frac{\chi_C}{2} \qquad \mathrm{tr}(\chi_A\, \chi_B) = 2\delta_{AB}.$$

It is well-known that in electrodynamics the physical coupling constant e, defined by the large distance behaviour of the electric potential (Thomson limit), is smaller than the effective coupling constant e_{eff}, one would measure at small distances, due to the presence of vacuum polarization effects. Since a "bare" electron is surrounded by an infinitely extended sea of electron-positron pairs, it will attract the virtual positrons contained in the vacuum polarization cloud, and thus shield part of its electric charge.

The behaviour of the effective coupling constant as a function of the distance can easily be calculated. One finds, using perturbation theory

$$\alpha^{eff}_{(t)} = \frac{\alpha_o}{1-\frac{2\alpha_o}{3\pi}\,t} \; ; \; \alpha^{eff} = \frac{e^2_{eff}}{4\pi} \; ; \; t = \log\,(^{\ell_o}/_{\ell}) \qquad (3.2)$$

where ℓ_o is a certain reference length at which α^{eff} is normalized to α_o, and ℓ the length at which the effective coupling constant is given by $\alpha^{eff}(t)$.

One realizes from (3.2) that at very small distances ($\sim e^{-\pi\cdot 137}\cdot\ell_o$) the function $\alpha(t)$ exhibits a pole. This is simply a sign that the perturbation theory result (3.2) breaks down at such very small distances (QED is unstable in the ultraviolet region); the behaviour of $\alpha(t)$ at such extremely small distances cannot be deduced from lowest order perturbation theory.

The essential difference between QCD and QED is that the nonabelian gluons carry color charges themselves, while the photon in QED is neutral. Consequently quarks surround themselves not only by a cloud of virtual $q\,\bar{q}$ pairs but also by a gluon cloud. Calculating the effect in lowest order perturbation theory, one finds that the gluon cloud acts in a very different way as the $q\bar{q}$ cloud: It leads to an increasing instead of a decreasing coupling constant. (Something similar would happen in QED, if equal sign charges would attract each other, and not charges of different signs). In QCD one obtains for the effective coupling constant $\kappa = g^2/_{4\pi}$:

$$\kappa(t) = \frac{\kappa_o}{1+2B\kappa_o t} \qquad (3.3)$$

($B = \frac{1}{4\pi}$ (11-2/3 number of quark flavors)). Note the opposite sign (compared to the result in QED) of the coefficient of t in the denominator, provided, the number of flavors is less than 16.

Due to the sign change the perturbative result (3.3) does not break down at small distances: the coupling constant becomes smaller, and perturbation theory becomes better at decreasing distances [6] ("asymptotic freedom"). The singularity in t occurs now at large distances: the critical length at which perturbation theory breaks down is of order

$$\ell_{crit} \simeq \ell_o \cdot \exp \frac{1}{2B\kappa_o}$$

Thus QCD is unstable in the infrared region. This, and the property of QCD to be asymptotically free, have given rise to the hope that QCD is the correct theory of hadron dynamics. Indeed these properties are very desirable. Due to asymptotic freedom the coupling constant is small at sufficiently small distances, and quarks become almost structureless objects (scaling in the deep inelastic region is almost exactly fulfilled, it is broken only by small logarithmic corrections). Due to the infrared instability the long distance properties of the theory, in particular the particle content (spectrum) of the theory, are unclear. Since the coupling constant increases with increasing distances, the hope is justified that in such a theory confinement occurs, and only color singlets appear in the physical spectrum.

The appearance of a critical length in QCD is an interesting phenomenon since a physical mass scale is introduced via the renormalization length ℓ_o, which is independent of the effective quark masses. If confinement is true in QCD, one is tempted to associate this scale with the typical hadronic mass scale introduced by the spectrum of hadrons (Sizes of hadrons, slopes of Regge trajectories, meson decay constants etc.). In particular it would be natural to associate the critical length ℓ_{crit} with the typical hadron size of 1 Fermi. This implies according to eq. (3.3) that the quark gluon coupling constant cannot be very small at distances like 10^{-15} cm, i.e. at distances where scaling in the deep inelastic region starts to become relevant.

The best estimates [7)] of κ at an energy of 2...3 GeV center around $\kappa \simeq 1/3$, i.e. the strong interactions are still much stronger than e.g. the electromagnetic interaction (despite asymptotic freedom). Note that higher order corrections to the free field theory results for deep inelastic scattering are in general κ/π, i.e. of order 10%. Recently violations of scaling of this order of magnitude have been observed at Fermilab,[8)], and it remains to be seen, if those violations are indeed the ones predicted in QCD.

Thus far the color confinement hypothesis has remained a conjecture. It is obvious from the discussion above that confinement must be a strong coupling phenomenon and cannot be obtained in perturbation theory. Until recently there was at least a hope that a signal for confinement can be obtained in perturbation theory by summing the leading infrared divergencies of the theory. The hope was that QCD shows in perturbation theory a qualitatively different behaviour in the infrared region as QED where the infrared divergencies can be treated e.g. by the Block - Nordsieck method. This hope has not substantiated itself; on the contrary, it seems that the situation in QCD is qualitatively the same as in QED[9].

As argued by Wilson, Susskind, and others,[10] the easiest way to attack the confinement problem as a strong coupling problem may be to treat the QCD Lagrangian on a lattice. What emerges from such a treatment, is intuitively easy to understand. For example let us imagine a quark - antiquark system (color singlet), where the quarks are very near to each other, say 10^{-15}cm. In this case lowest order perturbation theory should be applicable, and the force between the quarks would be basically a $1/r^2$ force, according to Coulomb's law, generalized to QCD. If one tries to pull the quarks away from each other, the attractive force between them would follow at the beginning Coulomb's law. However at distances larger than $\ell_{crit} \simeq 1$ Fermi the vacuum polarization effects due to the gluon selfcoupling squeeze the color flux lines into a narrow tube of radius ℓ_{crit}, and the force becomes constant, like the force between two condensator plates. As soon as there is enough energy in the system to create a new quark - antiquark pair, the system will break up into two separate color singlet systems (see Fig. 2)

If this picture of the confinement turns out to reflect the actual situation in nature, quarks and gluons would be fields without free particles associated to them. Color would be confined, and the spectrum of hadrons would emerge as the spectrum of the color singlet states.

Let me stress finally: If confinement is true, a new phenomenon has entered physics, namely a new type of realizing a symmetry. The two traditional ways of symmetry realization are a) the direct realization (realization in the spectrum, e.g. isospin) and b) the Nambu-Goldstone realization (realization via massless quanta, e.g. chiral symmetry). The present ideas about

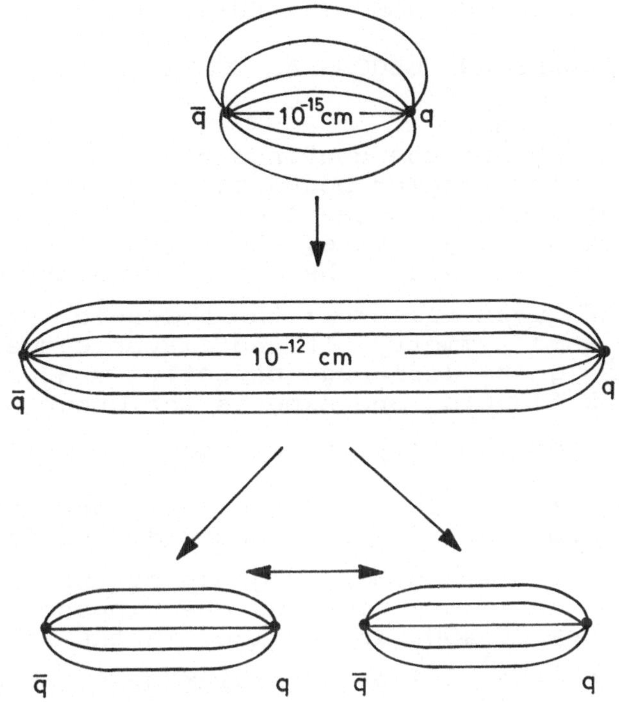

Fig. 2 : Schematic picture of the color flux lines in QCD in case of a separating $q\bar{q}$ system breaking up into two mesons.

QCD indicate a third way to realize a symmetry ("hidden realization"): Only singlets of the symmetry group exist as physical states. Note that this third possibility requires the symmetry to be an exact symmetry.

We should like to remind the reader, that thus far no internal symmetry in nature has turned out to be an exact symmetry (isospin is broken, and so is chiral SU_2xSU_2). Perhaps also the third type of symmetry (color) is not exactly conserved, but only approximately. This would imply that confinement is not absolutely true, and colored states can eventually be produced in hadronic reactions[11]. At present there is a strong tendency among theoreticians to believe in the forceful arguments brought up in favor of absolute confinement. However, no such argument is yet fully convincing, and we should like to stress the importance of experiments designed in order to look for colored states and in particular for quarks in the physical spectrum.

4. THE HADRON SPECTRUM IN QCD:

QUARKS AND GLUONS AS CONSTITUENTS

a) General remarks

According to QCD and the confinement hypotheses the hadrons are color singlet bound state systems composed of quarks and gluons. In the limit where all effective quark masses are degenerate, the Lagrangian of the system is invariant under the flavor symmetry group SU_f (f: number of quark flavors); the hadronic spectrum will exhibit the same symmetry. If the flavor symmetry is broken by assigning different effective masses to the various quark flavors, one expects the group SU_f still to be useful for classifying the hadronic multiplets.

In the limit where all effective quark masses vanish, the group $SU_f \times SU_f$ is conserved. For free massless quarks this is a symmetry group of the spectrum. In QCD where quarks are confined one supposes that the chiral symmetry $SU_f \times SU_f$ is realized in the Nambu - Goldstone way such that f-1 pseudoscalar mesons become massless; all other hadrons stay massive.

b) Low - lying meson states

The low lying meson states can be described by the wave function $\bar{q}q = \frac{1}{\sqrt{3}}(\bar{q}_r q_r + \bar{q}_g q_g + \bar{q}_b q_b)$, where q denotes the various quark flavors.

In a simple nonrelativistic approach one can separate the space and spin parts of the wave function. The parity of a $\bar{q}q$ - state with orbital angular momentum L is

$$P = (-1)^{L+1} \quad (4.1)$$

since q and $\bar{q}$ have opposite intrinsic parity. The charge conjugation of those $\bar{q}q$-states which are C-eigenstates is

$$C = (-1)^{L+1}$$

(s = spin of the $\bar{q}q$ system: S=O (antiparallel),
s=1 (parallel)).

It is useful to treat ordinary spin and flavor spin on the same footing. In this case one is dealing, in a very crude nonrelativistic picture, with the symmetry group $U_{2f} x U_{2f} x O_3$, where the first factor acts on the quark indices, the second one on the antiquark indices, and the third one on angular momentum.

For the lowest - lying mesons, where the three flavors u,d,s are relevant, the group is $U_6 x U_6 x O_3$. The lowest representation is $(6,\bar{6}), L^P = 0^-$. These 36 states represent the pseudoscalar and vector meson nonets.

The next higher representation is $(6,\bar{6}), J^P = 1^+$. These 9 x 4 x 3 = 108 states represent the tensor meson nonet, two axial vector meson nonets (with opposite charge conjugation), and the scalar meson nonet.

c) <u>Low - lying baryon states</u>
The low lying baryons are described by the color singlet configuration $\frac{1}{\sqrt{6}} \varepsilon_{ABC} q_A q_B q_C$ (A,B,C: color indices).
The simplest representation with baryon number one is $[(2f,2f,2f)_{symm.},1]$, $L^P = 0^+$, i.e. a configuration totally symmetric in the spin and flavor indices. It describes $\frac{2f\cdot(2f+1)(2f+2)}{1\cdot2\cdot3}$ states.

In the case of the three flavors u,d,s the representation is $(56,1), L^P = 0^+$ which describes the baryon octet and the baryon decimet. The excitation of the three quark systems to an $L^P = 1^-$ state changes the coordinate space wave function from a symmetric one to one with mixed symmetry. Fermi statistics requires that also the spin and flavor part of the wave function exhibits a mixed symmetry, and one obtains the representation $(70,1), L^P=1^-$.

The excitation of the three quark system by two units in angular momentum allows again a symmetric coordinate space wave function, i.e., $(56,1), L^P = 2^+$.

This pattern of the baryon spectrum, which agrees with observation, supplies one of the strongest arguments in favor of the color quantum number. Without color one expects (within the same line of thought) the representation $(20,1), L^P = 0^+$ to be the lowest baryon configuration, in disagreement with experiment.

d) Glue balls and Zweig's rule

In QCD one has, besides the tricolored quarks, eight formally massless vector gluons as dynamical degrees of freedom. In a naive approach to QCD one expects that there exist, besides the color singlet hadrons composed of quarks, also color singlet hadrons composed of gluons[4]. These states are SU_f singlets. Taking the gluons as massless vector quanta, one constructs the following set of color singlet configurations ("glue balls"):

Two gluons: $J^{PC} = \underline{0^{++}, 0^{-+}}$

$2^{++}, 2^{++}, 2^{-+}$

$3^{++}, 3^{++}, 3^{-+}$

$\vdots \quad \vdots \quad \vdots$

Three gluons: $J^{PC} = 1^{--}, 1^{+-}$, plus higher spin configurations

Thus far there is no evidence for the existence of a glue spectrum. However, if QCD is the correct theory of the hadronic world, glue states must exist. To find them is one of the challenges for the experimentalists.

Glue states can either be realized in the spectrum as well-defined resonances, in which case they must be relatively heavy in order to have escaped observation thus far, or as multiparticle states. In either case: Provided QCD is the correct theory of hadrons, there exists a peculiar asymmetry between quarks and gluons such that the low-lying part of the hadron spectrum is populated by bound quark systems, but there is no evidence for bound glue. A solution of the color confinement problem must, at the same time, also provide a solution for this peculiar fact.

What are the typical physical properties of glue balls, and how can one find them in the hadron spectrum? They are, of course, SU_3 singlets and electrically neutral. Consequently specific decay modes, e.g. $K\bar{K}$, are forbidden. Moreover electromagnetic decays like $(1^{--})_{glue} \rightarrow e^+e^-, \mu^+\mu^-$ should be highly suppressed. The leptonic decay width of a 1^{--} $\bar{q}q$-state is typical of the order of a few keV. Glue balls are expected to have a much smaller leptonic decay width ($\lesssim 0.1$ keV), since they can decay electromagnetically only via the $\bar{q}q$ - parts of their wave functions, e.g. via their mixing with $\bar{q}q$ states.

As we shall see below, the dynamics of "Zweig's rule" is closely related to the existence of glue ball states. Since the hadronic decay of the ψ(J)- particle (in the charm picture) proceeds via the violation of this rule, one expects the final state in the ψ(J) decay to be populated by glue ball states, which then decay strongly into the conventional $\bar{q}q$ - mesons ($\pi, \rho, \omega, \ldots$). Thus the best source of glue ball states nature may have provided us is the hadronic ψ(J) decay, and I would like to urge the experimentalists to look for new types of resonances in this decay.

The gluon dynamics is closely related to the "Zweig-rule". This rule states that hadron vertices should be represented by formal diagrams in which quark-antiquark lines do not annihilate inside a hadron (see Fig. 3).

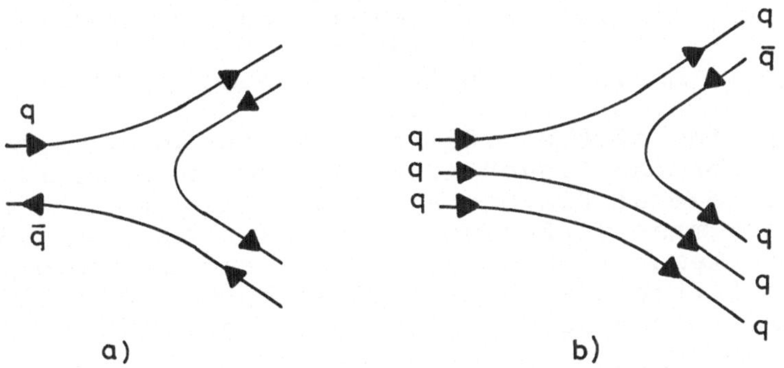

Fig. 3 a) Meson-meson-meson vertex according to Zweig's rule.
b) Baryon-meson-baryon vertex according to Zweig's rule.

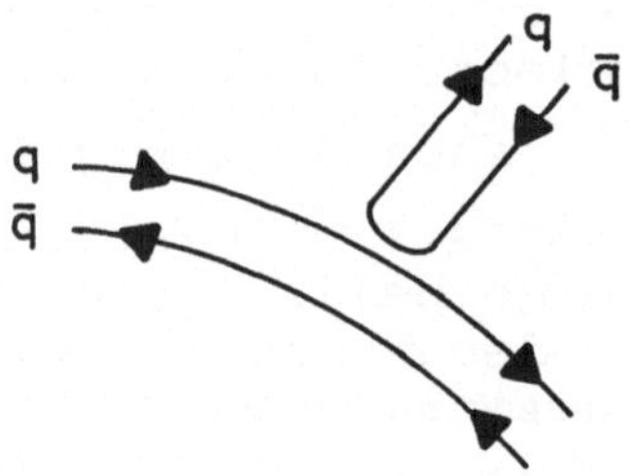

Fig. 4. Meson vertex, forbidden by Zweig's rule.

Zweig's rule forbids, for example, the decay $\phi \to \rho\pi$, since this decay requires the annihilation of the strange quark-antiquark system (ϕ).

Another way of looking at Zweig's rule is the statement that flavor neutral mesons tend to be relatively pure with respect to their quark content. For example, the ϕ-meson is essentially a pure $\bar{s}s$-system, with only a very slight admixture of $\bar{u}u$ and $\bar{d}d$. The same seems to be the case for the f'-meson, while it is definitely not true for the low-lying pseudoscalar η and η'. Here a relatively large mixing between the different $\bar{q}q$-combinations occurs such that the η-meson is nearly an SU_3-octet: $\eta \sim \frac{1}{\sqrt{6}}(\bar{u}u + \bar{d}d - 2\bar{s}s)$, while the η'-meson is nearly an SU_3-singlet. $\eta' \sim \frac{1}{\sqrt{3}}(\bar{u}u + \bar{d}d + \bar{s}s)$. Thus the Zweig rule is badly broken in the pseudoscalar channel.

The mixing of the different $\bar{q}q$-configurations in eigenstates of the flavor neutral meson mass matrix is strong, if several flavors are nearly degenerate, i.e. if the quark mass difference is smaller than the mixing term in the meson mass matrix. For example, the u-d-quark mass difference is negligible compared to the mixing term in the vector meson mass matrix. Therefore the 1^- mesons segregate into SU_3-singlet ($\omega = \frac{1}{\sqrt{2}}(\bar{u}u + \bar{d}d)$ and SU_2-triplet ($\rho_0 = \frac{1}{\sqrt{2}}(\bar{u}u - \bar{d}d)$. Furthermore, the sign of the mixing term must be such that the mixing raises the ω-mass above the ρ-mass.

A different behavior is exhibited by the 2^{++}-mesons. Here the mixing lowers the mass of the f-meson ($f=\frac{1}{\sqrt{2}}(\bar{u}u + \bar{d}d)$) relative to the A_2-meson mass $A_2=\frac{1}{\sqrt{2}}(\bar{u}u - \bar{d}d)$. Thus the sign of the mixing term is different in the 1^{--} and in the 2^{++} channel. Furthermore, one has $(\Delta M)_{1^{--}} = (M_\omega - M_\rho) < (\Delta M)_{2^{++}} = (M_{A_2} - M_f)$, i.e., the mixing is stronger in the 2^{++}-channel than in the 1^{--}-channel.

Conclusion: The purity of flavor neutral meson states, i.e., the quality of Zweig's rule, is channel dependent. It is relatively good in the 1^{--}-channel, less good in the 2^{++}-channel, and badly broken in the 0^{-+}-channel.

In QCD the mixing of the different flavor neutral $\bar{q}q$-configurations in mass eigenstates occurs through the annihilation of the $\bar{q}q$-systems into gluons, e.g. in the 1^{--}-channel the mixing is provided by the annihilation into an odd number of gluons (3,5,...) (see Fig. 5). Thus Zweig's rule is a statement about the gluon dynamics.

A different, but equivalent approach to Zweig's rule can be found by using intermediate physical glue states. According to Zweig's rule the mixing of flavor neutral $\bar{q}q$-meson states via glue states must be small, except in the pseudoscalar channel. The mixing involving glue states with the quantum numbers of two gluons seems to be stronger than the mixing involving three gluon states. For example, the A_2-f segregation occurs by the

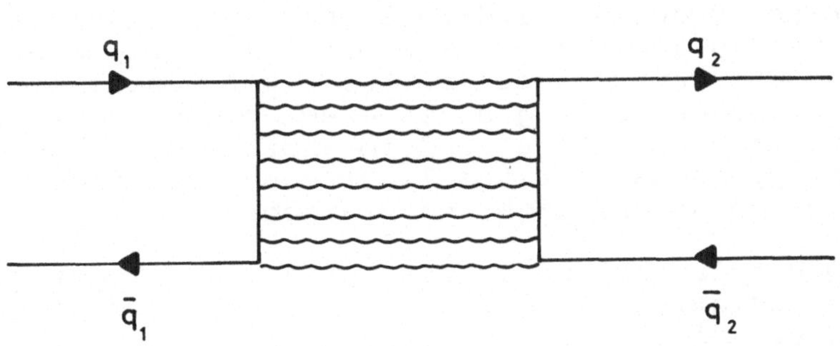

Fig. 5. Mixing of different $\bar{q}q$-configurations in the 1^{--}-channel.

mixing of the $\bar{q}q$-system and a 2^{++} glue system, while the ω-ρ segregation is caused by the mixing involving a 1^{--} glue state.

Thus far no satisfactory explanation of Zweig's rule has been given; it remains one of the mysteries of the strong interactions. It is likely that this rule is related to the absence of glue states in the low energy part of the hadron spectrum, as discussed above.

e) Exotic resonances

QCD, supplemented by the confinement hypotheses, predicts also the existence of hadron states containing more than two quarks (mesons) or more than three quarks (baryons) in their wave function. The simplest configurations of such exotic states are of the type $\bar{q}\bar{q}$ qq (they are obtained, e.g., if a diquark ($\bar{3}$,6 in color) and an antiquark (3,$\bar{6}$ in color) join to give a color singlet), and of the type $\bar{q}$qqqq (it is obtained, if we replace in an antibaryon configuration two of the antiquarks by diquarks). These exotic states can very easily decay into either a meson - meson or a baryon - meson system, i.e. they are expected to be relatively broad resonances. Thus far no exotic resonances have been isolated in the hadron spectrum, however it is quite possible that the very rich spectrum of broad exotic states sets in at an energy of $\sim$1.5 GeV.

5. QUARKS NEAR THE LIGHT CONE

The observed large cross sections as well as the scaling behaviour of the various amplitudes in electro and neutrino production indicate that quarks act at small or nearly lightlike distances like free pointlike particles. One way to describe the experimental situation is to assume scaling as an exact property of the leptoproduction amplitudes and to abstract the behaviour of current products near the light cone from free quark theory (parton model, light cone algebra).

In QCD this picture is not quite, but almost correct. There the deviations from the free quark commutators are small logarithmic "fine structure corrections" of order κ/π : due to the smallness of the quark - gluon coupling constant these corrections can be calculated in lowest order perturbation theory (for detailed calculations of the scale breaking effects with respect to deep inelastic scattering see e.g. the references 7,13).

In particular the ratio of the longitudinal and transverse cross sections in electroproduction $R^{el} = \sigma_{long}/\sigma_{trans.}$ which in the naive quark model approach is predicted to vanish like $1/q^2$ in the deep inelastic limit, approaches in QCD a function proportional to κ/π:[14)]

$$R^{el} \to \kappa/\pi \quad \phi(\xi)$$

where ϕ is a scaling function of order 1: the ratio R^{el} vanishes only logarithmically (since proportional to κ), and not like a power of q^2. The new experimental results[15)] on R^{el} are consistent with $R^{el} \simeq 0.1$, indicating $\kappa/\pi \simeq 0.1$, at $q^2 \simeq 10$ GeV^2. Note that the value of κ obtained here is consistent with the value $\simeq 1/3$ obtained previously from theoretical speculations about confinement.

Similarly the ration $R = \sigma_{e^+e^- \to hadrons}/\sigma_{e^+e^- \to \mu^+\mu^-}$ is predicted to approach the value $R \underset{s\to\infty}{\longrightarrow} \sum_{quark:} Q_e^2 (1+\kappa/\pi)$, i.e there is a shift $\sum Q^2 \cdot \kappa/\pi$ in R compared to the naive quark model result. Since κ approaches zero logarithmically, the naive quark results are obtained at extremely small distances. At energies at which the present experiments in e^+e^--annihilation are performed one has $\kappa/\pi \simeq 0.1$, i.e. typically a 10% correction to the naive quark result.

PART II: "STANDARD" THEORY

6. WEAK INTERACTIONS; CHARM; FOUR FLAVORS OF QUARKS AND LEPTONS

In this and the following sections I shall concentrate on phenomenological aspects of the various lepton - quark schemes to be discussed. The special field-theoretic aspects and in particular details of the spontaneous symmetry breaking are treated only superficially, and we ask the reader interested in these problems to consult one of the numerous review articles on gauge theories.

The conventional weak currents, which mediate the weak interactions among the observed leptons and the three flavors u,d,s, are known to be of the structure V-A, i.e., they couple only to the left-handed fermions. The observed universality of the weak interactions can be described by introducing the weak isospin SU_2^W and by assigning the left-handed leptons and quarks to SU_2^W doublets:

$$\text{Leptons} \quad \begin{pmatrix} \nu_e \\ e^- \end{pmatrix}_L, \quad \begin{pmatrix} \nu_\mu \\ \mu^- \end{pmatrix}_L$$

$$\text{Quarks} \quad \begin{pmatrix} u \\ d_c \end{pmatrix}_L \qquad \begin{aligned} d_c &= d\cos\theta_c + s\sin\theta_c \\ s_c &= -d\sin\theta_c + s\cos\theta_c \end{aligned}$$

(θ_c: Cabibbo angle, L: left-handed, R: right-handed)

The right-handed fermions are SU_2^W singlets, the weak charges are interpreted as generators of SU_2^W.

The weak and electromagnetic interactions can be described by the gauge theory based on the gauge group $SU_2 \times U_1$, where the group SU_2 is identified with the weak isospin defined above, while the U_1-generator Y (weak hypercharge) is needed in order to describe the electric charge:[16)]

$$Q^e = T_3 + 1/2\, Y \tag{6.1}$$

(T_3: neutral SU_2-generator, $T_3 = 1/2[T_+, T_-]$;

Y: weak hypercharge)

Since T_3 acts only on the left-handed fermions, the weak hypercharge Y is a mixture of a V + A current and a vector current.

According to eq. (6.1) the electric charge is a superposition of the two neutral generators of the group $SU_2 \times U_1$. Correspondingly the other (orthogonal) linear combination of these two generators must also be coupled to a gauge boson (neutral Z boson), giving rise to the exsistence of a neutral weak current. This neutral current will, in particular, couple to the neutral SU_2 charge T_3. If only the doublet $\binom{u}{d_c}_L$ were present (S_c:SU_2 singlet), the neutral current would have a $|\Delta S| = 1$ term proportional to $\sin\Theta_c \cos\Theta_c$. Such a term is known to be absent (it would e.g. lead to the decay $K_{oL} \to \mu^+\mu^-$, which is observed to be strongly suppressed). The most economical way to avoid the $[\Delta S = 1]$ neutral current is to introduce a new quark flavor (charm)[17] of electric charge 2/3, and to place it together with S_c into a SU_2 doublet: $\binom{C}{S_c}_L$. In this case the neutral current is flavor diagonal (the linear combination $\bar{d}\, d_L + \bar{s}s_L$ is invariant under rotations in the d - s space).

Table 2. The various quantum numbers of the fermions in the $SU_2 \times U_1$ scheme

Particles	T_3	Y
ν_e, ν_μ	1/2	-1
e_L^-, μ_L^-	-1/2	-1
e_R^-, μ_R^-	0	-2
u_L, c_L	1/2	1/3
d_{cL}, s_{cL}	-1/2	1/3
u_R, c_R	0	4/3
d_{cR}, s_{cR}	0	-2/3

Altogether one has sixteen elementary fermions, twelve colored quarks (four quark flavors) and four leptons; the sum of all electric charges is zero. The Lagrangian of the $SU_2 \times U_1$ theory in given by

$$L = -1/4\ \vec{W}_{\mu\nu}\ \vec{W}^{\mu\nu} - 1/4\ B_{\mu\nu}\ B^{\mu\nu} -$$

$$-\bar{f}\ \gamma^{\mu}(\partial_{\mu} - ig\ \vec{T}\vec{W}_{\mu} - ig'\ \frac{Y}{2}\ B_{\mu})f \qquad (6.2)$$

+ symmetry breaking terms,

where $\vec{W}_{\mu\nu} = \partial_{\nu}\ \vec{W}_{\mu} - \partial_{\mu}\vec{W}_{\nu} - g\ \vec{W}_{\nu} \times \vec{W}_{\mu}$

$$B_{\mu\nu} = \partial_{\nu}\ B_{\mu} - \partial_{\mu}\ B_{\nu}.$$

f : fermion fields

The $SU_2 \times U_1$ theory leads to the following expression for the phenomenological charged and neutral current interactions:

$$H^{weak} = \frac{4G_F}{\sqrt{2}}\ (j^{+}_{\mu}j^{-}_{\mu} + \rho\cdot\ j^{n}j^{n}) \qquad (6.3)$$

where

$$j^{+}_{\mu} = \bar{\nu}_e\gamma_{\mu}\ \frac{1+\gamma_5}{2}\ e^{-} + (e\rightarrow\mu) + \bar{u}\gamma_{\mu}\ \frac{1+\gamma_5}{2}\ d_c + \bar{c}\gamma_{\mu}\frac{1+\gamma_5}{2}\ s_c$$

$$j^{n}_{\mu} = \frac{1}{2}\ (\bar{\nu}_e\gamma_{\mu}\ \frac{1+\gamma_5}{2}\ \nu_e - \bar{e}^{-}\gamma_{\mu}\ \frac{1+\gamma_5}{2}\ e^{-}) + (e\rightarrow\mu)$$

$$+ \frac{1}{2}(\bar{u}\gamma_{\mu}\ \frac{1+\gamma_5}{2}\ u - \bar{d}\gamma_{\mu}\ \frac{1+\gamma_5}{2}\ d) + (u\rightarrow c,\ d\rightarrow s) - \sin^2\theta j^{e}_{\mu}$$

$$(6.4)$$

(j^{e}_{μ}: electromagnetic current, θ: weak mixing angle)

$$\rho = \frac{m_W^2}{\cos^2\theta\ m_Z^2}$$ (Z: neutral massive vector meson
$W^{\pm}$: charged massive vector mesons)

$$M_W = \frac{37.3\ \text{GeV}}{\sin\ \Theta}$$

(For a derivation of this formula see Appendix I, pg. 228).

The parameter ρ is introduced in order to keep the mass of the neutral intermediate boson a free parameter. Weinberg [16] proposed to describe the symmetry breaking by a doublet of scalar fields, in which case one obtains the specific prediction

$$M_Z = \frac{37.3}{\sin\theta_W \cos\theta_W} \quad (\rho=1). \tag{6.5}$$

The relation above rests on the specific doublet breaking mechanism for which we see no particular reason except simplicity. Since the whole subject of spontaneous symmetry breaking is thus far not well understood, it is unclear if this type of simplicity is the one nature may have chosen. Thus we prefer to introduce ρ as a free parameter to be measured by experiment.

There is one requirement, a renormalizable theory of the weak and electromagnetic interactions has to fulfil: It must be free of triangle anomalies,[18] i.e. the trilinear couplings of the gauge bosons (B, W_3) induced by fermion triangle diagrams have to vanish. (The fermion triangles lead to anomalous divergencies of the involved axialvector currents, which spoil the renormalizability). The constraint which follows for the fermion content of any $SU_2 \times U_1$ theory can be derived if we require the following triangle diagram to vanish (Fig. 6).
This diagram vanishes if

$$\mathrm{tr}\ T_3^2 \cdot Y = \mathrm{tr}\ T_3^2(Q_e - T_3) = 0. \tag{6.6}$$

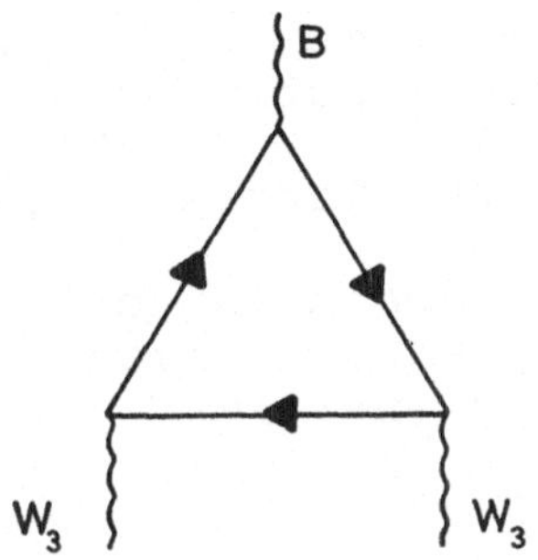

Fig. 6

In particular in case of a scheme containing only SU_2 doublets and singlets ($T_3^2 = \frac{1}{4}$, 0 respectively), the condition of the absence of anomalies reduces to the condition that the sum of all charges is zero:

$$\mathrm{tr}\, Q_e = 0. \tag{6.7}$$

We emphasize that the condition (6.7) does not ensure the absence of anomalies in more general cases where besides singlets and doublets other representations (triplets etc.) are present. In such cases the more general constraint (6.6) has to be applied.

In the four quark - four lepton scheme considered above the condition (6.7) is fulfilled, since $\sum_{\text{quarks}} Q_e = -\sum_{\text{leptons}} Q_e$ (Note: the color degree of freedom is essential for the argument).

Although it is not excluded that the new effects discovered during the last years in e^+e^--annihilation, neutrino production and other processes are due to the existence of several new quark flavors, it seems at present that at least the spectrum of particles belonging to the ψ-family can be explained by adding just one new flavor. The recent discovery[19] of new particles just below 2 GeV suggests strongly that this flavor is indentical with the charm flavor. In this case the neutral particle of mass 1.86 GeV can be identified with the D meson (quark composition $\bar{c}u$), and the charged particle of mass ~1.87 GeV with the D-meson of quark composition $\bar{c}d$. The F meson (quark composition $\bar{c}s$) is then expected to have a mass of 1.95...2 GeV, while the spectrum of charmed baryons is expected to set in at about 2.2 GeV.

Our preliminary conclusion is: the charm degree of freedom has been found; the effective mass of the charmed quark is much higher (~1.5 ... 2 GeV) than the effective masses of the three quark flavors u,d,s ($\lesssim$ 100 MeV).

Below we summarize the main consequences of the four quark scheme for e^+e^--annihilation and neutrino production at very high energies.

e^+e^--annihilation:
The electromagnetic current is

$$j_\mu^e = 2/3\ \bar{u}\gamma_\mu u - 1/3\ \bar{d}\gamma_\mu d + 2/3\ \bar{c}\gamma_\mu c - 1/3\ \bar{s}\gamma_\mu s, \tag{6.8}$$

and one predicts

$$R = \sigma_{e^+e^- \to hadrons} / \sigma_{e^+e^- \to \mu^+\mu^-} \xrightarrow[s\to\infty]{} 3\,\frac{1}{3}\ (1+\kappa/\pi) \tag{6.9}$$

If we take the value of κ relevant for energies $\gtrsim$ 6 GeV to be 1/4 ... 1/3, one finds $R \simeq 3.6 \ldots 3.7$.

The measured value of R at energies of 6 ... 8 GeV is [20] $R \simeq 5$, which indicates that besides the charm degree of freedom there might be other degrees of freedom excited at an energy above 4 GeV (new types of leptons, new flavors of quarks, other new particles?).

Neutrinoproduction off nucleons
Kinematic
We consider the reaction $\overset{(-)}{\nu}$ + hadron→charged lepton + hadrons (see Fig.7). We introduce the scaling variables $\xi = -q^2/2\nu$

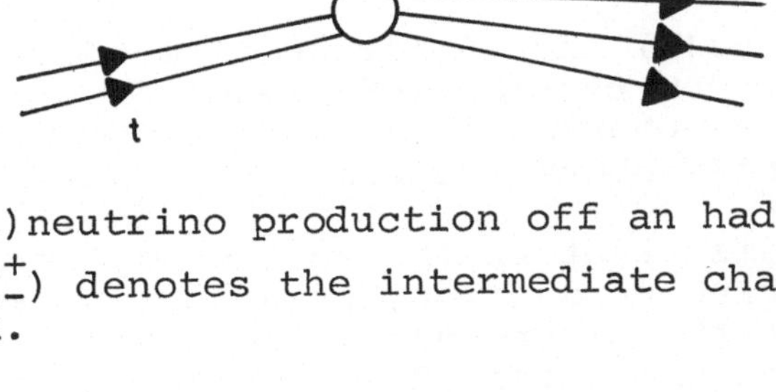

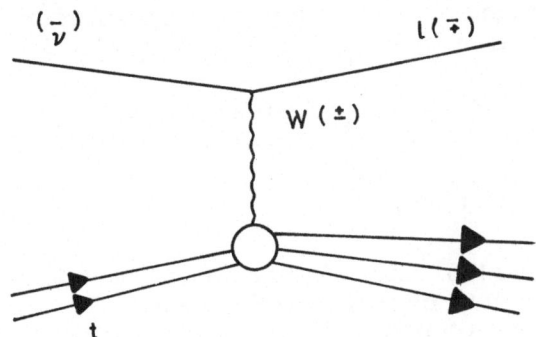

Fig. 7. (Anti)neutrino production off an hadronic target t. $W(\pm)$ denotes the intermediate charged weak boson.

and $y = \frac{\nu}{M_t E_\nu} = E_h/E_\nu$ (q: four-momentum transferred to the target, $\mu = p\cdot q$, p: target four-momentum, M_t:target mass, E_ν: energy of incident neutrino in lab-system, E_h:final hadronic energy in lab-system). Assuming scaling and the spin 1/2 character of the quarks, one finds for the differential neutrino production cross section

$$\frac{d^2\sigma^\nu}{dyd\xi} = 2K\cdot\xi\left[d\cos^2\theta_c + s\sin^2\theta_c + \bar{u}(1-y)^2 + \underline{d\sin^2\theta_c + s\cos^2\theta_c + \bar{c}(1-y)^2}\right] \quad (6.10)$$

$$\frac{d^2\sigma^{\bar{\nu}}}{dyd\xi} = 2K\cdot\xi\left[\bar{d}\cos^2\theta_c + \bar{s}\sin^2\theta_c + u(1-y)^2 + \underline{\bar{d}\sin^2\theta_c + \bar{s}\cos^2\theta_c + c(1-y)^2}\right]$$

with

$$K = \frac{G^2\cdot M_t\cdot E_\nu}{\pi} \quad . \quad \text{(G: Fermi constant)}. \quad (6.11)$$

In the valence quark approximation for nucleons or nuclei $\bar{u} = \bar{d} = s = \bar{s} = c = \bar{c} = 0$ one finds

$$\frac{d^2\sigma^\nu}{dyd\xi} = 2K\xi\ (d\cos^2\theta_c + \underline{d\sin^2\theta_c}) \quad (6.12)$$

$$\frac{d^2\sigma^{\bar{\nu}}}{dyd\xi} = 2K\xi\cdot u(1-y)^2 \ .$$

In eqs. (6.10) and (6.11) we have underlined the contributions from the charmed pieces of the weak current. Those are due to charmed final states and will be present only sufficiently above the charm threshold.

We emphasize that in the valence quark approximation only the neutrino production cross section acquires a new term above the c-threshold. The anti-neutrino cross section remains unchanged. However, the new contribution to the neutrino cross section is suppressed by the factor $\sin^2\theta_c \cong 0.04$.

A relatively big contribution to both the neutrino and the antineutrino cross section above c-threshold may come from the $s\bar{s}$-cloud in the nucleon, according to eqs. (6.10). However, the s- and $\bar{s}$-distribution functions in the nucleon are small and confined to small values of ξ. Thus one does not expect to see a sizable effect due to the production of charmed particles in the total (anti)-neutrino production cross sections.

Weak decays of charmed particles

The structure of the weak current in the conventional charm scheme requires that particles with C = +1 decay mainly (96%) into a state with S = -1. In a naive approach neglecting the strong interactions, one expects 40% of all decays to be semileptonic and 60% to be hadronic, provided the particle is relatively heavy ($M \simeq 2$ GeV). This follows simply from the counting of degrees of freedom. The virtual weak boson, emitted by the decaying charmed quark, can decay either into a lepton pair ($W^+ \rightarrow \nu_e e^+, \nu_\mu \mu^+$) or into a quark pair (e.g. $W^+ \rightarrow (\bar{d}u)_r$, $W^+ \rightarrow (\bar{d}u)_y$, $W^+ \rightarrow (\bar{d}u)_b$). It is well-known that these naive quark counting rules are wrong for the decay of strange particles. There the nonleptonic decays are enhanced in the $|\Delta I|=1/2$ channel such that the semi-leptonic decay modes contribute only ~1‰ to the total decay rates (except in cases where the $|\Delta I|=1/2$ decay is forbidden, e.g. in the decay of charged kaons). Thus the question arises: Is there a similar nonleptonic enhancement for charmed particles, or not?

Recently one has observed dimuon events[21)] in neutrino - nucleus scattering: $\nu_\mu + \text{nucleon} \rightarrow \mu^- + \mu^+ +$ hadrons. The kinematical properties of these events are consistent with the production and subsequent semileptonic decay of charmed particles. Since charm production is expected to contribute not more than $\sim$ 10% to the total neutrino production cross section (see above), and since the observed dimuon rate is of the order of 1% of the total cross section, one concludes that the semileptonic branching ration for the decay of charmed particles must be rather big $>20\%$, i.e. not far from the naive quark- counting value.

A similar conclusion can be drawn if one interprets the electron-hadron events observed recently in e^+e^- -

annihilation at DESY,[22] as due to semileptonic weak decay of charmed particles. A large semileptonic branching ratio for charmed particles is surprising, since in this case charmed particles behave quite differently as the strange particles, where the semileptonic decay modes are suppressed. This, of course, brings up the old problems of the nonleptonic weak decays ($|\Delta I|=1/2$ rule, etc.). Although these problems are unsolved, it is at least encouraging to see that the nonleptonic weak decays of charmed particles may be free of all those complications one has to cope with in case of strange particles.

PART III: VECTORLIKE THEORIES

7. TWELVE COLORED QUARKS AND FOUR LEPTONS, AND NO MORE?

This is the main question I shall address in the remaining parts of these lectures: Is the four lepton - four quark scheme discussed above attractive enough both theoretically and phenomenologically that it could be regarded as the final word in lepton - quark physics, or would it be worthwhile to have even more quark flavors and leptons? Of course, the introduction of any new quark flavor or lepton is a painful step: It complicates the subject and leads us further away from the dream that some day all of physics can be understood in terms of a few truly elementary objects and notions. However the world explored thus far already reveals many, if not too many, elementary fermions. The "minimal" four quark-four lepton scheme is composed of two "eightfold" units:

$$\begin{pmatrix} u_r & u_g & u_b & \nu_e \\ d_r & d_g & d_b & e^- \end{pmatrix} \qquad \begin{pmatrix} c_r & c_g & c_b & \nu_\mu \\ s_r & s_g & s_b & \mu^- \end{pmatrix}$$

Counting both fermions and antifermions, one has altogether 32 elementary fields, i.e. a number, already very large.

Let me first mention two phenomenological aspects which indicate that the four lepton - four quark scheme is incomplete. The presence of events of the type "$e^+e^- \to e^{(\mp)} \mu^{(\pm)}$ + missing energy" observed at SPEAR[23)] indicates the existence of a new lepton of mass ~2 GeV. If such a lepton ideed exists, it may imply the existence of yet further neutral and perhaps also charged leptons, and/or new types of quarks.

In (anti)-neutrino - nucleus scattering one has observed[24)] recently at very high energies an anomalous behaviour of the $\bar{y}$ - distributions, and an increase of the ratio $\sigma^{\bar{\nu}}/\sigma^{\nu}$ from ~0.38 (at low energies) to ~0.7 (at energies above 50 GeV). This phenomenon may be a pure strong interaction effect, caused by an increase of the quark - antiquark pairs in the nucleon wave function (i.e. scaling violations of the type predicted by QCD). However it may also be due to a new quark flavor being produced at energies above 30 GeV. Arguments based on QCD perturbation theory indicate that the effect is too large to be explained by the generation of $\bar{q}q$ pairs via radiative gluon corrections to the nucleon wave function,[25)] supporting the hypotheses that a new flavor is being produced. New experiments such as the one being prepared at the CERN-SPS are needed in order to resolve this issue.

On the theoretical side there are several problems one has to deal with in the four flavor scheme. Most serious is the charge quantization problem. It is well-known that gauge theories of the electromagnetic and weak interactions based on the gauge group $SU_2 \times U_1$ are not truly unified theories: They cannot explain why the proton and positron charge are equal, or in other words, why the quark charges are definite units (2/3,-1/3) of the positron charge. The only possibility to obtain quantized charges is to understand the group $SU_2 \times U_1$ as a subgroup of a larger simple (or perhaps semisimple) group. Models of this type which also include the strong interactions have been studied,[26,27,28)] they all have serious problems. For example one is led to a decay of the proton into leptons in second order of the gauge coupling which can only be stopped by choosing the masses of the associated gauge bosons to be of the order of 10^{16} GeV (the appropriate name for these bosons was introduced by S. Meshkov: intermediate vector baseballs). One way to avoid the proton decay problem is to enlarge the number of quarks and leptons[29]. In this case no need exists to introduce superheavy gauge bosons of the type mentioned above.

One aspect of the four flavor scheme is that there exists no natural way to introduce CP violation [30]. New weak currents perhaps associated with new quark flavors are required for this task. (Note, however, that the origin of CP violation might be the interaction of the leptons and quarks to the scalar Higgs fields[31]. If one introduces several weak doublets of scalar fields interacting with the fermions, CP violation can be introduced.)

A last problem of the four flavor scheme is that there exists no possibility to calculate the mass differences among the quarks and the leptons, or at least to understand the physical origin of these mass differences. For example the large mass difference between the effective charmed quark mass and the "light" quark masses m_u, m_d, m_s remains a mystery. Schemes involving further leptons and quarks may finally provide us with a pattern simple enough such as to render at least some of the mass differences calculable.

Recently many people became interested in vectorlike theories, which are a particular set of weak interaction theories containing several new types of leptons and new quark flavors besides charm[29,31,32,33,34,35]. The speculation is that the weak interactions are basically a vector theory, like the strong and electromagnetic theories, and the preference of the observed weak interactions for V - A type currents is viewed as a low energy effect, due to parity violating terms in the fermion mass matrix. This possibility is especially attractive, since it would render the weak interactions as a much closer relative of the electromagnetic and strong interactions as in the usual approach, and the idea of a unified theory of all interactions becomes a very reasonable one.

8. WHAT ARE VECTORLIKE THEORIES?

Let me first emphasize that there exist many different vectorlike schemes. Unfortunately many people identify with vectorlike theories particular schemes consisting of six quarks and six leptons which were studied by several authors last year[31...34]. In these particular models the neutral current is a vector current, a possibility which seems meanwhile excluded by experiment. However there exist many vectorlike theories for which the neutral current is not a vector current and which are in agreement with present observations.

A vectorlike theory is a theory in which the parity violation is not intrinsic, but due to the presence of parity violating terms in the fermion mass matrix. In the limit where the fermion masses are set to zero the theory is a pure vector theory, and the gauge currents are vectorial. If the fermion mass matrix is turned on, it acquires both scalar and pseudoscalar terms, and parity is violated. The unitary transformation in the fermion space which diagonalizes the fermion mass matrix and eliminates its pseudoscalar part (in the absence of CP violation) introduces apparent axial vector currents.

In the conventional gauge theory framework the fermion masses are generated by the coupling of the fermion fields to scalar fields which develop non-zero vacuum expectation values ("spontaneous symmetry breaking").

In a vectorlike theory the violation of parity is part of the spontaneous symmetry breaking, in contrast to, for example, the Salam-Weinberg model where the parity violation is an intrinsic property of the field equations.

Realistic vectorlike theories can only be constructed if there exist new leptons, and more than four quark flavors. A simple vectorlike scheme is, for example, given by the following lepton scheme:

$$\begin{pmatrix} \nu_e & N \\ e^- & E^- \end{pmatrix}_L \qquad \begin{pmatrix} N & \nu_e \\ e^- & E^- \end{pmatrix}_R \tag{8.1}$$

and an analogous scheme involving the muon. Here N is a new massive neutral lepton ($m_N > M_K$), and E^- a new charged lepton.

Introducing the fields $N_1 = (\nu_{eL}, N_R)$, $N_2 = (N_L, \nu_{eR})$, we can rewrite this scheme in a vector form:

$$\begin{pmatrix} N_1 & N_2 \\ e^- & E^- \end{pmatrix}_L \quad \begin{pmatrix} N_1 & N_2 \\ e^- & E^- \end{pmatrix}_R \tag{8.2}$$

while the mass term of the neutral fermions (neglecting a possible ν_e mass) can be rewritten as

$$\begin{aligned} m_N \bar{N}N &= m_N(N_L^* N_R + N_R^* N_L) \\ &= m_N(\bar{N}_2 \frac{1-\gamma_5}{2} N_1 + \bar{N}_1 \frac{1+\gamma_5}{2} N_2) \\ &= \frac{1}{2} m_N \ (\bar{N}_2 N_1 + \bar{N}_1 N_2 + \bar{N}_1 \gamma_5 N_2 - \bar{N}_2 \gamma_5 N_1) \end{aligned} \tag{8.3}$$

The scheme (8.1) reproduces the correct V-A pattern of the observed weak interactions, provided $m_N > M_K$. It shows the following general features of vectorlike theories:

A: Parity is violated only by pseudoscalar terms in the fermion mass matrix.

B: The numbers of left-handed currents and right-handed currents are equal. This implies in particular the absence of anomalies.

One can easily see that a vectorlike theory of the hadronic weak currents based on SU_2^{weak} can only be constructed if one introduces more than four quark flavours. Let us try to do it with just four. The left-handed currents are $\binom{u}{d'}, \binom{c}{s'}_L$. In order to construct a vectorlike scheme one has to introduce two right-handed currents. There is only one possibility to do so, namely $\begin{pmatrix} u & c \\ d'' & s'' \end{pmatrix}_R$ ($d_R'' = d_R \cos\theta'' + s_R \sin\theta''$, $s_R'' = \perp$; θ'': right-handed analogue of the Cabibbo angle). The V-A character of the weak current entering in β decay requires $\theta'' \doteq 90^o$, which gives $\begin{pmatrix} u & c \\ s & -d \end{pmatrix}_R$. This pattern is correct for β decay, however, wrong for Λ decay (both in strength and chirality). Thus we conclude:

<u>A vectorlike theory of the hadronic weak interactions can only be constructed if there exist more than four quark flavours and more than four leptons</u>.

9. VECTORLIKE MODELS BASED ON SU_2^W DOUBLETS

A radical possibility to construct vectorlike schemes is to place all leptons and quarks in SU_2^{weak} doublets. The smallest vectorlike scheme of this sort is one based on six lepton flavours and six quark flavours[31...34)

$$\begin{pmatrix} u & t & c \\ d' & b & s' \end{pmatrix}_L \qquad \begin{pmatrix} u & t & c \\ b & d'' & s'' \end{pmatrix}_R$$

$$\begin{pmatrix} \nu_e & N_E & \nu_\mu \\ e^- & E^- & \mu^- \end{pmatrix}_L \qquad \begin{pmatrix} \nu_e & N_E & N_\mu \\ E^- & e^- & \mu^- \end{pmatrix}_R \tag{9.1}$$

Here t and b are new flavours of quarks ("top" and "bottom") with charges 2/3 and -1/3, respectively[36)]. The left-handed quarks are rotated by the Cabibbo angle Θ_c into $d' = d\cos\Theta_c + s\sin\Theta_c$, and $s' = -d\sin\Theta_c + s\cos\Theta_c$. The observed universality of the weak interactions requires that there is no (or only little) mixing of d' with b. There may, however, be substantial mixing between b and s', i.e., the fields b and s' in the scheme above may be replaced by the rotated fields $\tilde{b} = b\cos\tilde{\theta} + s'\sin\tilde{\Theta}$ and $\tilde{s}' = -b\sin\tilde{\theta} + s'\cos\tilde{\theta}$. For simplicity we have not displayed this rotation in the scheme above.

Among the right-handed quarks, the observed V-A structure of the weak currents connecting the u quark with the d' combination forbids any appreciable coupling of u_R to either d_R or s_R. Consequently its partner must be pure or nearly pure b. The linear combinations $d'' = d\cos\Theta'' + s\sin\Theta''$ and $s'' = -d\sin\Theta'' + s\cos\Theta''$ are, for the moment, left arbitrary. Below we shall argue in favour of $\Theta'' \approx 0$.

The scheme contains, in addition to the conventional leptons, one new charged lepton M^-, one new massive neutral lepton N_M and the right-handed partner $(\nu_e)_R$ of the electron neutrino as well as the massive Majorana partner of the muon neutrino ν_μ. We assume that the neutral lepton N_M is a Fermi-Dirac particle (its mass term conserves lepton number). The fields $(\nu_e)_L, (\nu_e)_R$ and $(\nu_\mu)_L$ are assumed to be massless in the absence of the weak interactions. The violation of lepton number conservation occurs only via the N_μ mas term. Thus neutrino-less double β decay does not occur in lowest order of the weak interaction[37]. However, there exists the decay $K^- \to \pi^+ + \mu^- + \mu^-$, which occurs in second order of the weak interactions. [This decay was estimated in Ref. 38]. The expected branching ratio $K^- \to \pi^+ + \mu^- + \mu^- / K^- \to$ all is of the order 10^{-14} for $M_{N_\mu} \sim$ several GeV, i.e., more than ten orders of magnitude smaller than the experimental limit. It remains to be seen, if the accuracy of the experiments can be improved such that an experimental test of the lepton number violation of the kind we are discussing becomes feasible.

In the scheme (9.1) all lepton and quark fields take part in the weak interaction. Consequently the neutral current is a vector current:

$$j^{\text{neutral}} = (-\tfrac{1}{2}+z)\bar{e}^-\gamma_\mu e^- + (\tfrac{1}{2}-\tfrac{2}{3}z)\bar{u}\gamma_\mu u + +(-\tfrac{1}{2}+\tfrac{1}{3}z)\bar{d}\gamma_\mu d \qquad (z = \sin^2\theta) \tag{9.2}$$

+ terms involving the other fermions (θ: $SU_2 x U_1$ mixing angle).

A direct generalization of the six-flavour scheme (9.1) is the following eight-flavour scheme [29,39]:

$$\begin{pmatrix} u & t & c & v \\ d' & b & s' & h \end{pmatrix}_L \quad \begin{pmatrix} u & t & c & v \\ b & d & h & s \end{pmatrix}_R$$

$$\begin{pmatrix} \nu_e & N_E & \nu_\mu & N_M \\ e^- & E^- & \mu^- & M^- \end{pmatrix}_L \quad \begin{pmatrix} N_E & \nu_e & N_\mu & \nu_\mu \\ e^- & E^- & \mu^- & M^- \end{pmatrix}_R . \tag{9.3}$$

where we have not indicated the possibility to have weak angles between the various leptons and quarks. The eight-flavour scheme contains two new charged leptons E^-, M^-, and two new massive leptons N_E, N_M (Fermi-Dirac spinors), and five new quark flavours. The essential difference between the six-flavour and the eight-flavour scheme is that there is no need to introduce massive Majorana neutrals. The lepton mass matrix need not be lepton number violating.

10. VECTORLIKE THEORIES BASED ON SU_2 DOUBLETS AND SINGLETS

The schemes considered above are only specific examples of a large class of vectorlike theories. In general one may have, besides SU_2 doublets, also singlets, triplets etc.. For example the scheme (8.3) is still a vectorlike scheme if we abolish the lefthanded doublet $\binom{N}{E^-}_L$ and the righthanded doublet $\binom{N}{e^-}_R$; one obtains the scheme

$$\begin{pmatrix} \nu_e \\ e^- \end{pmatrix}_L; \quad \begin{pmatrix} \nu_e \\ E^- \end{pmatrix}_R; \qquad (E_L^-, e_R^-). \tag{10.1}$$

Similarly one can start from the 6 quark scheme (9.1) and cross out certain currents. Since the lefthanded currents $\bar{u}d_c$ and $\bar{c}s_c$ have to be kept for obvious phenomenological reasons, one is led to the following schemes:

$$\begin{pmatrix} u & c \\ d' & s' \end{pmatrix}_L; \quad \begin{pmatrix} u & c \\ b & s \end{pmatrix}_R \qquad (b_L, d_R) \tag{10.2}$$

or

$$\begin{pmatrix} u & c \\ d' & s' \end{pmatrix}_L; \quad \begin{pmatrix} c & t \\ s & d \end{pmatrix}_R. \tag{10.3}$$

Of course, the quarks entering in the righthanded currents need not be mass eigenstates, but there could be new types of weak angles. For simplicity we ignore such cases.

In the schemes considered above either d_R or u_R is a SU_2 singlet. There exist even vectorlike schemes for which both d_R and u_R are SU_2 singlets (as in the Salam-Weinberg scheme). The minimal scheme of this type is

$$\begin{pmatrix} u & c \\ d' & s' \end{pmatrix}_L ; \quad \begin{pmatrix} t & v \\ b & h \end{pmatrix}_R , \qquad (10.4)$$

where t,v;b,h are new quarks of charge 2/3, -1/3 respectively. In this case the righthanded currents needed for completing a vectorlike theory are reserved for the new quarks.

In Table 3, following, I have classified essentially all vectorlike theories one can construct with not more than eight quarks and leptons (for simplicity I did not consider new weak mixing angles). In the last column the neutral current is displayed in its coupling to electrons and the valence quarks of the nucleon u and d. The general expression can be derived easily according to the formula

$$j_\mu^n = j_\mu^3 - \sin^2\Theta \, j_\mu^e . \qquad (10.5)$$

11. VECTORLIKE LEPTONIC CURRENTS AND THEIR PHENOMENOLOGY

Any vectorlike theory implies the existence of right-handed weak currents. Typically those are relevant for the weak interactions of heavy leptons.

A) - Weak decays of heavy charged leptons

A heavy charged lepton can decay weakly via a right-handed weak current. For example, in the six-flavour scheme A the leptons E^- can decay via $E^- \to (\nu_e)_R$+(lepton pair + hadrons).

If $M_{E^-} \tilde{=} 2$ GeV, one can approximate the emission of hadrons by the emission of essentially massless free u or d quarks, in accordance with simple scaling ideas. Thus one predicts that the new lepton decays ~60% of the time into hadrons plus $(\nu_e)_R$, and 40% of the time into leptons. (In this estimate we have neglected the contributions of new flavours, which, however, will not matter, unless $M_{M^-} >> 2$ GeV.)

The observed dilepton events in e^+e^- annihilation[23) can be interpreted as the consequence of the production of a new lepton of mass $\simeq$ 1.9 GeV. According to the estimate above, one expects 8% of the time the heavy lepton pair to disintegrate into $e^+\mu^-$ or $e^-\mu^+$ (plus neutrinos). In this scheme the new neutrino $(\nu_e)_R$ is produced in the decay of E^-. This lepton may also decay into N_E provided $M_E>M_{N_E}$, which subsequently decays into e^- + lepton pair or e^- + hadrons. [This would lead to multilepton events, for example $e^+e^-\to\mu^++(e^-+e^-+\mu^+)$ + neutrinos.] The angular distribution of the dilepton events in e^+e^- annihilation depends on the weak coupling of E^-. When the statistics of the dilepton events improves in the near future, it will be possible to check whether a right-handed current is involved or not[40)].

B) - Radiative decays of heavy charged leptons

Especially interesting is the possibility to have radiative decays of heavy charged leptons. For example in scheme A the decay $E^-\to e^-\gamma$ occurs with a branching ratio of order[41)] $(m_N/m_E)^2\cdot\alpha/\pi$. It would be interesting to look for the radiative decays in e^+e^- annihilation, where one suspects that part of the observed total cross-section is due to the contribution of a new charged lepton[23)].

C) - The physics of neutral massive leptons

Many of the vectorlike schemes (A,B,C,D,E,G) contain new neutral massive leptons ("heavy neutrinos") along with new charged leptons, which are coupled by charged weak currents to the electron and muon. In general they will decay by emitting a charged lepton pair and a neutrino, e.g., in the six-flavour scheme A one has, for example, the decay $N_E \to e^- + \mu^+ + \nu_\mu$. The present experimental limits on the existence of such objects are very poor: they must only be heavier than the K meson, in order not to show up in the disintegration of the kaon. How can one see and discover the new massive neutral leptons?

One possibility is to see them in the decay of the new charged leptons. Another one is the production of a heavy neutral lepton N in ν_μ-electron scattering by the charged weak current. Because of the constraint $m_N \gtrsim M_K$

Table 3

Model	SU_2 doublets	SU_2 singlets	Remarks	New quarks charge, New leptons charge	Neutral current in $SU_2 \times U_1$ theory $z=\sin^2\theta$ (θ: $SU_2 \times U_1$ mixing angle
A	$\begin{pmatrix} u & t & c \\ d' & b & s' \end{pmatrix}_L \begin{pmatrix} u & t & c \\ b & d'' & s'' \end{pmatrix}_R$ $\begin{pmatrix} \nu_e & N_E & \nu_\mu \\ e^- & E^- & \mu^- \end{pmatrix}_L \begin{pmatrix} \nu_e & N_E & N_\mu \\ E^- & e^- & \mu^- \end{pmatrix}_R$	– –	$d''_R = \cos\theta'' d_R + \sin\theta'' s_R$ $s'' = \perp$ θ'' small ($\Delta I = \frac{1}{2}$ rule)	$c[2/3]$, $t[2/3]$ $b[-1/3]$ $N_E[o]$, $N_\mu[o]$ $E^-[-1]$	$\frac{1}{2}\bar{\nu}\nu - \left(\frac{1}{2} - z\right)\bar{e}e$ $+\left(\frac{1}{2} - \frac{2}{3}z\right)\bar{u}u - \left(\frac{1}{2} - \frac{1}{3}z\right)\bar{d}d + \ldots$ (vector; ν stands for ν_e, ν_μ)
B	$\begin{pmatrix} u & t & c & v \\ d' & b & s' & h \end{pmatrix}_L \begin{pmatrix} u & t & c & v \\ b & d & h & s \end{pmatrix}_R$ $\begin{pmatrix} \nu_e & N_E & \nu_\mu & N_M \\ e^- & E^- & \mu^- & M^- \end{pmatrix}_L \begin{pmatrix} N_E & \nu_e & N_M & \nu_\mu \\ e^- & E^- & \mu^- & M^- \end{pmatrix}_R$	– –	new possible weak angles neglected	$c[2/3]$, $t[2/3]$, $v[2/3]$ $b[-1/3]$, $h[-1/3]$ $N_E[o]$, $N_M[o]$ $E^-[-1]$, $M^-[-1]$	For light fermions, same as in A.
C	$\begin{pmatrix} u & c \\ d' & s' \end{pmatrix}_L \begin{pmatrix} u & t \\ b & d'' \end{pmatrix}_R$ $\begin{pmatrix} \nu_e & \nu_\mu \\ e^- & \mu^- \end{pmatrix}_L \begin{pmatrix} \nu_e & N_E \\ E^- & e^- \end{pmatrix}_R$	(t_L, c_R), (b_L, s''_R) (N_{E_L}, ν_{μ_R}) (E^-_L, μ^-_R)	$d''_R = \cos\theta'' d_R + \sin\theta'' s_R$ $\theta'' = o$, or $\theta'' = \frac{\pi}{2}$	$c[2/3]$, $t[2/3]$ $b[-1/3]$ $E^-[-1]$, $N_E[o]$	$\frac{1}{2}\bar{\nu}\nu - \left(\frac{1}{2} - z\right)\bar{e}e$ $+\left(\frac{1}{2} - \frac{2}{3}z\right)\bar{u}u - \left(\frac{1}{2} - \frac{1}{3}z\right)\bar{d}d_L - \left(\frac{1}{2} - \frac{1}{3}z\right)(\bar{d}''d'')_R$ $+ \ldots$

D	$\begin{pmatrix} u & c \\ d' & s' \end{pmatrix}_L \begin{pmatrix} u & c \\ b & s'' \end{pmatrix}_R$ $\begin{pmatrix} \nu_e & \nu_\mu \\ e^- & \mu^- \end{pmatrix} \begin{pmatrix} \nu_\mu & N_\mu \\ E^- & \mu^- \end{pmatrix}_R$	(b_L, d''_R) $(N_{\mu L}, \nu_{\mu R})$ (E^-_L, e^-_R)	$s''_R = \cos\theta'' s_R + \sin\theta'' d_R$ $\theta'' = 0$, or $\theta'' = \frac{\pi}{2}$	$c[2/3]$, $b[-1/3]$ $N_L[0]$, $E^-[-1]$	$\frac{1}{2}\bar{\nu}\nu - \left(\frac{1}{2} - z\right)\bar{e}e_L$ $+\left(\frac{1}{2} - \frac{2}{3}z\right)\bar{u}u - \left(\frac{1}{2} - \frac{1}{3}z\right)\bar{d}d_L$ $-\left(\frac{1}{2} - \frac{1}{3}z\right)\bar{s}'' s''_R + \ldots$,
E	$\begin{pmatrix} u & c \\ d' & s' \end{pmatrix}_L \begin{pmatrix} t & c \\ d'' & s'' \end{pmatrix}_R$ $\begin{pmatrix} \nu_e & \nu_\mu \\ e^- & \mu^- \end{pmatrix}_L \begin{pmatrix} \nu_e & \nu_\mu \\ E^- & M^- \end{pmatrix}_R$	(t_L, u_R) (E^-_L, e^-_R) (M^-_L, μ^-_R)	θ''small ($\Delta I = \frac{1}{2}$rule, see also: A)	$c[2/3]$, $t[2/3]$ $N_e[0]$, $N_\mu[0]$	$\frac{1}{2}\bar{\nu}\nu - \left(\frac{1}{2} - z\right)\bar{e}e$ $+\left(\frac{1}{2} - \frac{2}{3}z\right)\bar{u}u_L - \frac{2}{3}z\,\bar{u}u_R$ $-\left(\frac{1}{2} - \frac{1}{3}z\right)\bar{d}d + \ldots$
F	$\begin{pmatrix} u & c \\ d' & s' \end{pmatrix}_L \begin{pmatrix} u & c \\ b'' & h'' \end{pmatrix}_R$ $\begin{pmatrix} \nu_e & \nu_u \\ e^- & \mu^- \end{pmatrix}_L \begin{pmatrix} \nu_e & \nu_\mu \\ E^- & M^- \end{pmatrix}_R$	$(b_L, d_R)(h_L, s_R)$ $(E^-_L, e^-_R)(M_L, \mu_R)$	$b'' = b\cos\theta'' +$ $h\sin\theta''$ $h'' = \perp$	$c\,[2/3]$, $b[-1/3]$, $h[-1/3]$ $E^-[-1]$, $M^-[-1]$	$\frac{1}{2}\bar{\nu}\nu - \left(\frac{1}{2} - z\right)\bar{e}e_L + z\,\bar{e}e_R$ $+\left(\frac{1}{2} - \frac{2}{3}z\right)\bar{u}u - \left(\frac{1}{2} - \frac{1}{3}z\right)\bar{d}d_L + \frac{1}{3}z\,\bar{d}d_R$
G	$\begin{pmatrix} u & c \\ d' & s' \end{pmatrix}_L \begin{pmatrix} t & v \\ b & h \end{pmatrix}_R$ $\begin{pmatrix} \nu_e & \nu_\mu \\ e^- & \mu^- \end{pmatrix}_L \begin{pmatrix} N_E & N_M \\ E^- & M^- \end{pmatrix}_R$	$(t_L, u_R)(v_L, c_R)$ $(b_L, d_R)(h_L, s_R)$ $(N_{e_L}, \nu_{e_R})(N_{M_L}, \nu_{\mu_R})$ $(E^-_L, e^-_R), (M^-_L, \mu^-_R)$	Possible new mixing angles not displayed.	Same as in B.	$\frac{1}{2}\bar{\nu}\nu - \left(\frac{1}{2} - z\right)\bar{e}e_L + z\,(\bar{e}e)_R$ $+\left(\frac{1}{2} - \frac{2}{3}z\right)\bar{u}u_L - \frac{2}{3}z\,\bar{u}u_R - \left(\frac{1}{2} - \frac{1}{3}z\right)dd_L + \frac{1}{3}z\,dd_R$ $+ \ldots$ (For light fermions, same as in Salam-Weinberg model).

the lowest possible threshold for such a reaction is $M_K^2/2m_e \simeq 250$ GeV. Since N decays partly via $e^-\mu^+\nu_\mu$, one could see the production of the new neutral lepton by the appearance of trilepton events in $\nu_\mu e^-$ scattering.

It seems to me that the easiest way to discover new massive neutral leptons is to look for them in e^+e^- annihilation at very high energies where the production of the neutral leptons via the neutral current sets in. I estimated (within the six-flavour scheme A) the pair production of the massive neutral leptons to contribute about 30% of one unit in R at energies of 40 GeV. Their weak decays would lead, e.g., to events like $e^+e^- \xrightarrow[Z]{} (\mu^+\mu^-)+(\mu^+\mu^-)$, where both lepton pairs are produced in two jets and carry away ~2/3 of the available energy, the rest being carried away by neutrinos. The N leptons can in general also decay via $N\to\mu^-+\pi^+$, $N\to\mu^-+\rho^+$, etc., which leads to the very interesting events like $e^+e^-\to(\mu^-+\pi^+,\rho^+,\ldots)_{jet}+(\mu^++\pi^-,\rho^-,\ldots)_{opposite\ jet}$ + no missing energy. Most interesting I find the pair production of massive Majorana neutrals, like the N_μ lepton in the scheme A. Since the mass term of this lepton violates lepton number conservation, it can decay both into $\mu^- + \pi^+,\rho^+\ldots$ and into $\mu^+ + \pi^-,\rho^-\ldots$. This leads to spectacular new events, e.g., $e^+e^-\to(\mu^-\pi^+)_{jet}$ + $(\mu^-\pi^+)_{opposite\ jet}$.

- <u>Massive neutrinos and neutrino oscillations</u>

In a vectorlike theory the neutrinos cannot be described by a two-component Weyl theory, as for example in the Salam-Weinberg scheme. Neutrinos are four-component spinors. Consequently the basic reason why neutrinos should be massless is lost, and it would be most naturally if neutrinos like all other fermions would have a mass. To require the neutrino to be massless becomes an artificial constraint in a vectorlike theory. For example in all models considered, except G, both the left-handed and right-handed neutrino field components enter in the neutral current. Thus the emission and absorption of Z bosons will produce a (formally logarithmically divergent) neutrino mass term. In some models (A,B,C,H) the charged weak interaction leads to a mixing between neutrinos and the massive neutral leptons, producing a neutrino mass term of the order of a few eV[31,38).

Neutrino masses of the order of 1 eV, although being much below the experimental limits, would be very interesting since in this case the "neutrino sea" in the universe suspected on the basis of the observed 2.7°K "photon sea" would be a "sea" of massive, non-relativistic neutrinos. This "sea" can provide a large, perhaps even dominant fraction of the mass of the universe, which may well be the mass necessary for a closed universe compatible with the observed red shift and deceleration parameters. Furthermore the missing mass in galactic clusters can be provided by a cloud of massive neutrinos[42]. Estimates made on the basis of the observed astronomical parameters suggest an average neutrino mass of ~2 eV.

If the neutrino masses are indeed in the vicinity of 1 eV, the sea of cosmic neutrinos will consist of very non-relativistic neutrinos (temperature ~10^{-4} °K), which will not be distributed uniformly in space (as the photons constituting the 2.7°K photon sea), but will be concentrated in the galactic clusters. Inside those clusters, e.g., also here on earth, the neutrino density will be very high (typically $10^{9}...10^{10}$ ν/cm^{3}). To prove the existence of such a dense "neutrino sea" is a challenge for experimentalists.

Especially interesting in case of massive neutrinos is the possibility of neutrino beam oscillations[43]. If for example, the electron and muon neutrino are superpositions of two mass eigenstates:

$$\nu_e = \cos\Theta\ \nu_1 + \sin\Theta\ \nu_2$$

$$\nu_\mu = -\sin\Theta\ \nu_1 + \cos\Theta\ \nu_2$$

(Θ : leptonic analogue of the Cabibbo angle), oscillations will occur with an oscillation length

$$\ell = \frac{p\ [\mathrm{MeV}]}{4\,[m_1^2-m_2^2]\ [\mathrm{eV}^2]}\cdot 10\mathrm{m}$$

(p:neutrino momentum). This formula is already made suitable for application by use of the relevant units. The interesting aspect of it is that the oscillation lengths for $|m_2^2-m_2^2| \sim 1\ \mathrm{eV}^2$ are well within the range of experiments which could be done in the laboratory.

Besides the $\nu_e-\nu_\mu$ oscillations there exist in general also oscillations between the electron or muon neutrino and the new neutrinos $\bar{\nu}_{eR}$ and $\bar{\nu}_{\mu R}$, e.g., $\nu_{\mu L} \leftrightarrows \bar{\nu}_{\mu L}$. These oscillations are especially interesting since they violate lepton number conservation. One may expect that the main part of the mass is generated by the lepton number conserving part of the weak interaction, and the lepton number violating part is small compared to it, in which case the $\nu_\mu-\bar{\nu}_\mu$ mass matrix takes the form

$$\begin{pmatrix} m & \varepsilon \\ \varepsilon & m \end{pmatrix}$$

where ε is the lepton number violating mixing term (note that in general in a vectorlike theory of leptons and quarks it is desirable to have a violation of lepton number conservation[31]). This mass matrix is very similar to the $K^o - \bar{K}^o$ mass matrix: the eigenstates are $\nu_1 = (\nu_\mu+\bar{\nu}_\mu)/\sqrt{2}$ and $\nu_2 = (\nu_\mu-\bar{\nu}_\mu)/\sqrt{2}$. Thus in case of the $\nu_\mu-\bar{\nu}_\mu$ or $\nu_e-\bar{\nu}_e$ mixing one has a reason why the mixing angle might be 45°; no such reason exists for the $\nu_e-\nu_\mu$ mixing.

In many vectorlike models the new helicity components of the neutrinos are coupled by weak currents to new charged leptons (see,e.g., the models A,B,C,D,E,F). Thus, at a distance where the oscillations start to become relevant, a new charged lepton (M^+) would be produced in high energy ν_μ nucleus scattering. This leads in particular to a "wrong" charge signal: $\nu_\mu + \text{hadron} \rightarrow M^+ + \text{hadrons} \rightarrow \mu^+ + \text{neutrinos} + \text{hadrons}$.

It remains to be seen if neutrino oscillations of the type discussed here take place in nature, and experiments should be performed designed to search for the oscillations.

13. PHENOMENOLOGY OF VECTORLIKE HADRONIC CURRENTS

A) - Weak decays of the new quark flavours

One way to see the new weak currents would be to look for them in the weak decays of the heavy quark flavours. For example, in the six-quark scheme A the charm flavour decays via a vector current. It seems feasible to measure the chirality of the charm changing weak current in the near future[44).

B) - New right-handed currents in neutrino scattering

In many vectorlike models (A,B,C,D,E,F) there exist new right-handed currents acting on the "valence" quarks of the nucleon u and d. These can be discovered in neutrino or antineutrino scattering. For example, the new current $(\bar{u}b)_R$ could be seen in antineutrino scattering, once the threshold for producing bottom flavoured hadrons is passed.

Neglecting the antiquark content of the nucleon, the cross-section for the reaction $\bar{\nu}_\mu$+nucleon$\to\mu^+$+hadrons can be written as $d\sigma/dy = \text{const. } E\ (1-y)^2$. The new current $(\bar{u}b)_R$ produces a new component in the y distribution:

$$1/E\ d\sigma/dy = \text{const.}\ \left[(1-y)^2 + (\text{threshold factor})\cdot 1\right]. \tag{13.1}$$

The ratio $\sigma^{\bar{\nu}\mu^+}/E$ increases by a factor 4, once all threshold factors are passed.

Recently it has been suggested[45] that a reliable way to take into account the relevant threshold factors is to replace the scaling variable x in the deep inelastic antineutrino nucleon scattering by the variable $\hat{x} = x + m_b^2/2\nu$ (provided $m_b^2 \gg M^2_{\text{nucleon}}$), in which case the threshold effects can be parametrized by only one parameter (m_b). In Appendix II we reformulate the arguments in favour of the variable $\hat{x}$ both within the parton and light cone approach. The use of the variable $\hat{x}$ leads to a rather gentle onset of the new process. If we assume that the observed y anomaly is due to a new $(\bar{u}b)_R$ current, one finds $m_b \simeq 5$ GeV. (Here we have set a possible mixing angle, e.g., the b-h mixing angle Θ'' in scheme F, to zero).

At present it is not clear if the y anomaly is due to a new weak current. If this turns out to the case, and if $m_b \simeq 5$ GeV, one expects the first $\bar{b}b$ state to show up in e^+e^- annihilation in the region 8...10 GeV, just outside the range of the present e^+e^--machines.

A new right-handed current, acting on the d quark, e.g., $(\bar{t}d)_R$, could be seen most clearly in ν_μ scattering. However, here the effect would be less conspicuous as the corresponding effect due to the $(\bar{u}b)_R$ current in $\bar{\nu}_\mu$ scattering [a term proportional to $(1-y)^2$ is added to a constant term]:

$$1/E \cdot d\sigma/dy = \text{const.}\ (1+(\text{threshold factor})\ (1-y)^2). \tag{13.2}$$

Thus far there exists no indication in the experiments for the existence of such a current.

D) - New right-handed weak currents and the $\Delta I = 1/2$ rule

New currents add new terms to the non-leptonic weak Hamiltonian which may be relevant for the non-leptonic weak decays of strange (and charmed) particles. However, I expect a four-quark operator like $\bar{s}h\bar{h}d$ (h:heavy quark) to contribute very little to the non-leptonic strange particle decay, due to the absence of heavy quark flavours in normal hadrons. Recently it has been emphasized that new right-handed currents involving heavy quarks can lead in QCD to the appearance of a new term in the non-leptonic Hamiltonian which may be important and even dominating, namely the quark bilinear $\bar{s}\sigma_{\mu\nu} G^{\mu\nu} d$ ($G^{\mu\nu}$: gluon field strength, summation over colour understood) [46,47]. For example, in schemes A,B,D,E the new right-handed current $(cs)_R$ is present, which generates the new term

$$H^{\text{nonlept.}} = -\, G/\sqrt{2}\, \sin\Theta' \cdot m_c \cdot \frac{g}{4\pi^2}\, \bar{s}\sigma_{\mu\nu} G^{\mu\nu} \frac{1+\gamma_5}{2} d + \text{h.c.} \tag{13.3}$$

(it corresponds to the decay s→d+gluon, g:gluon-quark coupling constant). The important features of this new term are the following:

(i) It contains no heavy quark fields. Thus its matrix elements between normal hadrons can be big.

(ii) It is enhanced by m_c.

(iii) It is pure $|\Delta I|=1/2$ and has all the desired properties for a non-leptonic weak Hamiltonian[47].

A consistent picture of all non-leptonic decays of strange particles emerges if one assumes that the term (13.3) dominates the decays[46]. Specific wave function calculations performed recently within the MIT bag model indicate that this assumption may be consistent[48].

We emphasize that the term (13.3) is enhanced by the large value of m_c. A similar enhancement would not occur in the case of charmed particle decay. Thus charmed particles are supposed to have a large semi-leptonic decay mode, in agreement with recent estimates on the basis of the observed dimuon events in ν_μ hadron scattering.

Let me add: It may be that the $|\Delta I| = 1/2$ rule and other mysteries of the non-leptonic weak decays are due to a new term of type (13.3). However, it may not be due to the $(\bar{c}s)_R$ current, and not due to charmed intermediate states as in (13.3), but due to new currents and other intermediate states [see, for example, Ref. (49)].

Which heavy quark is really involved depends on the specific scheme of the weak interactions, and future investigations will shed more light on this question.

E) - Problems with $|\Delta S|=1$ neutral currents

In all vectorlike models involving SU_2^W doublets and singlets there is always the threatening danger to generate $\Delta S = 1$ neutral currents. For example, in model C one has to impose (artifically?) the constraint s"=s or s"=d. Such constraints seem unnatural, and we conclude that this model can hardly be regarded as realistic. The situation is even worse for model D. Even if we set $\Theta''=0$ or $\pi/2$ in this model, the higher order weak interactions will always generate a mixing between d and s (or order α):

$$\Theta'' \simeq \Theta' \cdot \frac{\alpha}{\pi} \cdot \frac{m_d \cdot m_c}{m_s^2} \simeq \Theta' \cdot \frac{\alpha}{\pi}, \qquad (13.4)$$

implying a $|\Delta S|=1$ neutral current of order $\alpha/\pi \sin\Theta'$, which is too large to be tolerable. Thus we conclude: the models C and D are unrealistic. There are, however, no problems of this sort, if both d_R and s_R are SU_2^W singlets. This is, for example, the case in models F and G.

14. PHENOMENOLOGY OF THE NEUTRAL CURRENT

A) - Hadronic neutral currents

In schemes based on SU_2^W doublets only the neutral current is a vector current, provided the gauge group is SU_2xU_1. This implies in particular $\sigma^{\nu\nu}=\sigma^{\nu\nu}$. The experimental results reported recently[50] are in disagreement with this prediction. Thus SU_2xU_1 vectorlike models based on SU_2^W doublets only are ruled out.

Let us consider the various schemes described in the Table. Especially, we concentrate on the ratio $\hat{R} = \sigma^{\bar{\nu}\bar{\nu}}_{hadr.}/\sigma^{\nu\nu}_{hadr.}$, which is measured to be ~0.5...0.7. The effective Hamiltonian describing the weak interaction is

$$H^{weak} = \frac{4G}{\sqrt{2}} (j^+_\mu j^-_\mu + \rho\cdot j^n_\mu \cdot j^n_\mu), \quad \rho = \frac{M_W^2}{M_Z^2 \cos^2\Theta} \qquad (14.1)$$

($j^+_\mu = \bar{u}\gamma_\mu(1+\gamma_5/2)d+\ldots$, j^n_μ : see the Table). The parameter ρ describes the strength of the neutral current coupling.

In the simple doublet symmetry breaking scheme discussed by Weinberg[16] one has $\rho=1$. We assume a more general symmetry breaking mechanism and keep ρ as a free parameter. The ratio $\hat{R}$ is, of course, independent of ρ.

In the following discussion of models, we apply the naive scaling assumptions for neutrino scattering and set all antiquark densities in the nucleon to zero. We use the definition $R^\nu = \sigma^{\nu\nu}_{hadr.}/\sigma^{\nu\mu}_{hadr.}$ and $R^{\bar{\nu}}=\sigma^{\bar{\nu}\bar{\nu}}_{hadr.}/\sigma^{\bar{\nu}\mu^+}_{hadr.}$; $z = \sin^2\Theta$. The various results for $\hat{R}$ are shown in Fig. 8).

Models

$$\text{A,B:} \qquad \hat{R} = 1 \qquad \text{(excluded)}$$

$$C(\Theta''=0),\ D(\Theta''=\tfrac{\pi}{2}): \quad \hat{R} = 1 \qquad \text{(excluded)} \tag{14.2}$$

$$C(\Theta''=\tfrac{\pi}{2}),\ D(\Theta''=0);\ F:$$

$$R^{\nu} = \left(\frac{7}{12}-\frac{11}{9}z + \frac{20}{27}z^2\right)\rho^2 \qquad R^{\bar{\nu}} = \left(\frac{5}{4} - 3z + \frac{20}{9}z^2\right)\rho^2$$

$$\hat{R} = \frac{5/4 - 3z + 20/9\ z^2}{7/4 - 11/3\ z + 20/9\ z^2}$$

Remarks

These models are consistent with the data for $z=0\ldots0.6$; note the rather smooth dependence of $\hat{R}$ on z, in contrast, e.g., to the Salam-Weinberg model. For $z=3/8 = 0.375$ the predictions of these models for inclusive neutrino scattering off hadrons coincide with the Salam-Weinberg model for the same value of z. The reason for this is simple: the neutral currents for both models in case $z=3/8$ are equal for d quarks, while for u quarks they are either a vector (here) or axial vector (Salam-Weinberg case). If we require $\rho=1$, consistency with the data for R^{ν} and $R^{\bar{\nu}}$ is reached for $z\simeq0.3\ldots0.4$.

$$\text{E:} \qquad R^{\nu} = \left(\frac{7}{12} - \frac{10}{9}z + \frac{20}{27}z^2\right)\rho^2$$

$$R^{\bar{\nu}} = \left(\frac{5}{4} - 2z + \frac{20}{9}z^2\right)\rho^2$$

$$\hat{R} = \frac{5/12 - 2/3z + 20/27\ z^2}{7/12 - 10/9z + 20/27z^2} \tag{14.3}$$

Remarks

Consistent with data for $z<0.1$. In this case ρ must be much different from 1 ($\rho^2\sim0.3\ldots0.5$).

G:

Within our approximations, the predictions of this model coincide with the Salam-Weinberg model generalized for arbitrary ρ:

$$R^{\nu} = (\frac{1}{2} - z + \frac{20}{27} z^2)\rho^2; \quad R^{\bar{\nu}} = (\frac{1}{2} - z + \frac{20}{9} z^2)\rho^2$$

$$\hat{R} = \frac{1/2 - z + 20/9\, z^2}{3/2 - 3z + 20/9\, z^2} \tag{14.4}$$

Remarks

Consistent with data for z=0.3...0.4. The same values give $\rho \simeq 1$.

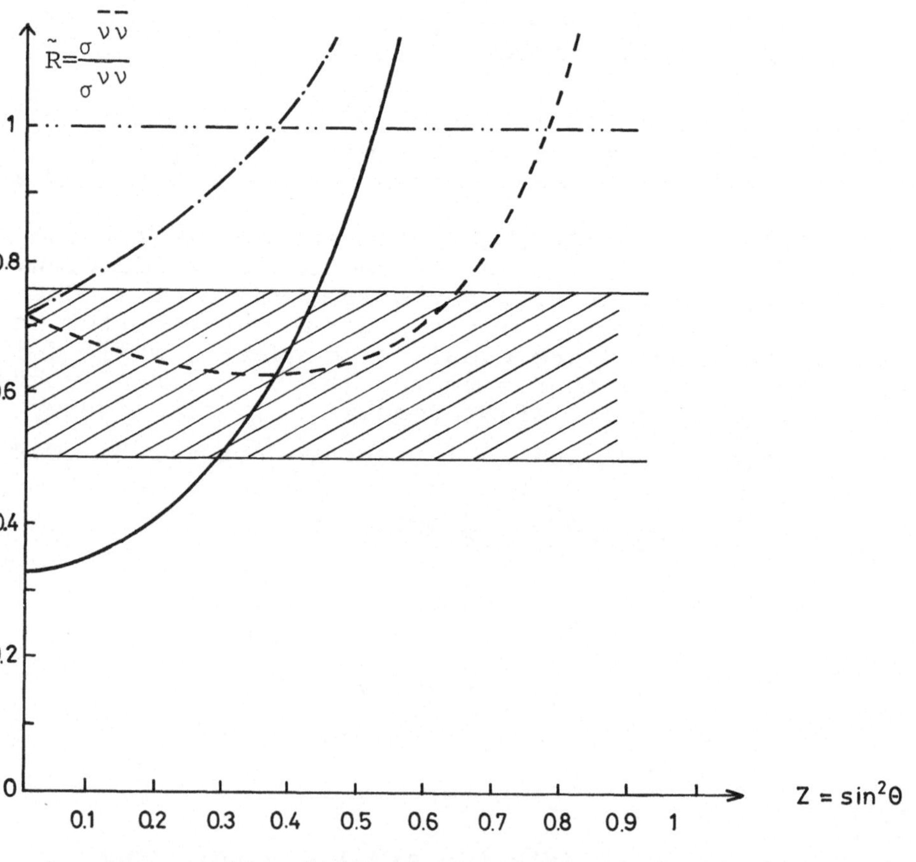

Fig. 8 : We use the following notation: (Dashed area judicates exp. data)

———: Salom Weinberg model and model G

-----: Models of the type $\begin{pmatrix} u & c \\ d_e & s_c \end{pmatrix}_L \begin{pmatrix} u \\ b \end{pmatrix}_R$ (see C,D,F)

-.--·-: Models of the type $\begin{pmatrix} u & c \\ d_c & s_c \end{pmatrix}_L \begin{pmatrix} t \\ d \end{pmatrix}_R$ (see E)

-··-··:Models with neutral vector current.

Conclusions

The hadronic neutral current data are consistent with either model G (no new right-handed couplings to the valence quarks u,d), or with the case E (new right-handed coupling to the d quark), or with the cases where a new right-handed coupling to the u quark enters (but none to the d quark).

15. PARITY VIOLATION IN ATOMIC PHYSICS

In the immediate future it will be feasible to test the possibly existing parity violation due to the neutral current in atomic physics[51]. Very soon one expects the first experimental results from the groups at Seattle and Oxford, investigating $^{209}_{83}Bi$.

The effective Hamiltonian relevant for atomic physics is in $SU_2 x U_1$ models:

$$H_{p.v.} = \frac{G}{\sqrt{2}} Qw \cdot \rho \{ \frac{\vec{\sigma} \cdot \vec{p}}{2m_e} \delta^3 (\vec{r}) \} (I^{weak}_{3(e_L)} - I^{weak}_{3(e_R)}), \quad (14.5)$$

where p is the electron momentum, and the effective charge Q_W is defined by $Q_W = 4\langle nucleus | \int j^n_o d^3x | nucleus\rangle$.

Table 4. In the various models, we find (Z:nuclear charge, N:neutron number).

Model	Q_W	$I^W_3(e^-_L) - I^W_3(e^-_R)$	Q_W for $^{209}_{83}BI$ and z=1/3
A,B,C (θ''=O)	Z(2-4z)-2N	O	-197
D($\theta''=\pi/2$)	Z(2-4z)-2N	-1/2	-197
D(θ''=O),F	Z(3-4z)	-1/2	+138
E	Z(-4z)-3N	-1/2	-489
G	Z(1-4z)-N	-1/2	-154

The predictions of model G (with $\rho=1$) for the parity violation in atomic physics are identical with the prediction of the Salam-Weinberg model. We characterize all models as follows. The models A,B,C($\Theta''=0$) give no effect. Model D($\Theta''=\pi/2$) and the models E and G give qualitatively the same results as the Salam-Weinberg model. The models D($\Theta''=0$) and F, i.e., those models involving the $(\bar{u}b)_R$ current, give roughly the same magnitude for the parity violation as the Salam-Weinberg model but the opposite sign. We emphasize the importance of the sign; it will be very important in discriminating among the various models.

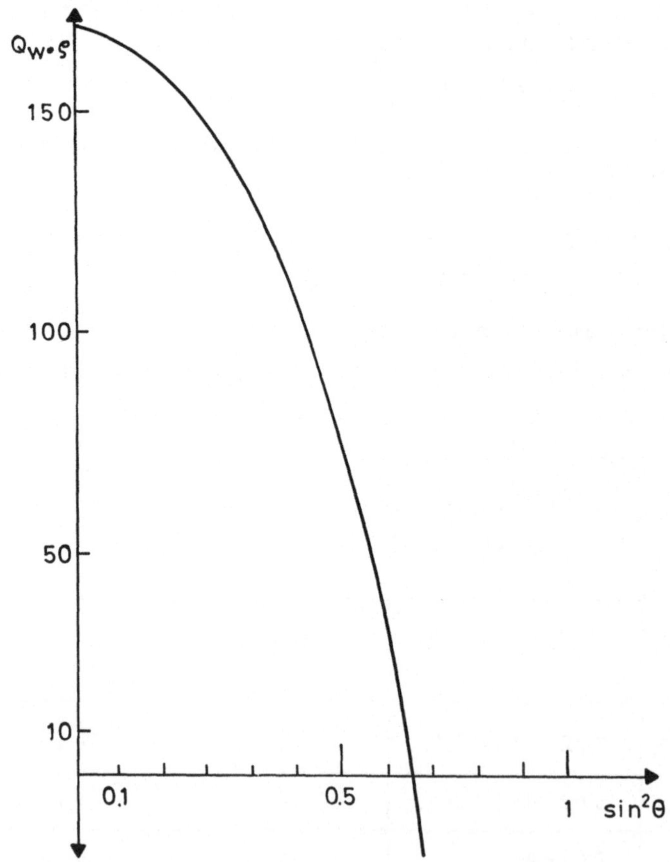

Fig. 9 : The quantity $Q_W \cdot \rho$ for models with the $(\bar{u}b)_R$ current, but without $(Fd)_R$ current. The strength parameter ρ is adjusted as a function of $\sin^2\Theta$ such that $R^\nu=0.28$.

In Fig. 9 we have displayed the quantity $Q_W \cdot \rho$ in the models of the type $\begin{pmatrix} u & c \\ \bar{d}^c & s^c \end{pmatrix}_L$, $\begin{pmatrix} u \\ b \end{pmatrix}_R \ldots$ (vectorlike models without the $(td)_R$ current). According to Fig (9) such a model is consistent with the neutrino production data for $0<\sin^2\theta<0.65$. The strength parameter ρ is adjusted such that R^ν agrees with the experiments: $R^\nu=0.28$. As displayed in Fig. (9), the quantity $Q_W \cdot \rho$ is very sensitive to $\sin^2\theta$. If we choose e.g. $\sin^2\theta=0.60$, one has $Q_W \cdot \rho=76$, i.e. a value significantly smaller in absolute magnitude as the value -154, obtained in the Salam-Weinberg model (or model G) for $\sin^2\theta=1/3$, and with the opposite sign. Qualitatively one can say that the addition of the $(\bar{u}b)_R$ current allows to choose $\sin^2\theta=0.6 \cdot\cdot 0.65$, in which case the atomic physics effect becomes suppressed by a factor 2...3 compared to the Salam-Weinberg model, and acquires the opposite sign.

The present experimental situation is rather unclear, but not inconsistent with such a situation[52].

16. SU_2^W TRIPLETS AND SU_3^W

The fact that the previously considered models D($\Theta''=0$) and F give the opposite sign for the parity violation in atomic physics as the Salam-Weinberg model depends crucially on the assignment of the right-handed electron to a SU_2^W singlet $(I_3^W(e_L^-)-I_3^W(e_R^-)=-\frac{1}{2})$. If we place the right-handed electron into a SU_2^W triplet by introducing new massive leptons:

$$\begin{pmatrix} \nu_e \\ e^- \end{pmatrix}_L ; \quad \begin{pmatrix} M^+ \\ N_- \\ e^- \end{pmatrix}_R \qquad (16.1)$$

(N, M^+: massive new leptons), the difference $I_3^W(e_L^-) - I_3^W(e_R^-)$ becomes $+\frac{1}{2}$, i.e., the sign of $H_{p.v}$ changes. In this case the sign of the parity violation in atomic physics is identical to the one obtained in the Salam-Weinberg model.

The triplet assignment for e_R^- will be forced upon

us within SU_2xU_1 models, if the sign of Q_W turns out to be negative and if the $(\bar{u}b)_R$ current exists.

If e_R^- belongs to a SU_2 triplet, the neutral current interaction of the electron is changed rather profoundly. In particular the differential cross-sections for ν_μ-e^- scattering are

$$\frac{d\sigma^\nu}{dy} = \frac{2G^2m_e}{\pi} E_\nu\cdot\rho^2 \left|(-\tfrac{1}{2}+z)^2+(-\lambda+z)^2(1-y)^2\right| \qquad (16.2)$$

$$\frac{d\sigma^{\bar{\nu}}}{dy} = \frac{2G^2m_e}{\pi} E_{\bar{\nu}}\ \rho^2 \left|(-\tfrac{1}{2}+z)^2(1-y)^2+(-\ \ +z)^2\right|$$

$z=\sin^2\Theta$, $\lambda=1$ (for comparison: the corresponding expressions in the generalized S-W model are obtained for $\lambda=0$). In order to free oneself from the parameter ρ, we consider especially the ratios

$$\overset{(-)}{r} = \frac{\sigma^{\overset{(-)}{\nu}_\mu e^-} / E_{\overset{(-)}{\nu}} \dfrac{2G^2m_e}{\pi}}{R^\nu_{hadr.}} \qquad (16.3)$$

For example, one finds, using $\frac{2G^2m_e}{\pi} E_\nu = 1.7\cdot10^{-41} E_\nu [\mathrm{GeV}]\,\mathrm{cm}^2$:

$$\bar{r} = \frac{(-1/2+z)^2\cdot1/3+(-1+z)^2}{7/12-11/9z + 20/27\ z^2}; \qquad (16.4)$$

z	0	0.1	0.2	0.3	0.4	0.5	0.6
$\bar{r}$	1.86	1.84	1.82	1.78	1.71	1.59	1.40

Experimentally, $R^\nu_{hadr.}$ is ~0.26. For example, for $R^\nu_{hadr.}=0.26$ and z = o.3...o.5, we obtain $\sigma^{\nu_\mu e^-} \simeq 7.10^{-42}$ $\mathrm{cm}^2E_{\bar{\nu}}$ |GeV|. Similarly one has $\sigma^{\nu_\mu e^-} \simeq 3.10^{-42}$ $\mathrm{cm}^2E_{\bar{\nu}}$[GeV]. Typically, these cross-sections are a factor 2.5

higher than the upper limits reported by the Gargamelle group[53)]. On the other hand they are not in disagreement with the cross-sections reported by the Aachen-Padova group[54)]. Thus at present the triplet possibility for e_R^- cannot be excluded.

Suppose the triplet hypothesis for e_R^- is correct. In this case, the following model H for leptons and quarks can be constructed.

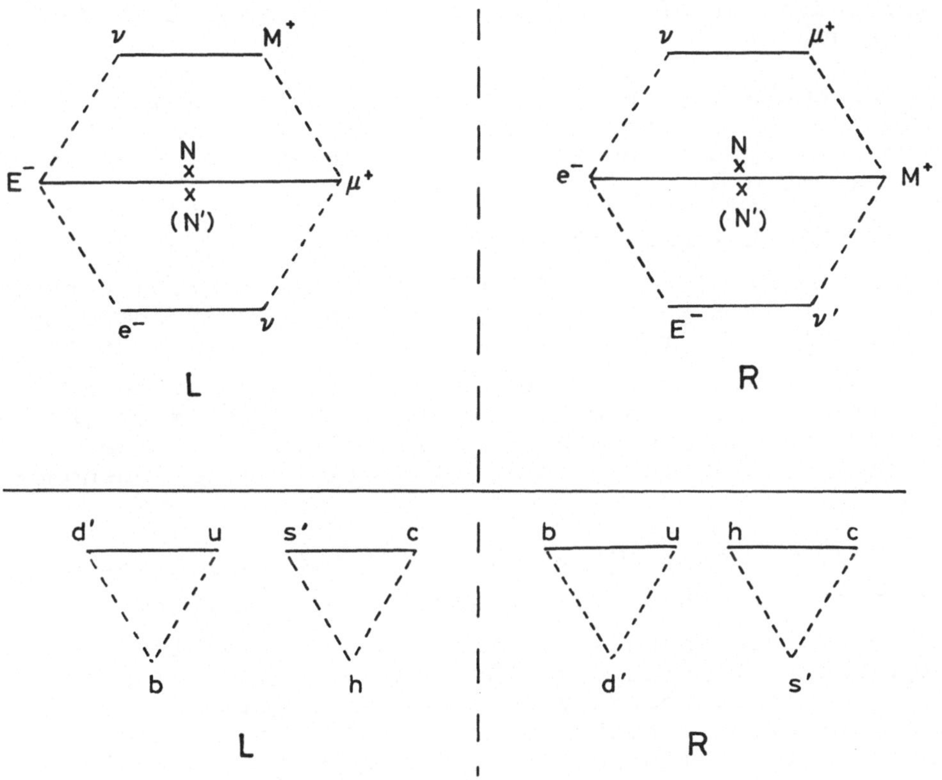

Fig. 10 : SU_3 scheme for leptons and quarks

This scheme is the smallest vectorlike extension of the scheme (16.1) incorporating the e-μ universality such that all leptons transform either as doublets or triplets. It contains two new charged leptons and two neutral ones (N,ν'). Note: $\nu_L = \nu_e$, $\nu_R = \bar{\nu}_\mu$.

The scheme H is very interesting, since it can immediately be incorporated into a SU_3 scheme, where the group SU_2xU_1 is regarded as a subgroup[49,55] of SU_3. If one adds one neutral lepton (in parenthesis), the leptons transform as an octet, while the quarks transform as triplets. The correct electric charges of the leptons and quarks follow as the consequence of the octet character of the leptons and the triplet character of the quarks.

We emphasize that in the scheme H the parity violation of the weak interaction is very simple. One notes easily that the scheme is invariant under a combined space and U spin reflection.

Within the SU_3 scheme the SU_2xU_1 mixing angle (unrenormalized) is 60° ($\sin^2\theta=3/4$); the neutral current would then be a pure axial vector. This is excluded by experiment; thus rather large renormalization effects must exist such as to give $\sin^2\theta<0.65$ as required by experiment. This can only happen if the masses of the unifying bosons are rather heavy implying that the interactions caused by them are negligible for phenomenology.

The SU_3 model (Fig. 10) is not the only model one can construct. There exists also the possibility to assign the righthanded electron to a SU_2 doublet[49] (Fig. 11). In this case the leptonic neutral current would couple vectorially to electrons, independent of renormalisation effects. There would be no parity violation in atomic physics induced by the leptonic neutral current (It can only be induced by the axial vector part of the hadronic neutral current).Since the latter does not act coherently on nucleons, the effect would be much smaller than the leading coherent effect, described by the Hamiltonian eq. (14.5).

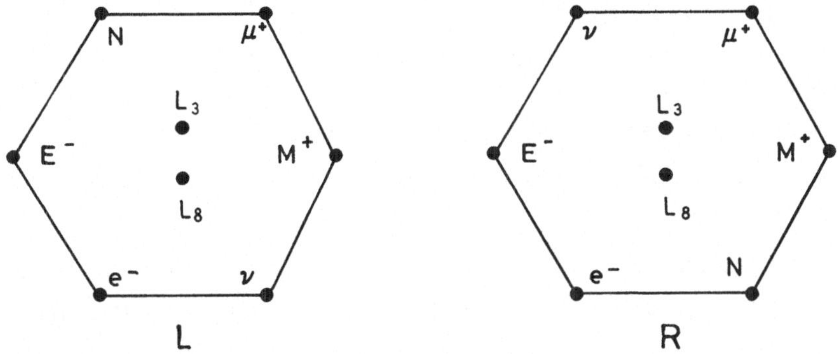

Fig. 11 . Another possible SU_3-octet assignment of the leptons.

17. UNIFIED THEORIES OF THE WEAK, ELECTROMAGNETIC AND STRONG INTERACTIONS

As we have seen above, it is possible to construct realistic models of the weak and electromagnetic interactions (e.g. the one based on the gauge group SU_3), which lead to a quantization of the electric charge. However even in such a case the situation is not yet satisfactory. One is still dealing with a gauge group of the type $SU_3^c \times G^{w-e}$ (w-e: weak-electromagnetic) consisting of at least two factors. It would be attractive to unify all interactions within one simple group. Several models of this type have been discussed by various authors in the past, and I ask the interested reader to consult the references (26,27,28). We emphasize one aspect of those theories, namely the problem of baryon number conservation or the stability of the proton. As one might expect from a theory which unifies the color and flavor group of the quarks and the flavor group of the leptons in one simple group, the conservation of baryon number or the stability of the proton with respect to its decay into leptons is a nontrivial aspect. Indeed, in many unified theories the decay "proton ⟶ leptons (e.g. $p \to \nu + \pi^+$) occurs, and one is forced to introduce very heavy gauge bosons (" intermediate vector baseballs") in order to keep the proton lifetime largen than the observed limit ($\sim 10^{30}$ years).

a) Is nature exceptional?

What kind of strategy should one follow in order to select a unifying group of all interactions? This group must contain the color group SU_e^c as one of its subgroup, likewise the gauge group of the weak and electromagnetic interactions acting on the various flavor indices (according to the discussion above this group might also be SU_3); thus one might look for gauge groups which provide "natural" embeddings of SU_3 groups. Such groups are the exceptional groups, and we shall study below various lepton - quark schemes based on the exceptional groups (G_2,F_4,E_6,E_r,E_8). The pioneering work in this direction has been done by the Yale group[56-59]. For details on the group theory aspects we refer to the mathematical literature[60]. A short discussion of the octonion algebra which is relevant for the structure of the exceptional groups can be found in the Appendix III. Some basic properties of the exceptional groups are denoted in Appendix IV.

b) <u>Lepton - Quark Schemes based on the exceptional groups</u>

<u>The group G_2:</u>

This group being of rank 2 is, of course, too small in order to be useful as a gauge group for a unified theory of the strong, electromagnetic, and weak interactions. Nevertheless a gauge theory based on G_2 is an interesting example since one can display here in a simplified manner some of the problems one is dealing with in case of the other exceptional groups.

The group G_2 has as its maximal subgroup of maximal rank the group SU_3. Let us identify this group with the color group SU_3^c. We suppose that the gauge group G_2 is broken such that SU_3^c is left as an unbroken subgroup:

$$G_2 \xrightarrow{\text{symmetry breaking}} SU_3^c . \tag{17.1}$$

The basic representation 7 transforms under SU_3^c as $7=3+\bar{3}+1$, i.e. it consists of a quark triplet q, antiquark triplet $\bar{q}$ and a lepton. (Majorana lepton). The adjoint representation decomposes as $14 = 8+3+\bar{3}$. Besides the color generators (coupled to the gluons) one has six generators which have both the quantum numbers of a diquark (qq) and a leptoquark ($\bar{q}\ell$), i.e. a diquark can annihilate into an antiquark and a lepton via the diagram displayed in Fig. 12.

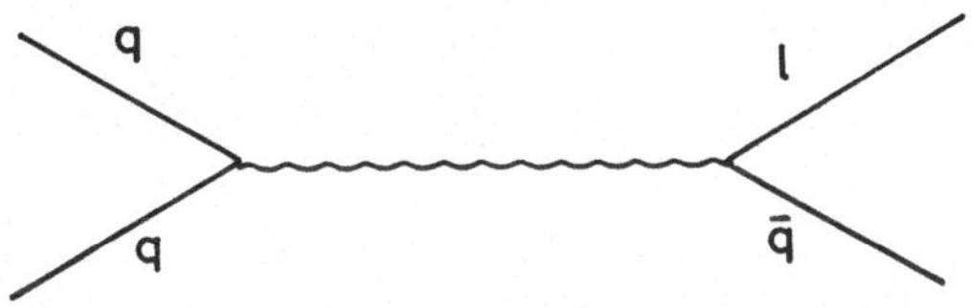

Fig. 12 . Annihilation of a diquark into an antiquark and a lepton in the G_2 scheme.

This feature (which is a general feature of the exceptional groups) implies not only that there exists no baryon number generator in the G_2 scheme, but that a three quark color singlet system like the proton can decay in second order of the gauge coupling. The only way to suppress this decay is to choose a very high mass M for the corresponding gauge boson; the experimental limit on the proton lifetime of $\sim 10^{30}$ years implies $M \gtrsim 10^{16}$ GeV.

The only way to avoid the second order proton decay in the G_2 scheme is to introduce besides the "light" quarks q a set of heavy quark Q as well as another lepton (fermion doubling). In this case the leptons can be interpreted as one Fermi - Dirac lepton, and we can form two 7 representations of G_2 as follows:

$(7)_1=(q,\bar{Q},\ell)$, $(7)_2=(Q,\bar{q},\bar{\ell})$. Since we can arrange the scheme such that the overall fermion number, i.e. the number of q states minus the number of Q states plus the number of leptons, is conserved, the second order proton decay as discussed above does not occur. Only the decay of the proton into three leptons is allowed; such a decay can only occur in rather high order of the gauge coupling, and essentially no constraints on the masses of the unifying gauge bosons follow.

Higher exceptional groups:

All exceptional groups contain the subgroup G_2, and therefore the color group SU_3^c. In Table 5, we display the decompositions of the various exceptional groups into SU_3^cx flavor group as well as indicate the transformation properties of the basic and adjoint representations.

In the previous section we have argued that a possible description of the strong, electromagnetic and weak interactions is one based on the group $SU_3^c \times SU_3^{w-el.}$. Needed for such a description were six quark flavors and eight leptons:

$$\text{fermions} = 2x(3^c,3)+(1,8)+2x(\bar{3}^c,\bar{3})+(1,8) \quad (17.2)$$

Below we shall discuss the embedding of the $SU_3 \times SU_3$ scheme into the various schemes based on the exceptional groups. We emphasize that all gauge theories based on the exceptional groups are anomaly free. This is easy to see for all exceptional groups except E_6, since those have only real representations. Some of the representations of E_6 are complex (e.g. 27,351); however also if these representations are involved as fermion representations, the corresponding theory is anomaly free[61].

The group F_4: F_4 is the smallest exceptional group which contains the desired flavor group SU_3. Since the basic representation 26 contains three quarks and antiquarks and a lepton octet, we need two 26 representations for a minimal realistic scheme (altogether six quarks and antiquarks, one lepton and one antilepton octet).

Table 5

Group G	$G \supset SU_3^C \times ?$	Basic representation and decomposition under subgroup	Adjoint representation and decomposition under subgroup
G_2	SU_3^C	$7 = 3^C + \bar{3}^C + 1$	$14 = 8^C + 3^C + \bar{3}^C$
F_4	$SU_3^C \times SU_3$	$26 = (3^C,3) + (\bar{3}^C,\bar{3}) + (8^C,1)$	$52 = (8^C,1) + (1^C,8) + (3^C,\bar{6}) + (\bar{3}^C,6)$
E_6	$SU_3^C \times SU_3 \times SU_3$	$27 = (3^C,3,1) + (\bar{3}^C,1,\bar{3}) + (1^C.\bar{3}.3)$	$78 = (8^C,1,1) + (1^C,8,1) + (1^C,1,8) + (3^C,\bar{3},\bar{3}) + (\bar{3}^C,3,3)$
E_7	$SU_3^C \times SU_6$	$56 = (3^C,6) + (\bar{3}^C,\bar{6}) + (1^C,20)$	$133 = (1^C,35) + (8^C,1) + (3^C,\overline{15}) + (\bar{3}^C,15)$
E_8	$SU_3^C \times E_6$	$248 = (1^C,78) + (8^C,1) + (3^C,27) + (\bar{3}^C,\overline{27})$	

Thus we obtain just the fermion content as required:

$$26 + 26 = (3^c,3)+(3^c,3)+(1.8) + \text{antiparticles}$$
$$= (u,d',b)+(c,s',h) + \text{leptons} + \text{antiparticles}. \quad (17.3)$$

As in the G_2 scheme, the unifying interactions in the F_4 scheme cause the decay of the proton into leptons in second order of the gauge coupling, thus superheavy bosons are necessary. Analoguously as above this decay can be avoided by doubling the fermion representation, i.e. by introducing four 26 representations (twelve quark flavors and two lepton octets).

The group E_6: Here the basic representation is 27 dimensional. The minimal scheme one can construct is essentially equivalent to the F_4 scheme discussed above, if we interpret the direct sum SU_3 of the two SU_3 groups in the flavor group $SU_3 \times SU_3$ as the gauge group for the weak and electromagnetic interactions. In this case the 27 representation decomposes into $(1^c,8)+(1^c,1)+$ $+(3^c,3)+(\bar{3}^c,\bar{3}^c)$; (one new neutral lepton (SU_3 singlet) has to be added to the fermion content of the F_4 scheme). Two 27 representations of E_6 provide us with six quarks and one lepton octet. A different possibility to construct a gauge theory based on the group E_6 has been discussed in Ref. (58).

The group E_7:[56,57) The fermion representation is 56 dimensional, the flavor group is SU_6. We interpret the group $SU_3^{w\text{-}elm}$ as the SU_3 subgroup of the flavor group, in which case the 56 representation decomposes under $SU_3^c \times SU_3^{w\text{-}elm}$ as follows:

$$56 = 2 \times \left[(3^c,\bar{3})+(\bar{3}^c,3)+(1,8)+2\cdot(1.1)\right]. \quad (17.4)$$

We obtain the desired fermion content (six quarks, one lepton octet) plus four neutral lepton states. The remarkable feature of the E_7 scheme is that here we need only one irreducible fermion representation. Of course, within this scheme superheavy gauge bosons are required in order to cure the problems of proton decay. The second order proton decay can be avoided by doubling the fermion representation.

The group E_8: E_8 is the only simple group, for which the basic and the adjoint representation coincide. Thus it is required that the boson and fermion representation of the E_8 gauge theory have the same structure with respect to the internal symmetry group, and E_8 is therefore especially suited for the formulation of a supersymmetric theory of quarks and leptons.

In case of E_8 the flavor group is E_6. The fermion representation decomposes under $SU_3^c \times E_6$ as follows:

$$248 = (1^c,78)+(8^c,1)+(3^c,27)+(\bar{3}^c,\overline{27}), \tag{17.5}$$

thus the fermion content of the theory consists of 78 leptons, 27 quark flavors and eight color octet "quarks". The latter are the fermion counterparts of the color octet gluons. Note that these color octet quarks are singlets under the flavor group, i.e. they would not participate in the electromagnetic and weak interactions.

The color octet quarks as well as 23 of the "normal" quark flavors must be relatively heavy: the observed hadron spectrum can be described in terms of only four quark flavors.

Since the flavor group E_6 is of rank 6, there are many possibilities to associate the generators of the electric charge with one of the selfadjoint generators of E_6. Consequently there exists a considerable amount of freedom to assign the electric charges to the various quarks and leptons. In order to be specific, let us consider a particular scheme for the electromagnetic and weak interactions. We start from the subgroup $SU_6 \times SU_2$ of E_6. The decomposition of the basic representation of E_8 under the subgroup $SU_3^c \times SU_6 \times SU_2$ is:

$$\begin{aligned}248 = {} & (3^c,6,2)+(\bar{3}^c,6,2)+(1^c,1,3)+(1^c,35,1)+ \\ & +(8^c,1,1)+(1^c,20,2)+(3^c,15,1)+(\bar{3}^c,\overline{15},1).\end{aligned} \tag{17.6}$$

Let us further assume that the SU_3^{w-elm} subgroup needed for the description of the weak and electromagnetic interactions is an SU_3 subgroup of SU_6. The 248 representation can be decomposed under $SU_3^c \times SU_3^{w-elm}$ as follows:

$$248 = 7\cdot[(3^c,3)+(\bar{3}^c,\bar{3})]+ 8\cdot (1^c,8) + 14\cdot(1,1) + (8^c,1) + (3^c,\bar{6}) + (\bar{3}^c,6). \qquad (17.7)$$

Thus we obtain:

21 quark flavors with the charges 2/3, -1/3, -1/3=

6 quark flavors transforming as the 6 representation of SU_3^{w-elm}: these quarks have unconventional electromagnetic and weak charges - in particular the charge 4/3 appears.

14 SU_3 singlet leptons: they are neutral and uncoupled from the "normal" weak interaction.

8 SU_3^{w-elm} octets of leptons, among those 32 leptons of charge -1.

We emphasize that within the E_8 scheme the second order proton decay still occurs; in order to avoid it we have to double the representation (introducing another 248 representation of fermions).

Obviously the E_8 scheme is a very complex system, containing about eight times more fermions as the 32 fermions, needed for the contemporary phenomenology of hadrons and leptons. At the present time it might look ridiculous to view such a scheme as a realistic one. However if future experiments should reveal the existence of many new quark flavors and leptons as well as of new interactions, the E_8 scheme might have to be regarded as a serious possibility to describe the pattern of leptons and quarks in nature.

Outlook:

In these lectures I could only discuss a few aspects of present day weak interaction theories and of their possible understanding within a larger scheme of interactions, including the electromagnetic and strong interactions. One of those aspects I tried to emphasize were the proliferation of the number of quarks and leptons, which is a feature of many schemes. How can a physicist who is trying to reduce the number of elementary constituents of nature to a minimal number live with this fact? One way to understand the large number of leptons and quarks would be to interpret them in some sense as composite objects. However at present there exists not evidence for a substructure of the fermions: no excitations of leptons and quarks with higher angular momenta and different parity have been observed. The fact that the anomalous magnetic moments of the electron and muon agree to very high accuracy with the values predicted in QED suggests that the electron and muon have no internal structure. If a substructure of quarks and leptons exists, it is presumably one occurring at a more sophisticated level than the kind of substructures studied in atomic, nuclear and hadron physics.

Another,and perhaps more convincing way to live with a large number of "elementary" fermions is to change our attitude towards the physics of leptons and quarks. It could well be as emphasized by Abdus Salam that even on the level of the "elementary" constituents nature does the same as anywhere else: It is rich in structure, but based on very few principles. Thus, if we regard principles, e.g. the gauge principle, a particular gauge group, etc. as the "elementary" constituents of nature, we need not to worry about how many leptons and quarks exist - be it 32 or 248.

APPENDIX I

The Effective Weak Hamiltonian in the $SU_2 x U_1$ Theory

The eigenstates of the vectormeson mass matrix are the photon A_μ and the massive neutral vectormeson Z_μ, which in terms of the $SU_2 \times U_1$ eigenstates can be written as:

$$W_{3\mu} = Z_\mu \cos\Theta + A_\mu \sin\Theta$$

$$B_\mu = -Z_\mu \sin\Theta + A_\mu \cos\Theta \qquad (\Theta: SU_2 x U_1 \text{ mixing angle})$$

The interaction of the massive vectormesons with the fermion is

$$L_{int} = -g \cdot \vec{j}_\mu \vec{W}^\mu - g'/2 \cdot j_\mu^Y B^\mu$$

$$\vec{j}_\mu = \bar{f} \gamma_\mu \frac{\vec{\tau}}{2} f_L \quad (f_L\text{: lefthanded fermion doublet, } \vec{\tau}\text{: Pauli matrices}).$$

The interaction of the charged vectormesons can be rewritten by introducing the eigenstates of the electric charge $W_\mp^\mu = \frac{1}{\sqrt{2}} (W_1^\mu \mp i W_2^\mu)$ as

$$L_{int} = -{}^g/_{\sqrt{2}}\, j_\mu^+ W_-^\mu + \text{h.c.} \quad (j_\mu^+ = (\bar{\nu} \bar{e})_L + \ldots)$$

which gives in the low frequency approximation

$$H_{weak} = \frac{4G}{\sqrt{2}} (j_\mu^+ j_\mu^- + \text{h.c.}) + \text{neutral current interaction}$$

(G: Fermi constant, $\frac{4G}{\sqrt{2}} = \frac{g^2}{2M_W^2}$).

The interaction of the neutral bosons is

$$-L_{int} = +g\, j_\mu^3\, W_3^\mu + {}^{g'}\!/2\; j_\mu^Y\, B^\mu = g\, j_\mu^3 (Z^\mu \cos\Theta + A^\mu \sin\Theta)$$

$$+ \frac{g'}{\sqrt{2}}\, j_\mu^Y (-Z^\mu \sin\Theta + A^\mu \cos\Theta)$$

$$= e \cdot j_\mu^e \cdot A^\mu + \ldots$$

which gives for the electric charges

$$e \cdot Q^e = g \cdot \sin\Theta \cdot T_3 + g' \cdot \frac{Y}{2} \cdot \cos\Theta = e \cdot (T_3 + \tfrac{1}{2} Y)$$

$$e = g \cdot \sin\Theta = g' \cdot \cos\Theta$$

$$\tan\Theta = g'/g.$$

The coupling of the Z-boson is

$$g . T_3 \cos\Theta - g' \cdot \frac{Y}{2} \sin\Theta = g \cdot T_3 \cos\Theta - g' (Q_e - T_3) \sin\Theta$$

$$= T_3 (g \cos\Theta + g' \sin\Theta) - g' \sin\Theta \cdot Q_e = \frac{e}{\sin\Theta \cos\Theta} (T_3 - \sin^2\Theta \cdot Q_e)$$

$$= g/\cos\Theta \cdot (T_3 - \sin^2\Theta Q_e).$$

Thus the neutral current is

$$j_\mu^n = j_\mu^3 - \sin^2\Theta j_\mu^e.$$

The weak interaction Hamiltonian in the low frequency approximation takes the form

$$H_{weak} = \frac{4G}{\sqrt{2}} (j_\mu^+ j_\mu^- + \rho \cdot j_\mu^n j_n^\mu)$$

where $\rho = M_W^2 / M_Z^2 \cos^2\Theta$.

APPENDIX II

We describe shortly the derivation of the scaling variable $\hat{\xi}$ both in the parton approach and by using the free quark model light cone commutators.

a) Parton approach

Consider the production of a b - quark by an incident antineutrino via the reaction $\bar{\nu}_\mu + u \longrightarrow b + \mu^+$. We neglect all masses besides the b - quark mass. Thus the kinematical restriction for the b quark production is:

$$(q+\xi p)^2 = m_b^2$$

(q: current four momentum, p: proton four momentum, $\xi \cdot p$ incoming quark four momentum). Thus the process scales in the variable

$$\hat{\xi} = - \frac{q^2 - m_b^2}{2(q \cdot p)}$$

Unfortunately the naive parton argument does not give any information as to the validity range of the scaling variable $\hat{\xi}$. This question can be studied more closely in the light cone approach.

b) Light cone approach

The relevant term in the current commutator

$$[(\bar{u}(x)\gamma_\mu b(x))_R, (\bar{b}(y)\gamma_\nu u(y))_R]$$

behaves near the light cone like

$$\partial_\mu \Delta_{m_b}(x-y)\ \bar{u}(x)\gamma^\mu u(y)$$

where Δ_{m_b} is the Fourier transform of the absorptive part of the b quark propagator $\frac{1}{q^2 - m_b^2}$. All powerlike corrections to the scaling result coming from less leading light cone singularities disappear in the deep inelastic region like $M_p^2/q \cdot p$ or faster. Thus, if we choose the b quark mass to be large compared to the proton mass: $m_b^2 >> M_p^2$, the relevant effect is the replacement of q^2 by $q^2 - m_b^2$ in the kinematics. In particular the scaling variable $\xi = -\frac{q^2}{2 \cdot p \cdot q}$ changes to $\hat{\xi}$. We find: The use of the scaling variable $\hat{\xi}$ is justified in case $m_b^2 >> M_p^2$, e.g. m_m 5GeV. It is questionable, however, to apply it to the problem of charm production (here m_c is of the same order as M_p).

APPENDIX III

Octonions (Cayley Algebra)

Octonions are generalized complex numbers with seven imaginary units $e_1, e_2, \ldots e_7$ fulfilling the multiplication law

$$e_a e_b = -S_{ab} + f_{abc}\, e_c.$$

Here the tensor f_{abc} takes the values 1,0, -1 and is totally antisymmetric; it is 1 for the following combinations (in rows):

1	2	4	3	6	5	7
2	4	3	6	5	7	1
3	6	5	7	1	2	4

In particular one has

$$e_1\, e_2 = e_3 \qquad e_7\, e_1 = e_4 \qquad e_7\, e_2 = e_5$$

$$e_7\, e_3 = e_6 \qquad e_4\, e_5 = -e_3 \qquad e_5\, e_6 = -e_1.$$

The conjugate elements are defined by $\bar{e}_a = -e_a$; the norm of an octonion $= \sum_{a=o}^{7} c_a e_a$ ($e_o = 1$, c_a real numbers) is given by

$$N^2(\omega) = \omega\bar{\omega} = \bar{\omega}\omega = \sum_{a=o}^{7} c_a^2.$$

It fulfills the composition law

$$N(\omega_1\omega_2) = N(\omega_2\omega_1) = N(\omega_1)\cdot N(\omega_2).$$

The algebra generated by octonions is a non-associative composition algebra. We remind the reader that a composition algebra is defined as an algebra with an identity element and with a nondegenerate quadratic form Q such that $Q(x\cdot y) = Q(x)\cdot Q(y)$. According to the theorem of Hurwitz there exist only four different composition algebras over the fields of the real or complex numbers:

1) Real numbers (dimension 1; commutative and associative)
2) Complex numbers (dimension 2; commutative and associative)
3) Quaternions (dimension 4; associative, not commutative)
4) Octonions (dimension 8; neither associative nor commutative).

APPENDIX IV

Basic Properties of the Exceptional Groups

The group G_2. The group G_2 is the group of all automorphisms of the octonion algebra (see App. III). Since the unit element is kept invariant under an automorphism, it is clear that the basic representation of G_2 must be seven - dimensional (namely the one given by the seven imaginary units). Below we summarize some properties of G_2:

Fundamental representations: 7, 14 (adjoint representation 14).

Rank: 2

$$7 \times 7 = 1+7+7+14+20$$

The other exceptional groups (F_4, E_6, E_7, E_8) are the automorphism groups of the Jordan algebras one can construct using the matrices of the type

$$\Omega = \begin{pmatrix} \alpha & \omega_3 & \omega_2 \\ \bar{\omega}_3 & \beta & \omega_1 \\ \bar{\omega}_2 & \bar{\omega}_1 & \gamma \end{pmatrix}$$

(α, β, γ: real, $\omega_1, \omega_2, \omega_3$: octonions).

We mention the following properties of these groups.

The group F_4. Fundamental representations: 26,52,273, 1274 (adjoint representation 52).

Rank: 4

26x26 = 1+26+52+273+324

The group E_6. Fundamental representation 27,78,351,351' 2925. (adjoint representation: 78)
Note: E_6 is the only exceptional group which has complex representations, e.g. 27,351

Rank: 6

$27 \times 27 = \overline{27}+351+351'$
$27 \times \overline{27} = 1+78+650$

The group E_7. Fundamental representations: 56,133,912, 1539, 8645, 27664, 355750.
(adjoint representation: 133)

Rank: 7

56x56 = 1+133+1539+1463

$E_7 \supset E_6 \times U_1$: $56 = 1+1+27+\overline{27}$

The group E_8. Fundamental representations: 248,3875, 30 380, 147 250, 2 450 240, 6 696 000, 146 325 270, 6 899 079 264.
(adjoint representation: 248)
Note: Basic and adjoint representation coincide.
Rank: 8

248x248 = 1+248+3875+30 380+27 000

$E_8 \supset E_7 \times SU_2$: 248 = (56,2)+(133,1)+(1,3)

REFERENCES

1. O.W. Greenberg - Phys. Rev. Letters 13, 598 (1964).
2. M.Han and Y. Nambu - Phys. Rev. 139, 1006 (1965).
3. W. Bardeen, H. Fritzsch and M. Gell-Mann - in "Scale and Conformal Symmetry in Hadron Physics", R. Gatto, Ed., p. 139 (John Wiley and Sons, New York, 1973).
4. H. Fritzsch and M. Gell-Mann, Proceedings of the XVI Int. Conf. on High Energy Physics, Chicago 1972, Vol.2.
5. H. Fritzsch, M. Gell-Mann and H. Leutwyler - Phys.Letters 47B, 365 (1973);
   S. Weinberg - Phys. Rev. Letters 31, 494 (1973).
6. D. Gross and F. Wilczek, Phys. Rev. Lett. 30, 1343, (1973)
   H.P. Politzer, Phys. Rev. Lett. 30, 1346 (1973)
   G. t'Hooft, unpublished.
7. See e.g., H. Fritzsch and P. Minkowski, Nucl. Physics B 76, 365 (1974).
8. See for example: J. Drees, Proceedings of the International Neutrino Conference, Aachen 1976.
9. T. Appelquist, J. Carrazzone, H. Kluberg - Stern, and M. Roth,Yale preprint 1976 ("Infrared Finiteness in Yang-Mills Theories").
10. For a recent discussion see:
    K.G. Wilson, Proceedings of the Coral Gables Conference, Florida 1976.
    L. Sussbind, Lectures given at the 1976 Les Houches Summer School in Theoretical Physics.
11. See for example:
    A. Salam, talk given at the Int. Neutrino Conference, Aachen 1976.
12. H. Fritzsch and P. Minkowski, Nuovo Cimento 30A, 393 (1975)
    K. Johnson, lectures given at this Sommer School.
13. D.J. Gross and F. Wilczek, Phys. Rev. D9, 980 (1974).
14. See also:
    D.V. Nanopoulos and G.G. Ross, Physics Letters 58B 105 (1975).
15. See e.g. J. Taylor, Proceedings of the Int. Neutrino Conference, Aachen 1976.
16. A. Salam and J.C. Ward - Phys.Letters 13, 168 (1964);
    S. Weinberg - Phys.Rev. Letters 19, 1264 (1967).
17. S.L. Glashow, J. Jliopoulos and L. Maiani, Phys. Rev. D2, 1285, (1970).
18. See e.g. C. Bouchiat, J. Iliopoulos, and Ph. Meyer, Phys. Lett. 38B, 519 (1972).

19. G. Goldhaber et al., to be published in Phys. Rev. Letters.
20. For a recent review see:
V. Lüth, Proceedings of the Int. Neutrino Conference, Aachen 1976.
21. See e.g.
A. Benvenuti, Proceedings of the Int. Neutrino Conference, Aachen 1976.
22. J. Burmester et al., DESY preprint 76150, (1976)
23. For a recent discussion see:
M. Perl, Proceedings of the Int. Neutrino Conference, Aachen 1976.
24. A. Benvenuti, Proceedings of the Int. Neutrino Conference, Aachen 1976.
B. Barish, ibid.
25. R. Barnett, H. Georgi and H.D. Politzer, Harvard preprint 1976.
see also:
G. Altarelli, G. Parisi and R. Petronzio, Rome preprint (2/76).
J. Kaplan and F. Martin, Paris preprint PAR/LPTHE 76/18 (5176).
26. J.C. Pati and A. Salam - Phys. Rev. D8, 1240 (1973)
27. H. Georgi and S.L. Glashow - Phys. Rev. Letters 32, 438 (1974).
28. H. Fritzsch and P. Minkowski - Ann.Phys. N.Y. 93, 193 (1974)
29 H. Fritzsch and P. Minkowski - Phys. Letter 56B, 69 (1975).
30. See e.g.:
M. Kobayashi and K.Maskawa, Progr. Theoret. Physics 49, 652 (1973)
31. S. Weinberg, Harvard preprint 1976.
31. H. Fritzsch, M. Gell-Mann and P. Minkowski - Phys. Letters 59B, 256 (1975).
32. R.L. Kingsley, S.B. Treiman, F.A. Wilczek and A.Zee - Phys. Rev. D12, 2768 (1975).
33. A. De Rujulà, H. Georgi and S.L. Glashow - Phys.Rev. D12, 3589 (1975)
N.V. Krasnikov, V.A. Kuzmin and K.G. Chetyrkin - JETP Letters 22, 47 (1975).
34. S. Pakvasa, W. Simmons and S.F. Tuan - Phys. Rev. Letters 35, 702 (1975)
35. For a review see:
H. Fritzsch, Proceedings of the Int. Neutrino Conference, Aachen 1976.
36. We use the notation of Harari.
H. Harari - Phys. Letters, 57B, 265 (1975).

37. See, in this respect:
A. Halprin, P. Minkowski, H. Primakoff and S.P.Rosen - Caltech Preprint CALT-68-533 (1976).
38. T.P. Cheng - to be published in Phys. Rev.
39. J.C. Pati and A.Salam - Proceedings of the 1975 Palermo Conference;
R.N. Mohapatra and J.C. Pati - Phys. Rev. D11, 2558 (1975).
40. Y.Park and A. Yildiz - to be published in Phys. Rev. Lett.
K. Fujikawa and N. Kawamoto - DESY Preprint 76-101 (1976).
41. H. Fritzsch and P. Minkowski - Caltech Preprint CALT-68-538 (1976);
K. Fujikawa - DESY Preprint (1976);
F. Wilczek and A.Zee - Princeton University Preprint (1976).
42. See, e.g.:
R. Cowsik and J.M. McClelland - Phys. Rev. Letters, 29, 669 (1972)
J. Gunn - unpublished;
G. Marx - Proceedings of the Int. Neutrino Conference, Aachen 1976.
43. B. Pontecorvo - JETP 26, 986 (1968);
H. Fritzsch and P. Minkowski - Phys.Letters 62B, 72 (1976);
M. Gell-Mann and J.B. Stephenson - unpublished;
S. Eliezer and A.R. Swift - Phys. Rev. D10, 3088 (1974).
44. A. Buras and J. Ellis, Nucl. Phys. B111, 341 (1976).
45. R.P. Feynman, unpublished.
R.M. Barnett, to be published in Phys. Rev.
46. H. Fritzsch and P. Minkowski - Phys.Letters 61B, 275 (1976).
R.K. Ellis - to be published.
47. See, in particular:
E. Golowich and B. Holstein - Phys. Rev. Letters 35, 83 (1975).
48. J.F. Donoghue, E. Golowich and B. Holstein - Amherst Preprint (1976).
49. H. Fritzsch and P. Minkowski, Phys. Lett. 63B, 99 (1976)
50. See the talks of B. Barish, A. Benvenuti, T. Hansl, W.Y. Lee, L. Sulak at the Int. Neutrino Conference, Aachen 1976.
51. See e.g.:
C. Bouchiat, Paris preprint 1976 ("Parity violation in atomic processes"), and references therein.
J. Bernabeu and C. Jarlskog , CERN preprint 1976.

52. For a preliminary report on the experimental situation see:
P.E.G. Baird, et al., contribution to the 5th Int. Conference on Atomic Physics (Berkeley, 1976).
53. See the various talks by members of the Gargamelle group at the Int. Neutrino Conference, Aachen 1976.
54. F. Bobisut, Proceedings of the Int. Neutrino Conference, Aachen 1976.
55. See also:
J. Kandaswamy and J. Schechter, University of Syracruse preprint 1976.
56. F. Gürsey and P. Sikivie - Phys. Rev. Letters 36, 775 (1976)
57. P. Ramond - Caltech Preprint (1976).
58. F. Gürsey, P. Ramond and P. Sikivie, Yale preprint 1976 ("Universal gauge theory model based on E_6")
59. M Günaydin and F. Gürsey, J. Math. Phys. 14, 1651 (1973).
60. R.D. Schafer, Introduction to Nonanociative Algebras (Academic Press, 1966);
N. Jacobson, Exceptional Lie Algebras (M. Dekker, 1971).
61. H. Georgi and S.L. Glashow, Phys. Rev. D6, 429 (1972)

THE M.I.T. BAG, 1976 EDITION

K. Johnson

Laboratory for Nuclear Science and
Department of Physics
M.I.T., Cambridge, MA 02139, U.S.A.

INTRODUCTION

We assume (1,2) that hadrons consist of colored quarks and gluons, and that the dynamics is described by the standard non-Abelian gauge field theory. However, we <u>also</u> assume that space exists in two phases. In the first phase, I, ordinary space, colored fields are not carried on the spatial points. In the second phase, II, which is the kind of space inside hadrons, the spatial points carry colored field variables. Further, we assume that the second kind of space carries an energy per unit volume, $B>0$, greater than the first kind and that the second kind of space can exist as bubbles in the first kind. This is done in a relativistically covariant way, by associating with the second kind of space a local stress-energy of the form, $-g^{\mu\nu}B$. The phase II space is called a bag. Thus, the local stress energy tensor for hadronic matter in our model has the form

$$T^{\mu\nu}(x)=\theta_B(x)\{T^{\mu\nu}_{Q+G}(x)-g^{\mu\nu}B\}, \tag{1.1}$$

where

$$\theta_B(x)=\begin{cases}0 \text{ on phase I space-time points}\\ 1 \text{ on phase II space-time points}\end{cases} \tag{1.2}$$

and where $T^{\mu\nu}_{Q+G}(x)$ is the stress tensor of the standard non-Abelian gauge theory of colored quarks and gluons.

In order that the stress-energy be conserved locally, we must have

$$\partial_\mu T^{\mu\nu}(x)=0. \tag{1.3}$$

If we differentiate (1.1), we find,

$$\partial_\mu T^{\mu\nu}(x)=\partial_\mu(\theta_B(x))\cdot(T^{\mu\nu}_{Q+G}(x)-g^{\mu\nu}B)$$
$$+\theta_B(x)\cdot(\partial_\mu T^{\mu\nu}_{Q+G}(x)). \tag{1.4}$$

The second term will vanish if

$$\partial_\mu T^{\mu\nu}_{Q+G}(x)=0 \text{ inside.} \tag{1.5}$$

This will follow as a consequence of the standard field theory equations of motion. The quantity

$$\partial_\mu \theta_B(x)=n_\mu \delta_s(x) \tag{1.6}$$

where n_μ is a local, space-like unit normal vector on the surface, and $\delta_s(x)$ is a surface delta function. Thus, in order that

$$\partial_\mu T^{\mu\nu}(x)=0 \tag{1.7}$$

we must have

$$n_\mu(T^{\mu\nu}_{Q+G}(x))=n^\nu B \tag{1.8}$$

on the surface. We may recognize this as the condition that the pressure of constituent fields on the surface be required to equal B in order to conserve energy and momentum at the boundary. Thus the term associated with B in the stress tensor acts as a confining pressure. The hadron is analogous to a bubble in the normal vacuum. In order that $n_\mu T^{\mu\nu}$ be proportional to n^ν on the surface, the fields must obey a boundary condition on the space-like surface of the hadron. For Dirac fields the appropriate condition is

$$i\gamma\cdot nq(x)=q(x). \tag{1.9}$$

On the boundary for a vector field, we have

$$n_\mu F^{\mu\nu}=0 \tag{1.10}$$

where $F^{\mu\nu}(x)$ is the field strength. For a colored field, (1.10) is analogous to the condition that the exterior space is classically "super-dichromatic". The color "electric" fields must be tangential to the surface, and vanish on the outside.

When (1.9) and (1.10) hold, it is easy to show that

$$n_\mu T^{\mu\nu}_{Q+G}=n^\nu P(x) \tag{1.11}$$

where

$$P(x)=\frac{1}{2}n\cdot\frac{\partial}{\partial x}(\bar{q}(x)q(x))-\frac{1}{8}\mathrm{Tr}(F_{\mu\nu}F_{\mu\nu}). \tag{1.12}$$

Thus, in this case the conservation equation (1.8), becomes

$$B=\frac{1}{2}n\cdot\frac{\partial}{\partial x}(\bar{q}(x)q(x))-\frac{1}{8}\mathrm{Tr}(F_{\mu\nu}F_{\mu\nu}) \tag{1.13}$$

on the surface,which may be regarded as a contour map of the surface. Thus, we now have a local, covariant description (at least classically) of a hadron including a local equation of the surface. We see that the boundary condition $n_\mu F^{\mu\nu}=0$, insures that a hadron is a color singlet since the color electric field lines cannot terminate on the surface. Accordingly, we have a theory of confined colors. Colored quarks and gluons are described as usual inside a hadron, but no hadrons will exist with a net color.

There are different views that one may take of our model. It may be a phenomenological version of the standard non-Abelian color gauge field theory. The problem would be to show that the vacuum has the character of the phase I and II vacuua described above. It may also be that the ultimate microscopic description of matter is completely novel, and that the theory above is a phenomological version which holds on a sufficiently long scale. It also may be that our model is simply a crude approximation, which is only useful to obtain the gross features of hadron structure. As we shall see in the remainder of these lectures, it is that, at the very least. In any case, since we now have a theory, we can explore its phenomenological consequences.

DEFORMED HADRONS

We have found in studying the theory that there exist two broad classes of hadrons. These are deformed hadrons and spherical hadrons. I will begin by discussing the deformed states, even though historically, we worked out the properties of the spherical ones first.

Let us imagine a meson consisting of a quark and antiquark in a color singlet state. If the state also has a high orbital angular momentum, we would expect that the quark and antiquark would be in a circular orbit rather separated from one another. In an ordinary system with short range forces, there would be no force sufficient to overcome the centrifugal repulsion, and hence such a state would be impossible. In our model, the quark and antiquark carry color fields. The quarks cannot separate since the color field lines from the quark must terminate on the antiquark. The field lines cannot fill an increasing volume of space since creating the space costs at least an energy per unit volume B. Thus, the color flux lines are squeezed by the confining pressure B and kept parallel to each other. One has the situation illustrated in Figure 1 (3,4).

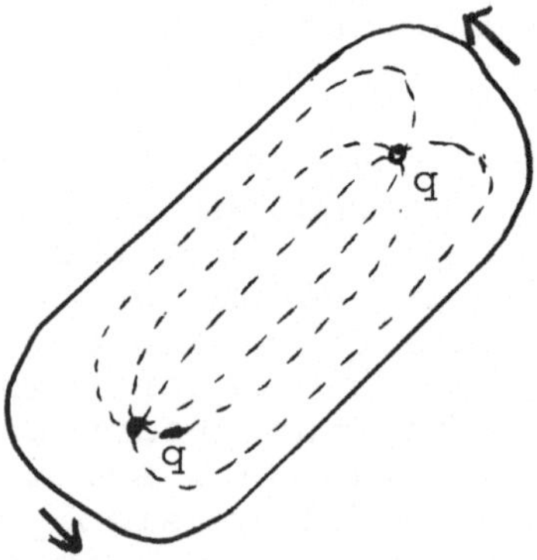

Fig. 1. A Deformed Meson.

Thus, the state is deformed. If the quarks are very light (we shall find that massless up and down quarks and a strange quark with mass ~280MeV will work the best), the energy and angular momentum will be carried mostly in the color field, and in this limit, we find that the ends will move with the velocity of light. In the instantaneous rest frame at a point between the quarks, let us suppose the color electric field strength is E^a_o, then from Gausses law, if A_o is the cross section,

$$E^a_o A_o = 2g_c(\lambda^1_a - \lambda^2_a)\frac{1}{2} \qquad (2.1)$$

where $2g_c$ is the color coupling constant* and λ^1_a is the color of the quark, λ^2_a is the color of the antiquark(1). In a color singlet $\lambda^1_a + \lambda^2_a = 0$, so

$$\frac{1}{2}\sum_a E^{a\,2}_o = [4g^2_c/A^2_o](C) \qquad C = \sum_a \lambda^2_a = \frac{4}{3} \qquad (2.2)$$

Here C is the color casimir operator of the quark on one end (or more generally, what ever color is on one end). The sides are determined by balancing the field pressure against B, (Eq. (1.12)) so

$$B = \frac{1}{2}\sum_a E^2_a = 4\frac{g^2_c}{A^2_o} C$$

or

$$A_o = 2\frac{g_c\sqrt{C}}{\sqrt{B}}$$

Consequently the energy per unit length in the rest system is

$$k = (\frac{1}{2}\sum_\alpha E^2_a)A_o + BA_o$$

$$k = 2BA_o = 4g_c\sqrt{C}\sqrt{B} \qquad (2.3)$$

and it is well known that such a system(3,5) (a relativistic string) on rotation gives a linear Regge trajectory,

$$J = \alpha' M^2$$

* $2g_c = g_s$ is the relationship to the convention used by some authors.

with

$$\alpha' = \frac{1}{2\pi k} = \frac{1}{8\pi g_c \sqrt{C}\sqrt{B}}$$
$$= \frac{1}{16\pi^{3/2}\sqrt{\alpha_c}} \frac{1}{\sqrt{C}} \frac{1}{\sqrt{B}} \qquad (2.4)$$

where $\alpha_c = g_c^2/4\pi$ is the chromostructure constant.

The quark and antiquark cannot fly apart but the state can decay by the spontaneous production of a quark-antiquark pair in the region between the ends. Thus, the dominate decay is associated with meson→meson+meson where the decay states have the same character. On the basis of this picture for mesons, we would anticipate that three quark baryons would have similar deformed high orbital angular momentum states. However, in this case we would have a single quark at one end and a di-quark with the antiquark color on the opposite end. Clearly the same slope would be obtained for the Regge trajectory and the decay channel would produce a meson and baryon again with the same quark configuration. However, this model predicts the existence of more states. If one considers the baryon-antibaryon system, we see that by annihilating a quark from one on an anti-quark from the other we could produce a new class of mesons, with two quarks (color $\bar{3}$) on one end and two antiquarks (color 3) on the other. Clearly the slope of this kind of trajectory is the same as that for the ordinary mesons. But in this class of mesons exotic flavors become possible. We shall return to discuss such exotic states later. Another kind of two-quark two-antiquark meson which is possible is to have the two quarks in a color $\underline{6}$ configuration, and the two anitquarks in a color $\bar{\underline{6}}$ configuration. These states have a lower slope than ordinary mesons, since the color Casimir of the $\underline{6}$ equals 10/3, hence according to (2.4),

$$\alpha'_6 = \left(\frac{4/3}{10/3}\right)^{1/2} \alpha'_3 = .63\alpha'$$

These higher mass states are not coupled strongly to either the meson-meson, or baryon-antibaryon states and hence would be difficult to produce.

Another interesting class of deformed states is possible because of the existence of heavy quarks(c).

If one considers a system consisting of a heavy quark and antiquark, it is not necessary to use a high orbital angular momentum to separate the colors. Indeed one would expect that if the quark is heavy enough so that its Compton size (1/M) is small in comparison to the bag size set by the pressure of its color fields ($\sim 1/\sqrt{A_o}$, A_o given above) then the state would be deformed, since the quarks would move slowly in comparison to the field lines.

Furthermore, the color field and bag energy is proportional to the distance, r, between the quarks yielding an effective potential

$$kr = \frac{1}{2\pi\alpha'} r$$

with k given by the calculation above. If the heavy quark carries the same color as a light quark, α' is the observed Regge slope $\sim .9\,(\text{GeV})^{-2}$, so

$$k \sim .18\,(\text{GeV})^2.$$

Indeed, this system is clearly the one that describes the J/ψ system, and the above slope for the linear potential is very close to that used in phenomenology(6).

Clearly, many other states are possible, a particularly interesting class are those which involve colored glue quanta instead of quarks. Some properties of these have been speculated about, but no detailed work has been done.

SPHERICAL HADRONS

If one considers states with light quarks (Compton size large in comparison to bag size), then one would expect that the bag should be spherical to minimize the kinetic energy of the quarks(7). Thus, consider a spherical bag with a given radius R, and imagine computing the energy of the constituents as a function of R. The total energy will be

$$E = E_{\text{constituents}}(R) + \frac{4\pi}{3} BR^3 \qquad (3.1)$$

where the last term is the volume energy. The constituent energy increases monotonically as $R \to 0$. We would therefore expect that if quantization of the complete system does not qualitatively change (3.1), that the surface wavefunction should be well localized around R_o, where

$$\frac{\partial}{\partial R}\left(E_{\text{constituents}}(R)+\frac{4\pi}{3}BR^3\right)\Bigg|_{R=R_o}=0 \tag{3.2}$$

(which is the same as B= pressure of constituents). Hence we would obtain for the mass of the particle

$$M=E_{\text{constituents}}(R_o)+\frac{4\pi}{3}BR_o^3. \tag{3.3}$$

This is the so-called "static bag" model result for the mass and radius of a spherical hadron.

We have assumed that the mass of the up and down quarks is very small. Why? Everyone knows that to get the gyro magnetic ratio for the proton to be 2.8, the quark magnetic moment must be

$$\mu_q=q/2m_{\text{quark}} \tag{3.4}$$

with $m_{\text{quark}}\simeq 330$ MeV for up and down quarks. The wavefunction has a form which also predicts that $\mu n/\mu p=-2/3$, which (neutron/proton) was one of the original successes of the non-relativistic quark model.

However, only a free Dirac particle has a magnetic moment given by (3.4). If the particle is bound in an S state, it is necessary that the momentum k be such that $k \ll m$ for (3.4) to be true. In a proton with radius R, the momentum of a non-relativistic quark is $\pi/R \sim 600$MeV (for R=1 Fermi). Hence, because $600 \not\ll 300$, formula (3.4) cannot be used.

If we compute the magnetic momenta of a free Dirac quark confined in a sphere of radius R we obtain the graph which is qualitatively illustrated in Fig. 2.

Therefore to get μ as large as possible, we put m=0, and have

$$\mu=.2R\cdot q$$

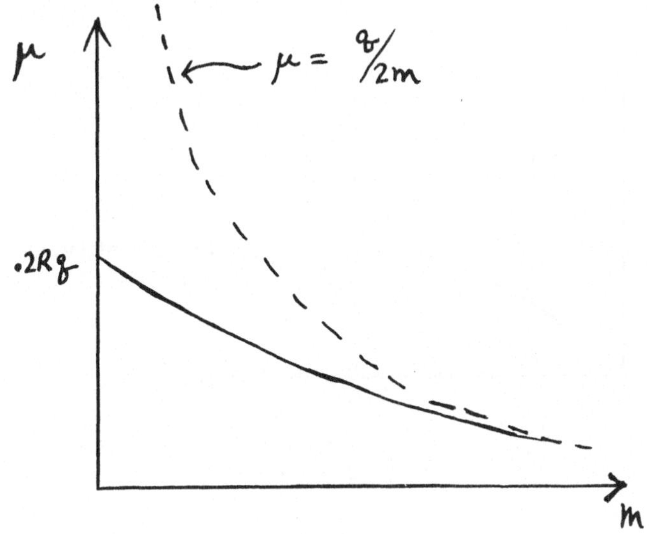

Fig. 2

To fit g_p we would then require that

$$.2R = \frac{1}{2 \cdot 330\text{MeV}}$$

or R=1.5 Fermi (3.5)

This is a large but at least conceivable proton. Another reason for taking the up and down quark masses to be very small is the ratio of the weak decay constants, GA/GV, which is observed to be ~1.24. In the non-relativistic model, GA/GV=5/3. We find for massless quarks, 1.1 .

For massless up and down quarks, the large and small Dirac wavefunctions have the form illustrated in Fig. 3.

It is still true that the large component dominates the small, so that in many cases the non-relativistic picture might work well. The momentum k of the massless quark in the spherical bag is 2.04/R. The momentum increases with mass until in the non-relativistic case, $k=\pi/R$.

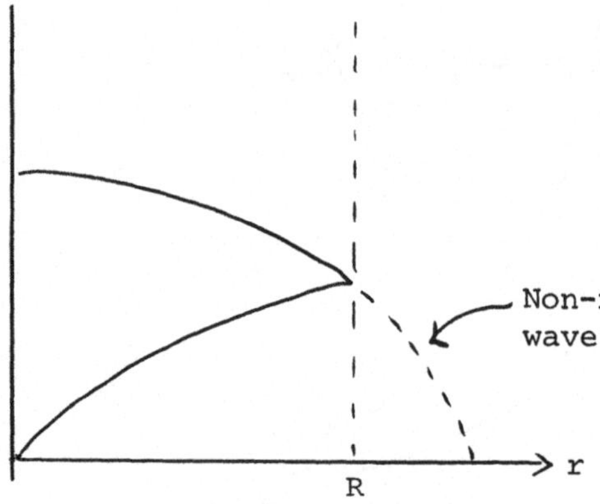

Fig. 3.

In the most naive model, where we omit the effects of the coupling, the energy of n quarks (or quarks and antiquarks) would be $n \cdot 2.04/R$ so

$$E=n\cdot\frac{2.04}{R} + \frac{4\pi}{3}BR^3 \tag{3.6}$$

and then $\partial E/\partial R=0$ gives,

$$R_o^4=(4\pi B)(n\times 2.04) \tag{3.7}$$

and

$$M=\frac{4}{3}\,\frac{2.04}{R_o}\,n \tag{3.8}$$

Therefore

$$MR_o=\frac{4}{3}(n\cdot 2.04)=8.16 \qquad (n=3) \tag{3.9}$$

Thus, the size is predicted in units of the mass, is independent of B, and we have for a proton with mass 1GeV,

$$R_o=\frac{8.16}{5}\ f\cong 1.6\ f \tag{3.10}$$

If we compare this with (3.5), we see that we have calculated the magnetic moment of the proton. We can now estimate α_c, the chromostructure constant since that appears in the Regge slope (3.4) with the value of $B^{1/4}$ which makes m_p~1 GeV ($B^{1/4}$~.1 GeV). We find that α'=.9 $(\text{GeV})^{-2}$ if $\alpha_c \approx 1$. Consequently, it is inconsistent to ignore the effects of the coupling to color which we have done in the above calculation of spherical hadrons. Another reason that we must calculate (3.6) with more sensitivity is the dependence upon the quark number which we have found, namely $M \sim n^{3/4}$. If we consider a six quark color singlet, since $(6)^{3/4} < 2\times 3^{3/4}$ we would predict a deuteron unstable against collapse into six quarks. However, we do not predict an absence of saturation in the number of quarks, as the fractional power suggests, because the exclusion principle saves us. Thus, for a large number of up and down quarks

$$N/V = \underset{\substack{\text{spin}\cdot\text{color}\cdot\\ \text{flavor}}}{2\cdot 3\cdot 2} \int^{k_F} \frac{d^3k}{(2\pi)^3} \sim k_F^3$$

and

$$E/V \simeq 2\cdot 3\cdot 2 \int k \frac{d^3k}{(2\pi)^3} \sim k_F^4$$

So

$$E^{tot}(V) = BV + E_{quarks} \simeq BV + \frac{N^{4/3}}{V^{1/3}}\, C\,.$$

Consequently at equilibrium

$$0 = \frac{\partial E}{\partial V} - B \simeq 1/3\,(N/V)^{4/3} C \tag{3.11}$$

so the density becomes constant, and hence the total energy is proportional to the number of quarks. Consequently, we don't necessarily predict that the atomic nucleus is unstable against collapse into quarks.

The quarks interact by means of color gluon exchange(8). If we treat the effect of interaction by perturbation theory in the spherical hadrons, the color electric effects which dominate the structure of the deformed states, have little importance. When the quarks move freely in the same spatial states, the color charge density is zero, since the total wavefunction corresponds to a color singlet. However, gluon exchange is a vector interaction and the color magnetic effect produces an effective color-spin exchange interaction. The interaction between two quarks (or quark and antiquark has the form,

$$-C\lambda_1\cdot\lambda_2\sigma_1\cdot\sigma_2 \tag{3.12}$$

where C is proportional to the chromostructure constant α_C, and is positive $(-J^1_\mu\cdot J^2_\mu = J^0_1 J^0_2 - \vec{J}_1\cdot\vec{J}_2)$. Hadrons are color singlets. For mesons ($3\ \bar{3}=\underline{\underline{1}}+8$)

$$\lambda_1+\lambda_2=0$$

so

$$\lambda_1\cdot\lambda_2 = -\lambda_1^2 = -\frac{4}{3}$$

In baryons each pair of quarks is a color $\bar{3}$, so

$$(\lambda_1+\lambda_2)^2 = \frac{4}{3}$$

so

$$\lambda_1\cdot\lambda_2 = -\frac{2}{3}$$

Consequently, we have

$$E = \begin{cases} +\frac{4}{3}\,C\ \sigma_1\cdot\sigma_2 & \text{in mesons} \\ +\frac{2}{3}\,C\ \sigma_1\cdot\sigma_2 & \text{in baryons.} \end{cases} \tag{3.13}$$

We therefore predict that in both mesons and baryons, the states with highest spin have the highest mass, $\Delta>N$, $P>\pi$, $K^*>K$, etc. This is in accordance with the facts. Another success of the color-spin exchange interaction concerns the Σ,Λ states. Both have a single strange quark, but $\Sigma>\Lambda$. The color spin interaction is weaker with increasing quark since it is essentially a dipole-dipole interaction and the dipole moment decreases with mass. In the $\Sigma-\Lambda$ states

$$E \simeq C(\sigma_1\cdot\sigma_2)+C'(\sigma_1\cdot\sigma_3+\sigma_2\cdot\sigma_3) \tag{3.14}$$

where σ_3 is the spin of the strange quark, $C'<C$. The SU(56) wavefunction is symmetric in spin and isospin and therefore the up-down quarks are such that in Σ, $J=1$, in Λ, $J=0$. Hence according to (3.14)

$$E_\Lambda = -3C$$

and

$$E_\Sigma = C + C' \cdot \sigma_3 \cdot (\sigma_1 + \sigma_2)$$

$$= C + 2C' \cdot \{\frac{1}{4}(\sigma_1+\sigma_2+\sigma_3)^2 - \frac{1}{4}\sigma_3^2 - \frac{1}{4}(\sigma_1+\sigma_2)^2\}$$

$$= C + 2C' \cdot \{\frac{1}{2}(\frac{1}{2}+1) - \frac{1}{2}(\frac{1}{2}+1) - 1 \cdot (1+1)\}$$

$$= C - 4C'$$

Therefore

$$E_\Sigma - E_\Lambda = 4(C - C') > 0; \tag{3.15}$$

that is, we see that the color-spin interaction predicts the proper sign of the Σ, Λ mass splitting(9). These results are correct in sign, but what about magnitude? Since we have quark wavefunctions we may compute C for all states and express the total color-spin exchange energy for all states in terms of one parameter, α_c, the chromostructure constant. The results are shown in Fig. 4.

In the calculations tabulated in this figure, we have a scale parameter. $B^{1/4}$, a strange quark mass, m_s, the coupling constant, α_c, and a parameter z used phenomenologically to take care of quantum corrections to the field energy associated with unoccupied states. Using these we fit four states, $\bar{\Omega}$, Δ, N, ω. All other states then have the masses shown. The results (considering the simplicity of the calculation) we regard as impressive.

The parameters $B^{1/4}$ and α_c are determined by this fit and we can now use them to compute the Regge slope α' according to the formula obtained in the deformed hadron. We find(3) $\alpha' = .9\,(\text{GeV})^{-2}$. We also regard this with some pride. Since we have incorporated SU(3) flavor breaking through a strange quark mass we may compute magnetic moment ratio(8,10). In the case of the Λ, we find

$$\mu\Lambda/\mu p = -.26$$

to be contrasted with $-.33$ in the case where the strange quark has a mass the same as the up-down. The observed magnetic moment ratio is

$$-.24 \pm .03.$$

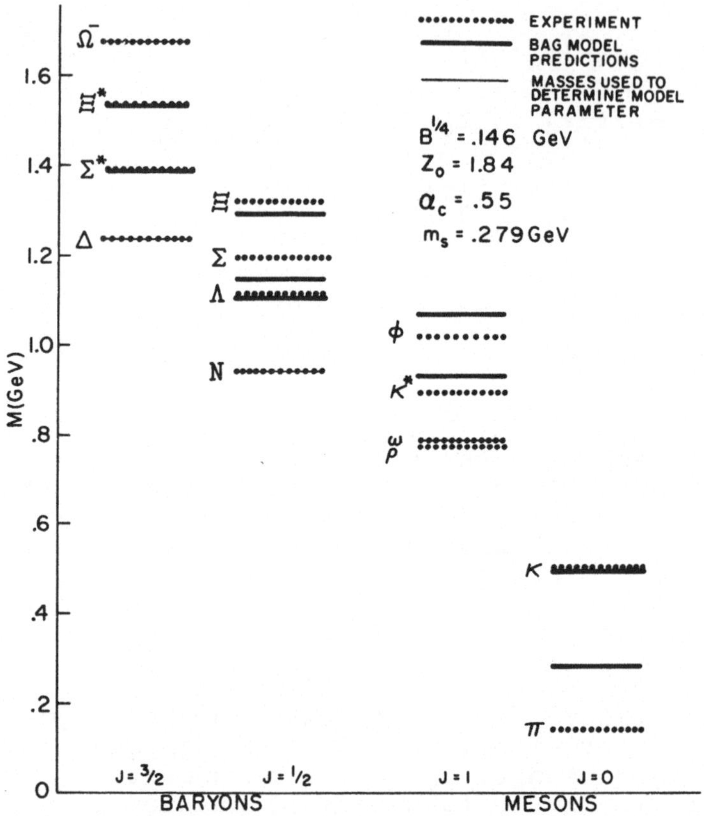

Fig. 4. Fit to the hadron masses with m_o=0, with B, α_c, Z_o, and m_s as shown. The actual masses are given by dotted lines for comparison. The masses of the N,Δ,Ω, andω were used to determine the parameters, all other masses are predicted.

While all of this is very nice, no calculation would be believable if there also were not some bad results. Otherwise there would be no room for improvement. The absolute magnetic moment of the proton was correctly calculated in our naive model, for g_p=2.8 one needed a proton radius, 1.5f. We now have R_p=1f, too small, g_p comes out 2/3(2.8). Also, not included in Fig. 4 are the masses of the 0^- flavorless states η, η'. These are also wrong. Whereas ϕ and ω are predicted to be pure $s\bar{s}$ and pure $u\bar{u}+d\bar{d}$, we would give the same result for η and η' with completely wrong masses. η ~$s\bar{s}$ with mass ~500 MeV and ' pure $u\bar{u}+d\bar{d}$ degenerate with π (as ρ,ω). These are familiar difficulties.

If there is an improvement in the calculation of the states, we believe that it is mixing with colored glue which must be the important ingredient. Such calculations are underway.

Another deficiency in the most naive static bag was the prediction that the deuteron would be more stable as six quarks than as two nucleons. The color spin exchange interaction also resolves this difficulty. With the color singlet restriction the Young diagram for color for six quarks is

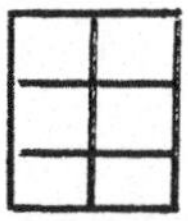

Total antisymmetry of the color-spin-flavor wavefunction means that the spin-flavor wavefunction is represented by the complementary diagram,

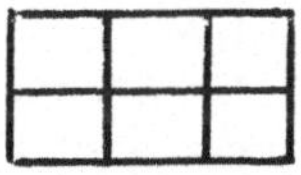

If we restrict our attention to up and down quarks, this diagram belongs to the fifty dimensional representation of $SU(4)=SU(2)\times SU(2)_{flavor}$ and decomposes with $(I,S)=$ $(3,0)+(0,3)+(2,1)+(1,2)+(1,0)+(o,1)$. These are the allowed spin isospin combinations for six quarks in a spherical bag with the same spatial wavefunction. The total color spin exchange energy is

$$-C\,\frac{1}{2}\sum_{i\neq j}\lambda_i\cdot\lambda_j\sigma_i\cdot\sigma_j \qquad (3.16)$$

To evaluate, we can use the permutation operators,

$$P^S_{12} = \frac{1}{2}(1+\sigma_1\cdot\sigma_2)$$

$$P^I_{12} = \frac{1}{2}(1+\tau_1\cdot\tau_2)$$

$$P^C_{12} = \frac{1}{3} + 2\,\lambda_1\cdot\lambda_2$$

where in the states of interest,

$$P^S_{12}P^I_{12}P^C_{12} = -1$$

for any pair. Since $P^2=1$,

$$P^S_{12}P^C_{12} = -P^I_{12}$$

or

$$-\lambda_1\cdot\lambda_2\sigma_1\cdot\sigma_2=\frac{2}{3}+\lambda_1\cdot\lambda_2+\frac{1}{6}\sigma_1\cdot\sigma_2+\frac{1}{2}\tau_1\cdot\tau_2$$

We can now easily evaluate (3.16) for n quarks,

$$-\frac{C}{2}\sum_{i\neq j}\sigma_i\cdot\sigma_j\lambda_i\cdot\lambda_j=[n(n-6)+S(S+1)+3I(I+1)]C/4$$

which is valid for any n quark color singlet. That is, n=3, 6, 9, 12. We see that if $n\geq 6$ the color spin exchange energy is always positive. It is lowest when for n=6, S=1, I=0, that is the deuteron quantum numbers. We know C from our three quark calculation. Hence we can calculate the mass of the lowest six quark state, and we find that it is 300 MeV greater than two nucleons. We have resolved the deuteron problem. The color spin exchange energy has been used to calculate many multiple quark states to test whether or not the same model used to compute the masses of the classic low mass hadrons, the SU(6) 36 and 56 also would predict the existence of exotic low mass hadrons. If we did, we would be in trouble. We don't(11,12). Indeed we do predict that certain low mass mesons namely the 0^+ nonet is better described as a two quark, two antiquark state than as an orbital excitation of the $q\bar{q}$ system. We have not time to discuss these questions here except to remark that an interesting prediction can be made concerning the types of meson trajectories which would couple strongly to the baryon, antibaryon system. On diagonalizing the color spin exchange interaction for the two quark, two antiquark mesons, one finds that the states have the form

$$\cos\theta|\bar{3},3\rangle+\sin\theta|6,\bar{6}\rangle$$

where in the first state the two quarks are either coupled up to form a color $\bar{3}$, and in the second to form a color 6. One would guess that the states where $\cos\theta \gg \sin\theta$, would be the ones which would lie on the Regge trajectory with the slope which couples to the baryon-antibaryon system. If we use the formula

$$J=\alpha' M^2+\alpha_o$$

and calculate α_o by using the mass for the spherical state, and J=0, we would then obtain a prediction for the masses and spins of higher resonances of this type. We would also obtain predictions for possible flavors. This project is underway.

CONCLUSIONS

In these lectures, I have stressed the spectroscopic successes of our model. I had had no time to go into many other areas in which work has been done. So far, we are pleased by the relative ease with which we have been able to extract the phenomenological consequences of our picture without the need of elaborate calculations or expedient assumptions. However, one must not under emphasize the need for a more profound technical development of the model. This area is the one which has been least stressed and which requires immediate attention. I invite the more technically proficient theorists to try their hand.

REFERENCES

1. A. Chodos, R. L. Jaffe, K. Johnson, C. B. Thorn, V. F. Weisskopf (1974) Phys. Rev. D9, 3471.
2. K. Johnson (1975) Acta Phys. Polonica, 136, 865. V. F. Weisskopf (1975) CERN Ref. Th., 2068.
3. K. Johnson, C. B. Thorn (1976) Phys. Rev. D13, 1934.
4. P. Gnädig, P. Hasenfratz, J. Kuti, A. S. Szalay Submitted to XVIII Conf. on High Energy Physics, Tbilisi.
5. P. Goddard et al.(1973) Nucl. Phys. 356, 109.
6. E. Eichten et al. (1975) Phys. Rev. Lett., 34, 369.
7. A. Chodos, R. L. Jaffe, K. Johnson, C. B. Thorn (1974) Phys. Rev. D10, 2599.
8. T. A. DeGrand, R. L. Jaffe, K. Johnson, J. Kiskis (1975) Phys. Rev. D12, 2060.
9. Independently worked out in the non-relativistic quark model, A. DeRújula et al. (1975) Phys. Rev. D12, 147.
10. E. Allen (1975) Phys. Lett. 57B, 263.
11. R. L. Jaffe, K. Johnson (1976) Phys. Lett., 60B, 201.
12. R. L. Jaffe, Phys. Rev., to be published; SLAC-PUB-1772, 1773 (July, 1976); (1977) Phys. Rev. Letters, 38, #5.

M.I.T. Bag Model "Complete" List (1976)

A (more or less) complete list of references to the M.I.T. bag model applications follows. Other kinds of bags (SLAC bag, etc) for which there is a much more extensive literature are not included because the emphasis has been on the attempt to derive bag-like behavior from more elaborate non-linear local field theories. These questions lie outside of the context of the M.I.T. model.

Bag Model; General. A. Chodos, R. L. Jaffe, K. Johnson, C. B. Thorn and V. F. Weisskopf, Phys. Rev. D9, 3471 (1974). K. Johnson (1975) Acta Phys. Polonica, B6, 865. V. F. Weisskopf (1975) CERN Ref. Th. 2068.

Static Bag. A. Chodos, R. L. Jaffe, K. Johnson, C. B. Thorn (1974) Phys. Rev. D10, 2599. E. Golowich (1975) Phys. Rev. D12, 2108. E. Allen (1975) Phys. Letters 57B, 263.

Hadron Spectroscopy. T. A. DeGrand, R. L. Jaffe, K. Johnson, J. Kiskis (1975), Phys. Rev. D12, 2060. T. A. DeGrand, R. L. Jaffe (1976) CTP 529. T. A. DeGrand (1976) CTP.

Multiquark States. R. L. Jaffe, K. Johnson (1976) Phys. Lett. 60B, 201. R. L. Jaffe (July 1976) SlAC-PUB-1772, SLAC-PUB-1773.
Motion of Surface. C. Rebbi (1975) Phys. Rev. D12, 2407. C. Rebbi (1976) CTP 551.
Hadron-Hadron-Elastic Scattering. F. E. Low (1975) Phys. Rev. D12, 163.
Charmed States. R. L. Jaffe, J. Kiskis (1976) Phys. Rev. D13, 1355.
Deformed Bags. C. DeTar, submitted to XVII Conf. on High Energy Physics, Tbilisi, CTP 546.
String-like Bags. K. Johnson, C. B. Thorn (1976) Phys. Rev. D13, 1934.
1+1 dimensional examples. C. B. Thorn, M. V. K. Ulehla (1975) Phys. Rev. D11, 3531. D. Shalloway (1975) Phys. Rev. D11, 3545. D. Shalloway (1976) Lab. of Nuc. Stud., Cornell, CLNS-331. V. Krapchev (1976) Phys. Rev. D13, 329.
Electromagnetic Mass. A. Chodos, C. B. Thorn (1974) Phys. Lett. 53B, 359. N. Deshpande, D. A. Dicus, K. Johnson, V. L. Teplitz, Phys. Rev., to be published.
Surface Tension Models and Deformed Bags. P. Gnädig, P. Hasenfratz, J. Kuti, A. S. Szalay, preprint. Central Research Inst. for Phys., Budapest, KFKI-75-67 preprint, submitted to XVIII Conf. on High Energy Physics, Tbilisi.
Structure Functions for Deep Inelastic Scattering. R. L. Jaffe (1975) Phys. Rev. D11, 1953. R. L. Jaffe, A. Patrascioiu (1975) Phys. Rev. D12, 1314.
Chiral Model. A. Chodos, C. B. Thorn (1975) Phys. Rev. D12, 2733.
Decays. J. F. Donoghue, E. Golowich, B. R. Holstein (1975) Phys. Rev. D12, 2875. Patrick Hays, Martin V. K. Ulehla (1976) Phys. Rev. D13, 1339 (improves with the addition of color).

CHARMED PARTICLES AND DRELL-YAN PHOTONS AS SOURCES OF DIRECT LEPTONS

K. Kajantie

Department of Physics and Research Institute

for Theoretical Physics, University of Helsinki

Finland

ABSTRACT

We discuss the possibility that the rapid increase observed in the single lepton yield when p_T decreases from 1 to o,25 GeV/c and the large (relative to Drell-Yan estimates) muon pair continuum at $1 < M < 3$ GeV might be due to a common source, semileptonic decays of charmed particles. This is possible if the total charm production cross section is of the order of 1oo μb at Fermilab energies.

1. INTRODUCTION

This lecture will address itself to the question of the origin of direct leptons (i.e., those not arising from the decay of pions and kaons) in proton-proton collisions. In particular, the following questions will be discussed:

(a) Can one learn something new by examining the consistency of single and double lepton distributions? The situation is somewhat analogous with the one in hadron production some time ago: many models gave a good description of single hadron spectra and a distinction could only be made on the basis double hadron distributions, correlations, etc.

(b) Where are the leptons coming from the semi-leptonic decays of charmed particles produced in proton-proton collisions? After all, there is strong evidence for the existence of charmed mesons from SPEAR [1] and charmed baryons from Fermilab [2] and these must also be produced in hadronic collisions.

(c) Are the leptons coming from semileptonic decays in charmed particles, in fact, already seen in data existing at present? An analogous case is the famous $pU \to \mu^+\mu^- X$ experiment at 29.5 GeV [3] in which the ψ appeared as a broad shoulder in the mass spectrum many years before it really was established as a narrow particle. More concretely, we shall discuss the possibility that the rapid increase observed as a function of p_T in the single electron spectrum at $\theta_{CM} = 30^o$, $s = 2800$ GeV^2 and $0,2 < p_T < 1,5$ GeV [4] and the excess of events (relative to Drell-Yan estimates) observed in $d\sigma/dM_{\mu\mu}$ for $1 < M_{\mu\mu} < 3$ GeV [5] might be related phenomena and, in fact, due to semileptonic decays of charmed particles.

The literature on direct lepton production is extensive. The problem is not the lack of models but their multitude and it may well be that different mechanisms will be needed in different parts of the lepton phase space. A general review is given in [6-7]. In addition to some less orthodox mechanisms [8-12] there are three obvious sources of direct leptons:

(a) Electromagnetic decays of strongly produced vector mesons [13-17]. Because the two-body decay $V \to \ell^+\ell^-$ leads to a maximum of the 90° single lepton spectrum at $p_T = m_V/2$, it is excluded that these leptons would give a sizeable part of the 90° lepton spectrum for $p_T < 2$ GeV. However, given our uncertainty of the vector meson distributions at large transverse momentum, it may well be that their decay leptons dominate the lepton spectrum at large p_T. The issue can only be settled by a large acceptance dilepton experiment, which makes it possible to verify whether there corresponding to each p_T lepton is an anti-lepton so that the lepton-antilepton invariant mass equals the mass of some vector meson. The solution is entirely in the hands of the experimentalists and in the following we shall only consider leptons with $p_T \lesssim 1$ GeV.

(b) Drell-Yan photons or the photon continuum [18-22]. This will be discussed in Section 2 below.

(c) Semileptonic decays of charmed particles [23-26]. This will be discussed in Section 3 below.

2. DRELL-YAN PHOTONS OR THE PHOTON CONTINUUM

The Drell-Yan mechanism (Fig. 1) leads to the following simple and elegant expression for the cross section of producing a $\mu^+\mu^-$ pair with invariant mass M and longitudinal momentum = x $\sqrt{s}/2$ in a hadron-hadron collision:

$$x_o \frac{d^2\sigma}{dxdM^2} = \frac{1}{3} \cdot \frac{4\pi\alpha^2}{3M^2} \frac{1}{M^2} \sum_i e_i^2 \left[f_i^h(x_+)\bar{f}_i^{h'}(x_+) + \bar{f}_i^{h'}(x_+) f_i^h(x_-)\right] \quad (1)$$

where the factor 1/3 arises from colour, $4\pi\alpha^2/3M^2$ is the cross section for $q\bar{q} \to \gamma \to \mu^+\mu^-$ at large M, $f_i^h(x)/x$ is the probability that the quark of type i carries the fraction x of the total longitudinal momentum of the hadron $x_o = \sqrt{x^2+4M^2/s}$ and $x_\pm = \frac{1}{2}(x_o \pm x)$. The condition for the validity of (1) is that M be larger than all other mass scales in the problem, like quark masses and transverse momenta, and that $M^2/s > o(1/\sqrt{s})$, to exclude the contribution of very slow partons ($x_\pm \geq M^2/s$). The magnitude of (1) is fixed by $4\pi\alpha^2 = 261$ nb$\cdot$GeV^2. In order to be able to say something about lepton distributions one also has

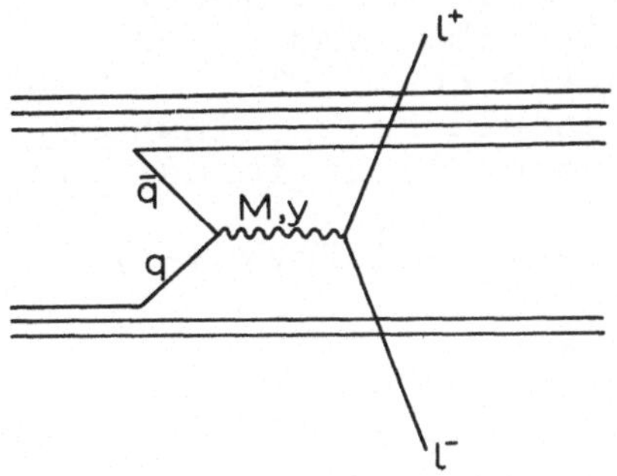

Fig. 1. The Drell-Yan mechanism for producing a lepton-antilepton pair of invariant mass M at rapidity y.

to know something about low M photons and of the transverse momentum distributions, of which the Drell-Yan mechanism as such says nothing. We shall treat the low-mass region and the transverse momentum dependence phenomenologically by writing the cross section for producing in a pp collision a $\mu^+\mu^-$ pair of invariant mass M, rapidity y and transverse momentum p_T in the form

$$\frac{d^4\sigma}{dM^2 dy d^2p_T} = \frac{\sigma(M^2)}{3M^2}\Big[V(x_+)S(x_-)+V(x_-)S(x_+) + \frac{4}{3}S(x_+)S(x_-)\Big]\frac{(n-1)m_o^{2n-2}}{\pi(p_T^2+m_o^2)^n} \qquad (2)$$

Here

$$x_+ = \sqrt{(M^2+p_T^2)/s}\, e^y, x_- = \sqrt{(M^2+p_T^2)/s}\, e^{-y}, \qquad (3)$$

$$\sigma(M^2) = \frac{4\pi\alpha^2}{3M^2}\sqrt{1-\frac{4m^2}{M^2}}\left(1+\frac{2m^2}{M^2}\right), \qquad (4)$$

where 2m is the threshold value of M. Since e and μ spectra are so similar, the effective threshold cannot depend on lepton type but is probably strongly affected by the quark masses. However, when calculating $\mu^+\mu^-$ spectra we shall set $2m = 2m_\mu$. The last factor in (2) describes the transverse momentum dependence of the pair. It is normalized to 1 and corresponds to $\langle p_T^2\rangle_{pair} = m_o^2/(n-2)$, where $m_o = 1$ and $n = 3$ in numerical examples later. Finally, in Eq. (2) we have performed the standard separation of the quark distributions to valence and sea parts. For an np collision the coefficient of the first term in square brackets should be 2/3 instead of 1. In numerical calculations the explicit forms [27]

$$V(x) = 1.21\sqrt{x}(1-x)^3 - 1.20\sqrt{x}(1-x)^5 + 4.93x^{0.85}(1-x)^9$$
$$S(x) = 0.101(1-x)^9 \qquad (5)$$

will be used. They fit the available electron and neutrino deep inelastic data.

Numerically, when $M > 1, \sqrt{s} \simeq 25$ and $y \simeq 0$, the factor in square brackets in Eq. (2) is about 0.03. The cross section is thus approximately given by (M in GeV)

$$\frac{d^2\sigma}{dM^2 dy} \simeq \frac{1}{M^4} \frac{nb}{GeV^2} \cdot \text{(threshold factors)} \tag{6}$$

This leads further to (taking the effective width of the rapidity distribution to be 4)

$$\frac{d\sigma}{dM^2} \simeq \frac{4}{M^4} \frac{nb}{GeV^2} \tag{7}$$

and to

$$\frac{d\sigma}{dy} \simeq 2o \text{ nb} \tag{8}$$

Assuming that the experimental result $\ell/\pi = 1o^{-4}$ is true everywhere and using the experimental value $d\sigma_{\pi^-}/dy$=3o mb

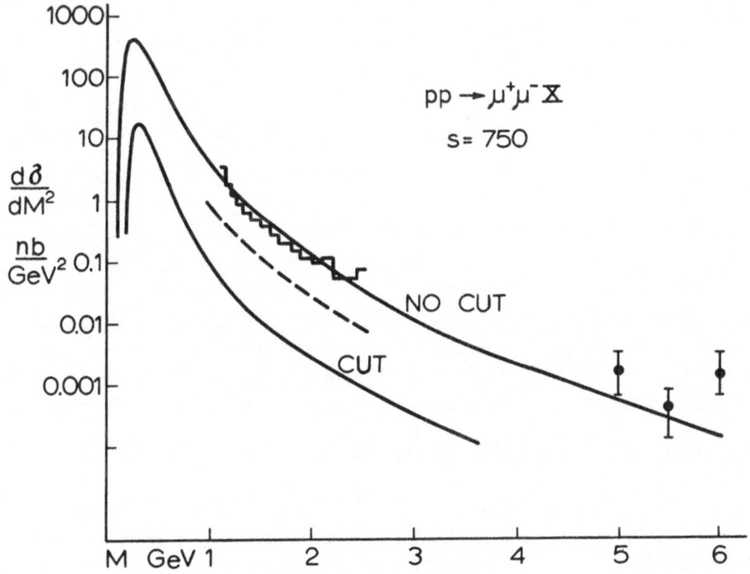

Fig. 2. The muon pair continuum calculated on the basis of Eqs. (2-5)(curve marked "no cut") compared with the data of ref.[29] (1 < M < 3) and of refs. [30-31] (M > 5; the data above M=6 agrees well with the curve marked "no cut"). The curve marked "cut" is the distribution obtained if both muons are required to have a laboratory longitudinal momentum > 60 GeV, the dashed curve corresponds to an experimental cut requiring the total longitudinal momentum of the pair in the laboratory system to be > 75 GeV.

one has that experimentally $d\sigma_\ell/dy \simeq 3\ \mu b$. This is two orders of magnitude above the Drell-Yan estimate in (8). Since the Drell-Yan mechanism, in fact, fits the large $M(M > 5\ GeV)$ part of the $\ell^+\ell^-$ spectrum (Fig. 2), one concludes that either the extrapolation (2) grossly underestimates the true yield of low M pairs or that there is a source of leptons contributing to the single but not to the double lepton spectrum. A possible enchancement mechanism for low M pairs is suggested in [21], but even this seems to be insufficient. Semileptonic decays, on the other hand, will offer an explanation of the latter type to increase the single lepton yield (Section 3).

Numerical evaluations of Eq. (2) are most conveniently performed with the aid of Monte Carlo techniques. In particular, these make it possible to take into account experimental cuts. The mass distribution for uncut events is, of course, most simply directly calculated from Eq. (1). Fig. 2 shows the mass distribution calculated on the basis of Eq. (2). There are essentially three experiments to compare with : $nBe \to \mu^+\mu^- X$ at 3oo GeV [28-29], $pBe \to e^+e^- X$ at 4oo GeV [3o] and $pCu \to \mu^+\mu^- X$ at 4oo GeV [31]. The observed mass intervals are approximately $1 < M < 3$, $5 < M < 10$, $8 < M < 11$ GeV, respectively. The mass distributions per nucleon extracted by using the measured A dependences are shown in Fig. 2 for $m < 6$ GeV. The higher mass points agree with each other and the Drell-Yan curve. The nBe experiment has an experimental cut eliminating events in which the muons have too small a momentum in the laboratory frame. The lowest curve in Fig. 2 is the mass distribution of muons produced according to Eq. (2) and satisfying $p_L^{\mu^\pm} > 60$ GeV in the lab frame.

According to these calculations and in agreement with other analyses it is quite probable that the events with $M > 5$ GeV are genuinely produced by the annihilation of a quark and an antiquark within the colliding hadrons. On the other hand, it is also probable that something else is producing the continuum below 3 GeV. In the following section we try to estimate to what extent charmed particles could leave their trace there.

3. SEMILEPTONIC DECAYS OF CHARMED PARTICLES

The contribution of semileptonic decays of charmed particles to the single lepton spectrum in hadron-hadron collisions was already discussed in the classic charm paper by Gaillard et al. [23]. However, the main emphasis was on large p_T leptons. Now we know that in this region there are many competing sources of leptons and if any single source dominates it is likely to be the electromagnetic decay of vector mesons. On the other hand, the semileptonic decay of a charmed hadron, being a three body decay, most clearly contributes to small p_T leptons [11,25]. It is thus tempting to associate these decays with the rise observed in the 3o^o lepton spectrum at $p_T < 1$ [4]. Here we shall pursue this idea further and, in particular, try to check whether it is consistent with the experimental $\mu^+\mu^-$ distribution discussed in Section 2 [26].

The charmed hadron spectrum is likely to be comlicated and up to now we have evidence for the mesons D(1,85), D(2.o2) and the baryon C(2.26). Here we shall, for simplicity, only include the D(1.85) and consider it as effectively representing the sum over all charmed hadron types.

In order to be able to calculate decay lepton distributions we have to make several assumptions about the strong production and weak decay properties of the D particle:

(a) Production cross section. The experimental numbers for ρ and ψ production at Fermilab energies ($\sqrt{s}$=25) are $\sigma_\rho \simeq$ 1o mb and $\sigma_\psi \simeq$ o.1 μb and it is obvious that these numbers bracket σ_D. The theoretical estimates (for $m_D = 2$) vary from 5o μb [32] to 5 μb [33]. There exists a very strong upper limit for charm production from an emulsion experiment [34] at Fermilab: $\sigma_D \leq 1$ μb for $\tau_D \geq 1o^{-14}$s. The upper limit from an ISR experiment [35] is $B(D \to K^+\pi^-)\sigma_D <$ o.41 mb. Accepting the former limit from the emulsion experiment would immediately exclude any sizeable contribution to lepton distributions from charmed particles. In the following we shall simply treat σ_D as a parameter and determine its value from the single lepton spectrum.

(b) x and p_T dependence of $f_{pp \to D}(x,p_T) \equiv E d^3\sigma_D/d^3p$. As a basis for assumptions one may here use the results for ψ production [3o] and for the production of massive muon pairs [36]. A suitable form for the x dependence is then $(1-x)^b$ with b = 3-4. As concerns the p_T distribution the main parameter is $\langle p_T \rangle$ or $\langle p_T^2 \rangle$. Interpolating between the proton ($\langle p_T \rangle$ = o.5 and $\langle p_T^2 \rangle$ = o.375) and the ψ ($\langle p_T \rangle$ = 1.25 and $\langle p_T^2 \rangle$ = 2.3) it is probably reasonable to use the forms

$$\frac{2}{\pi(p_T^2+1)^3} \quad \text{or} \quad \frac{A^2}{2\pi} \; e^{-Ap_T}, \quad A = 2.55 \tag{9}$$

with $\langle p_T \rangle$ = o.78 and $\langle p_T^2 \rangle$ = 1 (the former) and $\langle p_T^2 \rangle$ = o.92 (the latter), for the p_T dependence. In practice one, of course, has to test the sensitivity of the various distributions to changes in parameters.

(c) the semileptonic branching ratio B(D $\rightarrow \ell\nu$ hadrons). Since (for B<<1)

$$f_{pp\rightarrow D\rightarrow \ell} \propto B\sigma_D \tag{1o}$$

$$d\sigma_{pp\rightarrow D\bar{D}\rightarrow \ell\ell}/dM^2 \propto B^2\sigma_D$$

this parameter is important in fixing the relative scales of the single and double lepton rates. The theoretical estimates vary between 1% [37] to 2o% [38]. The semileptonic width can be estimated fairly reliably (it is believed) and the problem is how to estimate the width to purely hadronic channels. We shall use the value 2o% as the starting point. In decreasing order of believability the estimates are then:

$$\begin{aligned} \Gamma(D \rightarrow \ell\nu K) &= 2.1o^{11}s^{-1} \\ \Gamma(D \rightarrow \ell\nu \text{ hadrons}) &= 5.4\ 1o^{11}s^{-1} \\ \Gamma(D \rightarrow \text{hadrons}) &= 22\ \ 1o^{11}s^{-1} \end{aligned} \tag{11}$$

The semileptonic channels can be classified as follows [25]:

$$\begin{aligned} D \rightarrow\ & \ell\nu K \\ & \ell\nu K^* \\ & \ell\nu K(n\pi),\ n=1,2,\ldots \end{aligned} \tag{12}$$

As we shall see, a fit to the data of ref. [4] is possible only if the channels $\ell\nu K(n\pi)$ with small Q value dominate. Since there is no way of treating these channels in detail and since their only essential property in this connection is their small Q value, we shall compress all of the $K(n\pi)$ states into one effective particle of mass 1.4:

$$D \rightarrow \ell\nu\ 1.4 \tag{13}$$

The 1.4 could be anything between 1.3 and 1.6.

The leptons from the decay of D mainly go to small p_T and small x_{lepton}. For instance, Fig. 3 compares the input x dependence of $pp \rightarrow D$ with the x dependence of the leptons arising from the decay channels $D \rightarrow \ell\nu K$ and $D \rightarrow \ell\nu$ 1.4. It is obvious that any experiment excluding the small x region, i.e. slow leptons in the CMS, will be biased against seeing the effects of D production and decay (for instance, lepton polarization).

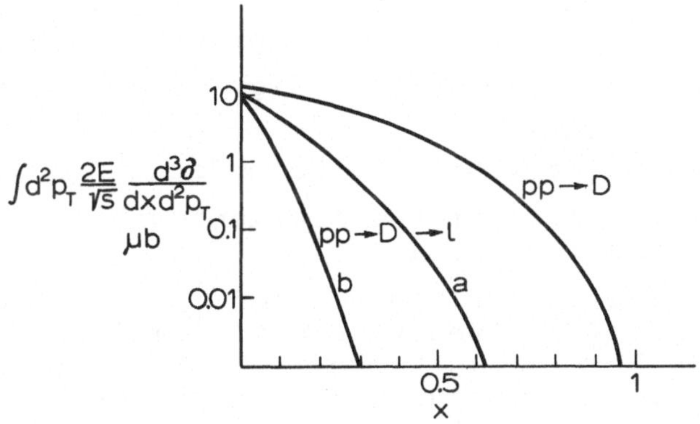

Fig. 3. The longitudinal momentum (scaled with $2/\sqrt{s}$) distribution of leptons produced via D decay at s=750. The D is produced according to Eq. (14) with N=16 μb/GeV^2, corresponding to $B\sigma_D$=36 μb. The curves marked a and b give the distribution of leptons coming from the channels $D \rightarrow \ell\nu K$ and $D \rightarrow \ell\nu$ 1.4, respectively.

Fig. 4 shows a fit to the single data at $\theta_{CM} = 30^o$ (which corresponds to fixed lepton rapidity = 1.3) and $s = 2800\ GeV^2$ [4]. This fit is obtained by assuming that the invariant D production amplitude is

$$f_{pp \to D}(x,p_T) = N(1-x)^3 \frac{2}{\pi(p_T^2+1)^3} \qquad (14)$$

where N is fixed so that

$$\begin{aligned} B(D \to e\nu K)\sigma_D &= 8\ \mu b \\ B(D \to e\nu K^*)\sigma_D &= 8\ \mu b \\ B(D \to e\nu 1.4)\ _D &= 60\ \mu b \end{aligned} \qquad (15)$$

(the numerical values of N are 2.4, 2.4 and 18 $\mu b/GeV^2$, respectively). The decay of D is treated with pure phase space; calculations with a $V \pm A$ matrix element do not significantly change the distributions discussed here. Eq. (15) implies that $B(D \to e\nu\ \text{hadrons})\sigma_D = 76\ \mu b$ and using the standard estimate 20% for the branching ratio we have $\sigma_D(\sqrt{s} = 53) = 760\ \mu b$. It is clear that there is much freedom in the fit and that the cross section numbers are at best good up to a factor 2 (for instance, the slightly different numbers given in ref. [26] are due to the fact that there the exponent of $(1-x)$ was 0 and $\langle p_T^2 \rangle$ was 0.5). However, if lepton production via charmed particles is to fit the data in Fig. 4, one sees that in any case

- σ_D must be at least of the order of 100 μb
- a dominant part of the semileptonic decays must have a small Q value

One may also note that a new version [39] of the experiment in ref. [4] gives values of $\ell/10^{-4}\pi$ following the same general trend as those in Fig. 4 but lying slightly lower. This would in direct proportion reduce the value of σ_D.

Now that one has fixed the magnitude of σ_D one may calculate the contribution of $D\bar{D}$ production to the muon pair continuum by assuming that the D's are always produced in pairs: $\sigma_D = \sigma_{D\bar{D}}$. To test the effects of $D\bar{D}$ production correlations we shall assume that the invariant two-particle distribution is given by

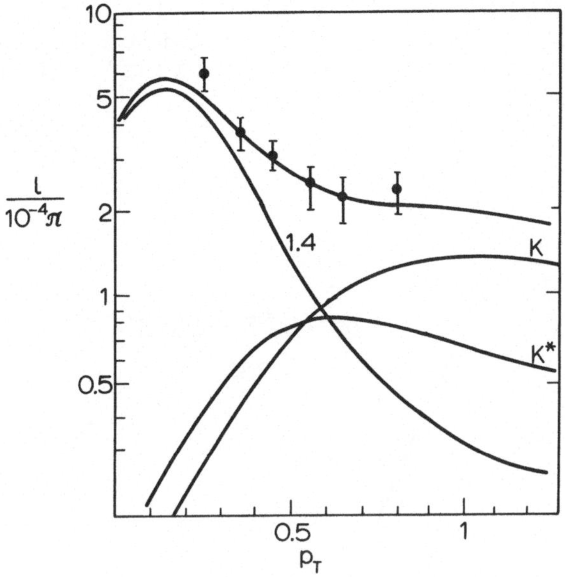

Fig. 4. A fit to the single lepton data at $\theta_{CM} = 30^o$ and $s = 2800$ of ref. [4] using Eqs. (14-15). The curves marked K, K^*, 1.4 correspond to leptons coming from the decay channels $D \to \ell\nu K$, $\ell\nu K^*$, $\ell\nu$ 1.4, respectively.

$$f_{pp\to D\bar{D}} = C_{D\bar{D}}\, f_{pp\to D} \cdot f_{pp\to \bar{D}} \tag{16}$$

where, apart from a normalization constant,

$$C_{D\bar{D}} = \begin{cases} 1, & \text{a} \\ \exp\,(-|y_{\bar{D}} - y_D|), & \text{b} \\ \exp\,(-M^2/4)\,(M^2/4 - m_D^2)^{1/2} & \text{c} \end{cases} \tag{17}$$

where $M = M_{D\bar{D}}$. Since the muon pair experiments are performed at Fermilab energies, we also must make an assumption about the energy dependence of σ_D. Although the different theoretical estimates of σ_D vary greatly in size they all agree in giving $\sigma_D(\sqrt{s} = 53)/\sigma_D(\sqrt{s} = 25) \simeq 4$. This is also the rate of variation of σ_ψ. We thus use the value $\sigma_D(\sqrt{s} = 25) = 190\ \mu b$. Calculating the muon pair distribution following from Eqs. (14-17) with the aid of Monte Carlo techniques (the calculation is equivalent with the evaluation of a 10 dimensional integral) we obtain the result shown in Fig. 5.

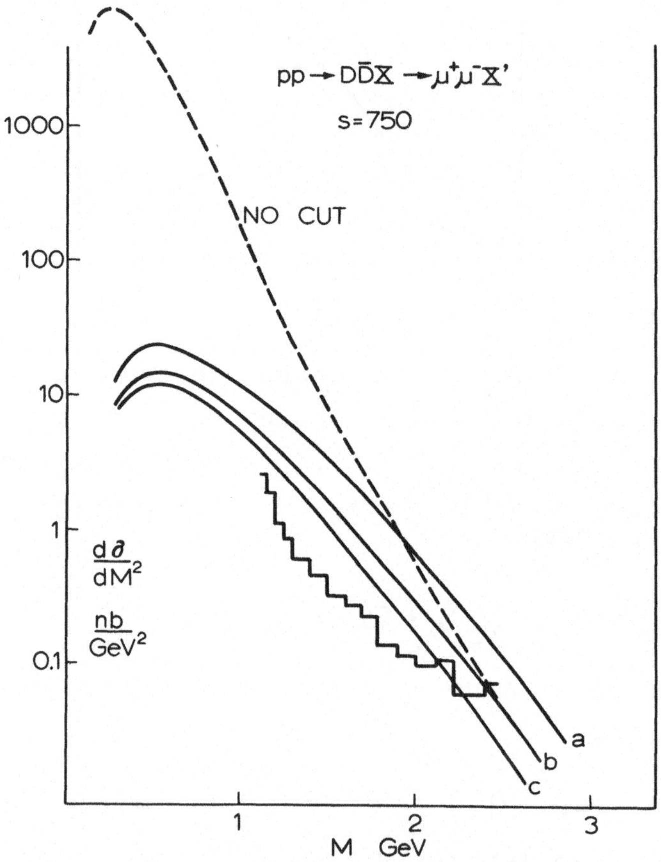

Fig. 5. The muon pair continuum arising from $D\bar{D}$ production and decay calculated on the basis of Eqs. (14-17). The solid curves marked a,b,c correspond to the $D\bar{D}$ production correlations in Eq. (17) and give the distribution of muons with $p_L^{lab} > 60$. The dashed curve marked "no cut" gives the distribution of all muons for correlation of type b, the data is from ref. [29].

4. CONCLUSIONS

On the basis of Fig. 5 we may conclude the following:

(a) If the main contribution to the muon pair continuum comes from some other source than charmed particles the calculated mass distribution must be reduced to considerably below the experimental curve, by a factor 100, say. Since the mass distribution is proportional to $B^2\sigma_D$, a fit to the single lepton spectrum, which is proportional to $B\sigma_D$ can only be maintained if B is reduced by 100 and σ_D increased by 100. These are quite unacceptable values. The only alternative then is to reduce only σ_D by a factor 100, in which case charmed particles have nothing to do with the increase of the small p_T lepton spectrum (nor with the muon pair continuum).

(b) If the muon pair continuum at $1 < M < 3$ is dominated by muons coming from semileptonic decays of charmed particles, one obtains a consistent fit to both the single and double lepton data by reducing the semileptonic branching ratio somewhat from the value 20% used in the calculation, in order to make the experimental and theoretical curves in Fig. 5 coincide. In view of the several assumptions made in the course of the calculation, it is hard to say what the reduction factor exactly should be.

In summary, it has been shown that the increase in the small p_T single lepton spectrum and the large muon pair continuum at small M may be due to a common source: semileptonic decays of charmed particles. The most evident experimental problem is the fairly large charm cross section required, of the order of 100 μb at Fermilab energies (this number refers to a sum over all charmed particle types). The experimental activity in this field is so high that the issue is bound to be resolved fairly soon. In fact, the measurement of $d\sigma/dM_{\mu e}$ would practically settle it.

REFERENCES

[1] G. Goldhaber et al., Phys. Rev. Lett. 37, 255 (1976), I Peruzzi et al., Phys. Rev. Lett. 37, 569 (1976).

[2] B. Knapp et al., Columbia, Hawaii, Illinois, Fermilab collaboration, submitted to Phys. Rev. Letters, August 1976.

[3] J.H. Christenson et al., Phys. Rev. Lett. 25, 1523

(1970), Phys. Rev. D8, 2016 (1973).
[4] L. Baum et al., Phys. Lett. 60B, 485 (1976).
[5] M. Binkley et al., Phys. Rev. Lett. 37, 574 (1976).
[6] J.D. Sullivan, Illinois Preprint ILL-(TH)-76-07, 1976.
[7] F.M. Renard, Montpellier Preprint PM/76/5, 1976.
[8] S. Pokorski and L. Stodolsky, Phys. Lett. 60B, 84 (1975).
[9] G.R. Farrar and S.C. Frautschi, Phys. Rev. Lett. 17, 1017 (1976).
[10] M. Fontannaz, Preprint LPTHE Orsay 75/29, 1975.
[11] Matts Roos, Phys. Lett. 61B, 457 (1976).
[12] J. Ranft and G. Ranft, Phys. Lett. 62B, 75 (1976).
[13] F. Halzen and K. Kajantie, Phys. Lett. 57B, 361 (1975).
[14] M. Bourquin and J.M. Gaillard, Phys. Lett. 59B, 191 (1975).
[15] G.R. Farrar and R.D. Field, Phys. Lett. 58B, 180 (1975).
[16] S. Chavin and J.D. Sullivan, Phys. Rev. 13D, 2990 (1976).
[17] N.S. Craigie and D. Schildknecht, CERN Preprint TH. 2193, 1976.
[18] S.D. Drell and T.-M. Yan, Phys. Rev. Lett. 25, 316 (1970).
[19] G.R. Farrar, Nucl. Phys. B77, 429 (1974).
[20] Minh Duong-van, Phys. Lett. 60B, 287 (1976).
[21] J.D. Bjorken and H. Weisberg, Phys. Rev. 13D, 1405 (1976).
[22] J. Finjord and F. Ravndal, Phys. Lett. 62B, 438 (1976).
[23] M.K. Gaillard, B.W. Lee and J.L. Rosner, Revs. Mod. Phys. 47, 277 (1975).
[24] M. Bourquin and J.M. Gaillard, CERN preprint, to be published.
[25] I. Hinchliffe and C.H. Llewellyn Smith, Phys. Lett. 61B, 472 (1976) and Oxford preprint, to be published.
[26] K. Kajantie, Helsinki Preprint TFT 11-76, Physics Letters, to be published, 1976.
[27] V. Barger, T. Weiler and R.J.N. Phillips, Madison preprint COO-474, 1975.
[28] T. O'Halloran, Proceedings of the 1975 International Symposium on Lepton and Photon Interactions at High Energies, ed. W.T. Kirk, Stanford, 1975.
[29] M. Binkley et al., Phys. Rev. Lett. 37, 574 (1976).
[30] D.C. Horn et al., Phys. Rev. Lett. 36, 1236 (1976).
[31] Kluberg et al., Princeton preprint CP 76-1, 1976.
[32] D. Sivers, Nucl. Phys. B106, 95 (1976), Rutherford preprint RL-76-026, 1976.

[33] A. Donnachie and P.V. Landshoff, CERN preprint TH. 2166, 1976.
[34] D. Davis, presented at Rutherford Informal Meeting on New Physics, February 1976.
[35] M.G. Albrow et al., CERN preprint, 1976.
[36] K.J. Anderson et al., Production of continuum muon pairs at 225 GeV by pions and protons, submitted to the XVIII Conference in High Energy Physics, Tbilisi, USSR, 1976.
[37] G. Altarelli, N. Cabibbo and L. Maiani, Nucl. Phys. B88, 285 (1975).
[38] John Ellis, M.K. Gaillard and D.V. Nanopoulos, Nucl. Phys. B100, 313, (1975).
[39] A. Staude, Rapporteur's talk at the XVIII Conference in High Energy Physics, Tbilisi, USSR, 1976.

THREE LECTURES ON LATTICE GAUGE THEORY*

J.B. Kogut**

Laboratory of Nuclear Studies
Cornell University
Ithaca, New York 14853, USA

Lecture 1: Motivation and Formalism of Lattice Gauge Theory

Quantum Chromodynamics. Why use a lattice? Gauge theory in zero dimensions. The lattice Hamiltonian in three dimensions. The classical continuum limit. The importance of being gauge invariant. Quark confinement in three dimensions.

Lecture 2: Two Dimensional Lattice Field Theories

Free fermions on a lattice. The massive Schwinger model. Strong coupling expansions. Padé approximants and the continuum limit. Irrelevant operators. Coupling constant renormalization and the SU(N) Thirring model.

* Lecture series presented at the International Summer School, McGill University, June 21-26, 1976 and at the International Summer Institute for Theoretical Physics University of Bielefeld, ZiF, August 23-September 4, 1976.

** Supported in part by the National Science Foundation.
A.P. Sloan Foundation Fellow.

Lecture 3: Spectrum Calculations in 3+1 Dimensional Gauge Theories.

Pure non-Abelian gauge fields. Abelian fields and a phase transition. Free fermions on a lattice. The ground state of lattice Quantum Chromodynamics. The low lying meson spectrum. Improved Hamiltonians.

MOTIVATION AND FORMALISM OF LATTICE GAUGE THEORY

Many physicists suspect that the underlying theory of strong interactions is Quantum Chromodynamics[1]-- the SU(3)' gauge theory consisting of colored quarks of various types coupled to eight colored gauge bosons. The continuum Lagrangian for the theory is

$$\mathcal{L} = \bar{\psi}\gamma^{\mu}(i\partial_{\mu} - gA^{\alpha}_{\mu}\lambda^{\alpha})\psi - \frac{1}{4} F^{\alpha}_{\mu\nu}F^{\mu\nu}_{\alpha} - \bar{\psi}M\psi \qquad (1.1)$$

where the notation is the usual. We label the fermion operator with its SU(3)' color label ψ_i and leave implicit its other quantum numbers such as isospin, strangeness, etc. One of the reasons behind the optimism for Eq. (1.1) is that the theory is asymptotically free[2] and probably explains the approximate Bjorken scaling of deep inelastic scattering. Asymptotic freedom means that the theory's running coupling constant vanishes in the deep Euclidean region of momentum space. This feature implies that at short distances the quarks and gluons of the theory are free. Therefore, one can use conventional perturbation theory to improve free quark model calculations of the short distance properties of hadrons.[3] The theory's simplicity at short distances is very intriguing and has led to a wealth of speculations concerning the unification of the ordinary interactions we observe at present-day accelerator energies.[4] What is the theory like at large distances? There is an interesting speculation that the running coupling constant grows without bound as its reference momentum becomes small. If one formulates the theory with a spatial cutoff "a", then the running coupling constant g(a) is <u>hoped</u> to have the behavior shown in Fig.1.1. An operational meaning to this curve might be the following. Consider Eq. (1.1) with the fermion mass matrix set to zero so that the theory has no explicit mass scale. It is not scale invariant,

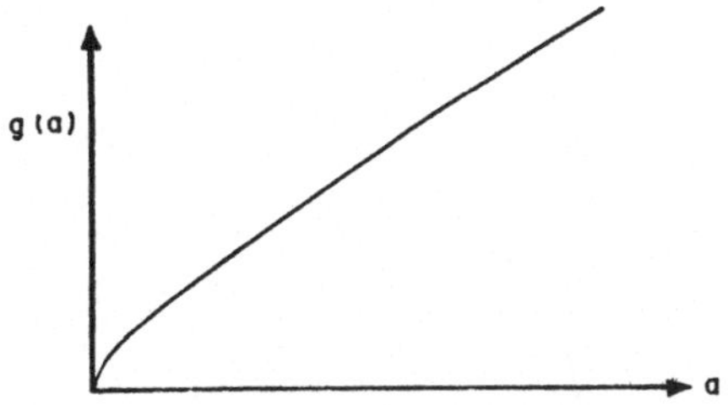

Fig. 1.1 The running coupling constant g(a) vs. a, the spatial cutoff. This is the conjectured one phase renormalization group trajectory of Quantum Chromodynamics.

however, because Fig. 1.1 distinguishes one frequency range from another. Studies of the hadron mass spectrum or the violations of Bjorken scaling should allow us to set $g^2(a)$ equal to a definite value for a definite, given cutoff $a = (1\ \text{GeV})^{-1}$, say. Then Fig. 1.1 states that in order to keep the physical predictions of the theory invariant to the choice of the cutoff "a" in the Lagrangian used in their computation, the coupling constant g(a) must be chosen to follow the curve. The possibility that g(a) diverges as $a \to \infty$ is called infrared slavery[5]. If the curve of Fig. 1.1 has no zeros except at the origin then the theory is infrared unstable -- the coupling constant literally vanishes at short distances but necessarily grows huge in the infrared region. If these speculations are true, then strong coupling methods will be necessary to discover the character of the theory's low energy mass spectrum and matrix elements. Perturbation theory of a conventional sort would be of little apparent help here. We need a new method and lattice gauge theory holds the promise of being just that.

To motivate the use of a lattice and the calculational scheme to be used, consider a very simple problem. Suppose we are interested in finding the energy spectrum of a particle in a potential V(r) which diverges as $r\to\infty$ as shown in Fig. 1.2. Obviously the wave functions of the stationary states which lie near the ground state are well localized in real space. But let's say that our only calculational tool for quantum mechanics is perturbation theory.[6] What would we then discover? In conventional perturbation theory the zeroth order wave function is a plane wave which gives a uniform probability density

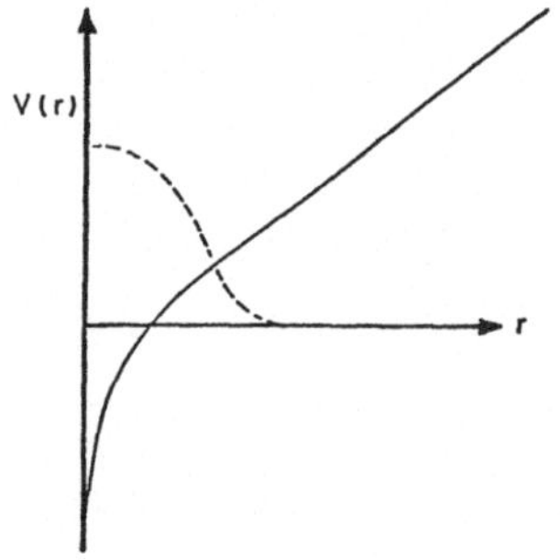

Fig. 1.2 The solid line is the confining potential as a function of distance r, and the dashed line is a low lying bound state wave function.

over all space. This state has a continuum normalization while the real stationary state has a discrete normalization. Therefore, we must expect a perturbation theory series which is infested with infinities--the zeroth order wave function must become qualitatively different as the presence of the potential is taken into account. Instead of doing a conventional treatment of this problem which first solves the kinetic piece of the Hamiltonian and then accounts for the potential iteratively, it seems more sensible to turn the procedure around. Clearly the potential V(r) is controlling the qualitative features of the low energy states and a better zeroth order wave function would have the particle near the origin in configuration space and certainly not spread everywhere. Since simplicity lies in configuration space, we might organize an approximate calculation of the spectrum into the following steps:

1. Divide space into cells (perhaps using a lattice)
2. Write the Hamiltonian as the sum of two terms

$$H = H^{static} + H^{kinetic}$$

   where H^{static} leaves the particle in one cell but $H^{kinetic}$ allows it to hop from cell to cell.
3. Solve H^{static} exactly.
4. Perturb the solutions to the H^{static} problem iteratively in $H^{kinetic}$.
5. Calculate to high enough orders that the artificial use of cells in configuration space does not seriously affect the analysis. In other words take a continuum limit (cell size $\rightarrow$ 0).

This procedure is essentially the attack we shall take in analyzing Quantum Chromodynamics. The lattice is introduced here only in order to handle the possible infrared enslavement of the theory. The lattice theory will be constructed so that its scaffolding disappears in the end and we have confidence that our cookbook prescriptions 1-5 yield an approximate, useful solution to the real theory.

We turn now to an introduction to the Hamiltonian form of lattice gauge theory. Lattice gauge theory was invented by K.G. Wilson[7] and independently by A.M. Polyakov[8] in the language of path integrals and Lagrangians. Such a formulation is well suited to making thorough field theoretic investigations using methods borrowed from statistical mechanics. Here we shall be interested in just a few aspects of the theory and the Hamiltonian with its continuous time development and spatial lattice is convenient.[9] Our objective is an understanding of the low energy spectrum and matrix elements of the theory and a Hamiltonian complete with canonical quantization exposes the low energy features of the theory in a particularly transparent fashion.

Our first task is to define and understand the conceptual basis of lattice gauge theory. To begin we choose a simple setting (following L. Susskind's Bonn Summer School[10] notes) and consider a world of just two spatial points in a time continuum. We wish to implement the requirement of local color gauge invariance as an exact symmetry for any value of the lattice spacing a. There will be a Hamiltonian H and we must determine its various elements. Let quarks live on the sites i=1,2. The fermion fields will be denoted $\psi_j(i)$ where j is a color index. Frequently the color index will be summed over and it will be left implicit. The precise way of placing fermion degrees of freedom on a lattice such that the Dirac equation is retrieved in the continuum ($a \to 0$) limit is rather tricky. It will be discussed in later lectures. For our purposes here the generic term $\psi_j(i)$ will suffice. We require that the H be locally color gauge invariant. A local color transformation is implemented on the fermions through a SU(3)' rotation matrix V(i) in the fundamental representation

$$\psi(i) \to V(i)\psi(i). \tag{1.2}$$

Therefore, we can think of fictitious color frames of reference at each lattice site i as shown in Fig. 1.3. Each term in H must be invariant to independent rotations of each frame. Clearly a mass term

$$\mu : \sum_i \psi^\dagger(i)\psi(i) : \qquad (1.3)$$

satisfies this criterion. However, the construction of a discrete form of the kinetic energy is not so simple because the obvious hopping Hamiltonian,

$$H_q = i[\psi^\dagger(2)\psi(1) - \psi(2)^\dagger\psi(1)] \quad \text{(tentative)} \qquad (1.4)$$

is not locally color gauge invariant. To patch up Eq. (1.4) we follow the original idea of Yang and Mills and introduce an SU(3)' rotation matrix U which relates the color frames at sites 1 and 2. In the fundamental representation we can write

$$U_3 = \exp(i\frac{\lambda^\alpha}{2} B_\alpha) \qquad (1.5)$$

where $B_\alpha (\alpha=1,2,\ldots,8)$ is an angular variable. It follows from the geometric definition of U that its color transformation law reads

$$U \to V(1)UV(2)^{-1}. \qquad (1.6)$$

If we write a new hopping Hamiltonian,

$$H_q = i[\psi^\dagger(1)U\psi(2) - \psi^\dagger(2)U^{-1}\psi(1)], \qquad (1.7)$$

then we have restored the local color symmetry. Later in these developments U will become a dynamical degree of freedom of the theory (B_α will be closely related to the conventional gauge fields) and these considerations will acquire more life.

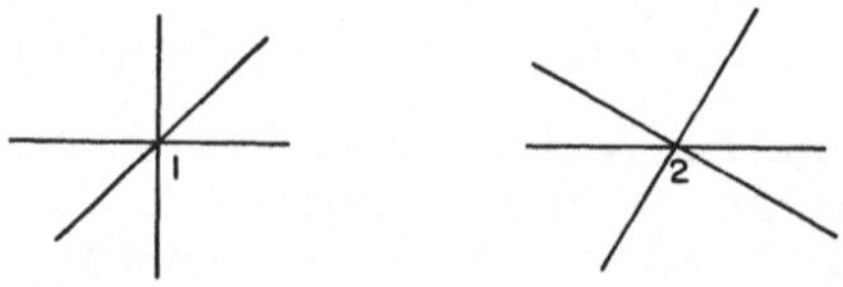

Fig. 1.3 Color frames of reference at adjacent lattice sites.

Let's study the U operators in isolation. These degrees of freedom relate adjacent color reference frames so we say that they occupy the links of the lattice. Let $E_\alpha(i)$ be the generators ($\alpha=1,\ldots,8$) of color rotations at each site. They satisfy the commutation relations

$$[E_\alpha(i), E_\beta(j)] = if_{\alpha\beta\gamma}\delta_{ij}E_\gamma(j) \qquad (1.8)$$

where $f_{\alpha\beta\gamma}$ are the structure constants of SU(3)'. The variables $E_\alpha(i)$ will be related to non-Abelian electric flux and will play an important role in subsequent developments. If we consider infinitesimal local color rotations so that

$$V(i) \simeq 1 + \frac{i\varepsilon^\alpha}{2}(i)E_\alpha(i) + \ldots \qquad (1.9)$$

where $\epsilon^\alpha(i)$ is an infinitesimal color vector, then Eq. (1.6) implies

$$[E_\alpha(1), U_3] = \frac{\lambda_\alpha}{2}U_3, \quad [E_\alpha(2), U_3] = -U_3\frac{\lambda_\alpha}{2}. \qquad (1.10)$$

Since Eq. (1.8) means that $E_\alpha(i)$ are color vectors and since U operator relates adjacent color frames of reference, we have the important result that

$$E(2) = -U_8E(1). \qquad (1.11)$$

Here U_8 means that U is expressed in the adjoint representation. This result can be obtained algebraically from Eq. (1.10), but its geometrical significance is our main concern here. It means that the "color flux" maintains its magnitude across links, but the gauge field (U or B) which lives on links rotates the flux in color space,

$$E(1)^2 = E(2)^2. \qquad (1.12)$$

This idea plays an important role in everything which follows.

Next we consider the space of states of the gauge field. At each site there is an SU(3)' group so we must label the states within an irreducible representation. Each irreducible representation is specified by two numbers, its quadratic and cubic Casimir operators (call them C_2 and C_3, respectively), and each state within a representation is labelled by two additional quantum numbers which are analogous to I_3, the third component of isospin and Y, the hypercharge, of ordinary SU(3). Denote these last two quantum numbers with a 2 dimensional vector $\underset{\sim}{m}$, so that a state can be written

$$|\underset{\sim}{m}(i);\ C_2(i), C_3(i)\rangle. \tag{1.13}$$

A complete set of states of the two sites can then be made from the products

$$\prod_{i=1}^{2} |\underset{\sim}{m}(i);\ C_2(i), C_3(i)\rangle. \tag{1.14}$$

However, not all of these possibilities are necessary since the quadratic Casimir operator is just $E(i)^2$ and we noted above that $E(1)^2 = E(2)^2$. Therefore,

$$C_2(1) = C_2(2). \tag{1.15}$$

Since C_3 is also an invariant constructed from the group generators, $C_3(1) = C_3(2)$. To explicitly construct the states in Eq. (1.14), we begin with the ground state $|0\rangle$ which is defined by

$$E(1)|0\rangle = E(2)|0\rangle = 0 \tag{1.16}$$

and apply the matrix elements of U. For example, the states

$$(U_3)^k_{\ \ell}|0\rangle \tag{1.17}$$

fill out the $(\bar{3},3)$ representation of SU(3)xSU(3). To check this claim, apply Eq. (1.10) and (1.16) to the computation,

$$\begin{aligned} E(1)^2 U_3|0\rangle &= (\tfrac{1}{2}\lambda^\alpha)(\tfrac{1}{2}\lambda_\alpha)U_3|0\rangle \\ &= C_2(3)U_3|0\rangle \end{aligned} \tag{1.18}$$

where $C_2(3)$ is the quadratic Casimir operator of the 3 dimensional representation, $C_2(3) = \frac{4}{3}$.

Before developing more aspects of the lattice theory, let's comment upon the physical meaning of E and U. We anticipate the continuum correspondences for these variables in the 3+1 dimensional theory,

$$\vec{E}_\alpha(r)\cdot\hat{n} = \frac{a^2}{g} F_\alpha^{0i}(r) n_i \tag{1.19}$$

where n is a unit vector pointing between lattice sites and F_α^{0i} is the usual field strength. In addition,

$$U_3(r,\hat{n}) = \exp(iag\,\vec{A}_\alpha(r)\cdot\hat{n}\,\frac{\lambda^\alpha}{2}) \tag{1.20}$$

in the class of gauges $A^0(r) = 0$. In the case of an Abelian gauge theory Schwinger[11] has taught us how to think about these operators. In Schwinger's point-separated currents one enforces gauge invariance in an exact form by considering objects like

$$\lim_{\underset{\sim}{\varepsilon}\to 0} \bar{\psi}(\underset{\sim}{r}+\underset{\sim}{\varepsilon})\gamma_\mu\gamma_5\, e^{ig\underset{\sim}{A}(\underset{\sim}{r})\cdot\underset{\sim}{\varepsilon}}\psi(\underset{\sim}{r}) . \tag{1.21}$$

By considering local gauge transformations one can show the gauge invariance of this fermion bilinear even when $\underset{\sim}{\varepsilon} \neq 0$. Alternatively, we can argue that the state which this current makes out of the vacuum satisfies Gauss' law and is therefore physically meaningful. To see this consider the state

$$|s\rangle = e^{ig\underset{\sim}{A}\cdot\underset{\sim}{\varepsilon}}|0\rangle . \tag{1.22}$$

Since $\underset{\sim}{E} = \dot{\underset{\sim}{A}}$ in this class of gauges, the exponential $\exp(ig\underset{\sim}{A}\cdot\underset{\sim}{\varepsilon})$ is a shift operator for $\underset{\sim}{E}$. Therefore,

$$\langle s|\underset{\sim}{E}|s\rangle = g\underset{\sim}{\varepsilon} . \tag{1.23}$$

The close analogy of these last two equations to the lattice Eq. (1.17) and (1.18) is important.

Now let's put the quark degrees of freedom back into the 2 site lattice and consider color rotations. First we wish to identify the generator of global color rotations, i.e. identical rotations at both sites

$$U \to VUV^{-1} \quad \text{and} \quad \psi(i) \to V\psi(i) . \tag{1.24}$$

For an infinitesimal rotation we have the generalization of Eq. (1.9),

$$V \simeq 1 + \frac{i}{2} \varepsilon^{\alpha} C_{\alpha} \qquad (1.25)$$

where C_{α} is the total color. The infinitesimal forms of Eq. (1.24) read

$$\delta U = [C_{\alpha}, U] = \frac{i}{2} \varepsilon^{\alpha} [\lambda^{\alpha}, U]$$

$$\delta \psi = [C_{\alpha}, \psi] = \frac{i}{2} \varepsilon^{\alpha} \lambda_{\alpha} \psi \ . \qquad (1.26)$$

These equations indicate that

$$C_{\alpha} = \sum_{i=1}^{2} \{\psi^{\dagger}(i) \frac{\lambda_{\alpha}}{2} \psi(i) + E_{\alpha}(i)\}. \qquad (1.27)$$

In the same way we find the generator of local rotations,

$$G_{\alpha}(i) = \psi^{\dagger}(i) \frac{\lambda_{\alpha}}{2} \psi(i) + E_{\alpha}(i). \qquad (1.28)$$

Now the requirement of local color gauge invariance can be stated in a precise form. A physically meaningful state in the lattice theory must be invariant to local color rotations and therefore must be annihilated by $G_{\alpha}(i)$,

$$G_{\alpha}(i)|\text{Physical State}\rangle = 0. \qquad (1.29)$$

Later we shall identify this statement as the lattice version of the non-Abelian Gauss' law. This oberservation should further motivate Eq. (1.29).

Now we can construct some examples of gauge invariant, physical states. The vacuum is locally invariant and is annihilated by $\psi(i)^{(-)}$, $\psi^{\dagger}(i)^{(-)}$ and $E_{\alpha}(i)$. A quark and an anti-quark at site i is clearly gauge invariant if the color indices are contracted

$$\sum_{j} \psi_{j}^{\dagger}(i)\psi_{j}(i)|0\rangle = \psi^{\dagger}(i)\psi(i)|0\rangle. \qquad (1.30)$$

If we place a quark at one site and an anti-quark on a neighboring site, then gauge invariance requires the presence of a U_3 operator between them,[12]

$$\psi^{\dagger}(1)U_{3}\psi(2)|0\rangle. \qquad (1.31)$$

The U_3 operator insures that the correct amount of electric flux flows between site 1 and 2 to satisfy Gauss' law. It is also true that Eq. (1.31) is constructed with all color indices contracted locally into scalars so the result is invariant to local color rotations.

Finally let's write a locally color gauge invariant Hamiltonian for the 2 site world. A fermion mass and a hopping term have already been constructed. In addition, we can assign energy to the pure gauge field excitations. Observe that the magnitude of the color vector $E_\alpha(i)$ is locally gauge invariant. We have indicated that this qunantity is closely related to the color electric field of continuum physics so it is reasonable to add

$$\frac{g^2}{2a} \sum_i E(i)^2 \tag{1.32}$$

to the Hamiltonian. The coefficient of this term will be discussed later. Now, an interesting 2 site Hamiltonian reads

$$H = \frac{g^2}{2a} \sum_i E(i)^2 + \frac{i}{a} \left[\psi^\dagger(1)U\psi(2) - \psi^\dagger(2)U^\dagger\psi(1)\right] + \\ + \mu \sum_i \psi^\dagger(i)\psi(i) \tag{1.33}$$

There are additional terms which can appear in H once we consider a cubic lattice in three spatial dimensions. For example, Eq. (1.33) has no magnetic field effects because the relevant term requires more than 2 sites for its construction. We turn to three dimensional lattices now.

Most of the concepts introduced for the 2 site model generalize in a trivial fashion to the realistic situation so this discussion will be brief. Label lattice sites with a triplet of integers $r = (r_x, r_y, r_z)$ and directed links with r and a unit vector specifying the first site's nearest neighbor as shown in Fig. 1.4. As before, fermion fields $\psi(r)$ occur at sites, gauge fields $U(r,\hat{n})$ occupy links and electric flux $\vec{E}_\alpha(r)\cdot\hat{n}$ lives on sites and points along the links. The algebra of $E(r)$ and $U(r,n)$ generalizes in the obvious fashion.

For example, the generalization of Eq. (1.11) reads

$$E(r+\hat{n})\cdot(-\hat{n}) = -U_8(r,\hat{n})E(r)\cdot\hat{n}. \tag{1.34}$$

The generator of local color gauge transformations reads

$$G_\alpha(r) = \frac{1}{2}\,\psi^\dagger(r)\lambda_\alpha\psi(r) + \sum_{\hat{n}} E_\alpha(r)\cdot\hat{n} \tag{1.35}$$

where the sum $\sum_n$ runs over the six unit vectors emanating from r. The condition that physical states be locally color gauge invariant reads as before[9]

$$G_\alpha(r)\,|> = 0 \tag{1.36}$$

and this will be interpreted below as the requirement that physical states satisfy the non-Abelian version of Gauss' law.

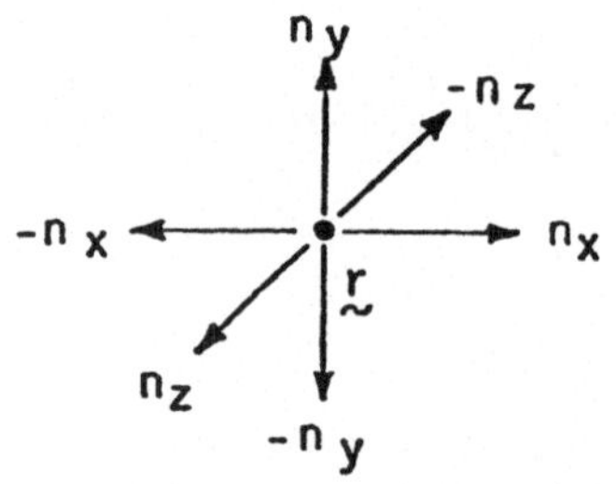

Fig. 1.4 Three dimensional spatial lattice showing one site r and the six directions to its nearest neighbors.

Next consider some examples of physical states. We obtain them by applying gauge invariant operators to the vacuum. Consider first pure gauge field examples. We can multiply strings of U operators together, but in order to make only locally color gauge invariant objects color indices must be contracted locally. In other words, let Γ be a closed path in the lattice as shown in Fig. 1.5 and label the links composing Γ with increasing integers as shown. Then an operator of the form

$$U(\Gamma) = \mathrm{Tr}\; U(1)U(2)U(3)\ldots U(n) \tag{1.37}$$

is locally color gauge invariant. The simplest such $U(\Gamma)$ occurs when we choose Γ to be a square made of four links. Note that if Γ is a non-intersecting, closed loop, then $U(\Gamma)|0\rangle$ has a particularly simple physical interpretation. If $U(\Gamma)|0\rangle$ is composed of U_3 matrices only, then each link of Γ has $E^2 = C_2(3) = \frac{4}{3}$. Thus $U(\Gamma)|0\rangle$ is a closed, continuous loop of color electric flux. We learn from this example that the flux in lattice gauge theory is quantized, i.e. on each link one can find

$$[E(r)\cdot\hat{n}]^2 = 0,\ \frac{4}{3},\ 3,\ \frac{10}{3},\ \frac{16}{3},\ \ldots \tag{1.38}$$

where the numbers on the right-hand side are the possible values of the quadratic Casimir operator for SU(3)'. The discrete character of the spectrum of this operator follows from the compactness of SU(3)'.

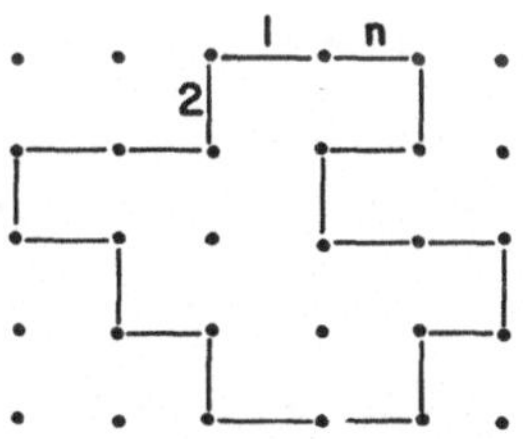

Fig. 1.5 A closed path in the lattice.

Now consider examples of physical states which also contain fermions. Meson excitations can be

$$\sum_k \bar{\psi}_k(i)\gamma\psi_k(i)|0\rangle = \bar{\psi}(i)\gamma\psi(i)|0\rangle \tag{1.39}$$

where the color indices k are locally summed so that the state is a color singlet and the γ indicates an appropriate Dirac matrix. Baryon excitations can be

$$\varepsilon_{k\ell m}\ \psi_k^\dagger(i)\psi_\ell^\dagger(i)\psi_m^\dagger(i)|0\rangle \tag{1.40}$$

since the $\varepsilon_{k\ell m}$ symbol makes Eq. (1.40) a color singlet. In addition, quarks and anti-quarks at different sites can be connected by U operators to make excited meson configurations. An excited baryon state shown in Fig.1.6 could be

$$\varepsilon_{k\ell m}\psi_p^\dagger(r_1)U_{pk}(1)\psi_q^\dagger(r_2)U_{q\ell}(2)\psi_r(r_3)U_{rm}(3)|0\rangle . \quad (1.41)$$

Many other examples in addition to these simple ones teach us that the physical space of states closely resembles that of the Dual Resonance Model. To make a real connection we must consider lattice dynamics. We turn to this topic and consider the construction of a Hamiltonian.

There are several demands placed on the Hamiltonian It must have local color gauge invariance. When its fields are treated classically (slowly varying on the scale of the lattice spacing a), then the conventional continuum Hamiltonian must be retrieved as $a \to 0$. All terms in the Hamiltonian must be local in real space. In practice, this locality requirement means that the $a \neq 0$ Hamiltonian should couple only fields separated by at most a few links. Our two site model suggested three terms which can appear in H. Another term which accounts for magnetic effects in the pure gauge field dynamics is constructed as follows. Consider the square shown in Fig. 1.7 and the related operator,

$$-\frac{1}{g^2 a}\,\mathrm{Tr}\,U(1)\,U(2)\,U(3)\,U(4) + \mathrm{h.c.} \quad (1.42)$$

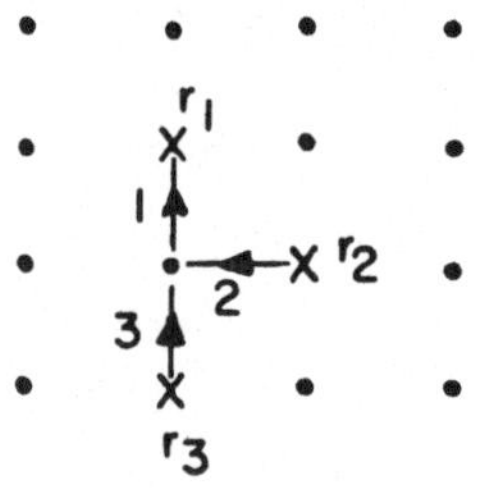

Fig. 1.6 An excited baryon state. The three links of flux tie the three quarks together in a gauge invariant fashion.

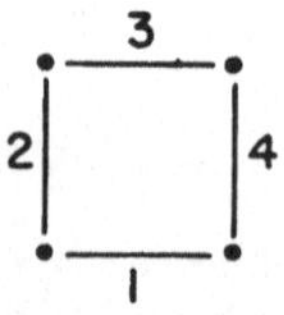

Fig. 1.7 Smallest square in the lattice associated with magnetic fields in the lattice Hamiltonian.

A term which can appear in H and which satisfies the requirements listed above is obtained by summing Eq. (1.42) over the entire lattice,

$$H_{\text{magnetic}} = -\frac{1}{g^2 a} \sum_{\text{boxes}} \text{Tr } UUUU + \text{h.c.} \qquad (1.43)$$

Let's take the classical a→0 limit here. That means we treat the dynamic variables as slowly varying on the scale of a and we ignore quantum effects such as coupling constant renormalization. We freely drop terms which are multiplied by extra powers of a when Eq. (1.43) is analyzed according to these rules. The product of four U operators is

$$UUUU = e^{iag \frac{\lambda^{\alpha}}{2} A_{\alpha}(r,n)} \, e^{iag \frac{\lambda^{\alpha}}{2} A_{\alpha}(r+n,m)}$$

$$e^{iag \frac{\lambda^{\alpha}}{2} A_{\alpha}(r+n+m,-n)} \, e^{iag \frac{\lambda^{\alpha}}{2} A_{\alpha}(r+m,-m)} \qquad (1.44)$$

where the quantities $A_{\alpha}(r,\hat{n})$ are the vector potentials at the site r which point along the link $\hat{n}$. The unit vectors $\hat{n}$ and $\hat{m}$ are directed around the square of Fig. 1.7. It will be convenient to write Eq. (1.44) as a single exponential, but this requires some care because the various factors of the equation do not commute. However, since we are taking the a→0 classical limit most of the non-commuting effects will not contribute Consider the formula

$$e^{\Theta_1} e^{\Theta_2} = e^{\Theta_1+\Theta_2} e^{1/2[\Theta_1,\Theta_2]} + \ldots \qquad (1.45)$$

where the + ... contains higher order terms. In the application here Θ_1 is of order a, Θ_2 is of order a and $[\Theta_1,\Theta_2]$ is of order a^2. The higher order terms in Eq. (1.45) will not contribute to the a→0 limit. The commutator of interest here is

$$\frac{1}{2}[iag \frac{\lambda_{\alpha}}{2} A_{\alpha}(r,\hat{n}), \; iag \frac{\lambda_{\beta}}{2} A_{\beta}(r+\hat{n},\hat{m})] = -$$

$$= -\frac{1}{2} a^2 g^2 A_{\alpha}(r,\hat{n}) A_{\beta}(r+\hat{n},\hat{m}) \, [\frac{1}{2}\lambda_{\alpha}, \frac{1}{2}\lambda_{\beta}]$$

$$= -\frac{1}{2} a^2 g^2 A_{\alpha}(r,\hat{n}) A_{\beta}(r+\hat{n},\hat{m}) \; i f_{\alpha\beta\gamma} \frac{1}{2} \lambda_{\gamma} \qquad (1.46)$$

where the $f_{\alpha\beta\gamma}$ are the structure constants of SU(3)'. Collecting everything,

$$\mathrm{UUUU} \simeq e^{iag \frac{\lambda^\alpha}{2} [A_\alpha(r,n)+A_\alpha(r+n,m)-A_\alpha(r+m,n) -A_\alpha(r,m)-agA_\beta(r,n)A_\gamma(r,m)f_{\mu\gamma\alpha}]} \quad (1.47)$$

where we have used the identity $A_\alpha(r,m)=-A_\alpha(r+\hat{m},-\hat{m})$. Identifying discrete gradients in Eq. (1.47) we have

$$\mathrm{Tr\ UUUU} \simeq \mathrm{Tr}\ e^{-ia^2 g \frac{\lambda^\alpha}{2} [\partial_m A_n^\alpha(r)-\partial_n A_m(r)+gf^{\alpha\beta\gamma} A_n^\beta(r)A_m^\gamma(r)]}$$

$$\simeq -a^4g^2 \frac{1}{4} (\partial_m A_n^\alpha - \partial_n A_m^\alpha + gf^{\alpha\beta\gamma}A_n^\beta A_m^\gamma)^2 \quad (1.48)$$

after using the identity $\mathrm{Tr}\ \lambda_\alpha\lambda_\beta = 2\delta_{\alpha\beta}$. Thus

$$-\frac{1}{g^2 a} \sum_{\text{boxes}} \mathrm{Tr\ UUU} + \text{h.c.} \simeq \frac{1}{g^2 a} \int \frac{1}{a^3} d^3x\ a^4 g^2 \frac{1}{2}(F_{mn}^\alpha)^2$$

$$\simeq \frac{1}{2} \int (F_{mn}^\alpha)^2\ d^3x \quad (1.49)$$

where we have identified the spatial (magnetic) components of the non-Abelian field strength tensor. We identify this result as the conventional continuum magnetic term in the class of gauges $A^0(r) = 0$. This exercise verifies that the coefficient $-\frac{1}{g^2 a}$ in the lattice Hamiltonian is correct. Similar exercises yield the other terms of H with the appropriate coefficients,

$$H = \frac{g^2}{2a} \sum_{r,\hat{n}} E^2(r,\hat{n}) + \frac{1}{a} \sum_{r,\hat{n}} \psi^\dagger(r) \frac{\underset{\sim}{\sigma}\cdot\underset{\sim}{n}}{i} U(r,n)\psi(r+n)+\text{h.c.}$$

$$-\frac{1}{g^2 a} \sum_{\text{boxes}} \mathrm{Tr\ UUUU} + \text{h.c.} + \mu \sum_{\text{sites}} \psi^\dagger(r)\gamma_0\psi(r) \quad (1.50)$$

As earlier in this lecture, our treatment of the fermion fields on the lattice is purely impressionistic In later lectures this tricky topic will be discussed in detail.

It is interesting to study the lattice quantization rules and the generator of local color rotations in the same classical limit. If we substitute Eq. (1.19) into the commutation relation Eq. (1.10) we have

$$\left[\frac{a^2}{g} F^{Oi}_{\alpha}(x),\ iagA^{j}_{\beta}(x)\right] = \delta_{\alpha\beta}\delta_{ij} \tag{1.51}$$

as $a\to 0$. Identifying the continuum Dirac delta function, we find

$$\left[F^{Oi}_{\alpha}(x), A^{j}_{\beta}(x)\right] = -i\delta_{\alpha\beta}\delta^{3}(x-y) \tag{1.52}$$

which is the usual canonical quantization condition in the axial gauge $A^{O}_{\alpha}(x) = 0$. Next consider the color carried by a link

$$\begin{aligned} C(r,\hat{n}) &= E(r)\cdot\hat{n} + E(r+\hat{n})\cdot(-\hat{n}) \\ &= (1-U_8(r,\hat{n}))E(r,\hat{n}) \end{aligned} \tag{1.53}$$

In the classical continuum limit,

$$C_{\alpha}(r,n) \approx -iagA^{n}_{\beta}f_{\alpha\beta\gamma}\cdot\frac{a^2}{g^2}F^{on}_{\gamma} = -ia^3(A^{n}x\xi^{n}) \tag{1.54}$$

where x refers to color space and ξ_{α} is the electric field. Therefore, the color density is

$$(A^{i}x\xi^{i})_{\alpha} \tag{1.55}$$

which is the correct answer in this gauge.

Next let's gain some familiarity with the Hamiltonian. Suppose that a is fixed and the coupling constant g is large. Then the static terms

$$\frac{g^2}{2a}\sum[E(r)\cdot\hat{n}]^2 + \mu\sum\psi^{\dagger}(r)\gamma_{o}\psi(r) \tag{1.56}$$

dominate the energetics. The vacuum state $|0\rangle$ must be fluxless, quark-free and locally gauge invariant. Next consider the energy of a single-quark. We shall argue that it is infinite, so in this static strong coupling approximation, the theory confines its fermions.[13] Since quarks carry color (they are in the 3 representation of SU(3)'), the requirement of local color gauge invariance (Gauss' law) means that a $U_3(r,\hat{n})$ operator must occupy a link emanating from the quark. But this operator has an open color index, so it must be connected to another $U_3(r+\hat{n},\hat{m})$, etc. Each U_3 operator costs an energy $\frac{g^2}{2a}\cdot C_2(3)$ and there are an infinite number of links in the string attached to the quark. More simply, the quark is a source of a flux tube of infinite extent and infinite energy--the energy of this configuration diverges like the linear dimensions of space. Note that this infinity is a large distance phenomenon and is called "confinement through soft forces". The fact that the large g form of lattice gauge theory embodies quark confinement in this natural fashion is one of its most interesting and promising features. Clearly the only finite energy states will be color singlets so many ideas of the color quark model[14] have a natural setting here. Hopefully, the lattice theory will lead to a quantitative, consistent theory of bound states of quarks and gluons. The bulk of these lectures describe the status of attempts to implement this hope.

Next we should gain a rough understanding of the other kinetic terms in H. Consider their effects in old-fashioned perturbation theory. As a first example retain only the pure gauge field pieces of the theory (no quarks now) and place an infinitely heavy source of color (in the 3 representation, say) at one point in the lattice and a heavy sink of color at another point as shown in Fig. 1.8. A line of U matrices must then lie between the source and sink in order to insure Gauss' law. We will refer to such configurations as a string or tube of flux.

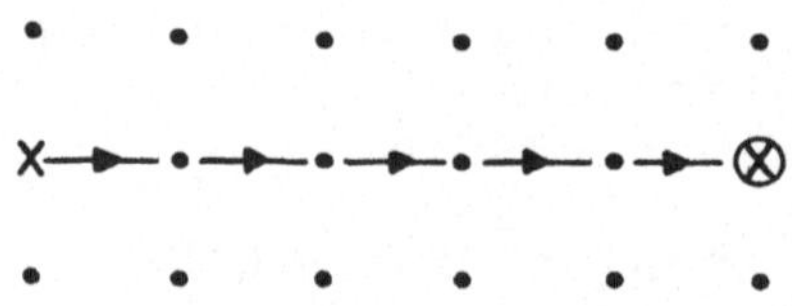

Fig. 1.8 A line of flux connecting a source and sink of color.

When $H_{magnetic}$ acts on this state one of its links can overlap with a link on the string and produce a string with wiggles as shown in Fig. 1.9. Of course, the overlapping U_3 matrices produce flux on one of the links which can either be in the singlet, $C_2(1) = 0$, or octet, $C_2(8)=3$, representations.

Now consider some simple effects of the quark hopping term H_q in Eq. (1.50). This term propagates quarks in color singlet states and it allows strings to break in a locally gauge invariant way. For example, H_q might act on the string in Fig. 1.8. The U_3 operator in H_q can annihilate a piece of flux in the initial state. However, when this happens a quark and an antiquark are also materialized by H_q so that Gauss' law is always fulfilled and no quarks are set free, Fig. 1.10. Thus, order by order in perturbation theory propagation occurs in the theory and since the dynamics is gauge invariant there is quark confinement order by order in this expansion method.

A simple and important strong coupling analysis can be done for Fig. 1.8 keeping only the pure gauge field pieces of H. The energy of this state is

$$\frac{g^2}{2a} C_2(3) \cdot N \qquad (1.57)$$

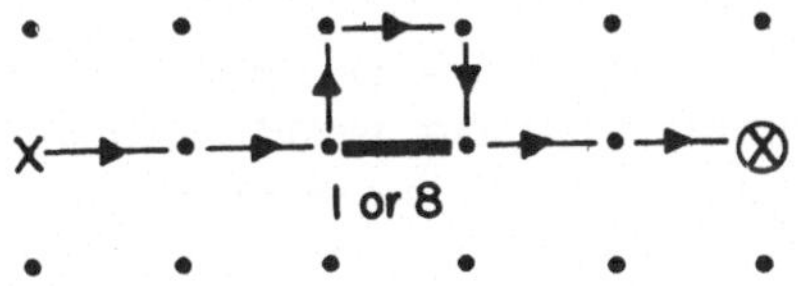

Fig. 1.9 A magnetic field fluctuation on Fig. 1.8.

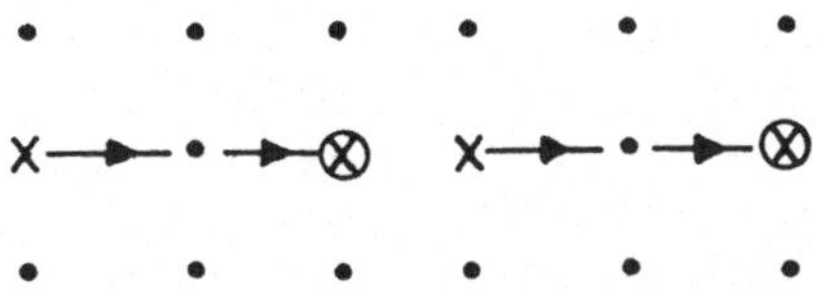

Fig. 1.10 A quark hopping fluctuation on Fig. 1.8 which allows the flux line to break in a gauge invariant, quark confining fashion.

where N is the number of links separating source from sink. Now write N in terms of physical units, GeV^{-1}, say. N equals L/a where L is the physical distance between the source and sink as measured in GeV^{-1} units. Thus, Eq. (1.57) becomes

$$\frac{g^2}{2a^2} \cdot C_2(3) \cdot L \tag{1.58}$$

But this energy should be physical, i.e. independent of the lattice spacing. Thus, the coefficient of L, the tension in the string, must be a physical number having units GeV^2,

$$g^2/a^2 = \text{constant} \quad \text{(large g)} \tag{1.59}$$

Therefore, the lattice coupling constant g must vary linearly with a. This is a nice conclusion because it shows that the strong coupling limit of the theory has coupling constant renormalization effects which favor confinement at large distances and free field behavior at short distances. Kinetic perturbation theory is expected to change Eq. (1.59) at small g^2 so that the renormalization group trajectory becomes asymptotically free, $g^2(a) \sim (\ell n\ a)^{-1}$, on fine-grained spatial lattices. Eq. (1.59) motivates the large a behavior of Fig. 1.1. One of the grand objectives of the calculational method to be described in the next two lectures is to extend Eq. (1.59) out of the strong coupling regime and prove or disprove the conjecture embodied in Fig. 1.1. The lack of zeros away from $g^2 = 0$ and singularities in the intermediate coupling region would show that quark confinement is a natural phenomenon in the asymptotically free theory of Eq. (1.1).

2. TWO DIMENSIONAL LATTICE FIELD THEORIES

Before attacking four dimensional gauge theories, we consider simple 1+1 dimensional models. These exercises demonstrate several general features of lattice theories and their continuum limits in relatively tractable, transparent settings. These lattice calculations will yield hard numbers which can be compared to the results of conventional continuum analyses. Our detailed discussion will be limited to the (massive) Schwinger model[15] although a few remarks will be made concerning the Thirring models.[16]

We begin by constructing a lattice Hamiltonian which reduces in the continuum limit to the conventional Schwinger model. First consider discrete forms of the free 1+1 dimensional Dirac equation,

$$i\dot{\psi} = i\alpha\partial_z\psi, \qquad \alpha = \gamma_5 = \begin{pmatrix} 0 & 1 \\ 1 & 0 \end{pmatrix} \tag{2.1}$$

An obvious attempt to place this equation on the lattice would be to (1) place a 2 component fermion field on each site and (2) interpret $\partial_z\psi$ as the difference $\frac{1}{2a}[\psi(n+1) - \psi(n-1)]$. However, we shall see that this attempt fails: in the continuum limit we find two mass-less Dirac particles from such a scheme.[17] To demonstrate this substitute plane waves into the discrete Dirac equation,

$$i\dot{\psi}(n) = \frac{i}{2a}\,[\psi(n+1)-\psi(n-1)] \tag{2.2}$$

and find the (ω) energy-momentum (ℓ=ka) relation,

$$\omega = \pm\,\frac{\sin\,\ell}{a}\,, \quad -\pi<\ell = ka < \pi \tag{2.3}$$

which is plotted in Fig. 2.1. This should be compared with the energy-momentum relation for the continuum Dirac equation,

$$\omega = \pm k \tag{2.4}$$

which describes right and left moving fermions and anti-fermions as shown in Fig.2.2. Now concentrate on those

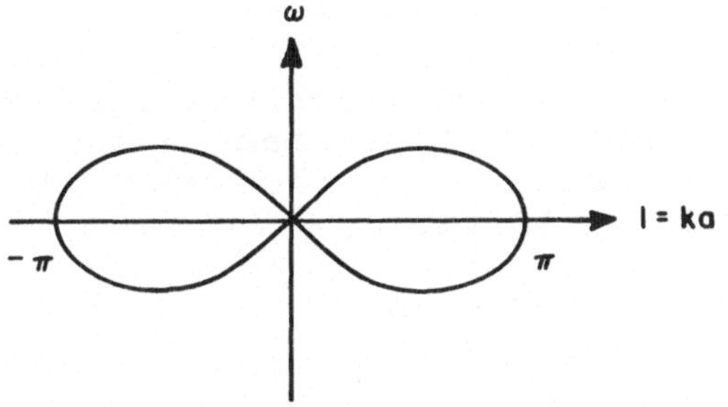

Fig. 2.1 The energy-momentum relation for the discrete Dirac equation in one spatial dimension.

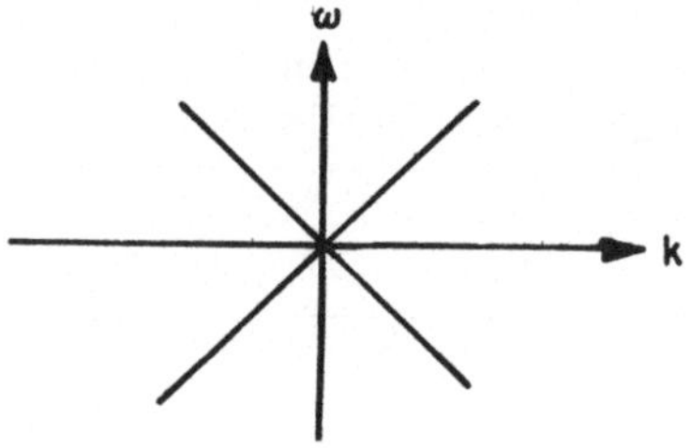

Fig. 2.2 The energy-momentum relation for the continuum Dirac equation.

energies of Fig. 2.1 which remain finite as $a \to 0$. This restricts us to the neighborhood of the origin $\ell \approx ka$ and the edges of the Brillouin zones, $\ell \approx ka \pm \pi$. But there are eight energies corresponding to given momenta in these regions and thus two species of fermions. The point is that the lattice itself allows momenta near the edge of the Brillouin zones to have low energy and these degrees of freedom survive in the continuum limit and give species doubling.[17]

These are many ways of correcting the unsuccessful approach just described. One way begins with the observation that there were too many degrees of freedom in the unit cell. By halving the number one can give a lattice description of a single fermion species. To do this place a single component fermion field $\phi(n)$ on each lattice site and let the hopping Hamiltonian be

$$H = \frac{i}{2a} \sum_n \left[\phi^\dagger(n)\phi(n+1) - \text{h.c.}\right] \tag{2.5}$$

The Heisenberg equation of motion which follows is

$$\dot{\phi}(n) = \left[\phi(n+1) - \phi(n-1)\right]/2a \tag{2.6}$$

That this lattice equation describes a 2 component fermion in the continuum limit can be seen by identifying $\phi(n)$ on even (odd) sites with the upper (lower) component of a conventional Dirac field,

$$\begin{aligned} \psi_1(n) &= \phi(n) \text{ for } n \text{ even} \\ \psi_2(n) &= \phi(n) \text{ for } n \text{ odd} \end{aligned} \qquad \psi = \begin{pmatrix} \psi_1 \\ \psi_2 \end{pmatrix} \tag{2.7}$$

Then Eq. (2.6) can be written

$$\begin{aligned} &\text{if } n = \text{even}, \quad \dot{\psi}_1 = \partial_z \psi_2 \\ &\text{if } n = \text{odd}, \quad \dot{\psi}_2 = \partial_z \psi_1 \end{aligned} \tag{2.8}$$

Since $\alpha=\gamma_5=\begin{pmatrix} 0 & 1 \\ 1 & 0 \end{pmatrix}$ we have reconstructed the Dirac equation $\dot{\psi}=\alpha\partial_z\psi$ as claimed.

This method of placing upper and lower components of a Dirac field on adjacent sites of a lattice occurs from time to time in models of statistical mechanics since many local spin-spin interaction systems can be rewritten in terms of fermion degrees of freedom.[18]

The fermion method of Eq. (2.5) and (2.7) has some novel features which stem from the fact that the fermion and anti-fermion parts of ψ are placed on different sites. For example a translation of the lattice by an even number of sites is clearly a symmetry of H and it corresponds to conventional translation invariance. However, a shift by an odd number of sites, which is also a symmetry of H, interchanges even and odd sites. This corresponds, according to Eq. (2.7), to an interchange $\psi_1 \rightleftarrows \psi_2$ which is implemented by the operator,

$$e^{i\gamma_5\pi/2} = i\gamma_5 \tag{2.9}$$

This is a discrete U(1) chiral transformation through the angle π. Thus a shift through an odd number of sites corresponds to the product of a conventional translation and a discrete chiral rotation.

The value of this fermion method becomes clearest when gauge interactions are incorporated into the Hamiltonian. The point is that the Hamiltonian of the massless lattice Schwinger model will also have the discrete γ_5 symmetry observed here and that this symmetry is sufficient to preclude the dynamical generation of a mechanical fermion mass. Thus, masslessness of the fermion fields is a natural symmetry of the theory even when it is placed on a lattice.

Next we place "free" gauge fields on the links of the lattice. Following the general discussion of Lecture 1, let there be U(1) rotation operators on each link,

$$U(n+1,n) = e^{iagA(n)} \equiv e^{i\Theta(n)} \tag{2.10}$$

where A(n) is the vector potential of the gauge field. As usual we work in the class of gauges $A^0(n)=0$ so that time independent gauge transformations can be implemented via unitary operators. In this gauge the electric field E(n) is $\dot{A}(n)$ so the canonical quantization condition reads

$$[A(n), E(m)] = i \frac{1}{a} \delta_{n,m} \tag{2.11}$$

It is more convenient to introduce the variables

$$\Theta(n) = agA(n) \tag{2.12}$$

and

$$L(n) = \frac{1}{g} E(n) \tag{2.13}$$

since then Eq. (2.11) becomes

$$[\Theta(n), L(m)] = i\delta_{n,m} \tag{2.14}$$

so that Θ and L are canonically conjugate. Since Θ enters the theory only in an exponential, its meaningful range of variation is compact,

$$0 < \Theta \leq 2\pi \tag{2.15}$$

It then follows from Eq. (2.14) that the spectrum of L is discrete and, in fact, runs over the integers. This is an important characteristic of the theory. Since $L = \frac{1}{g}E$ the electric field is quantized in units of size g. It is important to ask whether this result is a reflection of the lattice or whether it is characteristic also of the continuum model. Since electric flux cannot spread out in a world without transverse dimensions, the continuum model also has quantized flux. These facts will be important when we discuss the continuum limit of the lattice Schwinger model--the point is that lattice theory quantization leads to quantized electric flux which is also characteristic of the continuum theory.

Thus one can reasonably expect a smooth continuum limit of the lattice theory to exist.

The gauge field Hamiltonian H_G is obtained by the replacement,

$$H_G = \frac{1}{2}\int E^2(z)dz \to \frac{1}{2}\sum_n E^2(n)\cdot a = \frac{1}{2}g^2 a \sum_n L^2(n) \tag{2.16}$$

where the last form is most useful since the spectrum of L^2 consists only of squares of integers. The states of definite L^2 are obtained from the vacuum $|0\rangle (E(n)|0\rangle = 0)$ by applying the U operators,

$$L^2(n)\{e^{im\Theta}|0\rangle\} = m^2\delta_{n,m}\{e^{im\Theta}|0\rangle\} \tag{2.17}$$

Now we can couple the fermion and gauge field degrees of freedom together. The guide for doing this is the requirement of local gauge invariance: we insist that the fermion kinetic energy term be locally gauge invariant. Since

$$\phi(n) \to e^{ig\Lambda(n)}\phi(n) \tag{2.18}$$

and

$$U(n+1,n) \to e^{ig\Lambda(n+1)}U(n+1,n)e^{-ig\Lambda(n)} \tag{2.19}$$

under the discrete gauge transformation,

$$A(n) \to A(n) + \frac{1}{a}\left[\Lambda(n+1) - \Lambda(n)\right] \tag{2.20}$$

a gauge invariant quark hopping H_q reads

$$H_q = \frac{i}{2a}\sum_n \left[\phi^\dagger(n+1)U(n+1,n)\phi(n) - h.c.\right] \tag{2.21}$$

Finally, we need the discrete form of the fermion mass term, $m\int\bar{\psi}\psi\, dz$. Since $\gamma^0 = \begin{pmatrix}1 & 0\\ 0 & -1\end{pmatrix}$ there is the correspondence,

$$m\int\bar{\psi}\psi\, dz \to m\sum_n (-1)^n\phi^\dagger(n)\phi(n) \tag{2.22}$$

The massive Schwinger model lattice H is

$$H = \frac{1}{2} g^2 a \sum_n L^2(n) + \frac{i}{2a} \sum_n [\phi^\dagger(n+1)U(n+1,n)\phi(n) - h.c.]$$

$$+ m \sum_n (-1)^n \phi^\dagger(n)\phi(n) \tag{2.23}$$

It is more convenient to scale out the factor $\frac{1}{2}g^2 a$ and work with a dimensionless "Wamiltonian",

$$W \equiv \frac{2}{ag^2} H = W_o + x\, W_{int} \tag{2.24}$$

where

$$W_o = \sum_n L^2(n) + \mu \sum_n (-1)^n \phi^\dagger(n)\phi(n)$$

$$W_{int} = ix \sum_n [\phi^\dagger(n+1)U(n+1,n)\phi(n) - h.c.] \tag{2.25}$$

and the dimensionless parameters x and μ are

$$x = \frac{1}{g^2 a^2}\ , \quad \mu = \frac{2m}{g^2 a} = \frac{2m}{g} \sqrt{x} \tag{2.26}$$

In writing these formulas we have separated W into two terms. The first W_o involves operators all at single sites. It is a static operator. W_{int}, however, allows fermions to hop about the lattice in a gauge invariant fashion. It is multiplied by the dimensionless expanion parameter $x = 1/g^2a^2$. (Recall that the Schwinger model is super-renormalizable so g has dimensions of a mass and ga is therefore dimensionless.) According to our general method or analyzing lattice theories we shall do conventional perturbation theory in the variable x. The real challenge lies in a proper interpretation of such an expansion method since in the continuum limit ($a \to 0$) the expansion parameter x goes to infinity. Before discussing this issue, we should first understand some of the characteristics of the strongly cutoff theory. In particular, before calculating the spectrum of the theory we must find the theory's vacuum.

Consider the lattice theory with m=0. The vacuum state minimizes W_o so it must be fluxless, i.e. E(n)=0 for all n. However, the fermion content of the ground state is not determined by W_o. Thus all states $|0,\Psi\rangle$ which are fluxless and have an arbitrary fermion content are degenerate in zeroth order. This degeneracy is lifted in perturbation theory as we shall now argue. The shift of the vacuum energy due to W_{int} can be calculated using Raleigh-Schrödinger perturbation theory,

$$\delta W = \langle 0,\Psi|W_o+xW_{int} + x^2\, W_{int}\,\frac{1}{\omega_o+W_o}\,W_{int} +\ldots|0,\Psi\rangle \tag{2.27}$$

where ω_o is the zeroth order energy of $|0,\Psi\rangle$, i.e. zero. Now we shall evaluate Eq. (2.27) through second order. The first order shift in energy is zero since $U(n,n+1)|0,\Psi\rangle$ is a state with flux and does not project back onto $|0,\Psi\rangle$. We are left with

$$\delta W = x^2\langle 0,\Psi|W_{int}\,\frac{1}{-W_o}\,W_{int}|0,\Psi\rangle \tag{2.28}$$

The gauge field operator in W_{int} makes intermediate states with one excited link, so we can replace W_o by 1 in Eq. (2.28). Finally, the U operators can be eliminated and the problem of determining $|0,\Psi\rangle$ can be reduced to a static problem in fermion dynamics. Use the facts that

$$\begin{aligned} &\langle 0|U(n+1,n)U(m+1,m)|0\rangle = 0 \\ &\langle 0|U^\dagger(n+1,n)U(m+1,m)|0\rangle = \delta_{n,m} \\ &\qquad\text{etc.} \end{aligned} \tag{2.29}$$

to see that

$$\delta W = x^2\langle 0,\Psi|\sum_n \frac{[\phi^\dagger(n),\phi(n)]}{2}\,\frac{[\phi^\dagger(n+1),\phi(n+1)]}{2}\,|0,\Psi\rangle \tag{2.30}$$

Identify $\rho(n) = \frac{1}{2}\,[\phi^\dagger(n),\phi(n)]$. It is the local fermion number operator which takes the values $\pm 1/2$ depending on whether the site n is occupied or unoccupied by a fermion. Thus Eq. (2.30) becomes

$$\delta W = x^2 \langle O,\Psi| \sum_n \rho(n+1)\rho(n) |O,\Psi\rangle \tag{2.31}$$

We recognize that the minimization of δW is a standard problem in many body physics-- to find the ground state of the Ising anti-ferromagnetic chain. There are clearly two fermion configurations which minimize δW: occupied even sites and unoccupied odd sites or vice versa. Thus, the enormous degeneracy of W_o has been lifted and there is only a two-fold degeneracy left after doing second order perturbation theory. It is important to note that this remaining two fold degeneracy is a real property of the theory and not only of the approximation scheme, i.e. the degeneracy remains to all orders in perturbation theory. The reason for this is simply stated by observing the physical meaning of the twofold degeneracy. Clearly one vacuum is transformed into the second by making a lattice translation of a shift of one link. But as discussed earlier, this corresponds to a discrete chiral transformation $\psi \to \gamma_5 \psi$. Thus, once we choose one of the two vacua on which to base the lattice Schwinger model, we have spontaneously broken discrete chiral transformations. Since chiral transformations are known to be broken by the vacuum structure of the continuum model,[19] we learn that the strongly cutoff theory shares an important qualitative feature with the theory we eventually wish to make contact with.

Since we are interested in the massive Schwinger model, we should choose the vacuum of the massless model which smoothly changes as the fermion mass is increased from zero. With our γ matrix conventions, the mass term is $\mu \sum (-1)^n \phi^\dagger(n)\phi(n)$. For $\mu \geq 0$ this term picks out the vacuum in which odd sites are occupied. (Recognize this as the static limit of the Dirac sea). This is the zeroth order ground state on which we will base our strong coupling perturbation theory.

We can now calculate several low energy states and matrix elements of the lattice massive Schwinger model. First, one can generate a vector particle out of the vacuum by applying the lattice version of $\int \bar{\psi}\gamma_z \psi dz$. The integral over z insures us that the particle is at rest relative to the fixed lattice sites. Taking care of gauge invariance we have the lattice wave function of this state,

$$|V> = \sum_n [\phi^\dagger(n)U(n,n+1)\phi(n+1) + h.c.]|0> \qquad (2.32)$$

In addition a scalar particle can be made by applying the fermion kinetic energy to $|0>$,

$$|S> = i\sum_n [\phi^\dagger(n)U(n,n+1)\phi(n+1) - h.c.]|0> \qquad (2.33)$$

And, finally, various matrix elements such as the "vector particle sigma term", $<V|\int\bar{\psi}\psi dz|V>$, can be calculated. Using the Feynman-Hellmann Theorem[20], this is obtained from the dependence of the vector particle's mass on the bare fermion mass,

$$<V|\int\bar{\psi}\psi dz|V> = \frac{\partial}{\partial m} M_V\Big|_{m=0} = \frac{\partial}{\partial\mu} W_V\Big|_{\mu=0} \qquad (2.34)$$

A brief sketch of these calculations follows. The energy of the vector particle, for example, calculate in Raleigh-Schrödinger perturbation theory is

$$W_V = <V|W_0 + xW_{int} + x^2 W_{int}\frac{1}{\omega_0 - W_0} W_{int} + \ldots|V> \qquad (2.35)$$

where $|V>$ is the zeroth order wave function and $\omega_0 = <V|W_0|V>$. Clearly only the even powers in x contribute. We represent these terms diagrammatically as follows. The zeroth order contribution is shown in Fig. 2.3. The straight lines denote the fermions and the wiggly one the electric field between them. The zeroth order energy is $\omega_0 = 1 + 2\mu$. (Note that μ varies as $\sqrt{x}$ so the term "zeroth order" is somewhat misleading. However, one can incorporate μ into W_0 since it is a static operator and account for the x dependence of μ in the final formulas for masses and matrix elements.) To describe higher order terms in Eq. (2.35) we denote the action of W_{int} on the vacuum as shown in Fig. 2.4. Then at second order there are two distinct types of graphs as shown in Fig. 2.5. It is interesting that the vacuum structure does not allow graphs of the type shown in Fig. 2.6. Let us illustrate perturbation theory by calculating the contribution of Fig. 2.5a. The vacuum

(2.3)

(2.4)

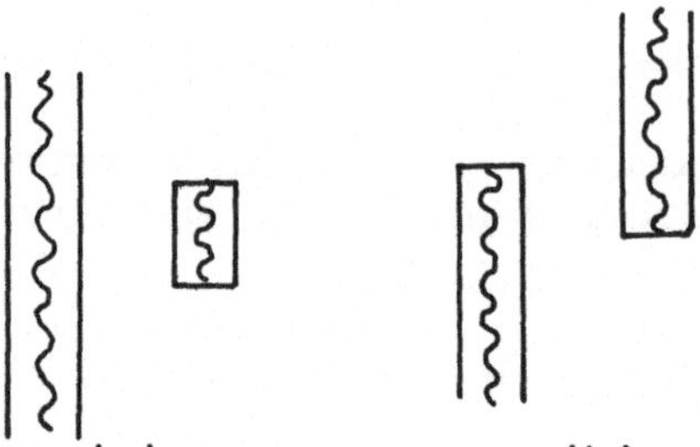

(a) (b)

(2.5)

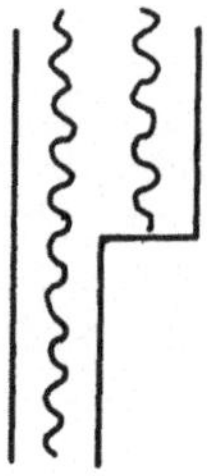

(2.6)

Fig. 2.3 Zeroth order vector particle wave function.

Fig. 2.4 Effect of the quark hopping Hamiltonian in the vacuum.

Fig. 2.5 Second order graphs for the mass of the vector particle.

Fig. 2.6 A disallowed graph.

fluctuation can occur on any of N-3 links and the energy denominator is $\omega_o - 2(1+2\mu) = -1 - 2\mu$, so this class of graphs contributes

$$(N-3)\,\frac{x^2}{-1\ -2\mu} \tag{2.36}$$

to Eq. (2.35). Of course the mass of the state must be measured relative to the vacuum energy which also shifts in second order. The graph responsible is shown in Fig. 2.7 and it has the value

$$N\,\frac{x^2}{-1\ -2\mu} \tag{2.37}$$

Fig. 2.7 A second order vacuum fluctuation.

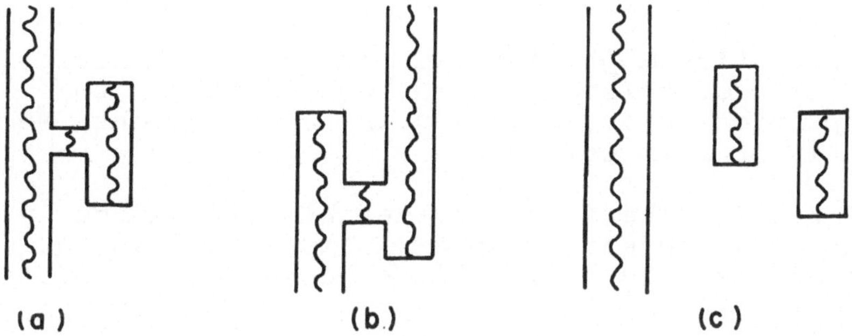

Fig. 2.8 Assorted fourth order graphs affecting the mass of the vector particle.

When Eq. (2.37) is subtracted from Eq. (2.36), the intensive character of the vector particle mass is re-established and a finite contribution is obtained. Before turning to results consider several fourth order graphs. In Fig. 2.8a one of the constituents of the vector particle propagates two links in the "linear confining potential". In Fig. 2.8b the vector particle propagates two lattice sites and in Fig. 2.8c the presence of the vector particle excludes certain vacuum fluctuations from a small part of the lattice. The calculation of these graphs is straightforward. Computer methods, initiated by A. Carroll[15], have allowed us to push many calculations to eighth order. So, we have expansions of the form

$$2\sqrt{x}\,\frac{M_{V,S}}{g} = \sum_n C_{2n}^{V,S}(\mu)\,x^{2n} \tag{2.38}$$

for the vector and scalar masses. In addition, the energy of these states for non-zero momentum p has also been calculated,

$$2\sqrt{x}\,\frac{E_{V,S}(p)}{g} = \sum_n d_{2n}^{V,S}(\mu,p)\,x^{2n} \tag{2.39}$$

Let's inspect a numerical example of these series. Consider the scalar particle in the massless (m=0) case,

$$2\sqrt{x}\,\frac{M_S}{g} = 1+6x^2-26x^4+190\,\frac{2}{3}\,x^6-1756\,\frac{2}{9}\,x^8 + \ldots \tag{2.40}$$

Since continuum physics lies at $x\to\infty$, a sensible interpretation of this expansion, which appears to be an asymptotic series, will require some thought. We turn to this problem now.

Luckily there is considerable experience in other branches of physics with expansions such as Eq. (2.40) Consider the high temperature expansion of statistical mechanics.[21] One can develop the partition function in powers of $1/kT = \beta$,

$$Z(\beta) = \mathrm{Tr}\ e^{-\beta H} = \mathrm{Tr}\ 1+\beta\frac{1}{1!}\ \mathrm{Tr}\ H+\frac{1}{2!}\beta^2\mathrm{Tr}H^2+\ldots \tag{2.41}$$

and obtain expansions for various quantities of interest such as the free energy, susceptibilities (correlation functions), etc. However, the high temperature region is often of little interest--one really wants to extrapolate from high T and explore the low T region for phase transitions. Calculations of critical temperatures and indices have been very successful using this method. Extrapolating the high temperature expansions away from $\beta=0$ is often done using Padé approximants[22] (or close relatives thereof). The point of view is that Eq. (2.41) defines a function with possible poles and cuts in the complex β plane and that this function is well determined by the series expansion itself. The Padé approximant extrapolates the series beyond its radius of convergence and gives a (sometimes) reliable guide to the qualitative and quantitative features of the complex function $Z(\beta)$. In practice one implements the extrapolation method by guessing (or determining by other arguments) what the form of the function of interest is near the critical point. For example, one expects various susceptibilities χ to diverge as $T\to T_c$ according to a power law,

$$\chi(T) \sim \frac{B}{(T-T_c)^{\gamma}} \text{ (regular function)} \qquad (2.42)$$

Given this hypothesis the high temperature expansion as extrapolated via Padé approximants allows one to determine the parameters T_c, γ and B.

It is interesting that the physics of the approach from large T to the critical temperature has many qualitative features in common with the extrapolation in lattice gauge theory from samll to large x. For $T \gg T_c$ the statistical mechanical system possesses only short range correlations. (For a model H with short range spin-spin forces,this is the fact that the graphs describing the low order terms in Eq. (2.41) cover only a few lattice sites.) The masses of eigenstates of the system are all of order a^{-1} in this extreme. However, as $T \to T_c$ the correlation lengths increase and in fact diverge at the critical temperature. Then the details of the lattice become unimportant because the important fluctuations which determine the character of the thermodynamic functions are infinite in size as measured in lattice units. The close analogy of these remarks with the continuum limit of field theory models should be clear--for large x the correlation lengths are small (masses of the particles are of order a^{-1}) but as $x \to \infty$ one expects (on the basis of known features of the continuum theory) to find bound states whose sizes and masses attain values independent of the lattice spacing a. In the statistical mechanics problem there is a smooth change in the physics as T decreases to T_c so that the high temperature expansion, although it must be continued outside its radius of convergence, contains the physics of the critical point. (It is important here that the phase transition at T_c be second or higher order because otherwise there will be no connection between the high and low temperature regions-- the analyticity assumption implicit in the continuation hypothesis appears to be false for a first order transition). If the function one is trying to approximate beyond the region of convergence of the Taylor series satisfies some rather weak regularity assumptions, then one can prove that the sequence of Padé approximants does in fact converge to it.[22] In many applications, however, one simply doesn't know if the hypotheses of the convergence theorems are satisfied, but a numerical investigation of Padê approximants for larger and larger order usually demonstrates convergence in a practical sense. Aside from a wealth of mathematical theorems and applications of principle, another reason Padê approximants enjoy their popularity is that low order approximants are often much better approximations to the function of interest than one has any right to expect. An example of this behavior will be examined below. In the field of critical phenomena, the high temperature expansion coupled with extrapolation methods such as Parê approximants still yield the best available estimates of parameters characterizing the critical points for many physical systems.

Now let's apply the Padê extrapolation method to the lattice Schwinger model. Consider the expansion for $\langle V|\int\bar{\psi}\psi dz|V\rangle$,

$$\langle V|\int\bar{\psi}\psi dz|V\rangle = \frac{\partial m_V}{\partial m}\Big|_{m=0} = 2(1-2x^2+28x^4-374.444x^6+ +4971.481x^8) \tag{2.43}$$

We can form the [2,2] Padê approximant here. We write the rational function

$$\frac{1+\alpha x^2+\beta x^4}{1+Ax^2+Bx^4}, \tag{2.44}$$

expand the dominator to order x^8, and match coefficients with Eq. (2.43). This gives

$$\langle V|\int\bar{\psi}\psi dz|V\rangle = 2\left(\frac{1+13.054x^2+21.538x^4}{1+15.054x^2+23.646x^4}\right) \tag{2.45}$$

which can now be continued to $x = \infty$. Note that the [2,2] Padê is singularity-free over the positive real axis--thus a smooth continuation from large to small lattice spacing is possible. The continuum value of the matrix element is obtained by letting $x\to\infty$ in Eq. (2.45),

$$\langle V|\int\bar{\psi}\psi dz|V\rangle = 1.822 \tag{2.46}$$

which should be compared to the exact answer,

$$\langle V|\int\bar{\psi}\psi dz|V\rangle = e^{\gamma} = e^{.522\ldots} = 1.781 \tag{2.47}$$

so Eq. (2.46) lies within 2.3% of the exact answer. A continuation to $x=\infty$ is also possible if one had only calculated the first three terms of Eq. (2.43). Then

$$\langle V|\int\bar{\psi}\psi dz|V\rangle = 2(1-2x^2+28x^4)$$

$$\overset{\text{Padé}}{=} 2\left(\frac{1+12x^2}{1+14x^2}\right) \rightarrow 1.71 \tag{2.48}$$

which is 3.6% below the correct value. This calculation gives us some hope that the method is reliable and pactical. Note that even the zeroth order fixed lattice spacing result (2.00) for this matrix element is only 12,3% above the exact answer. Thus the particular way we have chosen to set up the lattice theory is quite efficient.

Now consider the mass ratio m_S/m_V. We concentrate on mass ratios because these dimensionless quantities should approach constants in the continuum limit. Given this boundary condition we know that the diagonal Padé approximant should be used to extend the series expansion away from x=0. The ratio of the series expansions for the scalar and vector particles reads

$$\frac{m_S}{m_V} \cong \frac{1+6x^2-26x^4+190.667x^6-1756.667x^8}{1+2x^2-10x^4+78.667x^6-236.444x^8} \tag{2.49}$$

which should be expanded in a Taylor series which is meaningful through eighth order,

$$\frac{m_S}{m_V} \approx 1+4x^2-24x^2+200x^6-1975.11x^8 \tag{2.50}$$

and this series is extended beyond its region of convergence by replacing it with the [2,2] Padé approximant,

$$\frac{m_S}{m_V} = \frac{1+13.84x^2+64.41x^8}{1+13.84x^2+33.05x^8} \tag{2.51}$$

which gives a singularity-free extrapolation from $x=0$ to $x=\infty$. Taking $x\to\infty$ in Eq. (2.51) we have

$$\frac{m_S}{m_V} = 1.95 = 2(.975) \tag{2.52}$$

The exact answer for this ratio is 2-- in the massless Schwinger model the "scalar state" consists of 2 free vector particles. The zeroth order, [1,1] Padé and [2,2] Padé results for this mass ratio are plotted in Fig. 2.9. We have apparent convergence to the exact answer and considerable improvement even in the [1,1] Padé approximant beyond the crude static limit in which $m_S = m_V$.

It is interesting to study the mass ratio m_S/m_V as a function of the bare fermion mass m. Continuum theory calculations using the equivalent boson representation of the theory show that the scalar particle is a legitimate bound state when $m \neq 0$ and that as $m\to 0$, $m_S\to 2m_V$ and the state disappears into the continuum.[15] Thus it is interesting to study the quantity $(1-m_S/2m_V)$

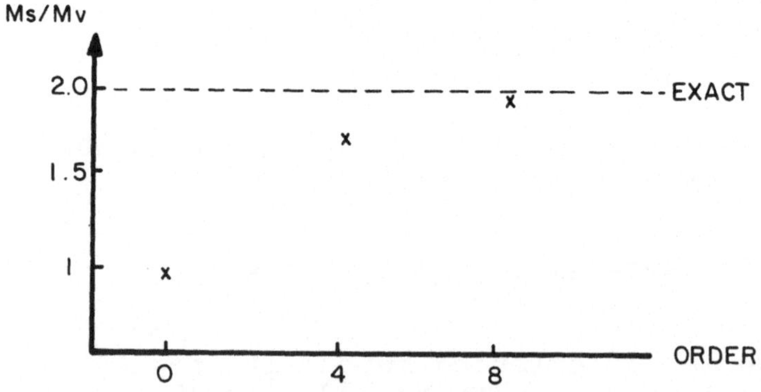

Fig. 2.9 Convergence of the lattice calculations of M_S/M_V to the continuum result.

as a function of the dimensionless measure of the fermion mass (4m/g). Lattice calculations with the Padé extrapolation yield the curve shown in Fig. 2.10. The region of the curve near $4m/g \approx 0$ lies within .05 of the exact answer (0 at $4m/g = 0$), and the region of the curve near $4m/g \approx 1$ has been checked against continuum calculations using infinite momentum variational methods. The agreement there was within a percent.[23]

Another interesting calculation concerns the propagation properties of the physical states. The lattice theory is set up so that we should retrieve all the consequences of relativity when the lattice spacing is taken to zero. One of those consequences which can be studied numerically is the energy-momentum relation. For $p \ll m$, the energy of a relativistic particle is

$$E(p) = \sqrt{p^2+m^2} \approx m + p^2/2m + \ldots \tag{2.53}$$

Thus, mass can be viewed as either the gap in the E-p relation at zero momentum (call this mass $m_{static}=E(0)$), <u>or</u> the curvature of the E-p relation (call this mass

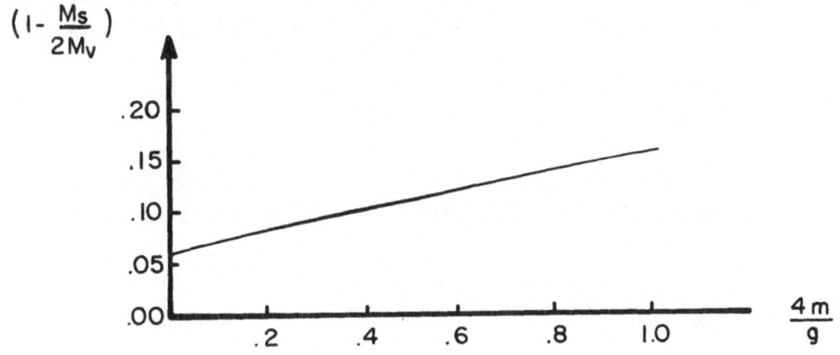

Fig. 2.10 Binding energy of the scalar particle as a function of the bare fermion mass.

$m_{kinetic} = (2dE/dp^2\big|_{p^2=0})^{-1})$. In lattice gauge theory one must check that these two definitions are equivalent ($m_{static}=m_{kinetic}$). This is non-trivial since the perturbation is in the kinetic part of the Hamiltonian. (The reader should appreciate just how irrelevant relativity is in the $x\simeq 0$ form of the theory. Consider just the free-massless Dirac equation on the lattice. In zeroth order the fermions do not move although they are massless. Thus, we have "infinitely massive-massless" fermions in this approximation. Of course, kinetic perturbation theory cures this crude zeroth order starting point.) In the Schwinger models one can form states of definite momentum, calculate E(p) for the vector particle, say, and consider the mass ratio $m^{static}/m^{kinetic}$. In zeroth order there is no meaning to this ratio, however, doing sixth order perturbation theory, forming a [1,1] Padé and taking the continuum limit yields .81 for this ratio.[15] Thus, relativity is re-instated order by order x.

Many other lattice calculations have been done and the lesson one learns is that the calculational method yields numbers which quickly converge to their known continuum limits, that the method is practical and programmable, and that it is not limited to simple 1+1 dimensional models.

Since Padé approximants play such an important role in these and the calculations to be described in the next lecture, it pays to gain some familiarity with this method. We begin with the Padé approximator's favorite function,

$$f(x) = \sqrt{\frac{1+2x}{1+x}} \qquad (2.54)$$

which varies smoothly from 1 at x = 0 to $\sqrt{2}$ at $x=\infty$ and has a cut between -1/2 and -1. The Taylor series about x=0,

$$f(x) = 1+\frac{1}{2}x-\frac{5}{8}x^2+\frac{13}{16}x^3-\frac{141}{128}x^4 +-\ldots \qquad (2.55)$$

has a radius of convergence $|x| = 1/2$ where the cut begins. However, the regions of convergence of Padé approximants can be more interesting than circles and in this example the approximants will, in fact, converge uniformly to f(x) in the entire complex plane except on the cut itself. Consider the [1,1] Padé for Eq. (2.55),

$$[1,1] = \frac{1 + \frac{7}{4}\cdot x}{1 + \frac{5}{4}\cdot x} \tag{2.56}$$

which equals 7/5 = 1.40 at x=∞. Next the [2,2] Padé,

$$[2,2] = \frac{1 + \frac{13}{4}\, x + \frac{41}{16}\, x^2}{1 + \frac{11}{4}\, x + \frac{29}{16}\, x^2} \tag{2.57}$$

equals 41/29 = 1.4138... at x=∞. So, even these low order approximants are excellent approximations to f(x) in the right hand plane.

It is interesting to see how they mock up the zero and cut in f(x). The [1,1] Padé has a pole at x=-.8 and a zero at -.57. But, the [2,2] Padé has poles at x -.60 and -.91 and zeros at -.52507 and -.74322. The trend is that higher order Padé approximants reproduce the zero of f accurately and replace the cut by a series of poles and zeros.

A more conventional approach to this extrapolation problem is to find a mapping $x \to \omega$ which takes the point at infinity to a point within the radius of convergence of the new power series of $f(\omega)$. Thus, the mapping should move the region of singularities (the cut in this example) far enough from the origin in the ω-plane. Consider the mapping

$$\omega = \frac{x}{1 + 2x} \tag{2.58}$$

and the function Eq. (2.54) in the ω-plane,

$$f(\omega) = \frac{1}{\sqrt{1-\omega}} = 1 + \frac{1}{2}\omega + \frac{3}{8}\omega^2 + \frac{5}{16}\omega^3 + \frac{35}{128}\omega^4 + \ldots \tag{2.59}$$

The point $\omega=1/2$, the image of infinity under the mapping, now lies within the radius of convergence of the series in Eq. (2.59) because the image of the cut extends from $\omega=1$ to ∞. Truncating the series in Eq. (2.59) after the first, second, etc., terms and evaluating these polynomials at $\omega=1/2$ give the approximations,

$$1,\ 1.25,\ 1.34,\ 1.38 \tag{2.60}$$

to $f(x=\infty) = \sqrt{2}$. So, the method converges to the correct answer but much slower than the Padé approximants.

There is an intimate relation between Euler transformations, $\omega = Ax/(1+Bx)$, as in this example, and the properties of diagonal Padé approximants. The relevant theorem is that diagonal Padé approximants are invariant to Euler transformations.[22]

Theorem: If $P_m(x)/Q_m(x)$ is the $[m,m]$ Padé approximant to $f(x)$, then $P_m(\omega)/Q_m(\omega)$ is the $[m,m]$ Padé approximant to $f(\omega)$, interpreted as a power series in X.

In certain cases, as in the example above, the point $x = \infty$ will be mapped to a point within the radius of convergence of the Taylor series for $f(\omega)$. Then the success of the Padé approximant to extrapolate from $\omega=0$ to $\omega=A/B$ is not mysterious at all. Since $\omega=A/B$ corresponds to $x=\infty$,

$$[m,m] \text{ Padé of } f(\omega=A/B) = [m,m] \text{ Padé of } f(x=\infty) \tag{2.61}$$

this theorem is the rationale behind the success of the Padé approximant to sum the x series itself in many cases.

In the applications of these ideas to field theory one is faced with the problem that little is known about the analyticity of masses and matrix elements in the complex $1/g^2$ plane. The theories of most interest to us are renormalizable one which are asymptotically free (the invariant charge in the Gell-Mann-Low sense vanishes logarithmically in the deep Euclidean region). If these theories have no intrinsic mass scale (bare quark masses, say), then the physical masses of the theory have an essential singularity[24] at $1/g^2=\infty$. This lack of analyticity precludes the uniform convergence of Padé approximants for masses themsleves. This is the reason we concentrate on mass ratios and dimensionless matrix elements which should approach constants as $g^2\to 0$ (this is the continuum limit for an asymptotically free theory, see later discussions) and constants are simple functions.

In potential theory a great deal is known about the analytic behavior of energy levels as a function of the strength of the potential. In many of these cases the Padé approximant has been proven to be a uniformly convergent, practical approximation method although perturbation theory may not converge at all.[25] Since field theory is qualitatively different from potential theory, these rigorous results are of no help. However, the problem is interesting and worth studying.

Now we return briefly to 1+1 dimensional field theories. The first issue we wish to consider concerns the uniqueness of the lattice Hamiltonian. The only requirement on lattice Hamiltonians is that they give the same continuum physics. Thus, operators can be freely added to H if they preserve the symmetries of the terms in H which contribute to the continuum limit and if they do not survive when $a \to 0$. Let's consider several examples in the context of the massive Schwinger model. A mass term in H which survives when $a \to 0$ is

$$m \sum (-1)^n \phi^\dagger(n)\phi(n) \tag{2.62}$$

An "irrelevant"[26] mass term is

$$\frac{1}{2}\lambda(g^2 a) \sum (-1)^n \phi^\dagger(n)\phi(n) \tag{2.63}$$

where λ is a dimensionless parameter. (Eq. (2.63) is dimensionally correct because ga is dimensionless.) Consider a second example drawn from the Schwinger-Thirring model. A four fermi interaction which survives the continuum limit is

$$\frac{G}{a} \sum_n \rho(n)\rho(n+1) \tag{2.64}$$

where G is a dimensionless coupling constant and $\rho(n) = \frac{1}{2}\left[\phi^\dagger(n),\phi(n)\right]$. By comparison the operator

$$Gg^2 a \sum_n \rho(n)\rho(n+1) \tag{2.65}$$

is irrelevant in the continuum limit. This term can be added to the H in Eq. (2.23) because it respects all the symmetries of the lattice Schwinger model H. We shall see in the four dimensional calculations that it is often convenient to add irrelevant operators to H in order to simplify the strong coupling perturbation theory. In that case the theory is renormalizable and the irrelevant operators one plays with are classified as non-renormalizable.[26]

How does the irrelevance of these terms show up in our calculational scheme? Clearly they affect the details of the $a \neq 0$ calculation greatly and the coefficients of the Taylor series show strong dependence on irrelevant parameters such as the λ in Eq. (2.63) and the G in Eq. (2.65). However, the irrelevance appears in the continuum limit, i.e. form the $[N,N]$ Padé approximant and let $x \to \infty$. The λ or G dependence in the result should be very weak and should disappear entirely when $N \to \infty$. Let's study a numerical example to confirm this. Consider the vector and scalar particles in the Schwinger model with the irrelevant mass term of Eq. (2.63) included. The expansions now read

$$2\sqrt{x}\,\frac{m_V}{g} = 1 + 2\lambda + \frac{2}{(1+2\lambda)}\,x^2 - \frac{(10+4\lambda)}{(1+2\lambda)^3}\cdot x^4 + - \ldots \tag{2.66}$$

$$2\sqrt{x}\,\frac{m_S}{g} = 1 + 2\lambda + \frac{6}{(1+2\lambda)}\,x^2 - \frac{(26+4\lambda)}{(1+2\lambda)^3}\cdot x^4 + - \ldots \tag{2.67}$$

So, defining the variable $y=x/(1+2\lambda)$,

$$m_V/m_S \cong \frac{1 + 2y^2 - (10+4\lambda)y^4}{1 + 6y^2 - (26+4\lambda)y^4} \tag{2.68}$$

which must be written as a Taylor series,

$$m_V/m_S \cong 1 - 4y^2 + 40y^4 \tag{2.69}$$

and converted to a $[1,1]$ Padé approximant

$$m_V/m_S = \frac{1 + 6y^2}{1+ 10y^2} \tag{2.70}$$

The point is that Eq. (2.69), and necessarily Eq. (2.70), have no explicit λ dependence. So, when $x \to \infty$ and y follows, the <u>same</u> continuum limit is found for m_V/m_S as in the case $\lambda=0$ which was plotted in Fig. 2.9. This example illustrates a general result: the inclusion of irrelevant operators into the lattice Hamiltonian can be absorbed into a change of the subtraction procedure for the theory's coupling constant. Since the coupling constant (ga in this example) vanishes in the continuum limit, the dependence on irrelevant parameters disappears.

It is gratifying that the approximate methods used here respect the general character of irrelevant operators. Of course, in examples different from Eq. (2.63) one does not find exact irrelevance in a low order Padé approximant. One can expect, however, that the dependence of the continuum limit on irrelevant parameters will be much less than the dependence in the strong coupling Taylor series. More illustrations of these ideas occur in the next section.

Our final 1+1 dimensional discussion describes the non-Abelian SU(N) Thirring model,[16]

$$\mathcal{L} = \sum_{\alpha=1}^{N} i\psi_\alpha \not\partial \psi_\alpha + \frac{g^2}{2} \left(\sum_{\alpha=1}^{N} \bar{\psi}_\alpha \psi_\alpha \right)^2 \qquad (2.71)$$

The four fermi interaction term appears with the "wrong" sign. This choice renders the theory asymptotically free[27], and causes the vacuum state to violate the γ_5 symmetry ($\psi_\alpha \to \gamma_5 \psi_\alpha$) of the Lagrangian. The theory also has a rich mass spectrum[28] so it is a nice setting to test lattice methods.

Since this theory is renormalizable, it has an interesting problem of coupling constant renormalization. An approximate calculation of the running coupling constant g(a) can be made using mean field theory. In this approximation the Hamiltonian is diagonalized in terms of free massive fermions. The requirement that the fermion mass M_F be a physical quantity, independent of the lattice spacing a then determines how the coupling constant g must vary as a is changed. The result of this analysis states that for small cutoffs, $a \ll 1/M_F$,

$$g^2(a) = -\frac{2}{(2N-1)\ln(aM_F)} \qquad (2.72)$$

which agrees with the exact continuum analysis[27] (lowest order Gell-Mann-Low equation), and for large cutoffs $a \gg 1/M_F$,

$$g^2(a) = \frac{4aM_F}{(2N-1)} \tag{2.73}$$

which agrees with the lowest order strong coupling calculation of the fermion mass. If we trust this mean field analysis, then we have the result that there is no phase transition between large and small a since the exact curve which connects Eq. (2.72) and (2.73) is smooth and has no zeros. This creates the possibility of a smooth continuation from the strong coupling expansion for masses and matrix elements to the continuum limit. The continuum limit is defined now in terms of Eq. (2.72)-- letting a go to zero means that g, the lattice coupling constant, also vanishes.

The mass spectrum calculations will not be discussed here. For the loosely bound multi-fermion states very good agreement was found between the DHN[28] results and the [1,1] Padé approximants.[16]

3. SPECTRUM CALCULATIONS IN 3+1 DIMENSIONAL GAUGE THEORIES

Now we discuss and summarize some of the calculations which have been done in 3+1 dimensional gauge theories. We begin with pure gauge fields and seek the low lying gauge invariant spectrum. The lowest energy excitations of the strong coupling limit are flux loops of the smallest possible perimeter. Flux is an oriented quantity, so the primitive building blocks for the states are

$$\mathrm{Tr}\ U(1)\ U(2)\ U(3)\ U(4)\,|0\rangle \tag{3.1}$$

and

$$\mathrm{Tr}\ U^\dagger(1)\ U^\dagger(2)\ U^\dagger(3)\ U^\dagger(4)\,|0\rangle \tag{3.2}$$

as shown in Fig.3.1. Given these objects one can make states which are irreducible with respect to permissible rotations on the cubic lattice. First, there is a scalar state made by superposing boxes of both orientations in all of the three planes as shown in Fig. 3.2a. This state clearly is unchanged by any rotation which takes the lattice into itself. The boxes shown in Fig. 3.2a are summed over the entire lattice in order that the scalar state have zero momentum. Next, one can make an axial

Fig. 3.1 Oriented boxes of flux.

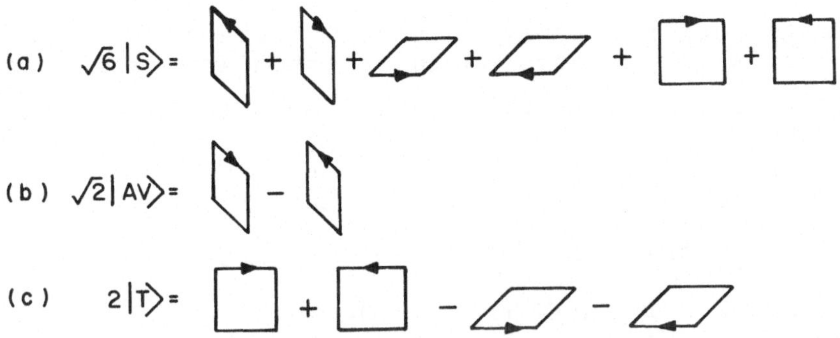

Fig. 3.2 The amplitudes giving the scalar, axial vector and tensor states in zeroth order.

vector giving the polarization of the state. For example, superpose an oriented loop in the x-y plane minus its hermitian conjugate as shown in Fig. 3.2b. The axial vector character of the state follows from the fact that it changes sign under a rotation through 180° about an axis along a link in the x-y plane and the fact that a reflection through the center of the box leaves the state unchanged. Finally, a few members of a tensor multiplet can be made in the way indicated in Fig. 3.2c as the reader can verify. Given these zeroth order states $|S\rangle$, $|AV\rangle$ and $|T\rangle$, one calculates their ω energies in the usual way. The Wamiltonian for this problem reads

$$W = \frac{2a}{g^2} H_{gauge} = \sum_{links} [E(r,\hat{n})]^2 - y \sum_{squares} (\mathrm{Tr}\ UUUU + h.c.) \tag{3.3}$$

$$= W_o - y\, W_{kinetic}$$

where

$$y = 2/g^4 \tag{3.4}$$

is the dimensionless expansion parameter.

Let's illustrate some pieces of the perturbation calculations which have been carried out to fourth order in y. We use Raleigh-Schrödinger perturbation theory as discussed in Lecture 2. In zeroth order all the 4 link states are degenerate and have a ω energy,

$$\omega^{(0)} = 4\cdot C_2(3) = \frac{16}{3} \tag{3.5}$$

In first order perturbation theory there is the possibility that $W_{kinetic}$ acts on an oriented loop and reverses its orientation. This is possible because of the SU(3) relation

$$3 \otimes 3 = 6 \oplus \bar{3}.$$

The mixing amplitude is

$$\langle 0|\mathrm{Tr}\ UUUU\ \mathrm{Tr}\ UUUU\ \mathrm{Tr}\ UUUU|0\rangle \tag{3.6}$$

where each product of four U operators has the same orientation around the same square. On each link we have

$$\langle 0|U^i{}_j U^k{}_\ell U^m{}_n|0\rangle = \frac{1}{6}\,\varepsilon^{ikm}\varepsilon_{j\ell n} \tag{3.7}$$

where we have used some elementary knowledge of the Clebsch-Gordon series for SU(3) to extract the singlet piece of the product $3 \otimes 3 \otimes 3$. Putting everything together the mixing amplitude in Eq. (3.6) is just unity. So through this order,

$$\begin{aligned} \omega\ (\text{scalar}) &= \frac{16}{3} - y \\ \omega\ (\text{axial vector}) &= \frac{16}{3} + y \\ \omega\ (\text{tensor}) &= \frac{16}{3} - y \end{aligned} \tag{3.8}$$

The first order term enters with a plus sign for the axial vector because the state's polarization vector changes sign when the orientation of the loops is reversed.

Now consider some second order graphs. There are disconnected graphs as shown in Fig. 3.3. The intermediate state contains four extra links of flux so the energy denominator is $-4\cdot\frac{4}{3}$. If N is the number of

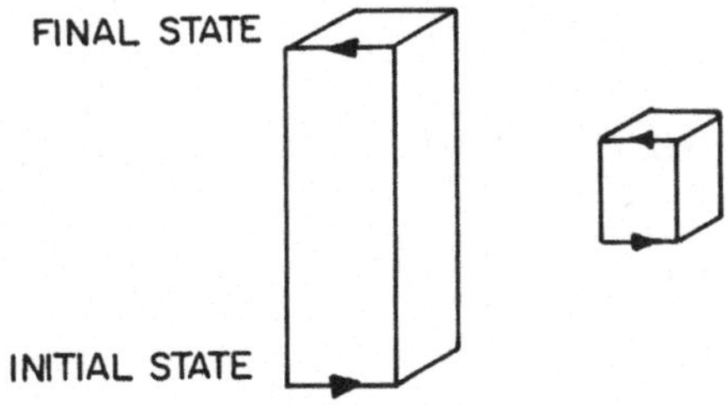

Fig. 3.3 Disconnected second order contribution to boxciton masses.

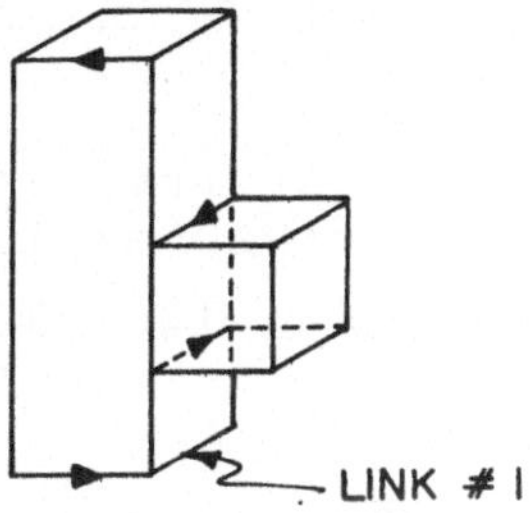

Fig. 3.4 Connected second order contribution to boxciton masses.

squares on the lattice, then the box fluctuation can occur on any N-13 squares. Therefore, the contribution of this set of graphs to the ω energy is

$$\frac{2(N-13}{-4\cdot\frac{4}{3}} \cdot y^2 \tag{3.9}$$

where the overall factor of 2 accounts for the two possible orientations of the box fluctuation. As in the 1+1 dimensional example, when the energy shift of the vacuum is accounted for the N dependence of Eq. (3.9) is subtracted away. Another class of graphs occurs when the perturbation creates and subsequently destroys a box which has a link in common with the through-going particle box as in Fig. 3.4. The link on which the U operators act four times is labelled ≠ 1 in the figure. In the intermediate state of link # 1 flux exists in either the $\underset{\sim}{6}$ or $\bar{3}$ representations (because $3 \otimes 3 = 6 \oplus \bar{3}$). For the 6 ($\bar{3}$) intermediate state the ω energy denominator is $4\cdot\frac{4}{3} - 6\cdot\frac{4}{3} - \frac{10}{3}\,(4\cdot\frac{4}{3} - 7\cdot\frac{4}{3})$.

Finally, we must compute the relative weights of these two intermediate states. It is not difficult to compute the Clebsch-Gordon series for $3 \otimes 3 = 6 \oplus \bar{3}$,

$$U^i{}_j U^k{}_\ell = \frac{1}{2}\varepsilon_{mik}\varepsilon_{nj\ell}(U^*)^m_n + \frac{1}{4}(\delta_{ir}\delta_{ks}+\delta_{rk}\delta_{is}) (\delta_{jt}\delta_{\ell u} + \delta_{\ell t}\delta_{ju})\, U_6^{\{rs\},\{tu\}} \tag{3.10}$$

Using this and the vacuum expectation value,

$$\langle 0|U^n_{\alpha\beta}U^{\bar{n}}_{\gamma\delta}|0\rangle = \frac{1}{n}\delta_{\alpha\gamma}\delta_{\beta\delta} \tag{3.11}$$

where n is the dimension of the representation of SU(3)' we conclude (after some algebra) that the weight for the $6(\bar{3})$ intermediate state is $\frac{2}{3}$ $(\frac{1}{3})$. Thus, this set of graphs equals

$$12y^2 \left\{\frac{\frac{1}{3}}{-3\cdot\frac{4}{3}} + \frac{\frac{2}{3}}{4\cdot\frac{4}{3} - 6\cdot\frac{4}{3} - \frac{10}{3}}\right\} \tag{3.12}$$

where the factor 12 counts the number of squares the box fluctuation can occupy.

The third and fourth order graphs have also been calculated. There are many fourth order graphs. In Fig. 3.5 we show some of the simpler varieties. The group theory considerations for some of these involve Racah coefficients (6-j symbols generalized to SU(3)'). The resulting expansions are[30]

$$\begin{aligned}
\omega_S &= \frac{16}{3}(1-\bar{y}-.569\bar{y}^2+17.393\bar{y}^3-95.206\bar{y}^4)\\
\omega_A &= \frac{16}{3}(1+\bar{y}+.097\bar{y}^2-3.112\bar{y}^3-51.840\bar{y}^4)\\
\omega_T &= \frac{16}{3}(1-\bar{y}+.583\bar{y}^2-7.650\bar{y}^3-15.357\bar{y}^4)
\end{aligned} \tag{3.13}$$

where $\bar{y}=y/(\frac{16}{3})$. These expansions are extrapolated to the continuum limit assuming asymptotic freedom ($g\to 0$ as $a\to 0$). Forming [2,2] Padé approximants,

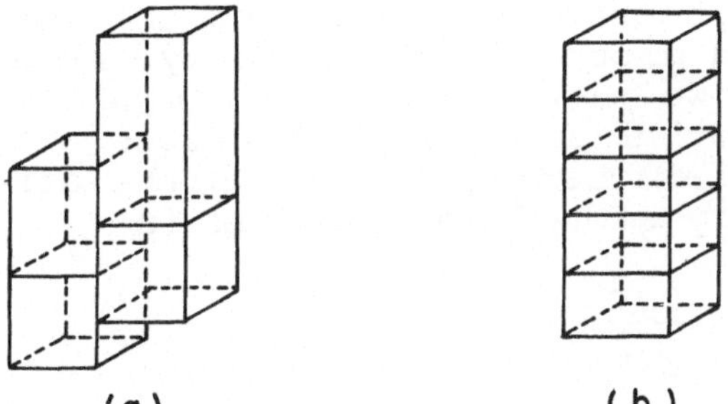

Fig. 3.5 Various fourth order contributions to boxciton masses.

$$\frac{M_A}{M_S} = \frac{1 + 2.773\,\bar{y} + 11.532\,\bar{y}^2}{1 + .773\,\bar{y} + 7.320\,\bar{y}^2} \quad \text{(Padé)} \tag{3.14}$$

and

$$\frac{M_T}{M_S} = \frac{1 + 20.739\,\bar{y} + 382.104\,\bar{y}^2}{1 + 20.739\,\bar{y} + 380.952\,\bar{y}^2} \quad \text{(Padé)} \tag{3.15}$$

we see that a singularity-free extrapolation from $\bar{y} = 0$ to $\bar{y} = \infty$ is possible. Taking $y \to \infty$ gives

$$M_A/M_S = 1.575, \qquad M_T/M_S = 1.003 \tag{3.16}$$

These results are very reasonable (although the calculation was so lengthy that I consider the fourth order coefficients in Eq. (3.13) tentative) and can be compared against a non-relativistic model of confined gluons. In such a model the tensor and scalar states consist of two S-wave gluons while the axial vector state consists of two gluons in a P-wave. The level ordering and splittings of Eq. (3.16), therefore, seem reasonable although the near degeneracy of the tensor and scalar states is somewhat surprising.

Another calculation which is conceptually interesting but very crude in practice, concerns the force law (renormalization group) in this theory. As explained in Lecture 1 if we place a colorful impurity (3) at one site and another impurity ($\bar{3}$)N sites away, then a flux line must form between them. In the strong coupling limit, the energy stored in the flux is

$$E = \frac{2g^2}{3a} N = \frac{2g^2}{3a^2} L \tag{3.17}$$

where L is the physical distance between the source and sink and is measured in GeV^{-1} units, for example. The requirement that the string tension be a physical quantity then implies how g must be varied with a in order that the low energy phenomena of the theory be independent of the renormalization procedure. It is amusing to let the string tension be $1/2\pi$ as in the Dual Model and to calculate Eq. (3.17) to higher orders in y in order to extend these considerations into the intermediate coupling region. Treating Fig. 1.8 as the zeroth order state we can calculate its energy in lattice perturbation theory

$$E = \left[\frac{4}{3} - .288\left(\frac{1}{g^2}\right)^4 - 7.04\left(\frac{1}{g^2}\right)^8 + \ldots\right] \frac{g^2}{2a^2} L \tag{3.18}$$

Requiring that the tension be $1/2\pi$ in GeV units implies that

$$\frac{4}{3} - \frac{a^2}{\pi}\left(\frac{1}{g^2}\right) - .288\left(\frac{1}{g^2}\right)^4 - 7.04\left(\frac{1}{g^2}\right)^8 = 0 \tag{3.19}$$

which relates g and a. A very rough plot is shown in Fig. 3.6. The dotted line is the expected asymptotically free curve and the dashed line is the zeroth order Eq. (3.17). The solid curve of Eq. (3.19) cannot be continued to the origin because the highest order $\left(\frac{1}{g^2}\right)^8$ term begins to dominate when $a \lesssim 1\ \mathrm{GeV}^{-1}$. However, the direction of the curve's deviation from super-renormalizable behavior is correct. It would be interesting to carry this simple calculation procedure to high order and verify asymptotic freedom directly. Probably this will require more sophisticated methods than strong coupling

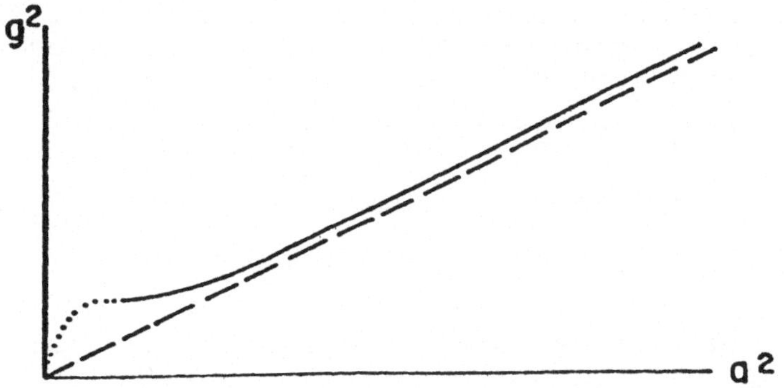

Fig. 3.6 Present status of the renormalization group calculation for pure non-Abelian fields. The calculation, the solid curve, is not accurate near the origin. The dotted line is the conjectured behavior.

expansions. Migdal[29] has produced an approximate recursion formula for non-Abelian gauge fields in the context of lattice theories and he has verified asymptotic freedom. However, it is not known whether his ingenious approximation methods are reliable.

Some interesting theoretical questions occur when these spectrum calculations are repeated for the lattice theory of pure Abelian gauge fields. Of course, the continuum form of this theory is free so we do not expect to be able to reach it from the lattice theory. In other words one expects there to be a barrier (singularities) between g=0 and large g so that an extrapolation from large g can never reach the origin. Explicit calculations discussed below support this point of view. The same question occurs again when one considers Abelian lattice gauge theory with quarks on the one hand and continuum Quantum Electrodynamics on the other. There are several reasons to suspect that these theories are also not related. First, in the lattice theory the natural renormalization effects favor strong coupling at large distances. But in continuum weak field electrodynamics screening favors decreasing couplings as distances increase. Second, lattice gauge theory quantization leads to quantized electric flux on links. This is natural in a non-Abelian theory with a compact group. But in Abelian quantum electrodynamics one naively expects flux to subdivide into units of arbitrary smallness.

Repeating the spectrum calculations of Eq. (3.13) for pure Abelian gauge fields on a lattice gives the following expansions for the flux loop states,[30]

$$\begin{aligned} \omega_S &= 4(1 - .771\, x^2 - 5.191\, x^4) \\ \omega_{AV} &= 4(1 - .729\, x^2 - 1.222\, x^4) \\ \omega_T &= 4(1 - .271\, x^2 + .483\, x^4) \end{aligned} \tag{3.20}$$

where $x = 1/g^4$. If we form the diagonal Padé approximant,

$$\frac{M_S}{M_T} = \frac{1 - 12.588\, x^2}{1 - 12.108\, x^2} \quad \text{(Padé)} \tag{3.21}$$

we find a zero at a critical value of x,

$$x_c^2 = 1/(12.588) \tag{3.22}$$

corresponding to a critical coupling constant of

$$g_c^2 = 1.884 \tag{3.23}$$

Clearly Eq. (3.21) cannot be extended beyond this point -- there is indeed a natural boundary. Since the scalar particle becomes massless at $g=g_c$, we have probably found a second order phase transition--the lattice theory defined for $g = g_c$ should be scale invariant with a conventional continuum limit. These comments on the Abelian theory are being tested by pushing the expansions of Eq. (3.20) to higher orders.[31]

Now we begin our discussion of the Real Thing--Quantum Chromodynamics. Our local gauge group will be SU(3)' and there will be an iso-doublet of massless quarks. We concentrate on the SU(2) Flavor group for several reasons. First, we are interested in the lowest lying part of the hadron spectrum. Heavy quarks are not expected to be important here. In addition, the success of current algebra sum rules and PCAC strongly suggest that chiral symmetry is spontaneously broken and the pion appears as a Goldstone boson of strong interactions. It remains an important challenge to see if this physical picture can be obtained from a field theory employing only quarks and gluons. This challenge motivates our choice of degrees of freedom and our lattice fermion method. We use a fermion technique in which the absence of a mechanical quark mass follows from a natural symmetry of the lattice theory (discrete chiral symmetry in this case).

To begin, we must generalize the fermion method of the previous lecture to 3+1 dimensions. We anticipate that there will be species-doubling (or worse) unless the components of the fermi field are placed thinly on the lattice. Following Susskind[17], place a single component per site. If we write the Dirac equation in discrete form and choose a conventional representation for the γ-matrices we see the sensible way of doing this. The equations are

$$\begin{aligned}\dot{\psi}_1 &= \Delta_z\psi_3 + \Delta_x\psi_4 + i\Delta_y\psi_4 \\ \dot{\psi}_2 &= -\Delta_z\psi_4 + \Delta_x\psi_3 - i\Delta_y\psi_3 \\ \dot{\psi}_3 &= \Delta_z\psi_1 + \Delta_x\psi_2 + i\Delta_y\psi_2 \\ \dot{\psi}_4 &= -\Delta_z\psi_2 + \Delta_x\psi_1 - i\Delta_y\psi_1\end{aligned} \tag{3.24}$$

where

$$\Delta_z\psi = [\psi(r+\hat{n}_z) - \psi(r-\hat{n}_z)]/2a \tag{3.25}$$

Consider the first line of Eq. (3.24). Place ψ_1 on a site. Then moving one site in the $\pm z$ direction couples ψ_1 to ψ_3 and moving one site in the $\pm x$ direction couples ψ_1 to ψ_4, etc. Therefore, the components of ψ should be arranged as in Fig. 3.7 to give a sensible geometric interpretation of Eq. (3.24). But now we see a problem. The unit cube contains each component of ψ twice -- once on one y = const. plane and again on the next y = const. +a plane. Therefore, there must be two fermion fields implicit in this scheme in the continuum limit. To identify them label successive y-planes differently -- denote the fields on the lower y-plane in Fig. 3.7 f_i, i=1,...,4, and in the upper y-plane g_i, i=1,...,4. Then, in this language the discrete Dirac equation, Eq. (3.24), becomes

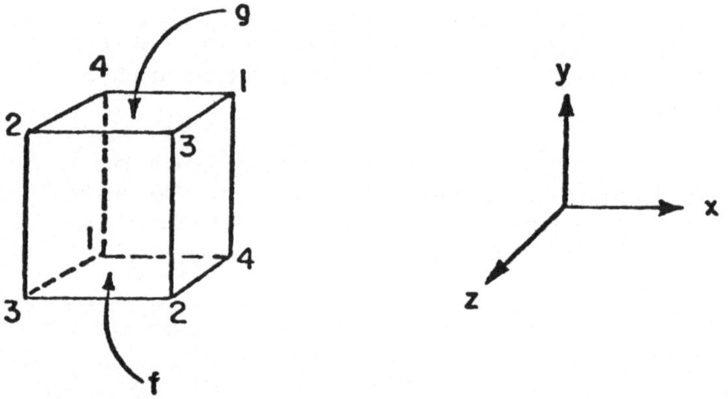

Fig. 3.7 Placement of the components of the fermion spinor on the spatial lattice. Fields on the lower y=const. plane are labelled f_i (i=1,...,4) and fields on the upper y=const. + a plane are labelled g_i (i=1,...,4).

$$\dot{f} = (\alpha_z\Delta_z + \alpha_x\Delta_x)f + (\alpha_y\Delta_y)g \qquad (3.26)$$

$$\dot{g} = (\alpha_z\Delta_z + \alpha_x\Delta_x)g + (\alpha_y\Delta_y)f$$

Now it is easy to identify the two species. The equation of motion for the sum u=f+g satisfies

$$\dot{u} = \underset{\sim}{\alpha}\cdot\underset{\sim}{\Delta}u \qquad (3.27)$$

and produces one fermion in the continuum limit. The difference $\tilde{d} = f - g$ satisfies

$$\partial_t\tilde{d} = (\alpha_z\Delta_z + \alpha_x\Delta_x - \alpha_y\Delta_y)\tilde{d} \qquad (3.28)$$

which is not a Dirac equation because of the sign error in the last term. But this can be removed via a unitary transformation,

$$d = \begin{pmatrix} i\sigma_y & 0 \\ 0 & -i\sigma_y \end{pmatrix} \tilde{d} \qquad (3.29)$$

which gives

$$\dot{d} = \underset{\sim}{\alpha}\cdot\underset{\sim}{\Delta}\ d \qquad (3.30)$$

In summary, this fermion method generates two massless fermion fields in the continuum limit. We are, therefore, free to interpret u and d as the usual members of an isodoublet,

$$\begin{pmatrix} u \\ d \end{pmatrix}$$

This possibility is a pleasing feature of the fermion method.

It is important to note that the long wavelength modes of Eq. (3.27) and (3.30) are not sensitive to the fine-grained details of the lattice. These modes, which are the only relevant ones in the continuum limit, satisfy the usual continuum Dirac equation. Therefore, although the lattice construction has less symmetry for $a \neq 0$ then one might hope the usual continuous space-time and isospin symmetries are retrieved in the finite energy spectrum when $a \to 0$.

Now let's write the lattice fermion Hamiltonian for Eq. (3.24). It is convenient to denote the single component fermion field at each site $\phi(r)$. Then the equation of motion is

$$\begin{aligned} \phi(r) = -\frac{1}{2a}\{&[\phi(r+n_z)-\phi(r-n_z)](-1)^{x+y} + \\ &+ [\phi(r+n_x)-\phi(r-n_x)] \\ &+ i\ [\phi(r+n_y)-\phi(r-n_y)](-1)^{x+y}\} \end{aligned} \tag{3.31}$$

and it can be obtained by canonical methods from the Hamiltonian

$$\begin{aligned} H_q = -\frac{i}{2a}\sum_n \{&[\phi^\dagger(r)\phi(r+n_z)-h.c.](-1)^{x+y} + \\ &+ [\phi^\dagger(r)\phi(r+n_x)-h.c.] \\ &+ i\ [\phi^\dagger(r)\phi(r+n_y) + h.c.]\ (-1)^{x+y}\} \end{aligned} \tag{3.32}$$

One can now read off some of the simpler symmetries of this fermion method by inspection of Eq. (3.32). Note that it is invariant under the replacement

$$\phi(r) \to \phi(r+n_z) \tag{3.33}$$

What symmetry does this correspond to in the familiar u and d quark language? Inspecting Fig. 3.7 shows that a shift by one unit in the z direction makes the replacements

$$f_1 \to f_3,\ f_4 \to f_2,\ g_1 \to g_3 \text{ and } g_4 \to g_2. \tag{3.34}$$

In terms of u and d spinors,

$$\begin{pmatrix} u_1 \\ u_2 \\ u_3 \\ u_4 \end{pmatrix} \to \begin{pmatrix} u_3 \\ u_4 \\ u_1 \\ u_2 \end{pmatrix}, \quad \begin{pmatrix} d_1 \\ d_2 \\ d_3 \\ d_4 \end{pmatrix} \to \begin{pmatrix} -d_3 \\ -d_4 \\ -d_1 \\ -d_2 \end{pmatrix} \tag{3.35}$$

which can be written more compactly as

$$\begin{pmatrix} u \\ d \end{pmatrix} \to \gamma_5 \tau_3 \begin{pmatrix} u \\ d \end{pmatrix} \tag{3.36}$$

which is identified as a discrete element of $SU(2) \times SU(2)\Big|_{\text{chiral}}$. Similar analyses for shifts in the x and y directions (some care with phases is necessary in these cases), show that they are symmetries which correspond to

$$\begin{pmatrix} u \\ d \end{pmatrix} \to \gamma_5 \tau_1 \begin{pmatrix} u \\ d \end{pmatrix}, \quad \begin{pmatrix} u \\ d \end{pmatrix} \to \gamma_5 \tau_2 \begin{pmatrix} u \\ d \end{pmatrix} \tag{3.37}$$

Given Eq. (3.36) and (3.37) we interpret discrete isospin transformations geometrically also. By combining x,y and z shifts in pairs, we find the symmetries $(\gamma_5 \tau_i)(\gamma_5 \tau_j) \sim \tau_k$. Therefore, shifts across face diagonals on the unit cube correspond to discrete isospin transformations,

$$\begin{pmatrix} u \\ d \end{pmatrix} \to \tau_i \begin{pmatrix} u \\ d \end{pmatrix} \tag{3.38}$$

In addition, one can compose an x shift, a y shift and a z shift, $\gamma_5\tau_1 \cdot \gamma_5\tau_2 \cdot \gamma_4\tau_3 \sim \gamma_5$, to produce a discrete γ_5 transformation. Therefore, shifts across a major body diagonal of the cube are identified as γ_5 transformations. The reader should consult the literature for the interpretations of other symmetry operations of the cube.[17] We learn from this that for $a \neq 0$ the Hamiltonian possesses only discrete pieces of the $U(2) \times U(2)|_{\text{chiral}}$ symmetry group.

These free field considerations are especially important because they generalize to the interacting case. It is important that the discrete pieces of $U(2)x(U(2)|_{\text{chiral}}$ which H possesses are sufficient to preclude the dynamical development of a fermion mechanical mass. Since the theory is presumed to be asymptotically free, the fact that the long wavelengths of the free u and d quarks possess full $SU(2) \times SU(2)|_{\text{chiral}}$ symmetry should generalize to the real theory. (This point will be discussed in greater detail later in this lecture.) Thus, in the continuum limit the lattice theory should have the same flavor symmetries as Eq. (1.1) with M=0. As in the 1+1 dimensional exercises, it is the fact that these symmetries should appear naturally in the continuum limit that motivates our use of this fermion method. We are in a position here to attack the popular idea that axial SU(2) symmetries are realized in the Nambu-Goldstone mode and that axial U(1) symmetry is destroyed by the character of the vacuum.[19] We shall see that the calculations have not proceeded far enough to really investigate these interesting problems, but our formulation may yet (with the help of more cleverness and labor) bear fruit.

Now let's put interactions back into the Hamiltonian and turn to computations. The ϕ formulation of the fermion method is quite ugly and unsymmetrical. These annoyances can be removed by defining another single component fermion field $\chi(r)$ which is related to $\chi(r)$ by some spatially dependent phase factors,

$$\chi(r) = (\text{phase factors})\,\phi(r) \tag{3.39}$$

The reader should consult the literature[17] for the details of Eq. (3.39). In this language a simple lattice Hamiltonian reads

$$H = \frac{g^2}{2a} \sum_{r,n} (\vec{E}_\alpha(r)\cdot\hat{n})^2$$

$$+ \frac{1}{2a} \sum_r \{(-1)^y \chi^\dagger(r) U(r,n_z)\chi(r+n_z) + (-1)^z \chi^\dagger(r)$$

$$U(r,n_x)\chi(r+n_x) + (-1)^x \chi^\dagger(r) U(r,n_y)\chi(r+n_y) + h.c.\}$$

$$- \frac{1}{g^2 a} \sum_{boxes} (\mathrm{Tr}\, UUUU + h.c.)$$

$$+ \ldots \tag{3.40}$$

The last line of Eq. (3.40) is meant to indicate that there is freedom in setting up the lattice H. We can always add operators to H which preserve the desired symmetries and which vanish in the continuum limit. These irrelevant operators were discussed in Lecture 2 and we will take advantage of them again. For the moment, however, ignore such possibilities and organize Eq. (3.40) in the following way. Introduce a dimensionless Wamiltonian,

$$W = \frac{2a}{g^2} H = W_e + xW_q + 2x^2 W_m \tag{3.41}$$

where

$$W_e = \sum_{r,n} (E(r)\cdot n)^2$$

$$W_q = \sum_r (-1)^y \chi^\dagger(r) U(r,n_z)\chi(r+n_z) + \text{etc.} \tag{3.42}$$

$$W_m = -\sum (\mathrm{Tr}\, UUUU + h.c.)$$

and

$$x = 1/g^2 \tag{3.43}$$

is the dimensionless expansion parameter for our analysis.

Before considering the mass spectrum and matrix elements of Quantum Chromodynamics, we must discover the character of the vacuum of W. These considerations will closely parallel the same discussion for the massless Schwinger model. We shall find that the discrete axial symmetries of $SU(2) \times SU(2)_{chiral}$ which W has are spontaneously broken by the theory's vacuum. Note that all states which are fluxless but have fermion content are degenerate in zeroth order. Call such states $|0,\Psi\rangle$. This degeneracy is lifted in perturbation theory as in the 1+1 dimensional models. An exercise in Raleigh-Schrödinger perturbation theory leads us to the minimization of

$$\delta W_o = \frac{x^2}{4} \langle 0,\Psi| \sum_{r,\hat{n}} \rho(r)\rho(r+\hat{n}) |0,\Psi\rangle \tag{3.44}$$

where

$$\rho(r) = [\chi^\dagger(r), \chi(r)] \tag{3.45}$$

is the color singlet fermion density. Thus, as in 1+1 dimension the vacuum problem is exactly soluble since it reduces to a generalized Ising anti-ferromagnet computation. To solve this explicitly we must find the spectrum of $\rho(r)$. It consists of the integers $-3,-1,+1$, and $+3$ corresponding to the possibilities that site r is unoccupied, once-occupied, twice-occupied or triply-occupied by χ excitations. In detail,

$$\begin{aligned}
&\rho|-3\rangle = -3|-3\rangle && \text{where } \chi_i|-3\rangle = 0 \\
&\rho|-1,i\rangle = -1|-1,i\rangle && \text{where } \chi_i^\dagger|-3\rangle = |-1,i\rangle \\
&\rho|1,\bar{j}\rangle = +1\ |1,\bar{j}\rangle && \text{where } \tfrac{1}{2}\varepsilon_{ijk}\chi_j^\dagger\ \chi_k^\dagger|-3\rangle = |1,\bar{i}\rangle \\
&\rho|3\rangle = +3|3\rangle && \text{where } \tfrac{1}{6}\varepsilon_{ijk}\chi_i^\dagger\chi_j^\dagger\chi_k^\dagger|-3\rangle = |+3\rangle
\end{aligned} \tag{3.46}$$

where the color indices on the fermion field have been exposed. Note that the states $|-3\rangle$ and $|+3\rangle$ are color singlets. In order to minimize Eq. (3.44) we must have all the nearest neighbors of a site with $\rho=+3$ have $\rho=-3$. There are two ways to do this. Label sites with $(-1)^{x+y+z}$ positive "even" and those with $(-1)^{x+y+z}$ negative "odd". Then there are two degenerate vacua corresponding to the possibilities: $\rho=+3$ on odd sites and -3 on even sites or vice versa, as shown in Fig.3.8. How are these two vacua related? One is taken into another by a shift through one link in any of the three possible directions. But these shifts are symmetries of H and correspond to discrete axial isospin, $\gamma_5\tau_i$, transformations. The two-fold degeneracy pictured in Fig. 3.8 persists to all orders of perturbation theory because of the $\gamma_5\tau_i$ symmetry of H. Note that translations by an even number of links leave each vacuum

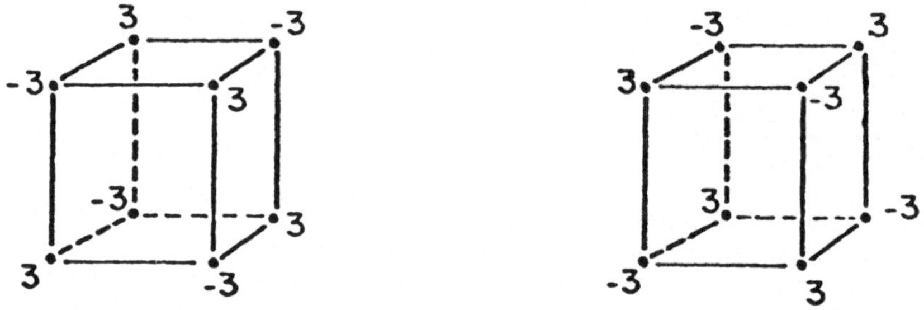

Fig. 3.8 The values of $\rho(r)$ in the two degenerate vacua.

invariant. These correspond to discrete isospin rotations, τ_i, and these symmetries are not broken by the theory's vacuum structure. And, finally, the vacuum is locally color gauge invariant. Thus, the vacuum preserves those symmetries we believe the vacuum of the real world has and it breaks the axial symmetries which also appear to be broken in the real world. This important result means that we can expect a smooth extrapolation of physical quantities from the strongly cutoff theory to the continuum limit. Since we use the Padé extrapolation method this possibility is crucial to the success of our scheme.

Conventional Quantum Chromodynamics has much more symmetry than its lattice version. It has full continuous $SU(2) \times SU(2)_{\text{chiral}}$ symmetry in its Lagrangian and presumably its vacuum spontaneously breaks the axial members of the symmetry group so that the isotriplet of massless pions appear as Nambu-Goldstone bosons. In a proper calculation of the continuum limit of lattice gauge theory these continuous symmetries should re-appear.

Roughly speaking, this occurs because the long wavelength ($\lambda >> a$) solutions of the free lattice Dirac equation have the continuous symmetries. In the continuum limit the coupling constant g(a) vanishes so that short distance lattice fermion-gluon interactions which respect only the discrete elements of $SU(2) \times SU(2)|_{chiral}$ become ineffective. These remarks can be made more precise by observing that one cannot construct operators which effect the continuum limit of the lattice Hamiltonian, which are compatible with the discrete symmetries of the lattice H, and which explicitly break $SU(2) \times SU(2)|_{chiral}$ in the continuum limit. Granting that the lattice theory is formulated so that it remembers the continuous symmetries when $a \to 0$, one must still verify that in this limit the axial SU(2) rotations are spontaneously broken symmetries of the vacuum. One must check, for example, that $\langle 0|\int\bar{\psi}\psi dx|0\rangle$ remains finite in the continuum limit when properly renormalized. If that is true, then the Nambu-Goldstone Theorem applies to the continuum limit of the lattice theory and we expect a massless pion triplet. Achieving this in a straightforward computation has still alluded us, as will be discussed below, but these general considerations give confidence that obtaining massless pions and the successful predictions of Current Algebra and P.C.A.C. may just be a matter of pushing strong coupling calculations to sufficiently high orders. This will be discussed after some lattice spectrum calculations have been presented.

Now consider a simple variation of the Hamiltonian in Eq. (3.40) which will simplify the strong coupling expansions. As discussed in Lecture 2 irrelevant operators can be added to H if they preserve its symmetries. Consider the term

$$W' = \mu \sum_{r,\hat{n}} \left[\rho(r)\rho(r+\hat{n})+9\right] \tag{3.47}$$

where μ is a dimensionless (irrelevant) parameter. This term is invariant under the discrete elements of $SU(2) \times SU(2)|_{chiral}$ which are symmetries of W. In addition, it leaves the two-fold defeneracy expressed in Fig. 3.8 intact to all orders, since W' has the same form as Eq. (3.44). The irrelevancy of W' is most easily seen by restoring its dimensions and writing it in a conventional form,

$$H' = \text{const.}\ \mu\cdot g^2 a^2 \int (\bar{\psi}\psi)^2 dx \tag{3.48}$$

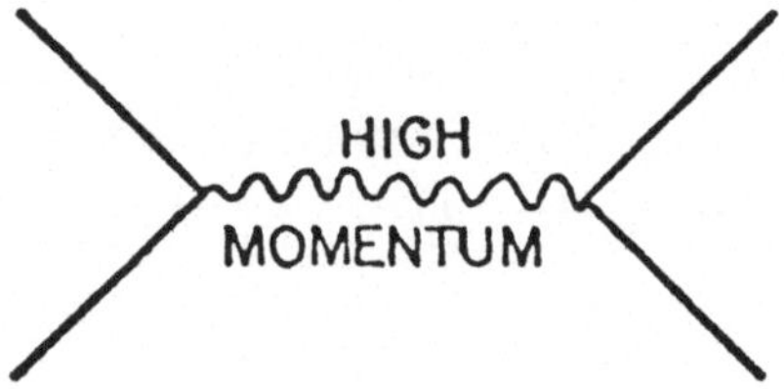

Fig. 3.9 Graph which contributes a four-fermi interaction in a cutoff effective Lagrangian.

Thus, it is a four-fermi interaction with a coupling $G\, g^2 a^2$. Hence, it disappears in the continuum limit faster than a^2 (because $g\to 0$ as $a\to 0$) and should leave continuum calculations unchanged. One could, in fact calculate its strength, i.e. compute μ as a function a. Consider a related calculation. Take the continuum form of Quantum Chromodynamics and define it with a gauge invariant cutoff (a Lorentz invariant regularization scheme, say). Then, if one integrates out the high frequency modes of that Lagrangian to produce an effective Lagrangian having a smaller momentum cutoff, four fermi terms would appear naturally from simple scattering graphs as in Fig. 3.9. Doing this in conventional, small g perturbation theory produces terms analogous but not identical to Eq. (3.48). Eq. (3.48) can occur because of the relatively small degree of natural symmetries in lattice spatially cutoff theories.

There are several reasons for incorporating W' into Eq. (3.40). Note that in its absence nucleon anti-nucleon pairs would cost no zeroth order ω energy, because there is no flux in their wave functions. This degeneracy is lifted by W' so using it frees us from applying the formalism of degenerate perturbation theory which becomes awkward at high orders. We also note that the zeroth order wave functions of mesons is simple only for μ sufficiently large. Include W' into the static, unperturbed piece of W. Then the zeroth order ω energy of a "quark-flux-anti-quark" meson is

$$\omega^{(0)}_{meson} = 4/3 + 68\mu \tag{3.49}$$

while the zeroth order ω energy of a baryon is

$$\omega^{(0)}_{nucleon} = 108\mu \tag{3.50}$$

The numbers 68 and 108 are easily obtained from counting arguments for the Ising-like W' term. Similarly, the zeroth order ω energy for a nucleon anti-nucleon pair separated by one link is

$$\omega^{(0)}_{N\bar{N}} = 180\mu \qquad (3.51)$$

The $N\bar{N}$ state can be constructed to have the same quantum numbers as a "q-flux-$\bar{q}$" meson state. Only for $\mu > (\frac{4}{3})/112$ will the q-flux-$\bar{q}$ state be less massive than the $N\bar{N}$ state in the static limit. For $\mu < \frac{4}{3}/112$ one must use the lighter $N\bar{N}$ state to find the mass of the lightest meson of those quantum numbers.For simplicity we restrict our attention to $\mu \geq \frac{4}{3}/112$. The approximate irrelevance of μ will be shown numerically over this range.

Our calculations of the low energy properties of Quantum Chromodynamics have focused on the low lying S-wave and P-wave mesons, the nucleon, and various matrix elements such as the nucleon's axial charge g_A. The zeroth order wave functions for mesons and nucleons are obtained by the same methods used in the 1+1 dimensional models. To make meson states from the vacuum we begin with the quark bilinear with the correct quantum numbers, write it in lattice language and apply it to the vacuum. For example,

$$|\pi_0\rangle \sim \int i\bar{\psi}\gamma_5 \frac{1}{2} \tau_3 \psi \, dx |0\rangle \qquad (3.52)$$

which reads in lattice form

$$|\pi_0\rangle = \frac{i}{\sqrt{3}} \left[\sum_r (-1)^x \chi^\dagger(r) U(r,n_z) \chi(r+n_z) - h.c.\right] |0\rangle \qquad (3.53)$$

Now we apply Raleigh-Schrödinger perturbation theory as usual. The elegance of the χ fermion method is pleasing in these lengthy tasks.

Now consider some results.[32] Strong coupling expansions have been carried out to fourth order in the expansion parameter $x=1/g^2$ for the mesons π,ρ,ω,σ (sometimes called the ε), B,f,A_1 and for the nucleon and its axial charge g_A. These calculations were done analytically for any value of μ. In Tables I and II we present the expansions using the notation

$$\omega_i = \omega_i^{(0)} + \omega_i^{(2)}\cdot x^2 + \omega_i^{(4)}\cdot x^4 + \ldots \qquad (3.54)$$

for the particles and

$$g_A = g_A^{(0)} + g_A^{(2)}\cdot x^2 + g_A^{(4)}\cdot x^4 + \ldots \qquad (3.55)$$

for the axial charge. These series are replaced by their [1,1] Padé approximants,

$$\frac{1 + \alpha x^2}{1 + \beta x^2} \qquad (3.56)$$

and the continuum limit $x\to\infty$ is taken, yielding α/β. As usual, mass ratios are computed and the nucleon is used as a common base. We find the results

$$\begin{aligned}
m_\rho^{(4)}/m_N^{(4)} &= .822 \qquad (.820)\\
m_\omega^{(4)}/m_N^{(4)} &= .824 \qquad (.834)\\
m_\pi^{(4)}/m_N^{(4)} &= .820 \qquad (.147)\\
m_\sigma^{(4)}/m_N^{(4)} &= .972 \qquad (820\text{-}1.10,\ \text{broad}) \qquad (3.57)\\
m_B^{(4)}/m_N^{(4)} &= 1.05 \qquad (1.32)\\
m_f^{(4)}/m_N^{(4)} &= 1.17 \qquad (1.35)\\
m_{A_1}^{(4)}/m_N^{(4)} &= 1.12 \qquad (1.17)
\end{aligned}$$

while the fourth order axial vector charge is

$$g_A^{(4)} = 1.81 \qquad (1.24) \qquad (3.58)$$

The experimental values for these quantities are the numbers listed above in parentheses.

Table I. Expansion coefficients, and continuum masses measured relative to the nucleon for various hadrons at difference choices of the irrelevant parameter ξ. The expansion coefficients occur in the expressions $\omega_i=\omega_i^{(0)}+\omega_i^{(2)}x^2+\omega_i^{(4)}x^4$, where i denotes the hadron of interest. At ξ=.61 degenerate perturbation theory must be applied. Non-degenerate perturbation theory gives the results at ξ=1.02 and 1.43. Quasi-degenerate perturbation theory lowers the ξ=1.02 results to those shown in the graphs. At ξ=1.43 the non-degenerate results are very accurate.

Hadron	ξ	$\omega^{(0)}$	$\omega^{(2)}$	$\omega^{(4)}$	m/m_N
ρ	.61	2.143	-3.512	66.222	.822
	1.02	2.693	- .292	12.483	.894
	1.43	3.237	- .161	4.075	.824
ω	.61	2.143	-3.507	66.435	.824
π	.61	2.143	-3.516	66.173	.821
σ	.61	2.143	- .880	41.963	.989
	1.02	2.693	1.955	6.994	1.11
	1.43	3.237	1.391	1.132	.995
B	.61	2.143	-1.821	55.891	.95
	1.02	2.693	1.560	11.109	1.10
	1.43	3.237	1.433	3.623	1.02
f	.61	2.143	-3.516	165.708	1.17
	1.02	2.693	- .301	76.698	1.14
	1.43	3.237	- .171	41.436	1.02
A_1	.61	2.143	- .988	47.502	1.00
	1.02	2.693	2.303	8.541	1.15
	1.43	3.237	1.943	1.943	1.07
Nucleon	.61	1.296	5.368	-15.029	--
	1.02	2.160	3.414	- 6.015	--
	1.43	3.024	2.341	- 3.194	--

Table II. Expansion coefficients and the continuum value of g_A for various choices of ξ.

ξ	$g_A^{(0)}$	$g_A^{(2)}$	$g_A^{(4)}$	g_A
0	3.00	-14.58	178.89	1.81
.31	3.00	- 7.22	55.02	2.10
.61	3.00	- 4.42	15.76	2.04
.92	3.00	- 2.96	6.47	1.91
1.22	3.00	- 2.11	3.62	1.70
1.53	3.00	- 1.58	2.49	1.31

Consider several interesting features of these calculations. For each mass ratio considered here the [1,1] Padé approximant exists with positive values for α and β. Therefore, an extrapolation from $y=0$ to $y=\infty$ is singularity-free in this approximation. This result gives us some confidence that the only phase of QCD is one which is strongly coupled at large distances. Of course, higher order calculations are essential to really argue this point.

Observe from the tables that the expansion coefficients for ω_{meson} and ω_N depend strongly on the irrelevant parameter μ, while the mass ratios do not. A meaningful measure of the size of μ is contained in the parameter $\xi = 68\mu/\frac{4}{3}$ which parametrizes the tables. ξ measures the amount of irrelevant energy on the scale of flux energy $(\frac{4}{3})$ in the unperturbed meson states. In the tables ξ varies from .61 to 1.43 (more than a factor of 2) and m^{ρ}/m_N, for example, varies only a few percent over this range. This is a good indication that the calculational and extrapolation method is respecting some general field theoretic principles. Note that the p-wave meson to nucleon mass ratios have more dependence on ξ. Since p-wave functions have more spatial variation than s-wave functions, this trend in the results is not unexpected. Higher order calculations will be necessary to approximate accurately the wave functions of excited states using lattice methods.

All of these calculations are being extended to higher orders using computer methods. Probably the only fair concerning the fourth order results is that they significantly improve the static calculations. For example, $g_A^{(0)} = 3$ because the nucleon consists of three static quarks in zeroth order. The value $g_A^{(4)} = 1.81$ which is found at $\xi=0$ is a considerable improvement over $g_A^{(o)}$, but it is nothing to brag about.

Our only real failure at this level is the calculation of the pion mass. The reason for this problem can be seen by inspecting low order graphs, and recalling that in quark models it is spin-spin forces which split the π and the ρ. The static H simply has not spin-spin effects and it is not until sixth order that they appear naturally. Therefore, we hope to achieve sizeable π-ρ splitting in our more ambitious computer calculations.

Besides pushing these calculations to higher orders in such a pedestrian fashion, we are also attempting to further exploit the + ... in Eq. (3.40). Renormalization group analyses suggest new terms which can appear in a cutoff Hamiltonian. For example, in addition to the very local terms of Eq. (3.40), next-nearest-neighbor coupling terms are generated naturally.[33] Some of these terms cause spin-spin forces in low orders of strong coupling expansions. We are optimistic that this work will yield a better understanding of lattice gauge theory and better numerical results than those reported here.

REFERENCES

1. The term "Quantum Chromodynamics" has been suggested by M. Gell-Mann. Advantages of this theory over previous field theoretic formulations of the quark model were pointed out by H. Fritzsch, M. Gell-Mann and H. Leutwyler, Phys. Lett. 47B, 365 (1973).
2. G. 't Hooft, Marseilles Conference on Gauge Theories, 1972 (unpublished); H.D. Politzer, Phys. Rev. Lett. 30, 1346 (1973); D.J. Gross and F. Wilczek, ibid. 30, 1343 (1973).
3. Applications to deep inelastic scattering: D.J. Gross and F. Wilczek, Phys. Rev. D8, 3633 (1973) and D9, 980 (1974), and H. Georgi and H.D. Politzer, Phys. Rev. D9, 416 (1974). Electron-positron annihilation: T. Appelquist and H. Georgi, Phys. Rev. D8, 4000 (1973) and A. Zee, Phys. Rev. D8, 4038 (1973). The $\Delta I = 1/2$ rule: M.K. Gaillard and B.W. Lee, Phys. Rev. Lett. 33, 108 (1974) and works by many other authors.

4. A recent interesting theory appears in F. Gürsey and P. Sikivie, Phys. Rev. Lett. 36, 775 (1976). See also H. Georgi, H. Quinn and S. Weinberg, Phys. Rev. Lett. 33, 451 (1974).
5. This speculation was made simultaneously by many authors. See, for example, S. Weinberg, Phys. Rev. Lett. 31, 494 (1973).
6. Ugh.
7. K.G. Wilson, Phys. Rev. D10, 2445 (1974).
8. A.M. Polyakov (unpublished) and Phys. Lett. 59B, 82 (1975).
9. J. Kogut and Leonard Susskind, Phys. Rev. D11, 395 (1975).
10. L. Susskind, Lectures presented at the Bonn Summer School, 1974.
11. J. Schwinger, Phys. Rev. Lett. 3, 296 (1959).
12. Exposing the color indices which are locally contracted, Eq. (1.31) is:
$\bar{\psi}_i (U_3)^i{}_j \psi^j |0\rangle$.
13. A similar argument implies that colorful gluons do not exist in the finite energy spectrum of the theory.
14. Colored quark models are discussed by H. Fritzsch and M. Gell-Mann, in Proceedings of the Sixteenth International Conference on High Energy Physics, The University of Chicago and National Accelerator Laboratory, 1972, edited by J.D. Jackson and A. Roberts (National Accelerator Laboratory, Batavia, Illinois, 1973), Vol. 2, p. 135.
15. T. Banks, L. Susskind and J. Kogut, Phys. Rev. D13, 1043 (1976). A. Carroll, J. Kogut, D.K. Sinclair and L. Susskind, Phys. Rev. D13, 2270 (1976).
16. J. Shigemitsu and S. Elitzur, CLNS-333 (May 1976), to appear in Phys. Rev. D. The dynamical symmetry breaking and mass gap equation for large N were obtained by A. Zee, Phys. Rev. D12, 3251 (1975).
17. Leonard Susskind, "Lattice Fermions", Ecole Normale Superieure preprint, December 1975 (submitted for publication).
18. Most recently, see A. Luther, Nordita preprint, June 1976, and the references therein.
19. J. Kogut and L. Susskind, Phys. Rev. D11, 3594 (1975).
J. Lowenstein and A. Swieca, Ann. Phys. (N.Y.) 68, 172 (1971).
20. See E. Merzbacher, Quantum Mechanics (John Wiley and Sons, Inc., New York, 197o), 2nd Ed., Chap. 17, Sec. 8, for a discussion of the Feynman-Hellmann Theorem.

21. See, for example, Chap. 8 of D. Mattis, The Theory of Magnetism (Harper and Row, New York, 1965) and the references to the original publications cited there.
22. Background on Padé approximants may be gleaned from G.A. Baker, Jr., Essentials of Padé Approximants (Academic Press, New York, 1975). Other reviews which discuss many applications include J.L. Basdevant, Fortschritte der Physik 20, 283 (1973 (and J. Zinn-Justin, Physics Reports 1C, No. 3 (1973).
23. H. Bergknoff, "The Physical Particles of the Massive Schwinger Model", CLNS-341, August 1976 (submitted to Nucl. Phys. B).
24. K. Lane, Phys. Rev. D10, 1353, 2605 (1974). See also D.J. Gross and A. Neveu, ibid. D10, 3235 (1974).
25. Rigorous results for the anharmonic oscillator were obtained by J.J. Loeffel, A. Martin, B. Simon and A.S. Wightman, Phys. Lett. 30B, 656 (1969). See also the general discussions of ref. 22.
26. K.G. Wilson and J. Kogut, Phys. Reports 12C, 75 (1974).
27. See the second article of ref. 24.
28. R.F. Dashen, B. Hasslacher and A. Neveu, Phys. Rev. D12, 2443 (1975).
29. A.A. Migdal, "Phase Transitions in Lattice and Gauge Systems", Chernogolovka reprint, 1975), and "Recursion Equations in Gauge Theories", ibid., 1975.
30. J. Kogut, D.K. Sinclair and Leonard Susskind, CLNS-336 (June 1976), to appear in Nucl. Phys. B.
31. R. Fredrickson, in progress.
32. The C.O.T.Y. Collaboration, "Strong Coupling Calculations of the Hadron Spectrum of Quantum Chromodynamics", CLNS-339 (July 4, 1976).
33. K.G. Wilson, unpublished.

CLUSTERING AND CLUSTERS

André Krzywicki

Laboratoire de Physique Théorique et

Particules Elémentaires

Orsay, France*

1. SELECTED TOPICS IN THE THEORY OF STOCHASTIC POINT PROCESSES

1.1 Systems of Random Points

When one is lecturing in a centre for interdisciplinary research, it is, I imagine, appropriate to start by emphasizing the interdisciplinary character of the topic one is going to discuss. In my case this does not present any difficulty. Clustering is a pictorial term for positive correlations in an ensemble of randomly distributed bodies or events, which can be represented in some space by a system of random points. Probabilistic techniques which enable one to study such systems constitute the theory of stochastic point processes[1)]

* Laboratoire associé au C.N.R.S. Postal address: LPTPE Bâtiment 211, Université de Paris Sud, 91405 Orsay, France.

(the word "process" is used because in applications one often considers time series of correlated events). The theory of stochastic point processes found applications in radio physics, in epidemiology, in studying certain aspects of road traffic, in quantum optics etc. In high energy physics one of the most familiar examples of a system of random points is a rapidity plot representing a multiparticle event:

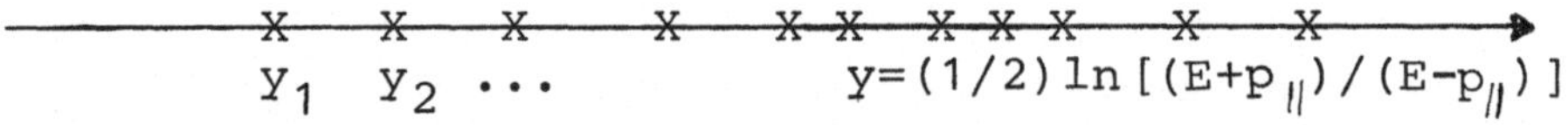

I shall stick to this one-dimensional example, but the generality of the results discussed in this chapter should be clearly borne in mind. The use of quantum mechanics language is very convenient for a high energy physicist but is by no means necessary. It is perhaps worth mentioning that most of the general results concerning systems of random points have been known since twenty years ago to people studying stochastic point processes in the context of radio physics.

1.2. Statistical Ensemble of Random Points

Statistical ensemble of random points [2)] can be described by a density matrix ρ. The probability of finding the system in a given state, or in a given set of states, is

$$\mathrm{Prob}(s) = \mathrm{Tr}\,(\rho P_s) = \langle P_s \rangle \tag{1}$$

where P_s is the operator projecting on the states in question and < > denotes the ensemble average. The random points discussed in the following are taken to be the rapidities of particles belonging to a rapidity interval R of length Y. However, R is not necessarily the whole rapidity interval accessible kinematically at a given energy: $Y \lesssim \ln(E_{lab}/E_o)$. If not stated otherwise the integrations are always over R.

When we speak about particles belonging to R, we mean particles detected by some experimental set-up. We do not assume that the set-up records all types of partilces which are actually produced. However, for simplicity, we assume that the set-up does not distinguish between different types of the recorded particles. In other words, our random points are indistinguishable.

The operator projecting on a state where there are n and only n particles in R, localized at $y_1 \ldots y_n$, is

$$P_{exclusive} = a^+(y_1) \ldots a^+(y_n) \; |0\rangle\langle 0| \; a(y_1) \ldots a(y_n) \tag{2}$$

where $a^+(y)$ and $a(y)$ are the operators, respectively creating and annihilating a particle with rapidity y and satisfying the standard commutation relations:

$$(a(y_1), a^+(y_2)) = \delta(y_1 - y_2), (a(y_1), a(y_2)) = $$
$$= (a^+(y_1), a^+(y_2)) = 0 \tag{3}$$

However (see the Appendix)

$$|0\rangle\langle 0| = :\exp(-\int N(y)dy) : \tag{4}$$

where $N(y) = a^+(y)a(y)$ is the particle number operator and the symbol : : denotes, as usual, the normal product, in which all creation operators stand to the left of all annihilation operators. Hence

$$P_{exclusive} = : N(y_1) \ldots N(y_n) \exp(-\int N(y)dy) : \tag{5}$$

The operator projecting on states containing at least n particles at $y_1 \ldots y_n$ is

$$P_{inclusive} = \sum_s a^+(y_1) \ldots a^+(y_n) |s\rangle\langle s| a(y_1) \ldots a(y_n) \tag{6}$$

Using the completeness relation $\sum_s |s\rangle\langle s| = 1$ one rewrites (6) in the form

$$P_{inclusive} = :N(y_1) \ldots N(y_n): \tag{7}$$

1.3 Exclusive and Inclusive Probabilities and the Generating Functional

From (1) and (5) one obtains the exclusive probability of finding n and only n particles at $y_1 \ldots y_n$:

$$e_n(y_1 \dots y_n) = \langle :N(y_1) \dots N(y_n) \exp(- \int N(y)dy) :\rangle \quad (8a)$$

Notice that e_o is the probability that the interval R is empty. Likewise, from (1) and (7) one obtains the inclusive probability of finding at least n particles in R, localized at $y_1 \dots y_n$:

$$i_n(y_1 \dots y_n) = \langle :N(y_1) \dots N(y_n) :\rangle \quad (8b)$$

Exclusive and inclusive probabilities can be written as functional derivatives of the same generating functional $g(\phi)$:

$$g(\phi) = \langle :\exp(\int N(y)\phi(y)dy) :\rangle \quad (9)$$

Indeed

$$e_n(y_1 \dots y_n) = \delta^n g(\phi)/\delta\phi(y_1) \dots \delta\phi(y_n) \Big|_{\phi=-1} \quad (10a)$$

$$i_n(y_1 \dots y_n) = \delta^n g(\phi)/\delta\phi(y_1) \dots \delta\phi(y_n) \Big|_{\phi=0} \quad (10b)$$

Of course

$$g(\phi) = \langle :\exp(\int N(y)(1+\phi(y))dy) \exp(- \int N(y)dy) :\rangle \quad (11)$$

Expanding the 2nd exponential in (11) and taking the functional derivative indicated in (10a) we find

$$e_n(y_1 \dots y_n) = \sum_{k=0}^{\infty} ((-1)^k/k!) \int i_{n+k}(y_1 \dots y_{n+k}) \, dy_{n+1} \dots dy_{n+k} \quad (12a)$$

Expanding the 1st exponential in (11) and taking the functional derivative indicated in (10b) we obtain

$$i_n(y_1 \dots y_n) = \sum_{k=o}^{\infty} (1/k!) \int e_{n+k}(y_1 \dots y_{n+k}) \, dy_{n+1} \dots dy_{n+k} \quad (12b)$$

which is the conventional definition of an inclusive probability.

1.4 Correlation Functions

Correlation functions are defined by the equation

$$c_n(y_1 \ldots y_n) = \delta^n \ln g(\phi)/\delta\phi(y_1)\ldots\delta\phi(y_n)\Big|_{\phi=0} \quad (13)$$

Thus

$$g(\phi) = \exp\left(\sum_{n=1}^{\infty}(1/n!)\int c_n(y_1\ldots y_n)\prod_{i=1}^{n}(\phi(y_i)dy_i)\right) \quad (14)$$

Comparing (13) and (10b) one easily checks that

$$c_1(y_1) = i_1(y_1) \quad (15a)$$

$$c_2(y_1,y_2) = i_2(y_1,y_2) - i_1(y_1)i_1(y_2), \text{ etc} \quad (15b)$$

It is suggested by the above equations and by the terminology we have adopted, that the vanishing of all multiparticle correlation functions implies that particles are emitted independently. Indeed, assume that

$$c_n(y_1\ldots y_n) = 0 \text{ for } n \geq 2 \quad (16)$$

In this case

$$g(\phi) = \exp\left(\int i_1(y)\phi(y)dy\right) \quad (17)$$

and therefore the inclusive probabilities factorize:

$$i_n(y_1\ldots y_n) = i_1(y_1)\ldots i_1(y_n) \quad (18)$$

1.5 Multiplicity Distribution

The integrals of the multiparticle distributions defined heretofore characterize the distribution of the multiplicity of points in R. The probability of finding exactly n particles in R is

$$P_n = (1/n!)\int e_n(y_1\ldots y_n)\, dy_1\ldots dy_n \quad (19)$$

The so-called binomial moments $\langle N(N-1)\ldots(N-n+1)\rangle$ are the integrals of the inclusive probabilities

$$\langle N(N-1)\ldots(N-n+1)\rangle = \int i_n(y_1\ldots y_n)dy_1\ldots dy_n \tag{20}$$

This is readily checked using (12b) and (19). The normalization of the inclusive probabilities is not restricted to be less than unity, because the integration in (20) involves multiple counting of equivalent configurations. The integrals

$$f_n = \int c_n(y_1\ldots y_n)\, dy_1\ldots dy_n \tag{21}$$

measure the strength of the inclusive correlations.

Notice that eq. (10a) is equivalent to

$$g(\phi) = \sum_{n=o}^{\infty} (1/n!) \int e_n(y_1\ldots y_n) \prod_{i=1}^{n} ((1+\phi(y_i))dy_i) \tag{22}$$

Setting $\phi(y) = z$, independent of y, one obtains the <u>generating function</u> for the multiplicity distribution

$$g(z) = \sum_{n=o}^{\infty} p_n(1+z)^n \tag{23}$$

The following equations are analogues of eqs. (10a), (10b) and (13)

$$p_n = (1/n!)\, d^n g(z)/dz^n \Big|_{z=-1} \tag{24a}$$

$$\langle N(N-1)\ldots(N-n+1)\rangle = d^n g(z)/dz^n \Big|_{z=0} \tag{24b}$$

$$f_n = d^n \ln g(z)/dz^n \Big|_{z=0} \tag{24c}$$

When all multiparticle correlations are absent, (17) implies that $g(z)=\exp(\langle N\rangle z)$ and the multiplicity distribution has the Poisson form

$$p_n = (\langle N\rangle^n/n!)\, \exp(-\langle N\rangle) \tag{25}$$

1.6 Rapidity Gap Distributions

Let $\varepsilon_o(y,y')$ denote the probability that the closed interval $[y,y']$ is empty. Thus $\varepsilon_o(y,y')$ is given by the equation (12a), with n=0 and R replaced by $[y,y']$:

$$\varepsilon_o(y,y')=\sum_{k=0}^{\infty} ((-1)^k/k!) \int_y^{y'} i_k(y_1 \ldots y_k) \, dy_1 \ldots dy_k \qquad (26a)$$

The probability $\varepsilon_2(y,y')$ that the open interval (y,y') is empty, but there are particles at y and y' is likewise obtained from (12a):

$$\varepsilon_2(y,y')=\sum_{k=0}^{\infty} ((-1)^k/k!) \int_y^{y'} i_{k+2}(y,y',y_1 \ldots y_k) \, dy_1 \ldots dy_k \qquad (26b)$$

Of course

$$\varepsilon_2(y,y') = -\partial^2 \varepsilon_o(y,y')/\partial y \, \partial y' \qquad (27)$$

The conditional probability p(r) that there is a rapidity gap of length r between a particle at y and its nearest right neighbor is

$$p(r) = (-1/i_1(y))\partial^2 \varepsilon_o(y,y')/\partial y \, \partial y' \qquad (28)$$

Notice, that $\varepsilon_o(y,y')$ is equal to $g(\phi)$, with $\phi=-1$ within the interval (y,y') and zero elsewhere. Using (14) one rewrites (26a) as follows

$$\varepsilon_o(y,y')=\exp(\sum_{k=1}^{\infty} ((-1)^k/k!) \int_y^{y'} c_k(y_1 \ldots y_k) \, dy_1 \ldots dy_k) \qquad (29)$$

1.7 The Short-Range Order* (SRO)

The short-range order (SRO) hypothesis consists in assuming that the density fluctuations in the system are qualitatively of the same nature as the density fluctuations in a non-critical fluid. One postulates, that the density moments $\langle N(y_1)\ldots N(y_n)\rangle$ are independent of Y and satisfy the following two conditions:

(a) <u>cluster decomposition:</u>

$$\langle N(y_1)\ldots N(y_n)\rangle = \langle N(y_1)\ldots N(y_k)\rangle\langle N(y_{k+1})\ldots N(y_n)\rangle \tag{30}$$

provided

$$\min |y_i - y_j| \gg \lambda, \quad i=1,\ldots,k \text{ and } j=k+1,\ldots,n \tag{31}$$

where λ is a certain characteristic distance.

(b) <u>stationarity or translation invariance:</u>

$$\langle N(y_1)\ldots N(y_n)\rangle = \langle N(y_1+a)\ldots N(y_n+a)\rangle \tag{32}$$

Thus the density moments depend on rapidity differences only. Strictly speaking (a) and (b) are meaningful only when $Y \gg \lambda$ and when one is many characteristic distances away from the boundaries of R.

A corollary to (b) is that the average particle density $\langle N(y)\rangle = i_1(y)$ is independent of y, except perhaps near the boundaries of R, and therefore the average multiplicity $\langle N\rangle$ is a linear function of Y:

$$\langle N\rangle = \int i_1(y)dy = a_1 Y + b_1 \tag{33}$$

It is obvious that the condition (b) is shared by the functions $i_n(y_1\ldots y_n)$ and $c_n(y_1\ldots y_n)$.

* It would be, perhaps, more appropriate to speak about short-range correlations and not short-range order. In mathematics, systems of random points satisfying SRO are called ergodic systems.

Using the identity (see the Appendix)

$$:\exp(a^+ a\phi): = \exp(a^+ a\ln(1+\phi)) \quad (34)$$

one obtains from (9) the generating functional for the density moments $<N(y_1)\ldots N(y_n)>$:

$$g(\phi) = <\exp(\int N(y) \ln(1+\phi(y))dy)> \quad (35)$$

Let R_1 and R_2 be two rapidity intervals, much longer than λ and separated by a distance much larger than λ. Let furthermore $\phi_i(y) \neq 0$ only when $y \in R_i (i=1,2)$. Set

$$\phi(y) = \phi_1(y) + \phi_2(y) \quad (36)$$

Then, eq. (35) and SRO imply that

$$g(\phi=\phi_1+\phi_2) = g(\phi_1)\, g(\phi_2) \quad (37)$$

Of course, the above condition is satisfied when $<:N(y_1)\ldots N(y_n):>$ has a cluster decomposition property analogous to (a). In other words:

(a') $i_n(y_1\ldots y_n)=i_k(y_1\ldots y_k)\, i_{n-k}(y_{k+1}\ldots y_n)$

provided (31) is satisfied.

Inserting (36) in (14) and requiring that (37) holds one finds the 3rd formulation of (a):

(a") a correlation function $c_n(y_1\ldots y_n), n\geq 2$, vanishes when at least one rapidity difference $y_i - y_j$, $i,j=1\ldots n$, becomes much larger, in absolute value, than the characteristic distance λ.

1.8 SRO and Multiplicity Distributions

Assume now that SRO holds exactly*. Then, for $Y>>\lambda$ one has

*This is an idealization, especially when one sets $Y = \ln(E_{lab}/E_o)$. Cf. the next chapter.

$$f_n = a_n Y + b_n \qquad n=1,2,... \tag{38}$$

since the integration in (21) extends effectively only over distances of order $O(\lambda)$ from the diagonal of R^n. Defining

$$\begin{pmatrix} a(z) \\ b(z) \end{pmatrix} = \sum_{n=1}^{\infty} (z^n/n!) \begin{pmatrix} a_n \\ b_n \end{pmatrix} \tag{39}$$

one can write the generating function in the form

$$g(z) = \exp(a(z)Y+b(z)) \tag{40}$$

Since $p_n \leq 1$ for any Y, one must have

$$a(-1) < 0 \tag{41}$$

This in turn means that, apart from a polynomial in Y, the probability p_n falls exponentially with increasing Y. This result is a particular case of a more general theorem, due to Le Bellac[3]:

Let $f_j \sim a_j Y$ for $j \leq 2M$ and $Y \to \infty$. Then p_n falls, with increasing Y, at least as fast as $(1/Y)^M$.

Replace the summation in (23) by an integration, use (40) and invert the equation so as to obtain an integral representation of p_n. Provided a(z), b(z) are regular enough at z=-1, the integral can be evaluated by saddle point techniques which, in the limit $Y \to \infty$ and n/Y fixed, yield[4,5]

$$(1/Y)\ \ln(n!p_n/Y^n) = a(-1)+c(n/Y) + O(1/Y) \tag{42a}$$

where c(u) is a priori unknown, depending on the details of the dynamics, but satisfies the constraint

$$c(0) = 0 \tag{42b}$$

1.9 SRO and Rapidity Gap Distributions

Assume again that SRO holds exactly, that y and y' are far from the boundaries of R and that

$$r = y - y' >> \lambda \tag{43}$$

Then, using (21) and (29) we find

$$\varepsilon_o(y,y') = \exp\left(\sum_{k=1}^{\infty} ((-1)^k/k!)(a_k r+b_k)\right)$$

$$\propto \exp(a(-1)r) \tag{44}$$

Eq. (28) further implies that

$$p(r) \propto \exp(a(-1)r) \tag{45}$$

for r large enough. It is remarkable that the same constant a(-1) appears in the above equation and in eq. (42). This is a general relation between two distinct features of the system [5].

1.10 Cluster Emission Models

Cluster emission models rest on the hypothesis that the observed particles are produced in two steps: first are created some intermediate objects, called clusters, which subsequently disintegrate into observed particles. By assumption, there is no interference between cluster decays.

Notation: we use the same symbols to represent quantities describing particle production, cluster production and cluster decay. However, a quantity refering to cluster production or decay has, respectively, one or two bars on top of it. The rapidity distribution of the decay products of a cluster depends, of course, on the cluster rapidity - this dependence is indicated by an argument appearing after a semicolon. For example, $\bar{\bar{i}}_n(y_1 \ldots y_n;\bar{y})$ is the inclusive probability that a cluster of rapidity $\bar{y}$ produces at least n particles with rapidities $y_1 \ldots y_n$.

The processes of cluster creation and disintegration are themselves random point processes and can be described by the generating functionals $\bar{g}(\phi)$ and $\bar{\bar{g}}(\phi;\bar{y})$. The functional $\bar{\bar{g}}(\phi;\bar{y})$ depends in general, on the cluster's rapidity $\bar{y}$. However, when $\phi(y) = z$, independent of y, $\bar{\bar{g}}$ becomes the generating function summarizing information about the particle multiplicity in the cluster decay. This decay multiplicity is independent of $\bar{y}$ and consequently the function $\bar{\bar{g}}(z;\bar{y})$ is independent of $\bar{y}$.

An easy calculation involving convolutions of exclusive probabilities (see the Appendix) leads to the result

$$g(\phi) = \bar{g}(\bar{\bar{g}}(\phi) - 1) \tag{46}$$

Functional differentiation of (46) yields equations, whose probabilistic interpretation is quite obvious:

$$i_1(y_1) = \int \bar{i}_1(\bar{y}_1)\, \bar{\bar{i}}_1(y_1;\bar{y}_1) d\bar{y}_1 \tag{47a}$$

$$i_2(y_1,y_2) = \int \bar{i}_1(\bar{y}_1)\, \bar{\bar{i}}_2(y_1,y_2;\bar{y}_1)\, d\bar{y}_1 + \int \bar{i}_2(\bar{y}_1,\bar{y}_2)\, \bar{\bar{i}}_1(y_1;\bar{y}_1)\, \bar{\bar{i}}_1(y_2;\bar{y}_2)\, d\bar{y}_1 d\bar{y}_2 \tag{47b}$$

etc.

A cluster model becomes a useful phenomenological tool provided one makes simplifying assumptions about cluster's production and decay. It is usually postulated that clusters are produced independently of each other (cf. (17)):

$$\bar{g}(\phi) = \exp\left(\int \bar{i}_1(\bar{y})\phi(\bar{y}) d\bar{y}\right) \tag{48}$$

with constant average density

$$\bar{i}_1(\bar{y}) = \text{const} \tag{49}$$

except perhaps near the boundary of R. One furthermore assumes, that cluster decay products occupy a finite rapidity interval, independent of Y, and that the inclusive probabilities $\bar{\bar{i}}_n(y_1 \ldots y_n;\bar{y})$ are functions of the differences of rapidities $y_1 \ldots y_n, \bar{y}$ only. The resulting picture is a simple, though reasonably flexible, mechanistic representation of SRO. In the following, unless stated otherwise, the expression "cluster model" will refer to models where clusters are produced independently, as described above [6].

Combining (46) and (48) one obtains

$$g(\phi) = \exp\left(\int \bar{i}_1(\bar{y})(\bar{\bar{g}}(\phi;\bar{y})-1)d\bar{y}\right) \tag{50}$$

The correlation functions are calculated using (13) and (10b):

$$c_n(y_1 \ldots y_n) = \int \bar{i}_1(\bar{y})\, \delta^n \bar{\bar{g}}(\phi;\bar{y})/\delta\phi(y_1)\ldots\delta\phi(y_n)\Big|_{\phi=0} d\bar{y} = \int \bar{i}_1(\bar{y})\bar{\bar{i}}_n(y_1 \ldots y_n;\bar{y})d\bar{y} \tag{51}$$

Since the above expression is a convolution of non-negative quantities, all correlation functions are positive or zero:

$$c_n(y_1 \ldots y_n) \geq 0 \tag{52}$$

Furthermore, if a cluster never decays into more than m secondaries, $c_n(y_1 \ldots y_n)=0$ for $n>m$. This can be easily understood: particles coming from distinct clusters are uncorrelated. However, when one finds a particle somewhere, there is an increased probability of finding nearby the rest of the progeny of the parent cluster. The maximum number of correlated particles equals the maximum multiplicity in the disintegration of a cluster.

The generating function for the multiplicity distribution is immediately found from (50):

$$g(z) = \exp(\langle\bar{N}\rangle(\bar{\bar{g}}(z)-1)) \tag{53}$$

Probability distributions with a generating function of this form are known in the literature under the name of compound Poisson distributions[7].

For very large r, eqs (23),(29) and (53) give [8]

$$p(r) \propto \exp(-r\bar{i}_1(1-\bar{\bar{p}}_o)) \tag{54}$$

where $\bar{\bar{p}}_o$ is the probability that a cluster decays into secondaries which are not recorded by our experimental set-up.

Comparing (45) and (54) we arrive at the following guess: the cluster model which mimics best a more complicated SRO dynamics is the one which satisfies the condition

$$\bar{i}_1(1-\bar{\bar{p}}_o) = -a(-1) \tag{55a}$$

In applications $\bar{\bar{p}}_o$ is usually fairly small and

$$\bar{i}_1 \approx -a(-1) \tag{55b}$$

2. CLUSTERING IN MULTIPARTICLE FINAL STATES

2.1 Introduction

The results presented in the preceding chapter are very general and apply to any one-dimensional random point process. Moreover, the generalization to spaces with more than one dimension is straightforward. We have only apparently limited our discussion to rapidity plots and we have done this for purely pedagogical reasons. Now, however, we switch to problems which are truly specific for high energy physics.

The plan of this chapter is the following: we first explain the dynamical meaning and the limitations, in the context of high energy physics, of the short-range order picture. We further introduce the related concept of local compensation of quantum numbers and briefly mention the consequences of this local compensation for hadronic diffraction. We then turn to cluster models and determine from general considerations the basic features of the cluster model, which is expected to offer the best representation of the "true" SRO dynamics at asymptotic energies. Next are discussed some recent comprehensive fits to data, where "realistic" cluster models have been used. We emphasize the close similarity between the asymptotic expectations and the features of those "realistic" cluster models, which are the most successful. This similarity is non-trivial because the "realistic" models are built so as to satisfy kinematic constraints, which are unimportant asymptotically. We argue that, in fact, Nature "conspires" to reduce the effect of the conservation laws. In particular, we discuss in some details the clustering associated with the leading particle effect.

2.2 Dynamical Meaning of SRO

The data on high energy collisions of hadrons provide a clear evidence for two prominent features of strong interactions: these interactions are soft, in the sense that large invariant momentum transfers are strongly suppressed, and they have an effective short range in rapidity space, provided one neglects diffractive effects. This is obvious from the observed behaviour of exclusive cross-sections for non-diffractive processes. This implies very long time scales for soft hadronic interactions and thereby enables one to understand, among others, some of the salient features of hadron-nucleus collisions. This is also, apparently, the essence of what we have learned about multiparticle production, etc. I insist on the point, that the short range, in rapidity space, of the effective hadronic forces is evidenced by a wide variety of phenomena and is something fundamental. In this manner I try to justify, from the outset, some biases which will become apparent in the further discussion.

The short range, in rapidity space, of the effective hadronic forces is analogous to the behaviour, in the co-ordinate space, of Van de Waals forces and has the same implications for correlations between particle densities. This is the dynamical idea behind the hypothesis of SRO.

It is an experimental fact that secondaries whose rapidities are close to the rapidity of one of the incident particles, and therefore close to the kinematic boundary, keep a memory of this incident particle. Consequently, the formulation of SRO, given in section 1.7, has to be supplemented with the following addendum:

$\langle N(y_1)\ldots N(y_n)\rangle$ is independent of the initial state provided $y_1 \ldots y_n$ are many correlation lengths away from the kinematic boundaries.

It is worthwhile mentioning that all the postulates which together form the SRO hypothesis are automatically satisfied by the Mueller-Regge model[9]. It is not the purpose of these lectures to review the latter model (cf., for example, ref. 10) and we make little use of it. Nevertheless, a reader conversant with the Mueller-Regge phenomenology will certainly follow our discussion with more ease.

We also have no time to review the experimental situation. Such reviews are periodically presented at conferences and the reviewers alternatively emphasize either the approximate consistence of the data with SRO or the importance of SRO breaking. This is a matter of the choice of perspective. The basic ideas of SRO are now quite old and were formulated for the first time and advocated when very different ideas were prevailing[11]. The data obtained during the late 60's and the early 70's were resembling the predictions and the SRO picture became, in turn, a rather commonly accepted theoretical framework. The observed importance of certain SRO breaking effects caused some disaffection, especially because the intrinsic limitations of the picture were not always fully realized, but the central position occupied by SRO and related concepts in the study of multiparticle production remains unchanged. Perhaps because other existing theories seem to have less prospects for a further development.

2.3 The Limitations of the SRO Picture

In spite of its attractiveness, SRO is bound to be only approximately true. First, the fluid analogy [12] has a serious limitation: in a genuine fluid, the coordinates of molecules do not have to fulfil anything like the energy-momentum conservation constraints satisfied by the rapidities of particles produced in a high energy collision. These conservation contraints involve all secondaries and correspond to correlations which are basically of long range in rapidity space. The constraints of energy-momentum conservation weaken progressively with increasing collision energy, especially when one moves away from kinematic boundaries, but cannot be safely neglected at energies accessible with present accelerators.

A second, much more serious limitation of SRO has to do with unitarity. A collision of two energetic hadrons usually results in their break-up: because of the short-range property of forces, the colliding systems cannot adjust themselves to the perturbation caused by the interaction of their wee constituents. Thus, particle production is expected to be the dominant process at high energies, in agreement with observation. However, these production processes correspond to absorption of the incident wave and necessarily generate diffraction, both elastic and inelastic. The observed diffractive cross-sections depend weakly on energy: the "force" corresponding to diffraction and conventionally represented by the exchange of a quasi-particle, the pomeron, has an extreemly long, presumably infinite range in rapidity. The observed inelastic diffraction is far from being negligible:we face a complicated bootstrap problem, since the inelastic diffraction is itself a cause of diffraction and so on. In other words, SRO is broken by unitarity corrections and appears, merely as the 1st order approximation in a perturbative approach to unitary constraints*. Whether this approach will be viable, or stated differently, whether SRO is a good starting point for a more complete theory, will depend crucially on our ability to calculate the unitary corrections to SRO. In this respect, the development of the reggeon field theory justifies some optimism.

The claim that departures from SRO are due to inelastic diffraction can be checked by selecting events, so a to reduce the importance of diffractive effects. In general, it does not make much sense to speak about "diffractive" and "non-diffractive" events. However, it is commonly believed that at present accelerator energies the dominant inelastic diffractive effect is the dissociation of one of the incident hadrons. A rejection of events which present the characteristic features of this class results in a sample where diffractive effects are expected to be reduced. Of course, selection demands considerable information about events and consequently requires time-consuming experiments. But once this information is available, many different tests can be performed. Indeed, SRO predictions have been extensively checked by studying rapidity gap distributions[13)], semi-inclusive data[14)], the pattern of charge fluctuations[15)] etc. We avoid entering into a discussion of these analyses, refering the reader to original publications or to appropriate reviews. Let us mention only, that the SRO structure of such "purified" samples of events is quite remarkable.

* A more precise argument concerning SRO breaking by unitarity rests on Le Bellac's theorem[3)], mentioned in section 1.8: because of kinematics, SRO cannot hold exactly at finite E_{lab}. However, if kinematic constraints were the only SRO breaking effect one would have, for all n, $f_n \sim a_n \log E_{lab}$ in the limit $E_{lab} \to \infty$. Hence, with $E_{lab} \to \infty$, $p_n = \sigma_n / \sigma_{inel}$ would fall faster than any inverse power of $\log E_{lab}$. This prediction is in variance with data.

2.4 Local Compensation of Quantum Numbers

Until now we treated particles as if they were all identical. This simplified considerably the writing and the loss of generality has been, in fact, not very serious. In reality, particles can be distinguished by their charges, transverse momenta etc. The purpose of this section is to discuss the distributions of quantum numbers among secondary particles.

Let Q be a conserved, additive quantum number. Define Z(y) as the amount of Q transferred across the rapidity y:

$$Z(y) = \sum_{j=1}^{n} Q_j \Theta(y-y_j) - Q_{target}\Theta(y-y_{min}) - Q_{beam}\Theta(y-y_{max}) \qquad (56)^*$$

The random function Z(y) vanishes outside the kinematically accessible interval (y_{min}, y_{max}) and oscillates around zero within this interval.

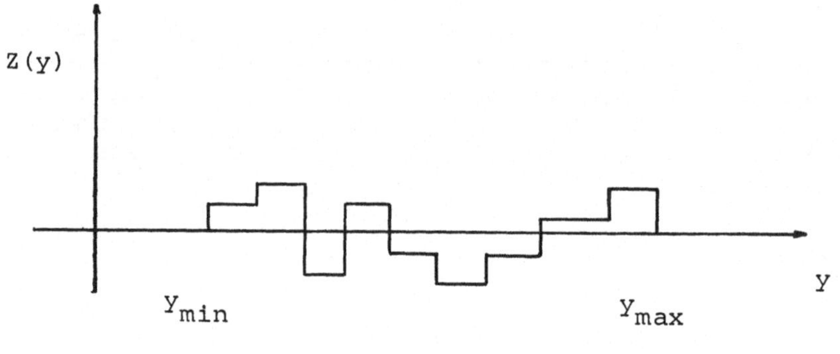

A connected interval where $Z(y) \neq 0$ is called a <u>zone</u> and the plot of Z(y) vs. y is called a <u>zone graph</u> of an event.

The quantum number Q is said to be locally compensated [16] in rapidity space if Z(y) has the following short-range order properties:

* In order to simplify further writing we choose a frame where $y_{min}>0$. When Q=transverse momentum, Z(y) is defined in a frame where $Q_{beam}=Q_{target}=0$.

(A) cluster decomposition:

$$\langle Z(y_1)\ldots Z(y_n)\rangle = \langle Z(y_1)\ldots Z(y_k)\rangle\langle Z(y_{k+1})\ldots Z(y_n)\rangle \tag{57}$$

provided

$$\min |y_i - y_j| \gg \lambda \quad i=1,\ldots,k \text{ and } j=k+1,\ldots,n, \tag{58}$$

where λ is a correlation length, which can depend on Q.

(B) stationarity:
the moment functions $\langle Z(y_1)\ldots Z(y_n)\rangle$ become independent of $Y = y_{max} - y_{min}$ and of the initial state, when $y_1 \ldots y_n$ are many correlation lengths from kinematic boundaries. Lorentz invariance then implies that $\langle Z(y_1)\ldots Z(y_n)\rangle$ depends, in the central region, only on the rapidity differences $y_i - y_j$, $i,j = 1\ldots n$.

Thus, in the central region, a secondary with $Q=Q_s$ is surrounded by a collection of particles, with total $Q=-Q_s$, contained in a rapidity interval whose average length, the average length of a zone, is of the order $O(\lambda)$*. The Q of an incident particle is carried by one or a few respective fragmentation secondaries.Finally the compensation of Q in a given rapidity interval is independent of what happens far away in rapidity.

This simple structure of multiparticle events enables one to understand some salient features of hadronic diffraction. Without entering into details, which were discussed at length elsewhere[16,17], we only state the principal result:

Let the Q exchanged in a two-body reaction $ab \to cd$ be equal to Δ. Assume, that only multiparticle states X, where Q is locally compensated, contribute to the right hand side of the unitarity equation

$$\text{Im}T(ab\to cd) = (1/2)\sum_X T(ab\to X)T^*(cd\to X) \tag{59}$$

* A random distribution of Q would result in zones with average length of the order $O(Y^{1/2})$.

Then one can prove that for $E_{lab} \to \infty$ and fixed Δ:

$$|ImT(ab \to cd)| \; / \; |ImT(ab \to ab) \quad ImT(cd \to cd)|^{1/2} \leq$$
$$\leq \beta(\Delta) \; E_{lab}^{\;\gamma(\Delta^2)} \tag{60}$$

where $\gamma(\Delta^2)$ is strictly negative and can be written as an explicit function of constant parameters characterizing zone correlations. Thus, diffractive parts of exchange amplitudes fall like some powers of energy compared to the diffractive parts of elastic forward amplitudes.

When Q is a transverse component of momentum, it is meaningful to speak about infinitesimal Δ's. For $\Delta \to 0$, the function $\gamma(\Delta^2)$ takes a particularly simple form

$$\gamma(\Delta^2) = -\Delta^2/(8\int d(y_1-y_2)\langle Z(y_1)Z(y_2)\rangle) + O(\Delta^4) \tag{61}$$

Eqs. (60), (61) correspond to the following lower bound for the slope of the pomeron trajectory:

$$\alpha'_p \geq 1/\;(8\int d(y_1-y_2)\langle Z(y_1)Z(y_2)\rangle) \tag{62}$$

Assuming that $\langle Z(y_1)Z(y_2)\rangle = \text{const.} \times \exp(-|y_1-y_2|/\lambda)$ and taking into account the constraints, analogous to eqs, (A.9) in the Appendix, which must be satisfied by $\langle Z(y_1)Z(y_2)\rangle$, one rewrites (62) in a particularly suggestive form:

$$\alpha'_p \geq 1/(4\lambda^2\langle p_t^2\rangle i_1) \tag{63}$$

The hypothesis of local compensation of quantum numbers has also its roots in the idea that hadronic forces have short range in rapidity. In fact, local compensation can be regarded as an aspect of the inclusive SRO. However, local compensation seems to hold under more general conditions and is therefore less sensitive to diffractive effects. We illustrate these points with a simple example.

We are consistently working in a representation where the number and the rapidities of particles are diagonal: in each event, N(y) has a well defined value

$$N(y) = \sum_{j=1}^{n} \delta(y-y_j), \tag{64a}$$

while n and y_j's vary from one event to another. Thus, N(y) can also be regarded as a random function defined on the ensemble of events. Assume, for definiteness that Q is the electric charge and that secondary particles carry charges 0 and ±1 only. Define, in analogy to (64a) the densities of + and - charges:

$$N_\pm(y) = (1/2) \sum_{j=1}^{n} (1 \pm Q_j)\delta(y-y_j) \tag{64b}$$

Then, for $y_{min}<y_1<y_{max}$

$$Z(y_1) = \int_{y_{min}}^{y_1} dy(N_+(y)-N_-(y))-Q_{target} \tag{65}$$

Assume that charge is locally compensated. For symmetry reasons one has $\langle Z(y_1)\rangle=0$, provided y_1 is far from kinematic boundaries. Using eq. (65) we find

$$Q_{target} = \int_{y_{min}}^{y_1} dy \sum_{j=\pm} j\langle N_j(y)\rangle, \quad y_{min}<<y_1<<y_{max} \tag{66}$$

Hence, the average charge of particles included in a rapidity interval beginning at y_{min} and extending far enough into the central region equals Q_{target}. This is guaranteed when SRO holds.

Using (65) and (66) we further write

$$\langle Z(y_1)Z(y_2)\rangle = \int_{y_{min}}^{y_1} dy \int_{y_{min}}^{y_2} dy' \sum_{j,k=\pm} jk \tag{67}$$

$$(\langle N_j(y)N_k(y')\rangle - \langle N_j(y)\rangle\langle N_k(y')\rangle)$$

Charge conservation implies that

$$\int_{y_{min}}^{y_{max}} dy \sum_{j=\pm} j(\langle N_j(y)N_k(y')\rangle - \langle N_j(y)\rangle\langle N_k(y')\rangle)=0 \tag{68a}$$

In the SRO regime, the integrand is a function of y-y' only and vanishes for $|y-y'|>>\lambda$. Therefore

$$\int_{y'-O(\lambda)}^{y'+O(\lambda)} dy \sum_{j=\pm} j(\langle N_j(y)N_k(y')\rangle - \langle N_j(y)\rangle\langle N_k(y')\rangle) = 0 \tag{68b}$$

This in turn guarantees that $\langle Z(y_1)Z(y_2)\rangle$ depends on y_1-y_2 only and vanishes for $|y_1-y_2|>>\lambda$, as required by local compensation. Again, local compensation constraints follow from SRO. Notice, however, that $\langle Z(y_1)Z(y_2)\rangle$ is entirely determined by a very particular combination of density moments: $\sum_{kj=\pm} jk\langle N_j(y)N_k(y')\rangle$. This combination is insensitive to any term in $\langle N_j(y)N_k(y')\rangle$ which does not depend on j and k. Such a term just cancels out. In particular, it appears that local compensation of charge is compatible with the presence of long range inclusive correlations, provided the latter are charge independent, in the sense obvious from the above discussion.

The experimental status of local compensation of charge and transverse momentum can be found in recent reviews[18]. One comment: the following sum rules are derived in the Appendix:

$$\langle Z(0)^2\rangle = \int (2c_2^{+-}(\tau)-c_2^{++}(\tau)- c_2^{--}(\tau))\,|\tau|\,d|\tau| \tag{69a}$$

$$i_1^{\pm} = \int (2c_2^{+-}(\tau)-c_2^{++}(\tau)- c_2^{--}(\tau))\, d|\tau| \tag{69b}$$

The last equation is nothing else but the charge conservation constraint . If the secondaries under consideration are pions, the combination of inclusive correlation functions which appears in the integrands corresponds, in the Mueller-Regge language, to isovector central exchange in the appropriate 4 to 4 forward amplitude. Empirically, the + - correlations are stronger than the ±± ones, the integrands in (69) are positive and therefore

$$\lambda_{eff} = \langle Z(0)^2\rangle / i_1^{\pm} \tag{70}$$

can be regarded as an effective length over which the two-particle correlation, or at least its isovector "part" is important. The observed values $i_1^{\pm} \approx 1$ and $<Z(0)^2> \approx 1$ imply that $\lambda_{eff} \approx 1$, in good agreement with other estimates based on data "purified" from diffractive "contaminations", like the semi-inclusive data, to give an example.

2.5 Clusters at Asymptotic Energies

We shall argue now, that when the dynamical significance of SRO is kept in mind, then there is little freedom left to cluster model builders, at least at asymptotic energies. Indeed, the following considerations are limited to the central region, safely distant from the kinematic boundaries.

The independent emission of clusters is, strictly speaking, incompatible with local compensation of transverse momentum, and a fortiori with SRO, when clusters have non-vanishing transverse momenta. As repeatedly stated in this text, the independent emission of clusters is, in my opinion, merely a recipe to mimic SRO. From this standpoint it is rather natural that we limit our discussion to an idealized world, where the motion of secondaries is purely longitudinal. With this idealization we also avoid entering into a rather technical discussion which, at the present time at least, does not seem to be very relevant for phenomenology (see the footnote after eq. (71)).

The effective forces between hadrons correspond to reggeized exchanges and a large rapidity gap in a multiparticle event is associated with the exchange of a reggeon. If the gap length is r, this exchange implies a propagation factor exp(αr) in the amplitude. Since SRO is expected to be broken by diffractive effects, we exclude pomeron exchanges and set $\alpha=\alpha_R \approx 1/2$. All normalization factors taken into account, one expects the following behavior of the gap distribution[19] at not too small r:

$$p(r) \propto \exp((2\alpha_R-\alpha_p-1)r), \qquad (71)$$

where $\alpha_p \simeq 1$ is the intercept of the pomeron*. Eq. (71) also follows from eqs (40) and (45) if one assumes that, apart from logarithmic factors, the cross-sections have the Regge behaviour:

$$\sigma_n \propto (E_{lab}/E_o)^{2\alpha_R-2} \quad \text{and} \quad \sigma_{inel} \propto (E_{lab}/E_o)^{\alpha_p-1}$$

Thus

$$a(-1) = 2\,\alpha_R - \alpha_p - 1 \qquad (72)$$

and, with the above mentioned values of the Regge intercepts, one finds $a(-1) \simeq -1$. As explained in section 1.10, the best choice of the density, in rapidity space, of independently emitted clusters is then

$$\bar{i}_1 = -a(-1) \approx 1 \qquad (73)$$

NB.: the prediction $a(-1) \simeq -1$ is in remarkably good agreement with the fall of p(r), for $r \gtrsim 1.5$, observed in Fermilab. This could be an accident, however. Some insight into the problem can perhaps be gained using

* A more realistic calculation, taking into account the transverse motion, gives

$$p(r) \propto \exp((2\alpha_R-\alpha_p-1)r)/(\text{const.} + \alpha'_R r),$$

where α'_R is the slope of the Regge trajectory. The r-dependence in the denominator comes from the integration with respect to the transverse momentum carried by the reggeon. This "shrinking" effect does not seem to be important for phenomenology.

One can wonder, why the Regge model form of p(r) is in variance with eq. (45) (pure exponential in r), derived from such general premises. The point is that SRO does not hold exactly: the right-hand-side of (40) should be regarded as the leading term of an asymptotic expansion of g(z). In the neighborhood of z=0 the next to the leading terms of this expansion are very small, for Y large enough. This is no longer true near z = -1. It is instructive to analyze this question in the framework of the multi-Regge exchange model, but I have no time to do this here.

the model independent relation between the gap distribution (45) and the multiplicity distribution (42). If SRO holds and when Y is identified with the total rapidity interval, the observable quantity $L(n/Y) = (1/Y)\ln(n!p_n/Y^n)$, extracted from data at different, sufficiently high energies, is expected to follow a scaling curve, approaching the value -1 at small n/Y. This is indeed the case (for details cf. ref. 5). Instead of claiming further accidents, we are inclined to take $a(-1) = -1$ as a reliable estimate, based not only on a theoretical prejudice but also on data.

Experience with two-body dynamics indicates that α_R is approximately the same for iso-singlet and iso-vector exchanges. Hence, the probability of transferring a charge 0 or ±1 through a large rapidity gap falls at the same rate with increasing length of the gap. The corresponding couplings are also of similar magnitude and therefore the two probabilities are expected to be comparable. This immediately implies that the fractions of charged and neutral clusters are comparable. If all non-leading clusters were neutral, only zero charge could be transferred through rapidity gaps much larger than the extension of a cluster[20).

A corollary: Random distribution of charges among clusters is incompatible with local compensation of charge. Therefore, the independent emission of clusters is a "reasonable" idealization as long as one does not distinguish between clusters of different charge.

The Mueller-Regge arguments lead to the prediction

$$c_2(y_1,y_2) \propto \exp((\alpha_R-\alpha_P)|y_1-y_2|) \tag{74}$$

The above equation holds when the effective mass of the two secondaries, almost always pions, is large enough to justify the use of the Regge representation, $M_{eff} \gtrsim 1$ to 2 GeV say. This corresponds, on the average, to $|y_1-y_2| \gtrsim 2$ to 3. With the conventional values of the intercepts, $\alpha_P \simeq 1$ and $\alpha_R \simeq 1/2$, eq. (74) implies that correlations are appreciable over a distance $\lambda \lesssim 2$ only. This can also be considered as a generous estimate of the rapidity extension of a cluster.

In order to find a better extimate of cluster's extension one has to make some extra assumptions. For example, it seems reasonable to guess, that strong final state interactions within an ensemble of mesons, belonging to a rapidity interval less than two units long, determine, to a large extent, the pattern of correlations between the mesons. But these intrinsically low-energy interactions are known to be dominated by resonance effects and the disintegrations of resonances tend to produce positive correlations between secondary particles. This is an important point, since the observed positivity of correlations - at least the two-particle correlations, those which are the best studied, are definitely positive for $E_{lab} \gtrsim 100$ GeV - has been the basic motivation for constructing cluster models. If correlations were negative, the whole discussion of this section would be without object.

Continuing along this line of reasoning, we expect that the identification of a cluster with a statistical mixture of bosonic resonances corresponds to a very rough but not unreasonable, representation of short-range rapidity correlations[21]. Now, the disintegration of an unpolarized body gives the following inclusive distribution of secondaries[22]:

$$\bar{\bar{i}}_1(y;0) = \int_{\text{fixed } y} f(p/\langle p\rangle)\, d_2p \qquad (75)$$

$$\propto \cosh^{-2} y \int_{((m/\langle p\rangle)\sinh y)^2}^{\infty} f(u)u\, du$$

The function $\cosh^{-2}y$ is well approximated, for not too large y, by a gaussian with dispersion $\delta \simeq 0.9$. The second factor on the right hand side of (75) reduces this dispersion: for pion emission one finds $\delta \simeq 0.7$. Assuming further that

$$\bar{\bar{i}}_2(y_1,y_2;0) \approx \bar{\bar{i}}_1(y_1;0)\, \bar{\bar{i}}_1(y_2;0) \qquad (76)$$

one obtains from eqs. (47)[23]:

$$c_2(y_1,y_2) \propto \exp(-(y_1-y_2)^2/4\delta^2) \qquad (77)$$

This corresponds to a correlation length, defined as the average of $|y_1-y_2|$, $\lambda = 2\ \delta/\sqrt{\pi} \approx 0.8$ for pions.

A rapid survey of the Particle Data Group compilation suffices to convince oneself that bosonic resonances with mass less than about 2GeV or so, decay on the average into approximately 2 charged particles. This should also be the average number of charged particles in a cluster. Together with the expectation that the average density of clusters is 1, cf. eq. (73), this implies that the average density of charged particles should be about 2, in excellent agreement with data at highest accelerator energies.

2.6 Realistic Cluster Models

The kinematic effects which break SRO are also, strictly speaking, incompatible with models where clusters are produced without any mutual correlation. In practice, one faces the following alternative:

--Neglect energy-momentum conservation. This often enables one to write elegant and instructive analytic formulae, but quantitative confrontation of model's predictions with data is a priori hazardous since the formulae in question are asymptotic.

--Forget about elegance and employ the MonteCarlo techniques, generating events following the uncorrelated jet model recipe, applied to cluster production. When one is careful enough, this "realistic" approach represents a simple and unambiguous scheme of SRO breaking and enables one to attack problems where analytic methods are almost powerless.

The second of the above mentionned approaches has been used in a recent, particularly comprehensive analysis by Arneodo and Plaut[24]. These authors have successfully discribed a wide variety of data: inclusive and semi-inclusive longitudinal correlations, data on zones, rapidity gap distributions, charge transfer correlations. We list below their conclusions, referring the reader to the original papers for details:

--The average rapidity density of clusters is $\bar{i}_1 \approx 1$.

--About 50 to 60% of clusters are charged. Arneodo and Plaut assume that clusters have charges 0,±1 and that the charge transferred between clusters is, in absolute value, less or equal to unity. This last constraint[20] ensures local compensation of charge and generates a charge-dependent short-range correlation between clusters (cf. the preceding section).

--With the assumption that the distribution of cluster decay products is gaussian, the best fit is obtained when the dispersion δ=0.6 to 0.7.

It is fair to mention that very similar conclusions have been reached earlier by other people[25)], but on the basis of a less thorough analysis. An example illustrating the usefulness of the model: Arneodo and Plaut have demonstrated that a troublesome energy dependence of the semi-inclusive correlations observed by the Pisa-Stony Brook ISR collaboration[26)] can be interpreted as a purely kinematic effect and is entirely compatible with energy independent cluster characteristics.

It is striking, that when one attempts to fit many different data, one is led to cluster production and decay characteristics which are almost identical to the asymptotic expectations. I do not say this to lessen the merit of people who have carried out systematic analyses of data - a semi-quantitative discussion is not a substitute for a careful analysis. However it can help to put cluster models in a proper perspective. In fact, a disagreement between "realistic" models and asymptotic expectations would be rather disastrous for cluster models: if kinematic effects were strong enough to conceal the underlying SRO dynamics, most of the motivation for constructing cluster models would disappear. In this respect, the analysis of the rapidity gap distributions is instructive. It is true that one can fit the Fermilab data with clusters decaying, on the average, into 4 charged secondaries and therefore with $\bar{i}_1 \approx 0.5$ or so. But such a cluster model is not "reasonable". The increase of $\bar{i}_1$ from 0.5 to the asymptotic value 1 would correspond to a 100% rise of the central plateau for E_{lab}>1000GeV, if cluster decay is indpendent of E_{lab}. On the other hand, a decrease of the cluster decay multiplicity with increasing E_{lab} would mean that there are important dynamical long range correlations. Also, a perusal of curves presented by Arneodo and Plaut enables one to verify that with such clusters the gap distribution is strongly energy dependent for E_{lab} between 400 and 1400 GeV (this energy dependence is practically absent when $\bar{i}_1$=1). Thus, paradoxically, the model does not show much trace of SRO at highest accelerator energies, although the successes of the SRO picture have been the basic reason for constructing cluster models. Somebody will perhaps argue that one should disconnect cluster models from the SRO dynamics. It is difficult to see the advantages of such an approach: the theoretical basis of the model as well as the connection with the rest of the strong interaction physics would be lost. Of course, one cannot exclude the possibility that SRO will turn out to be a poor starting point for a more complete theory and will eventually be abandoned. But in this case the survival of cluster models is unlikely. Cluster models emerge quite naturally in the context of random point processes with ergodic properties but are rather artificial when long range correlations are prevailing.

We have emphasized the close similarity between the asymptotic expectations of the preceding section and the features of the "realistic" cluster models: cluster production and decay is such that SRO is broken almost as little as possible, just enough to explain certain disturbing effects! Why is it so? A discussion of the clustering associated with the leading particle effect provides some insight into this question.

2.7 Clustering and the Leading Particle Effect.

2.7.1. Leading particle effect and cluster models. As is well-known, the kinematic reflections of the leading particle effect are particularly important. Let us forget about charge and transverse momentum distribution of clusters and let us assume that clusters are produced without mutual correlations. Following Stodolsky[27],one can calculate the distribution of the energy "left over" in the process of cluster production:

$$\mathrm{Prob}(x) = \bar{i}_1 (1-x)^{\bar{i}_1 - 1} \qquad (78)$$

where x is the fraction of the available momentum "left over" by the right moving clusters. It is obvious that the kinematic reflections of the leading particle effect are minimized when the shape of Prob(x) is close to that of the leading particle momentum distribution, after x has been identified with the scaled momentum of the leading particle. The observed leading particle distributions are fairly flat, apart from the diffractive peak at x=1. Thus, the cluster model which is the least sensitive to the leading particle effect is the one where $\bar{i}_1=1$. Nature apparently conspires to reduce the effect of the energy-momentum conservation constraints!

Notice, that local compensation of quantum numbers also results in a reduction of the effect of overall conservation laws. The dynamics forces secondary particles to choose configurations which are the least sensitive to the overall conservation constraints. A comment: models where clusters are produced independently over-estimate the breaking of SRO by overall conservation constraints, because in such models quantum numbers do not compensate locally. The remedy is simple in the case of dicrete quantum numbers - we already mentioned that Arneodo and Plaut[24] used the limited charge exchange ansatz of Pirila, Quigg and Thomas[20]. However, it is more difficult to ensure local compensation of transverse momentum.

2.7.2. Leading particle effect and SRO. We have shown in the preceding subsection that independent emission of clusters, with average rapidity density $\bar{i}_1$, is compatible with the existence of the leading particle effect, provided the momentum distribution of the leading particle has the form (78). One might feel that this compatibility between SRO and the leading particle effect is a welcome but accidental feature of an over-simplified model. After all, even when the diffractive peak near $x = 1$ is neglected, the leading particle distribution does not have exactly the form (78). The leading particle takes, on the average, a half of the momentum available for particles emitted in one hemi-sphere. The large momentum of the leading particle must be balanced by the momenta of the other particles. In the absence of any correlation between the left and the right movers, one expects that the leading particle effect is accompanied by a collective recoil of the non-leading secondaries. How is it possible that this collec-tive motion is compatible with SRO?

The key of the answer is the following empirical fact[28]: the multiplicity distribution of secondaries recoiling off a leading particle is very similar to the multiplicity distribution in a final state produced in a collision with c.m. energy equal to the effective mass of the recoiling secondaries.

We illustrate the point with a simple model[29]), which is constructed keeping in mind the above mentionned regularity. Assume, for simplicity, that all secondaries are identical. Assume further that in each final state there exists a particle which, when removed, leaves a configuration very similar to a final state produced in a collision at $E_{lab} = M^2_{eff}/2m$, where M_{eff} is the effective mass of the remaining particles. We call this "privileged" particle the leading one and we assume that its momentum distribution scales, in the Feynman sense. All these assumptions are summarized in the following equation:

$$p_n(E_{lab}) - \int_0^1 dx\; h(x)p_{n-1}(E_{lab}(1-x)) = R_n(E_{lab}) \tag{79}$$

where $R_n(E_{lab})$ falls very rapidly when $E_{lab}\to\infty$, and $h(x)$ is the leading particle distribution written as a function of the scaled momentum (remember that $M^2_{eff}=2mE_{lab}(1-x)$). Also,

$$\int_0^1 dx\; h(x) = 1 \tag{80}$$

Multiplying (79) by $(1+z)^n$ and summing over n we get

$$g(z,E_{lab})-(1+z)\int_0^1 dz\; h(x)g(z,E_{lab}(1-x)) =$$

$$= G(z,E_{lab}) \tag{81}$$

where $G(z,E_{lab})$ is again a rapidly falling function of E_{lab}. Eq. (81) is essentially the multiperipheral equation. It is obvious that

$$g(z,E_{lab}) \propto E_{lab}^{\;a(z)} \tag{82}$$

is the asymptotic solution of (81), provided the eigenvalue equation

$$1-(1+z)\int_0^1 dx\; h(x)(1-x)^{a(z)} = 0 \tag{83}$$

has a unique solution. It can be shown that this is the case provided

$$\left| \int_0^1 dx\, h(x)(\ln(1-x))^k \right| < \infty \quad k=1,2,... \tag{84}$$

The point is that a(z) is a holomorphic function of z in the neighbourhood of z = 0 when (84) holds: the successive dervatives of a(z) at z = 0 are readily obtained upon differentiating eq. (83) with respect to z.

Comparing (82) with (40), where we set $Y=\ln(E_{lab}/E_o)$, we notice that the multiplicity distribution in the model has the form characteristic for SRO. Thus, the correlations induced by the leading particle effect are, in the model, of the SRO type, provided the leading particle distribution is smooth enough at x = 1. When the condition (84) is not fulfilled, for example when apart from logarithmic factors $h(x) \propto (1-x)^{-1}$, SRO is broken by a combined effect of dynamics and kinematics: Events where one secondary, the leading particle, is produced very close to the kinematic boundary appear so frequently that the fluid analogy becomes inadequate.

Appendix

A.1 The operator identities (4) and (34) are easily proved[2)] using the overcompleteness property of the so-called coherent states

$$|z\rangle \underset{def}{=} \exp(za^+)|0\rangle \tag{A.1}$$

where z is a c-number. For more details and references cf.ref. 30.

One multiplies the left and the right hand side of (4) (resp. (34) by an arbitrary coherent state vector $|z\rangle$ and one checks that, indeed, the two multiplications give the same result. The calculations are extremely simplified by the fact that coherent states are eigen-states of the annihilation operator:

$$a^+|z\rangle = z\,|z\rangle \tag{A.2}$$

A.2 Generating functionals and cluster models. The probability $P_{nj}(y_1 \ldots y_n ; \bar{y}_1 \ldots \bar{y}_j)$ that j clusters with rapidities $\bar{y}_1 \ldots \bar{y}_j$ produce exactly n secondary particles with rapidities $y_1 \ldots y_n$ is given by the following convolution:

$$P_{nj}(y_1 \ldots y_n ; y_n \ldots y_j) = \sum_{k_1 \ldots k_j} \delta(n-k_1-\ldots-k_j) \times$$

$$\times \sum_{\text{perm.}} \bar{\bar{e}}_{k_1}(y_1 \ldots y_{k_1} ; \bar{y}_1) \ldots \bar{\bar{e}}_{k_j}(y_{k_1+\ldots+k_{j-1}+1} \ldots$$

$$\ldots y_{k_1+\ldots k_j} ; \bar{y}_j) \tag{A.3}$$

where the second summation is over $(n!/k_1! \ldots k_j!)$ non-equivalent permutations of the rapidities of the secondary particles. Remembering eq. (22), one readily finds from (A.3) that

$$\bar{\bar{g}}(\phi ; \bar{y}_1) \ldots \bar{\bar{g}}(\phi ; \bar{y}_j) = \tag{A.4}$$

$$\sum_n (1/n!) \int P_{nj}(y_1 \ldots y_n ; \bar{y}_1 \ldots \bar{y}_j) \prod_{i=1}^{n} ((1+\phi(y_i))dy_i)$$

Since

$$e_n(y_1 \ldots y_n) = \sum_j (1/j!) \int P_{nj}(y_1 \ldots y_n ; \bar{y}_1 \ldots \bar{y}_j)$$

$$\bar{e}_j(\bar{y}_1 \ldots \bar{y}_j) d\bar{y}_1 \ldots d\bar{y}_j \tag{A.5}$$

eq. (46) is easily obtained from eqs. (22) and (A.4).

A.3 The sum rules (69) are a consequence of the relation (68) between $\langle Z(y_1)Z(y_2)\rangle$ and the inclusive spectra

Far from kinematic boundaries, the zone moment function $\langle Z(y_1)Z(y_2)\rangle$ depends on $\tau = y_1 - y_2$ only. One has identically

$$d^2\langle Z(y_1)Z(y_2)\rangle/d\tau^2 = d^2\langle Z(y_1)Z(y_2)\rangle/d|\tau|^2 + \qquad (A.6)$$
$$2\delta(\tau)d\langle Z(y_1)Z(y_2)\rangle/d|\tau|$$

On the other hand, eq. (67) implies that

$$d^2\langle Z(y_1)Z(y_2)\rangle/d\tau^2 = -\sum_{kj=\pm} jk\langle N_j(y_1)N_k(y_2)\rangle \qquad (A.7)$$

Normal ordering the annihilation and the creation operators in $\langle N_j(y_1)N_k(y_2)\rangle$ one finds

$$\langle N_j(y_1)N_k(y_2)\rangle = i_2^{jk}(y_1,y_2) + i_1^j(y_1)\ \delta_{jk}\ \delta(y_1-y_2) \qquad (A.8)$$

Far from kinematic boundaries i_2^{jk} is a function of τ only and $i_1^+ = i_1^- =$ const. Comparing (A.6) with (A.7) and taking into account (A.8) one obtains

$$d^2\langle Z(y_1)Z(y_2)\rangle\ /d|\tau|^2 = 2c_2^{+-}(\tau)-c_2^{++}(\tau)-c_2^{--}(\tau) \qquad (A.9a)$$

$$d\ \langle Z(y_1)Z(y_2)\rangle\ /d|\tau|\Big|_{\tau=0} = -i_1^{\pm} \qquad (A.9b)$$

These differential equalities are readily converted into the sum rules (69), if one assumes that $\langle Z(y_1)Z(y_2)\rangle$, and its first derivative with respect to $|\tau|$, fall rapidly when $|\tau|\to\infty$.

REFERENCES

1. See e.g. R.L. Stratonovich, Topics in the Theory of Random Noise, Vol. 1, Gordon and Breach, New York 1973; also O. Macchi, Processus Ponctuels et Coincidences, Thesis, Université de Paris-Sud, 1972.
2. See Z. Koba, Proceedings CERN-JINR School of Physics Ebeltohoft 1973 and references therein; following Koba we employ the operator techniques first used in this context by K.J. Biebl and J. Wolf, Nucl. Phys. B 44, 301 (1972)

3. M. Le Bellac, Phys. Lett. B37, 413 (1971)
4. R.C. Arnold and G.H. Thomas, Phys. Lett. B47, 371 (1973); G.H. Thomas, Phys. Rev. D8, 3042 (1973).
5. A. Krzywicki, C. Quigg and G.H. Thomas, Phys. Lett. B57, 369 (1975).
6. An exhausive bibliography on cluster models would take much more space than we have at our disposal. The list of early papers, where the model has been applied to the phenomenology of multiparticle production, includes:
P. Pirila and S.Pokorski, Phys. Lett. B43, 502 (1973); C. Quigg and G.H. Thomas, Phys. Rev. D7, 2752 (1973); A. Bialas, K. Fialkowski and K. Zalewski, Phys. Lett. B45, 337 (1973); J. Ranft and G. Ranft, Phys. Lett. B45, 43 (1973); E.L. Berger and G.C. Fox, Phys. Lett. B471 162 (1973) S. Pokorski and L. Van Hove, Acta Phys. Polon. B5, 229 (1974); F. Hayot and A. Morel, Nucl. Phys. B68, 323 (1974); E.L. Berger, Nucl. Phys. B85, 61 (1975)...
Further references can be found in the following review papers:
J. Ranft, Proceedings of the 5th International Symposium on Many Particle Hadrodynamics, Leipzig 1974; K. Zalewski, Proceedings of the 17th International Conference on High Energy Physics, London 1974; P. Darriulat, Proceedings of the 6th International Colloquium on Multiparticle Reactions, Oxford 1975; L. Foa, Physics Reports C22, 1(1975); See also refs. 8, 20, 21, 23.
7. W. Feller, An Introduction to Probability Theory and its Applications, vols. I and II, Willey & Sons Inc. New York 1971.
8. C. Quigg , P. Pirila and G.H. Thomas, Phys. Rev. Lett. 34, 290 (1975). It is tacitly assumed in this paper that $\bar{p}_o$=0.
9. A.H. Mueller, Phys. Rev. D2, 2963 (1970)
10. See e.g. the following reviews: W.R. Frazer et al., Rev. Mod. Phys. 44, 284 (1972); Chan Hong-Mo, Proceedings of the CERN School of Physics, Grado 1972; P. Salin, Ecole d'Été de Physique des Particules, Gif 1973 and references therein.
11. The multiperipheral model of Amati, Bertocchi, Fubini, Stanghellini and Tonin is the prototype of SRO ideas: see the review by S. Fubini, Proceedings of the Scottich Universities Summer School 1963 and references therein; K.G. Wilson, Proceedings of the Scottish

Universities Summer School 1973* and references therein; R.P. Feynman, Photon Hadron Interactions, Benjamin Inc., Reading Mass. 1973; the most recent review is, to my knowledge, the one by M. LeBellac, CERN Academic Training Programme, CERN 76-14 (1976).
12. See the paper by K.G. Wilson quoted in our ref. 11.
13. Argonne/Fermilab/ Stony Brook Collaboration and Michigan/Rochester Collaboration as quoted in our ref. 20.
14. See the long list of references given by P. Darriulat in his review quoted in our ref. 6.
15. J. Derré et al., French-Soviet Collaboration, Saclay preprint M-12 (1974); C. Bromberg et al., Phys. Rev. D12, 1224 (1975); D. Fong et al, Phys. Lett. B61, 99 (1976); J. Lamsa et al., Phys. Rev. Let. 37,73 (1976).
16. A. Krzywicki and D. Weingarten, Phys. Lett. B50, 265 (1974);
A. Krzywicki, Nucl. Phys. B86, 296 (1974) and Proceedings of the 10th Rencontre de Moriond (1975)
D. Weingarten, Phys. Rev. D11, 1924 (1975) and Phys. Rev. D13, 1494 (1976).
17. See the review by A. Krzywicki, Proceedings of the 6th International Colloquium on Multiparticle Reactions, Oxford 1975 and references therein.
18. P. Darriulat, already quoted in our ref. 6; T. Ferbel, E. Majorana School of Subnuclear Physics, Erice 1976 (Rochester preprint UR-587, to be published in the Proceedings of the School).
19. See e.g. D.R. Snider, Phys. Rev. D11, 140 (1975).
20. P. Pirila, C. Quigg and G.H. Thomas, Phys. Rev. D12 92 (1975)
21. See e.g. F. Hayot, Lett. al Nuovo Cim., 12, 676 (1975).
22. E.L. Berger, G.C. Fox and A. Krzywicki, Phys. Lett. B43, 132 (1973).
23. A. Bialas, M. Jacob and S. Pokorski, Nucl. Phys. B75 259 (1974).
24. A. Arneodo and G. Plaut, Nucl. Phys. B107, 275 (1976) and Nice preprint NTH-76/3 (1976), to be published.
25. See e.g. the reviews quoted in our ref. 6.
26. S.R. Amendolia et al., Nuovo Cim., 31, 17 (1976)
27. L. Stodolsky, Phys. Rev. Lett. 28, 60 (1972); the importance of this argument for cluster models has been emphasized in the paper by Pokorski and Van Hove, quoted in our ref. 6.

* This very important paper has circulated a long time in the form of a Cornell University preprint CLNS-131 (1970) before being published.

28. See the review by J. Whitmore, Physics Reports, C27, 187 (1976) and references therein.
29. Cf. R. Jengo, A. Krzywicki and B. Petersson, Nucl. Phys. B65 319 (1973) and references therein.
30. J.R. Klauder and E.C.G. Sudarshan, Fundamentals of Quantum Optics, Benjamin Inc., 1968.

QUARK PARTONS

P.V. Landshoff*

CERN, Geneva

ABSTRACT

A description is given of some recent data from e^+e^- annihilation, deep inelastic e and μ scattering, lepton-pair and J/ψ production in hadronic collisions, and large p_T hadronic reactions. These data are compared with expectations derived from the quark parton model.

FIRST LECTURE: REACTIONS INVOLVING LEPTONS

I shall assume that you are generally familiar with the application of the quark-parton model to reactions that involve leptons [1)], and shall discuss how the model fits in with some recent experimental results that I have found interesting.

Before I start, I will just mention a measurement that has just been made [2)] at SLAC of the asymmetry in polarized electroproduction. The results fit nicely with quark-parton expectations [1)], where to a first approximation the spin configuration of the valence quarks is related to that of their parent nucleon in the same way as is given by an SU(6) wave function.

* On leave of absence 1975-1976 from DAMTP, University of Cambridge U.K.

e^+e^- Annihilation: Jets

The magnitude of the e^+e^- annihilation cross-section has been fully discussed by many people, so I will consider only the beautiful analysis of the jet structure of the final state that was recently reported by Gail Hanson [3].

The quark-parton model for e^+e^- annihilation into hadrons couples the virtual photon to a quark and an antiquark [4]. Each of these then fragments into a system of hadrons (Fig.1). There has to be some final-state interaction if no particles with quark-like quantum numbers are to appear in the final state. Little is known about the precise effects of this final-state interaction, though it is usually assumed that it does very little apart from sorting out the quantum numbers [5]. Probably, it influences only the slowest particles in the final state.

In particular, the final-state interaction should not remove the jet-structure that is implied in Fig.1. In the centre-of-mass frame the q and $\bar{q}$ emerge in opposite directions, and so there should be two bunches of hadrons moving in opposite directions. The experimental analysis [3] looks for the axis that minimizes the sum of the squares of the momentum components $p_\perp$ of the hadrons, transverse to that axis. It is found then that the distribution of $p_\perp$ falls off rapidly for large values of $p_\perp$ (Fig.2). The average value of $p_\perp$ turns out to be about 300 MeV/c. Notice that the shape of the distribution hardly varies with energy: even though at the higher values of E_{cm} there is more chance of finding a particle whose total momentum is rather large, still $\langle p_\perp \rangle$ remains at about 300 MeV/c. This feature of jets is also found in hadron-hadron collisions, as I shall describe in my next lecture, and for practical purposes it <u>defines</u> what we mean by a jet: a collection of particles whose components of momentum perpendicular to some axis, called the jet axis, are small.

The experiment detects only charged particles. An accurate determination of the jet axis would require also the detection of the neutrals. To minimize the error in the jet axis, the data of Fig.2, and also of Fig.3, are from events that contain at least one rather fast particle, having more than half the maximum momentum possible. The data of Fig.3 show the distribution, in such events, of momentum components $p_{/\!/}$ parallel to the jet axis. Because in these data one of the jets is required to have a rather fast particle, most of the par-

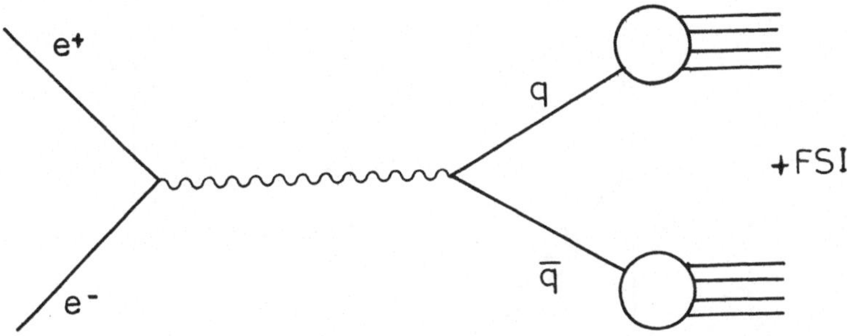

Fig.1 : The quark-parton model for e^+e^- annihilation into hadrons. The final-state interaction avoids appearance of particles having quark-like quantum numbers.

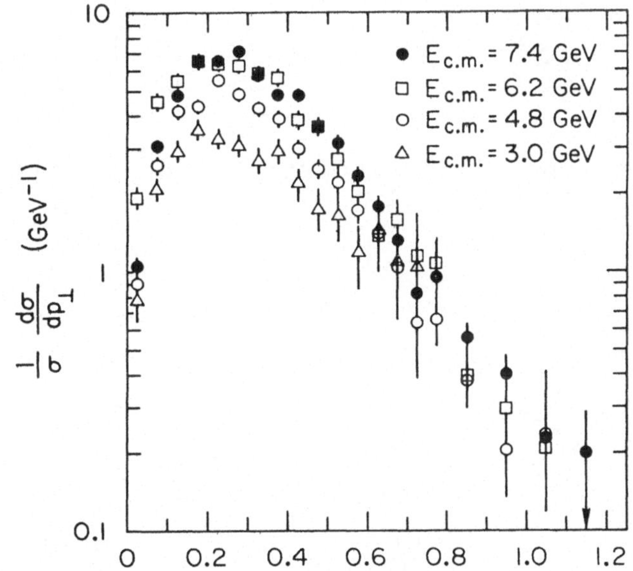

Fig.2 : e^+e^- annihilation: distribution of momentum components perpendicular to jet axis for events having a particle with $x=2p/E_{cm}$ greater than ½. Data from reference 3).

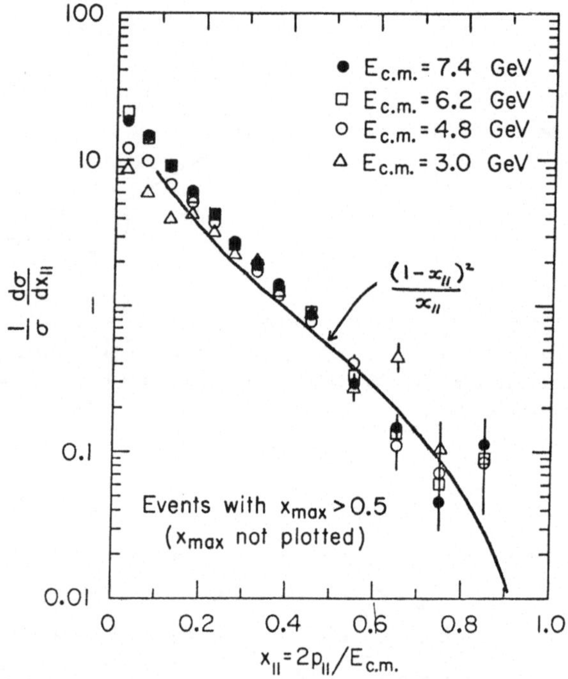

Fig.3 : e^+e^- annihilation: distribution of momentum components parallel to jet axis.

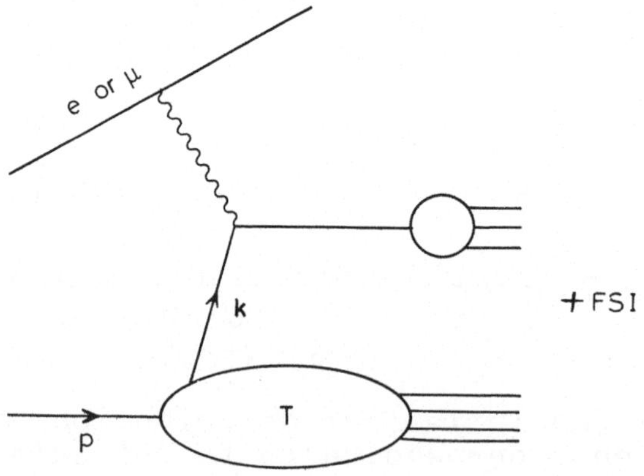

Fig.4 : Quark-parton model for deep inelastic e and μ scattering.

ticles in this plot are in the opposite-side jet, that is this plot approximately describes the fragmentation of just one of the jets. The plot is of $dN/dx_{//}$ against $x_{//}$, where $x_{//}$ is the ratio of $p_{//}$ to its maximum possible value. Notice that the data plotted in this way seem to scale well: there is little or no variation with energy. I have drawn the curve $(1-x_{//})^2/x_{//}$ against the data for use in the next lecture.

In the parton picture of Fig.1, the angular distribution of the jet axis must be of the form $(1+\alpha\cos^2\theta)$, because the virtual photon has spin-1. If the partons have spin-1/2, as is the case for quarks, α should be equal to 1. The measured $\alpha = 0.97 \pm 0.14$.

Deep Inelastic e and μ Scattering: Jets

In deep inelastic scattering on a nucleon, the parton picture corresponds [1] to Fig.4: the exchanged virtual photon knocks a parton out of the nucleon and the parton then fragments. There must again be some final-state interaction if there are to be no final-state particles having quark quantum numbers.

The fragmentation of the quark should be similar to that in e^+e^- annihilation, and a comparison has recently been made in a muon scattering experiment [6] at SLAC. In Fig.5 the e^+e^- data are similar to those of Fig.3, but multiplied by $x_{//}$. The agreement with the corresponding

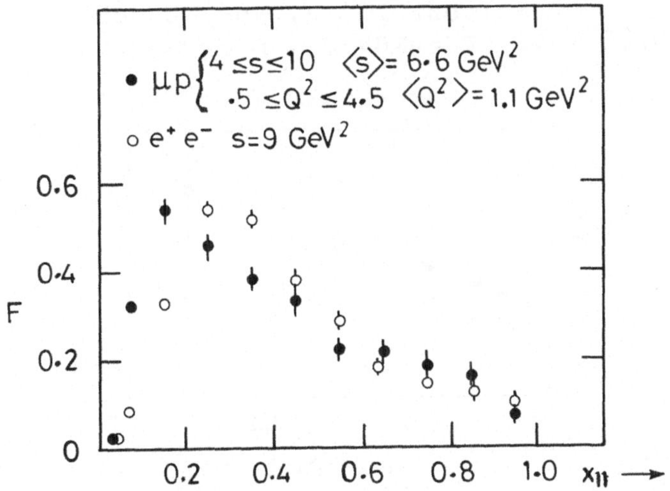

Fig.5 : Parton fragmentation in deep inelastic muon scattering and e^+e^- annihilation. This comparison is from reference 6).

data from deep inelastic muon scattering is very encouraging, though much more work must be done to establish whether one is really seeing the same jets in the two processes.

An interesting prediction of the quark-parton model first made by Pantin [7], concerns the ratio π^+/π^- for the very fastest pions produced in deep inelastic e or μ scattering. On a proton target, the ratio is predicted to be 8, if one makes the simplest assumptions and if $x=Q^2/2\nu$ is large enough for the contribution from scattering on valence quarks to dominate. This result is easy to derive. The very fastest pions take nearly all the momentum of the parent fragmenting quark, and in this situation it is generally believed that a π^+ is almost sure to come from a p quark rather than an n quark, and a π^- is almost sure to come from an n quark rather than a p quark. Since the proton has twice as many p valence quarks as n valence quarks, and the squared charge of a p quark is 4 times that of an n quark, the result follows. The corresponding prediction for a neutron target is 2. For very small x, where valence quarks do not contribute to the scattering, the predicted ratio is 1 for either target.

A recent SLAC electroproduction experiment [8] measures the π^+/π^- ratio on both proton and neutron targets (the latter data are, of course, deduced from scattering on deuterium). Only for the larger values of x shown in Fig.6 is the assumption of the dominance of valence quarks likely to be applicable. The data are not confined to the very fastest pions: they cover the range $0.4 < z < 0.85$ with presumably a preponderance of events

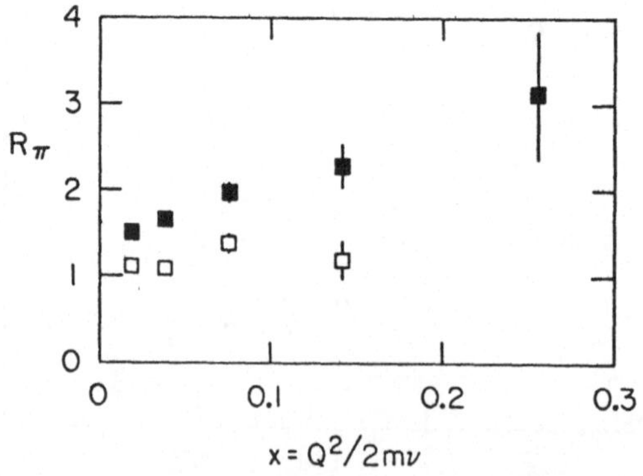

Fig.6 : π^+/π^- ratio for fast pions produced in deep inelastic electron scattering, on proton (■) and neutron (□) targets. Data from reference 8).

at the smaller values of z, where z is the fraction of momentum of the pion relative to the maximum possible. For the smaller values of z one does not expect anything like a 100% correlation between the charge of the pion and the charge of the parent quark, so that the ratio π^+/π^- should be nearer to 1 than the above predictions. Clearly, the most that one can say at present is that the data are sufficiently interesting to make further investigation well worth while.

Deep Inelastic Electron and Muon Scattering: Inclusive Processes

The Bjorken scaling prediction is that, for sufficiently large Q^2 and ν , the deep inelastic structure functions should depend only on $x=Q^2/2\nu$. However, a problem is that when Q^2 and ν are not so large, some other variable, which asymptotically is indistinguishable from x, might give better scaling. This has been investigated in the analysis of an experiment [9] at SLAC.

In Fig.7 the data for a proton target are plotted for fixed x, for fixed $x'=Q^2/(2\nu+1)$, and for fixed $x_s =Q^2/(2\nu+1.5)$. The variable x' has frequently been used to plot data previously, but the choice x_s clearly results in the least variation with Q^2. In terms of this variable, the measured structure function takes a particularly simple form:

$$2\,M\,W_1^{\text{proton}} = 3.2\,(1 - x_s)^4 \qquad x_s \gtrsim \frac{1}{5} \qquad (1)$$

This is demonstrated in Fig.8.

The result (1) is in violation of the Drell-Yan-West threshold relation [1], which predicts that the power should be 3 rather than 4 for values of the variable that are close to 1. However, since the threshold relation is based on a supposed relationship between inclusive and exclusive processes, it is deeply dynamical in character and its derivation involves assumptions additional to those needed elsewhere in the parton model. It seems that no other reasonable choice of variable, instead of x_s , will yield a power 3, and it is intriguing that the power in terms of the variable x_s works out to be so very closely equal to 4.

The same experiment[9] considers also electroproduction on a neutron target. Previous comparisons[10] of the scattering on neutron and proton targets at the same values of ν, Q^2 suggested a value for σ^N/σ^P close to 1/4 at large x. In the quark parton model, this would be

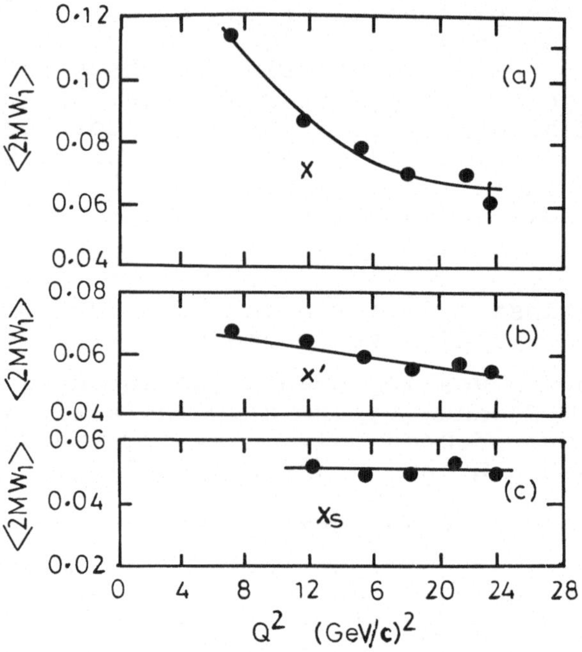

Fig.7 : Test of scaling in electroproduction using the variables $x=Q^2/2\nu$, $x'=Q^2/(2\nu+1)$ and $x_S=Q^2/(2\nu+1.5)$. In each case, the data are for values of the corresponding variables between 0.6 and 0.7 and are from reference 8).

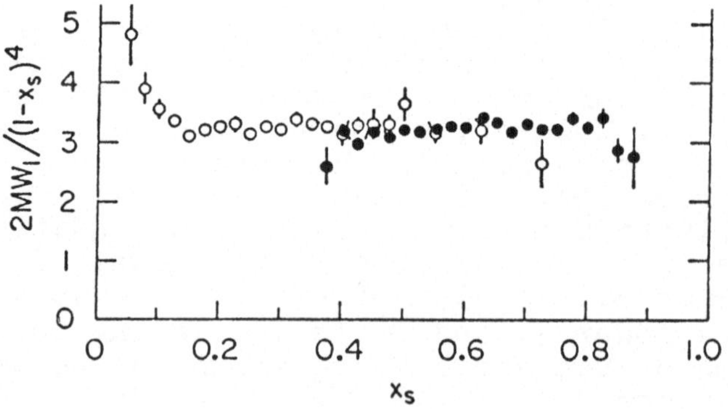

Fig.8 : 50^0 and 60^0 electroproduction data from reference 8), for W>2 GeV and Q^2>1 GeV^2 (hydrogen target).

interpreted as being the result of the momentum distributions of the p and n valence quarks in the proton having different shape, such that as $x \to 1$ the probability of finding a p quark is much larger than that of finding an n quark. However, it now seems that actually the ratio σ^N/σ^P does not scale all that well. In fact, for a neutron target (remember always that the data are actually extracted from a deuterium target, and that the nuclear corrections are large!) the variable that gives the best scaling is not the proton variable $x_s = Q^2/(2\nu + 1.5)$. Rather, it is $x_s' = Q^2/(2\nu + 0.6)$

The safest conclusion to draw from this is that one can say <u>nothing</u> about the asymptotic ratio σ^N/σ^P . Certainly, there is no longer good reason to believe from the data that the value 1/4 is significant. However, the fit to the neutron-target data in terms of x_s' is interesting:

$$2\,M\,W_1^{\text{neutron}} = 2.2\,(1 - x_s')^4 \qquad x_s' \gtrsim \frac{1}{5} \tag{2}$$

If one makes the bold assumption that the fits (1) and (2), which agree well with the present data, can be extrapolated to values of ν and Q^2 large enough for x_s and x_s' both to be indistinguishable from x, then σ^N/σ^P will be roughly constant for $x \gtrsim 1/5$, and close to 2/3. This is just what one gets if one supposes that the contribution from the quark-antiquark sea is small for $x > 1/5$, and takes the momentum distribution of the p and n valence quarks to have the same shape.

Some striking results on how good is Bjorken scaling have come [11] also from the scattering of 100 GeV/c and 150 GeV/c muons on hydrogen, at Fermilab. The proton structure function shows a very marked rise at small x as Q^2 increases (Fig.9a). Where they overlap in Q^2 , the results agree quite well with those from electroproduction at SLAC, and Fig.9b shows fits to the combined data. The plots here are against the variable x, so that the changes with Q^2 for the larger values of x can possibly be interpreted as arising because x_s is the more appropriate variable. However, the changes at small x cannot be explained in this way.

Do We Want Scaling ?

These results obviously raise a number of questions for theorists.

First, is it reasonable to suppose that x_s and x_s' are natural variables to use in analysing the data, and is it reasonable that these variables, for proton and neutron targets, respectively, turn out to be rather

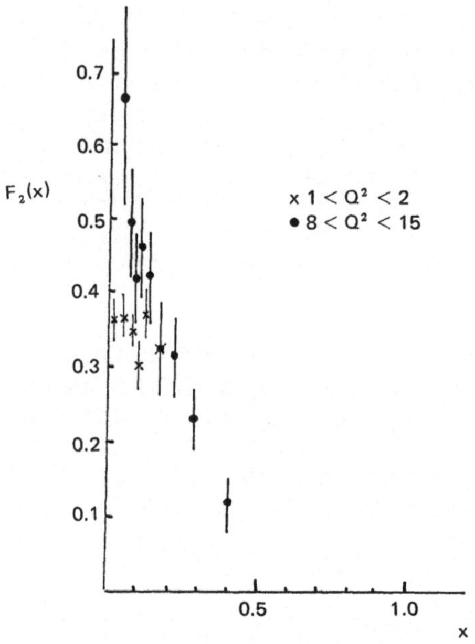

Fig.9 : (a) Data from reference 11 for scattering of 100 GeV/c and 150 GeV/c muons on hydrogen;

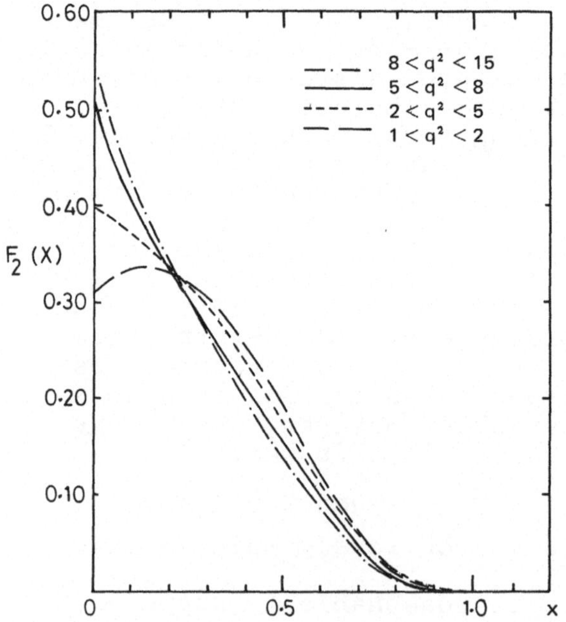

(b) suggested fits to these data and to SLAC electroproduction data.

different? My own feeling, admittedly a posteriori, is that the answer to both these questions is yes. If one supposes that asymptopia is not reached until the kinematical conditions are such that the invariant mass of the lower bunch of particles in Fig.4 is able to reach values of one or two GeV, one can show[12] that something like x_s or x_s' is the neutral variable, for reasonably large values of these variables. Since changing from a proton to a neutron target involves a change in the total quantum numbers of the lower bunch of particles, it is not astonishing that the necessary mass should change, and therefore also the appropriate variable.

This explanation, if it is correct, applies only to the variation with Q^2 at the larger values of x. The more dramatic variation at small x (Fig.9) would have a different origin. Since it is confined to $x \leq 1/5$, it is natural to blame it on the quark-antiquark pairs in the sea, rather than on the proton's valence quarks. The simplest explanation of the rise with Q^2, which at fixed x corresponds to a rise with the energy ν/M carried by the virtual photon, is that the sea contains some relatively massive quarks and antiquarks that need to absorb a rather energetic photon if they are to be fully excited. In terms of Fig.4, this is now a mass effect associated with the upper branch of hadrons, though remember that the final state interactions may prevent anything with the newly excited quantum numbers being present in the final state.

I think that this is the simplest explanation of the data, but there are, of course, other possibilities. One is that the need for quark confinement [5] actually breaks the scaling. However, it is widely believed [1] that this is not so: deep inelastic scattering is a short distance phenomenon and it is only after a comparatively long time, when the absorption of the virtual photon has been completed, that the quark begins to feel the effect of the long-range potential that confines it. But as it is essentially non-relativistic, maybe this argument applies only to the valence quarks and so it breaks down for small values of x.

Again, it may be that the breakdown of scaling that we are seeing is a manifestation of asymptotic freedom, or even of anomalous dimensions. The parton model may be expressed [1] in terms of field theory, in which case one finds that the behaviour of the theory at short distance must be suitably soft if one is to get scaling. Scalar ϕ^3 theory has this property, but no theory that has perturbation theory as its base and which involves fermions and is renormalisable is quite soft enough. The closest that such theories get to scaling is the case of asymptotic freedom[1]: for m = 0,1,2,...

$$\int_0^1 dx\ x^m\ F_2\ (x) \sim\ (\log Q^2)^{-\epsilon_m} \qquad \epsilon_0=0, \quad \epsilon_{m+1} > \epsilon_m \qquad (3)$$

Qualitatively, this behaviour corresponds just to the curves drawn in Fig.9b: the m = 0 moment scales, but the higher moments decrease with increasing Q^2. Actually, the decrease may be rather too rapid for (3) to apply with a reasonable mass scale inside the logarithm. If one replaces the inverse powers of log Q^2 in (3) by inverse powers of Q^2 , one is in the situation of anomalous dimensions, which seems to give a more reasonable fit to the data.

As I have said, for the present I prefer the simpler explanation of the data, that new degrees of freedom in the quark-antiquark sea are being excited.

Lepton-pair and J/ψ Production in Hadronic Collisions

Very shortly after the J/ψ was discovered, various authors [13)] proposed that its production in hadron-hadron collisions might arise through the fusion of a charmed quark c from the sea of one hadron with a $\bar{c}$ from that of the other (Fig.10). There are additional interactions in the initial and final states that can have an important effect on the quantum numbers of the particles found with the J/ψ in the final state, and [14)] which can avoid the need for charmed particles to be produced conjointly with the J/ψ .

The early calculations based on this model agree surprisingly well with the subsequent data, even though the momentum distribution of quarks and antiquarks in the sea was rather poorly known. Encouraged by this success, Donnachie and I were led to assume [14)] that this mechanism is actually the correct one, in which case the

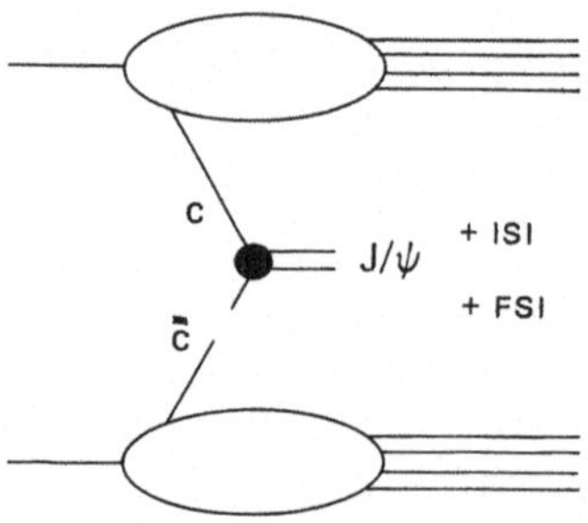

Fig.1o : J/ψ production through c,c̄ fusion. There are additional initial state and final state interactions.

J/ ψ production gives us the best information that we have yet about the sea. Choosing a momentum distribution that gives a good fit to the dramatic rise in the J/ψ production cross-section with increasing energy (Fig.11) gives also good fits to the distributions in Feynman's variable x_F at the different energies (Fig.12).

Notice in Fig.12 that the data for J/ψ from a pion beam have a distribution in x_F that is wider than for a proton beam. If one assumes, as we do, that the production is caused by quarks and antiquarks that belong to the sea of the beam particle, one must conclude that the x distribution of the sea of the pion is wider than that of the nucleon. It is widely believed that this is true for the x distributions of the valence quarks, so a simple way to achieve this is to use an idea of the Rome school [16]. This assumes that the sea of a hadron is

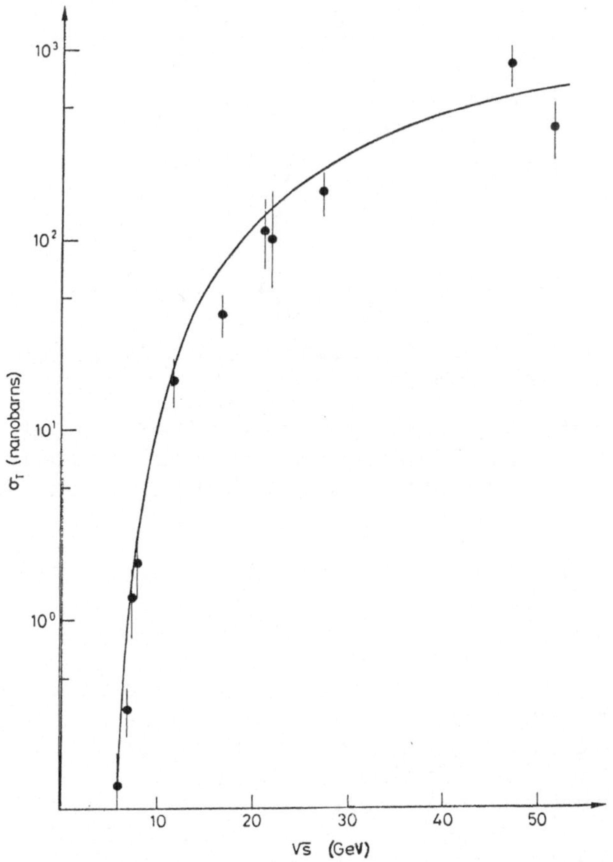

Fig.11 : Total cross-section for J/ψ production in nucleon-nucleon collisions.

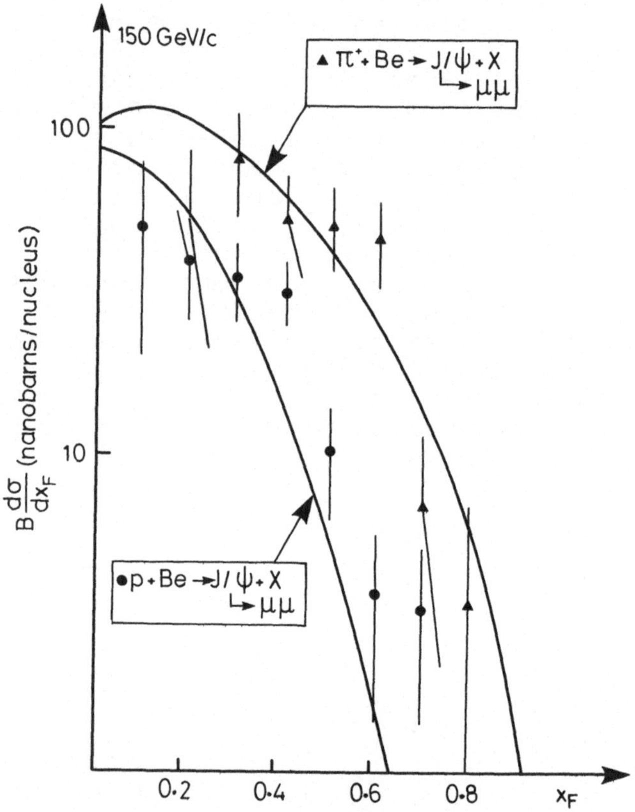

Fig.12 : Feynman x_F distribution for J/ψ production. The data are from reference 15).

made up of parts that are associated with each valence quark. If the valence quark momentum distribution is V(x), the sea quark distribution S(x) for the hadron is then

$$S(x) = \sum_{\substack{\text{valence}\\ \text{quarks}}} \int^1 dx_1\, dx_2\, V(x_1)\, \Sigma(x_2)\, \delta(x_1 x_1 - x) \qquad (4)$$

where Σ is the distribution of the sea quark relative to the valence quarks. The curves in Figs.11 and 12 correspond to taking the valence quark distribution in the proton as (1) (with a suitable matching form for $x < 1/5$), the valence quark distribution of the pion vanishing linearly in $(1-x)$ for $x \to 1$, and Σ constant. This means that the sea distributions for the proton and the pion vanish as $(1-x)^5$ and $(1-x)^2$, respectively, as $x \to 1$. Details are given in Ref.(14).

Many different authors have calculated the continuum lepton-pair production through the Drell-Yan mechanism [1], that is the fusion of a non-charmed quark and a corresponding antiquark to form a massive virtual photon. The results of the calculations differ a little in normalization and shape, according to the quark and antiquark distributions that are assumed, but there is general agreement that the fit with the available data is encouraging. Our own calculation [14], with some of the data, is shown in Fig.13. Notice that a possible theoretical problem arises in the calculation, depending on what is the explanation of the breaking of scaling in deep inelastic scattering. If the explanation is the excitation of new quarks and antiquarks in the sea, there is no difficulty. But if the ordinary quarks give contributions that vary significantly with Q^2, it is not clear what should be used in the Drell-Yan calculation.

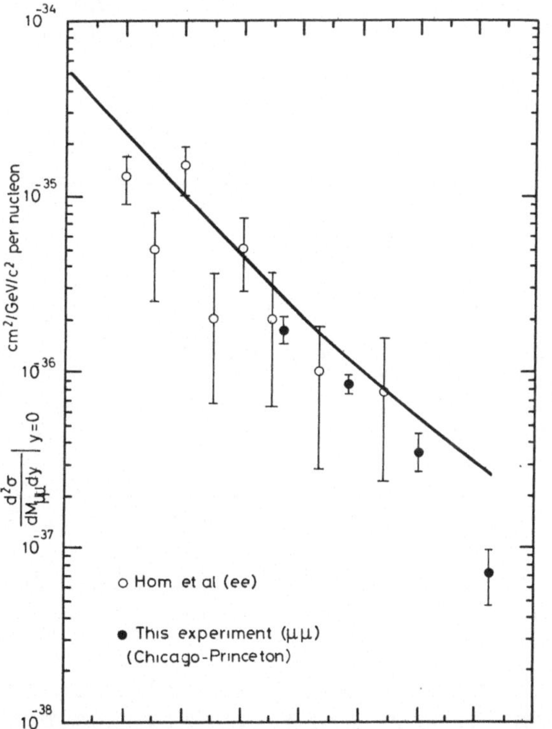

Fig.13 : Data for lepton pair production from reference 17), with calculation according to reference 14).

There is a hint, from deep inelastic muon scattering [18], that the average transverse momentum of partons in the proton increases with their fractional longitudinal momentum x. This idea is strongly reinforced by measurements [15] of the average transverse momentum of muon pairs produced in nucleon-nucleon collisions(Fig.14). In the Drell-Yan picture, this transverse momentum is just the sum of the transverse momenta of the fusing quark and antiquark. The larger the invariant mass $M_{\mu\mu}$ of the muon pair, the larger is the fractional longitudinal momentum of the quark and of the antiquark, so that the data of Fig.14 again suggest that the average transverse momentum of a parton increases with its fractional longitudinal momentum. Notice, however, that the Drell-Yan picture is supposed to apply only at the larger values of $M_{\mu\mu}$, greater than 1 GeV, say. A crucial question is whether, when the energy is changed, the average p_T of the muon pair is found to be fixed at fixed $M_{\mu\mu}/\sqrt{s}$. That is, at a given $M_{\mu\mu}$ the average p_T should decrease with increasing energy. Similarly, if the J/ψ is indeed produced by $c\bar{c}$ fusion, its average p_T should decrease with increasing energy.

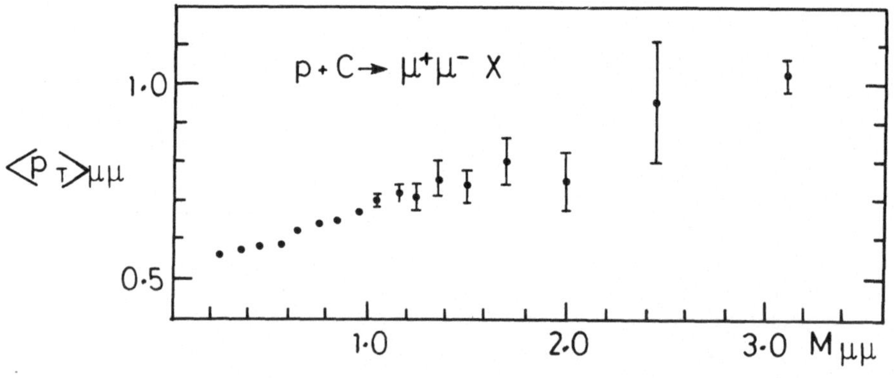

Fig.14 : Data for average transverse momentum of muon pairs, from reference 15).

On the theoretical level, the rise of the average transverse momentum of the parton with increasing x is quite natural [12], provided that one treats the parton model in a proper relativistic way [1]. A rather crude argument is the following. As x increases, the parton k in Fig.4 takes more and more of the longitudinal momentum of the incoming hadron p. However, it cannot take

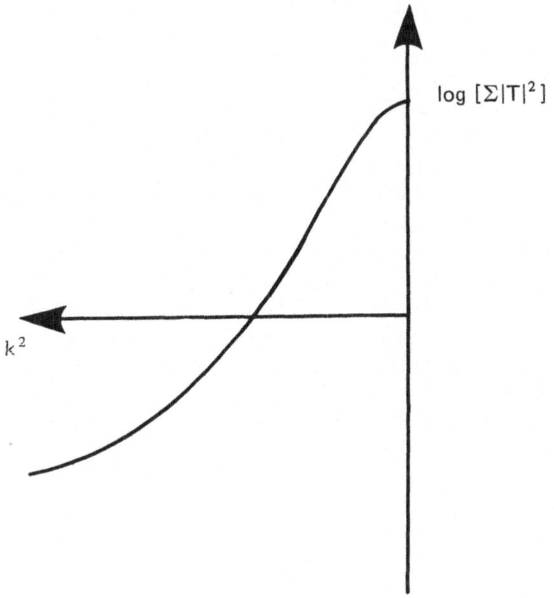

Fig.15 : Plot of $\sum|T^2|$ for figure 4 against k^2.

more and more of its energy, because then there would not be any energy left over to create the lower bunch of hadrons. So to avoid this, it must be [1] that k^2 becomes more and more negative (spacelike) as x increases. To calculate the parton distributions, one needs $|T|^2$(see Fig.4) summed over the possible configurations of hadrons in the lower bunch. To achieve scaling, or at least approximate scaling, it must be that $\sum|T|^2$ decreases sufficiently rapidly for large $(-k^2)$, and from the power behaviour of $F_2(x)$ as $x \to 1$ we know [1] that actually $\sum|T|^2$ falls like an inverse power of $(-k^2)$ at large $(-k^2)$. Thus the plot of log $(\sum|T|^2)$ against k^2 looks something like Fig.15. I have said that the maximum value of k^2 moves to the left as x increases. Evidently $(k^2)_{max}$ is achieved with zero transverse momentum, k_T=0, since including transverse momentum reduces k^2 by an amount k_T^2. So the inclusion of a transverse momentum results in a fall in the value of $\sum|T|^2$, which is why in fact not too much transverse momentum is allowed. However, because the curve in Fig.15 curves upward as one moves to the left, including a given k_T when one starts from a more left-most $(k^2)_{max}$ results in a smaller relative decrease in $\sum|T|^2$, that is the average value of k_T becomes larger.

SECOND LECTURE: LARGE p_T REACTIONS

Single-Particle Distributions

In this lecture I am going to talk about the production of large-transverse-momentum hadrons in hadron-hadron collisions. Most of the time I shall deal specifically with the reaction $pp \to \pi X$, with the pion emerging at 90° in the centre of mass.

Nearly always, the pions that are produced have rather small transverse momentum. Here we shall be concerned with the rare events where there is a large-transverse-momentum pion. This is used to trigger the detection apparatus in the experiments, and so is known as the "trigger particle".

For pions produced with small transverse momentum p_T the inclusive distribution fits well to

$$E\frac{d\sigma}{d^3p} \sim e^{-6p_T} \tag{5}$$

and increases fairly slowly with s. However, for $p_T > 1$ GeV/c the fall-off at given s is rather more gentle than the exponential (5), and the rise with s becomes more marked as p_T increases. This suggests the possibility that new dynamical mechanisms operate at large p_T.

There have been described in the literature a large number of models, differing greatly in character, each of which provides a rather good fit to single-particle distributions. However, most of the models do not naturally include the features that are revealed in large p_T correlation experiments. Many of the models can incorporate these features <u>a posteriori</u>, but there is a particular class of models that actually predicted them. These are the "hard scattering" models, which I shall describe below.

Jet Kinematics

The information that is available so far about correlations in high p_T reactions comes entirely from the CERN ISR, from a number of different experiments [19]. The information is almost entirely kinematic in nature, and the very important question of quantum number correlations is so far largely unexplored. Most of the recent and most important data come from three experiments[20-22] at the Split Field Magnet, and it must be realized that this facility is far from ideal for large p_T experiments. The magnetic field is very weak at 90°, where it is needed most, and it is so complicated in

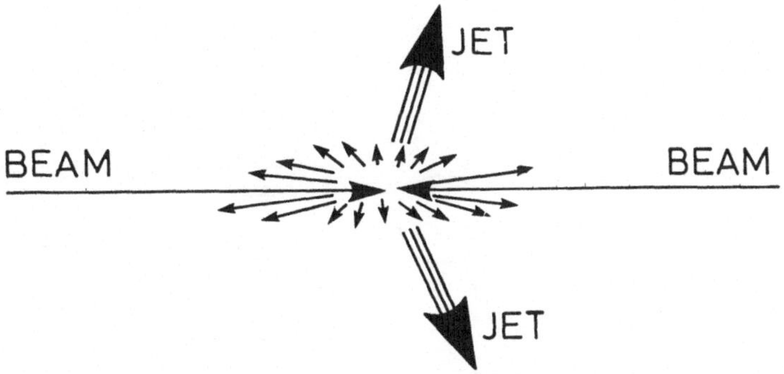

Fig.16 : Structure of a high-p_T event: a pair of jets together with a background of mostly small-p_T particles.

structure that event analysis is very difficult. Nevertheless, it is becoming clear that large p_T events, although superficially they seem to be far from simple, actually have a structure that is basically straightforward.

It seems that in the final state there are two jets of approximately equal and opposite large transverse momenta (Fig.16). One of the jets provides, among its fragments, the high-p_T trigger particle. Generally, the two jets do not have equal longitudinal momentum components, and when the angle θ_1 at which the trigger side jet emerges is fixed, the angle θ_2 at which the opposite-side jet emerges varies from event to event.

In addition to the jets, there is a background of mostly small-p_T particles. Not much is known yet about this background. The best guess is that, on average, it resembles the final state that is produced in ordinary, non-high-p_T events, with the distribution of the transverse momenta of the component particles concentrated at small values. However, there is probably a tail to the distribution, extending out to large values. The background is a nuisance, since both in practice and in principle it is hard to distinguish particles in the jets from particles in the background.

The kinematic configuration of Fig.1 is just what is predicted from hard scattering models [23]. These models have the structure shown in Fig.17. Some part A of one of the initial particles scatters at wide angle on some part B of the other initial particle, resulting in two objects C and D that have large transverse momentum.

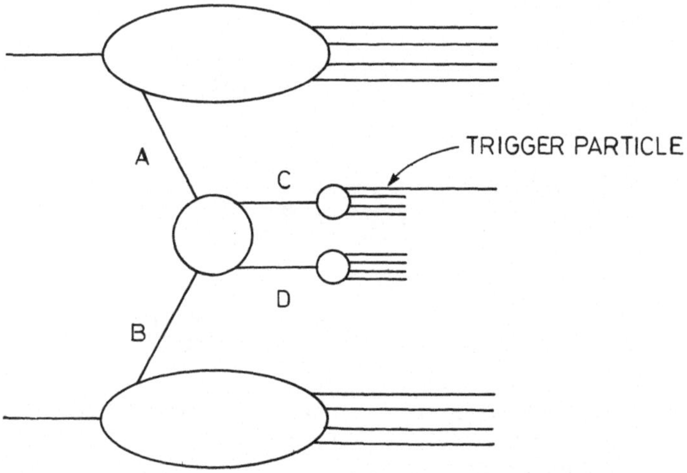

Fig.17 : Hard-scattering models: the central scattering is a wide-angle scattering. There are additional interactions in the initial and final state, not shown.

In addition to what is shown explicitly in the figure, there are initial and final state interactions; these generate the background [24]. There are a number of different models of this structure; they differ in the assumptions they make about the nature of A,B,C and D (I will discuss this briefly later), but they all have the same kinematic content. A and B each take some (variable) fraction of the longitudinal momenta of their parent particles. They can also have some transverse momentum, but this is small, perhaps 300 MeV/c on average. So the transverse momenta of C and D are nearly equal and opposite. After the central wide-angle scattering, C and D each fragment into a system of hadrons; these are the jets in Fig.16. As in the case of the SPEAR jets (see Fig.2), the jet fragments have small momentum components transverse to the jet axis, about 300 MeV/c on average and with a distribution that seems to be largely independent of the longitudinal momentum.

Because of the limited transverse momenta of A,B relative to their parent hadrons, and of the jet fragments relative to C,D, all the particles have moderately small momentum component p_{out} perpendicular to the plane defined by the incoming beams in the centre of mass frame and the trigger particle. Data [20] for the p_{out} distribution are shown in Fig.18. These data refer to particles on the side opposite to the trigger particle, with varying momentum component p_x in the direction in the beam/trigger plane perpendicular to the beams, and varying rapidity Y. The straight lines correspond to $e^{-2|p_{out}|}$

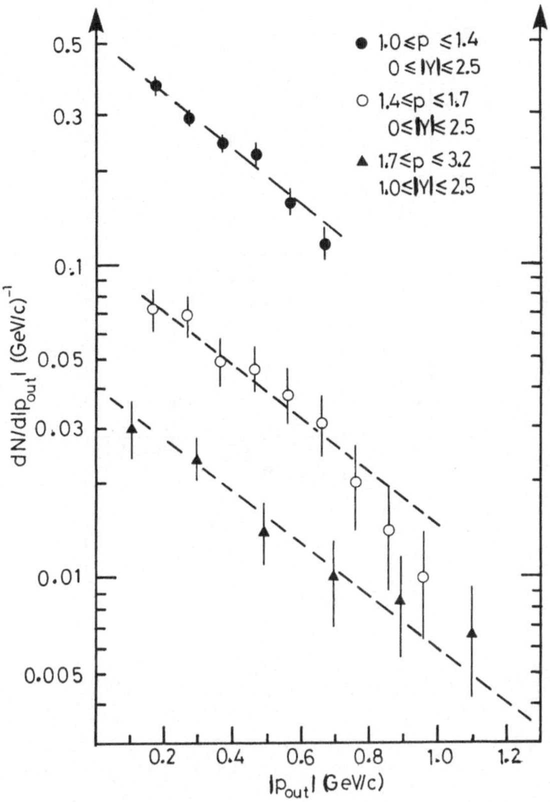

Fig.18 : Data from reference 2o) for distribution of momentum components of away-side particles perpendicular to the beam/trigger plane. The straight lines are $e^{-2|p_{out}|}$.

and seem as a first approximation to fit equally well to the data for the different values of p_X and Y. The average value $\langle p_{out}\rangle$ 500 MeV/c agrees well with the kinematical expectations that I have outlined in the previous paragraph.

Because the angle θ_2 at which the opposite-side jet emerges varies from event to event, the distribution [25] of opposite-side particles, summed over events, is quite wide (Fig.19). To discuss a jet structure in the rapidity distribution, it is necessary to look at each separate event in more detail. To minimize the problem of separating the jet from the background, it is necessary to consider those events that have on the side opposite to the trigger two particles with large p_T, at least 900 MeV/c in the plot of Fig.2o. This plot [21] shows the distribution of the difference in rapidity of the two

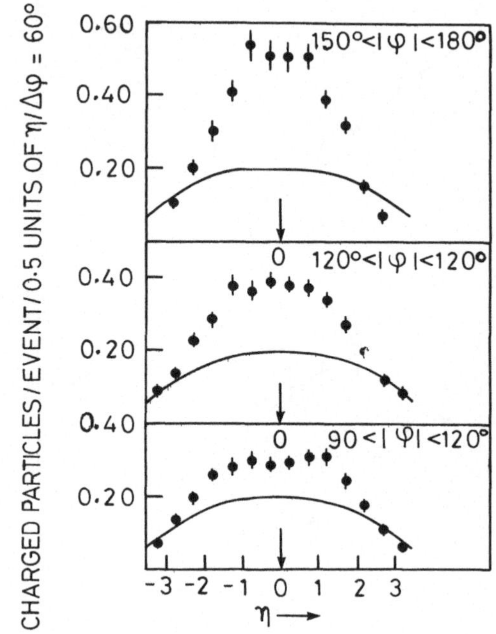

Fig.19 : Data from reference 25) for distribution of charged particles in pseudorapidity η in various intervals of azimuthal angle ϕ opposite to a π^0 trigger at $\eta = 0$. The solid lines correspond to small-p_T events.

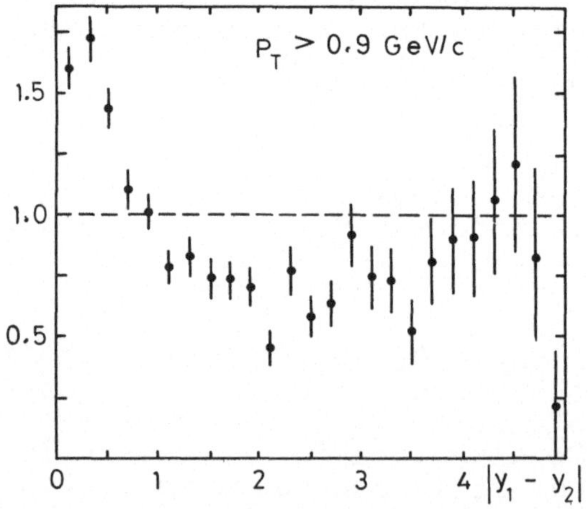

Fig.2o : Data from reference 21) for distribution of rapidity difference of two moderately large p_T particles opposite to the trigger, normalized to the distribution obtained by taking the two particles from different events.

particles, normalized by dividing by what is obtained by instead taking the two particles from different events. This latter distribution already has a peak at small $|y_1-y_2|$, because it seems that the angle θ_2 at which the away-side jet emerges is more or less limited to $45^o<\theta_2<135^o$(compare Fig.18), so that the additional peak in Fig. 19 seems to be clear evidence for jet structure. The same experiment 21)quotes a value of about 300 MeV/c for the average momentum component of the jet particles transverse to the jet axis.This agrees well with what is seen in the SPEAR jets (Fig.2),though the problem of identifying the jet axis is even more severe than at SPEAR.

Various experiments have measured the average multiplicity of charged particles in the jet, that is the average multiplicity seen on the side opposite to the trigger minus the average multiplicity of the background. Methods of estimating the latter vary a little, but the results agree quite well with each other and with multiplicity measurements at SPEAR. An early measurement by the Stony Brook-Pisa experiment is shown in Fig.21. Notice that the average multiplicity at a given p_T of the trigger is essentially independent of the total energy.

Knowing the average multiplicity (assuming that there are roughly half as many neutral as charged particles in

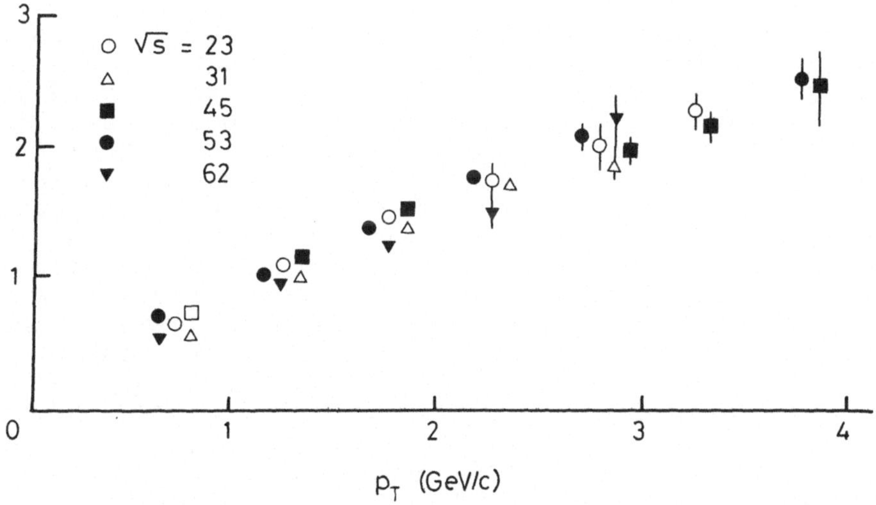

Fig.21 : Data from reference 26) for what is essentially the average multiplicity of charged particles in a jet, plotted against the transverse momentum of the trigger particle on the other side.

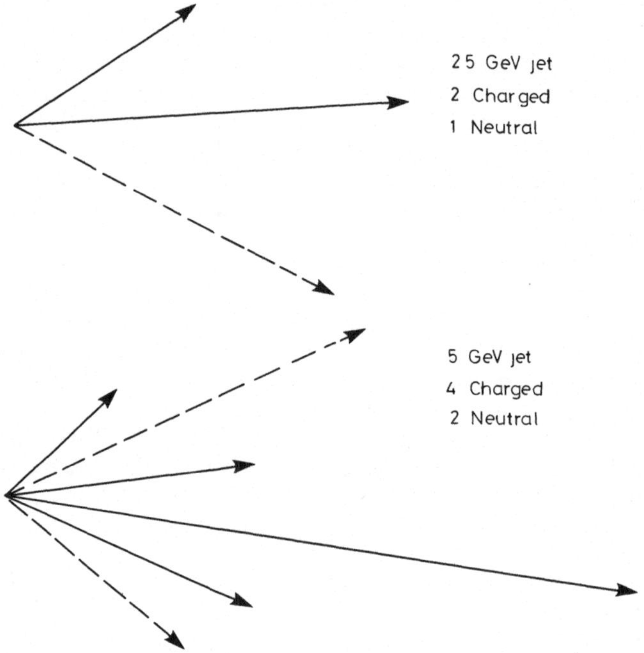

Fig.22 : Scale drawings of typical jets; the broken lines represent neutral particles.

the jet) and knowing that the average momentum transverse to the jet axis is 300 MeV/c, one can construct scale drawings of what a "typical" jet looks like (Fig.22). These drawings apply both to the jets produced at SPEAR, and to those produced at 90° at the ISR. They show that, at the relatively small momenta that have been accessible so far, the jets are not all that well collimated in angle: in this sense, they are not very jet-like. The drawings also demonstrate the problem of the background in the ISR experiments: when its transverse momentum is less than, say, 600 MeV/c, the problem of deciding whether a given particle belongs to a jet rather than to the background is particularly acute, and so the typical jets do not stand out very well from the background.

Correlation Analysis

A number of authors [27)] have analyzed the available large p_T correlation data in a two-jet picture, and it is agreed that, while many questions remain to be answered, there are no serious problems in understanding the data in such a picture.

The analysis is greatly simplified if one makes some or all of the following assumptions:
1. The way that a jet fragments into hadrons is independent of the way in which it is produced. In particular, it does not depend on the total energy of the process nor on the angle at which the jet emerges. This assumption enables one to make calculations in the jet picture without having any understanding of the dynamical mechanism that produces the jets.
2. The jet fragmentation scales. This means that the probability that a jet of total momentum P contains among its fragments a pion whose momentum component along the jet axis is yP is a function $F_1(y)$ of y only. This assumption can only be valid, if it is valid, for values of P that are large compared with the invariant mass of the jet. (This invariant mass does not appear explicitly in most of the calculations, which implicitly integrate over the unknown mass spectrum.)
3. The fragmentation of both jets is described by the same $F_1(y)$. This assumption is not valid in all of the explicit hard-scattering models described in the literature [23].

As I have said, not all of the theoretical papers make all of these assumptions. There is no very clear experimental information on whether or not each of them is valid, but they do simplify calculations considerably. Some doubt has been cast [21] on the experimental validity of the scaling property.

Trigger Bias[28]

Even if the two jets are described by the same fragmentation function $F_1(y)$, so that if one could look at all two-jet events one would usually find that the two jets looked similar, the trigger system will pick out events in which they look rather different. This is particularly true of events picked out using a single high-p_T particle as a trigger. This is easy to show in a simple calculation [28].

First, parametrise the cross section for the production of a pair of jets (of almost equal and opposite transverse momentum P_T) in the form

$$\frac{d\sigma}{dP_T}^{\text{Jets}} = \frac{A}{P_T^{\,n-1}} \qquad (6)$$

This is integrated over the angles θ_1, θ_2 at which the two jets emerge, and over their invariant mass etc. As I shall explain shortly, it seems that at a given energy $\sqrt{s}$ the parameters A and n vary only slowly with

P_T, so that it is a good approximation to treat them as constants at fixed $\sqrt{s}$. Consider now the production of a pion of large transverse momentum p_T. This comes from a jet of transverse momentum P_T, which has to be integrated over. Neglecting the momentum component of the particle transverse to the jet axis, one has $y=(p/P)$ approximately equal to p_T/P_T. So

$$\frac{d\sigma}{dp_T} = 2 \int dP_T \, dy \frac{A}{P_T^{\,n-1}} F_1(y) \, \delta(y-p_T/P_T) \tag{7}$$

where the factor 2 enters because either of the two jets can produce the particle. The P_T integration is trivial:

$$\frac{d\sigma}{dp_T} = \frac{2A}{p_T^{\,n-1}} \int dy \; y^{n-2} \; F_1(y) \tag{8}$$

So the power of p_T in the formula for pion production is the same as the power of P_T in the formula (6) for jet production: this is Bjorken's parent-daughter relation[29]. The data for $d\sigma/dp_T$ confirm that n is almost constant at a given energy. It varies from about 11 to about 9 over the ISR energy range. Notice that the parametrizations (6) and (8) are different from the one that is most directly suggested by hard-scattering models [23], namely

$$\frac{d\sigma}{dp_T} = \frac{1}{p_T^{\,n'-1}} \quad f(x_T), \quad x_T = \frac{2p_T}{\sqrt{s}} \tag{9}$$

We now know[30] that n' is rather constant, at a value of about 8.5, throughout the FNAL/ISR energy range.

To illustrate the effect of trigger bias, I will choose a form of $F_1(y)$ that actually seems [28] to result in quite good agreement with the various correlation data:

$$F_1(y) = K \, \delta(y-1) + L + B \frac{(1-y)^2}{y} \tag{10}$$

The first term corresponds to the possibility that the pion forms the whole of the jet, so that it takes all of its momentum. The second term corresponds to the possibility that the pion is one fragment of the ρ, or some other two-body resonance, which forms the whole of the jet. The last term can be thought of as the contribution from multibody resonances, together perhaps with a continuous component.

Inserting (10) into (8), one obtains

$$\frac{d\sigma}{dp_T} = \frac{2A}{p_T^{\,n-1}} \left[K + \frac{L}{n-1} + \frac{2B}{n(n-1)(n-2)} \right] \qquad (11)$$

The fit to the correlation data works well with $K \approx L$, and K/B a few per cent. So for a jet that is not subjected to trigger bias (which is normally more or less the case for the jet opposite to the trigger) there is a very small chance of finding that the jet consists just of a single π or a single ρ . However, because n is rather large ($n \approx 10$), according to (11) these possibilities are relatively enhanced on the trigger side: the denominator in the last term suppresses the continuum term, and about 75% of the contribution comes from the first term.

Notice also that the ratio of the jet cross-section (6) to the pion cross-section (11) at the same transverse momentum, $P_T = p_T$, is some two orders of magnitude [28,29]. This can be tested by triggering with a calorimeter, which in principle can pick up all the two-jet events without imposing any trigger bias.

What are the Jets?

We cannot yet give any definite answer to this question, but the most likely possibilities are

(a) both jets are the same as at SPEAR, and so basically quark-like (qq);

(b) both jets are a sequence of resonances (MM);

(c) one jet of each of these two kinds (qM).

Each of these possibilities has been proposed in explicit hard-scattering models [23].

While none of these possibilities can be eliminated at present, my strong impression is that the data are rather less favourable to (qq) than to (MM) or (qM), though there is still the attractive possibility that the case (qq) will emerge at rather higher energies and/or higher transverse momenta.

One factor in the argument is the value close to 8 observed for the parameter n^1 defined in (9). The value of n^1 depends on the nature of the central hard-scattering $AB \to CD$ in Fig.17. Hard-scattering models [23] of types $q\bar{q} \to MM$ and $qM \to qM$ predicted the value $n^1 = 8$, while the prediction of type $qq \to qq$ was $n^1 = 4$. However, it may well be that the theoretical basis for these predictions is not correct [31].

Some further evidence in the discussion is centered around the δ-function term that I have put in the expression for $F_1(y)$ in (1o). Such a term corresponds to the jet being a single pion. As I explained in my last lecture, it is generally thought that the SPEAR jets are basically quark-like, so that such a δ-function term would be hard to understand in a SPEAR jet. As we saw in (11), including the δ-function in the trigger-side jet results in the trigger-side jet rather often being just a single pion: other particles are present only in about one event in four. So on average the P_T of the trigger-side jet is only a little greater than the p_T of the trigger pion. On the other hand, without the δ-function term in the trigger-side jet one calculates[28] that $\langle P_T \rangle \approx 3/2\ p_T$.

Having $\langle P_T \rangle$ as large as $3/2 p_T$ obviously would lead to fairly large correlations on the same side as the trigger particle, and we[28] have argued that the data for these correlations rather seem to favour a smaller value of $\langle P_T \rangle$. The value of $\langle P_T \rangle$ also affects the momentum distribution of particles on the opposite side, if one assumes that the two jets have roughly equal P_T. A larger value of P_T evidently results in a greater likelihood of rather fast particles appearing on the opposite side. More specifically, if one calculates the distribution dN/dz, where z is the ratio of the transverse momentum of the opposite-side particles to that of the trigger, one gets (Fig.23) a good fit[28] to the data[2o] if one takes $P_T \approx p_T$ and supposes that the opposite-side jet fragments like at SPEAR. (See Fig.3; this assumption amounts to assuming that $F_1(y)$ for the opposite-side jet is given by just the last term in (1o)). On the other hand[28], if one assumes instead that the trigger-side jet also fragments as at SPEAR, the normalization of the calculated curve increases by about a factor of 2. It is possible that there are ways out of this conclusion[32].

One consequence of the trigger-side jet being meson-like would be that one would expect a relatively high rate of production of direct photons at large p_T. As I explained above, the jet-production cross-section is expected to be some two orders of magnitude greater than the pion-production cross-section. Any dynamical mechanism that produces a large p_T jet that has the quantum numbers of a neutral meson can surely instead produce a photon with that p_T; essentially, one just pays the price of introducing a factor $\alpha = 1/137$. The more specific calculation of Escobar[33] is shown in Fig.24, together with data from reference[2o]. Notice that the experiment is very difficult, because of calibration and background problems.

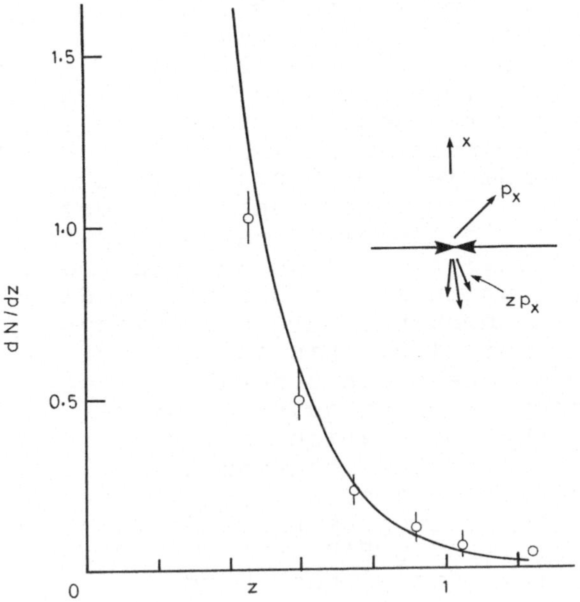

Fig.23 : Data from reference 2o) for momentum distribution of opposite-side particles, with calculation as described in the text.

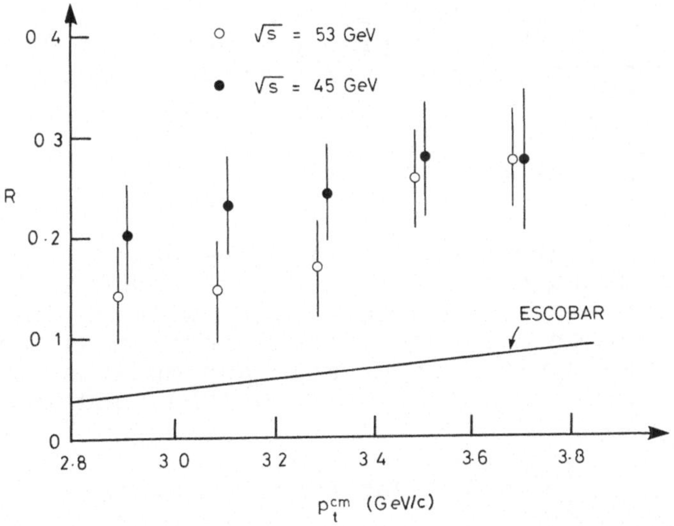

Fig.24 : Ratio of large p_T γ production to π production. Calculation from reference 33), data from reference 2o).

So far, then, my tentative conclusion is that the nature of the jets is more likely to be (MM) or (qM), than (qq). This conclusion is reinforced if one is willing to introduce more specific assumptions about the nature of the central scattering in Fig.17. For case (qq) where the central scattering is qq→qq, the dynamical mechanism responsible for this at wide angle is usually assumed to be the simple exchange of a vector gluon. However, the Bielefeld school[34)] calculates that spin 1 exchange is in disagreement with what is known about the rapidity distribution of the opposite-side jet. If the scattering goes through a pure spin exchange, the data seem to favour spin 1/2. This is the most natural assumption for a model where the central scattering is $q\bar{q}\to MM$, which could occur through quark exchange.

However, the correlation data so far exist only for $p_T \lesssim 3$ GeV/c. Even if the both-meson-jet case (MM) is what is seen at these values of p_T, it may be that at larger p_T there is a transition to the case (qM), where the opposite-side jet is of SPEAR type. The SLAC school has proposed counting laws[31)] that have the consequence that, even though the value of n in (9) is the same in each of these two cases, the behaviour of the corresponding functions f is such that case (qM) will dominate[35)] at sufficiently large x_T.

There is very indirect evidence that this may be happening. With simple assumptions about the central scatterings, and with the p and n valence quark distributions in the proton having the same shape (see my first lecture) we[36)] predicted that for pion production at 90° with x_T large enough for valence quarks to dominate, that is >0.1, the case (MM) gives

$$\frac{pp \to \pi^+X}{pp \to \pi^-X} \quad 1.2 \tag{12}$$

with very weak dependence on x_T. The data[30)] (Fig.25) show signs of a short plateau at about this value, but there is an abrupt rise beyond x_T=0.3. It will be interesting to learn whether a second plateau sets in at a value of π^+/π^- =2. Again with very simple assumptions, one can obtain this value in models of type (qM), and indeed also (qq).

Conclusions About Large P_T

There are no firm conclusions. We do not yet know what is happening, but we have a number of very strong hints that the dynamics are basically simple. It is important to have more information about kinematics, using

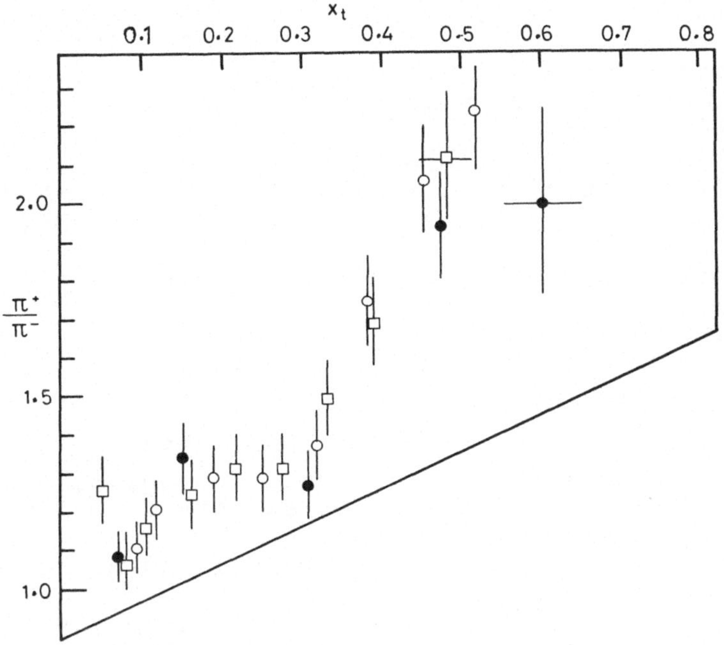

Fig.25 : Data for $(pp \to \pi^+ X)/(pp \to \pi^- X)$ from reference 3o. The production is at approximately 90° in the centre of mass, at various Fermilab energies.

single-particle triggers with transverse momenta greater than the 3 GeV/c so far studied in correlation experiments. Hopefully, calorimeter triggers will allow even larger transverse momenta to be explored. So far, very little is known about quantum number correlations at high p_T, and the study of these should provide valuable new information.

References

1) A recent theoretical review of the parton model is that by P.V. Landshoff and H.Osborn, CERN preprint TH 2157 (1976) (to be published in Electromagnetic Interactions of Hadrons, ed. A. Donnachie and G.Shaw, by Plenum Publishing Co.) This paper contains an extensive list of references.
2) M.J. Alguard et al., Phys.Rev.Lett. 37, 1261 (1976).
3) G. Hanson, Proc.7th International Colloquium on Multiparticle Reactions, Tutzing (June 1976).
4) N. Cabibbo, G. Parisi and M. Testa, Nuov.Cim.Lett.4, 35 (197o).
5) For a discussion of quark confinement, see K.Johnson's lectures at this Institute.

6) K. Bunnell et al., Phys.Rev.Lett. 36, 772 (1976).
7) C.F.A. Pantin, Nuclear Physics B46, 2o5 (1972).
8) J.F. Martin et al., Phys.Lett. 65B, 483 (1976).
9) W.B. Atwood, Report no. SLAC-185 (1975) and seminar at CERN (1976);
W.B. Atwood et al., Phys.Lett. 64B, 479 (1976).
1o) See, for example, R.E. Taylor, Proc.VII International Symposium on Electron and Photon Interactions, Stanford (1975).
11) H.L. Anderson et al., submitted to the 1976 Tbilisi Conference.
12) P.V. Landshoff, Phys.Lett. to appear.
13) See, for example, J. Gunion, Phys.Rev. D12, 1345(1975) M. Green, M. Jacob and P.V. Landshoff, Nuov. Cim. 29A, 123 (1975).
14) A.Donnachie and P.V.Landshoff,Nucl.Phys.B112,233(1976)
15) K.J. Anderson et al., Phys.Rev.Lett. 36, 237 (1976). See also their various papers submitted to the 1976 Tbilisi Conference.
16) G. Altarelli, N. Cabibbo, L. Maiani and R. Petronzio, Nuclear Physics B69, 531 (1974).
17) D.C. Hom et al., Phys.Rev.Lett. 36, 1236 (1976); L.Kluberg et al., submitted to the 1976 Tbilisi Conf.
18) L. Mo, Proc.VII International Symposium on Electron and Photon Interactions, Stanford (1975).
19) See the review by P. Darriulat at the 1976 Tbilisi Conference.
2o) CERN Experiment 412: P. Darriulat et al., Nuclear Physics B1o7, 429 (1976) and B11o, 365 (1976).
21) CERN-Collège de France-Heidelberg-Karlsruhe collaboration: M. Della Negra et al., Nuclear Physics B1o4, 365 (1976) and M. Della Negra, Proc.VII International Colloquium on Multiparticle Reactions, Tutzing (1976).
22) British-Scandinavian-French collaboration: R. Møller, Proc.VII International Colloquium on Multiparticle Reactions, Tutzing (1976).
23) For more extensive reviews of hard-scattering models, see the talks by P.V. Landshoff and S.D. Ellis at the 1974 London Conference, and D. Sivers, S.J. Brodsky and R. Blankenbecler, Physics Reports 23C, 1 (1976).
24) C.E. DeTar, S.D. Ellis and P.V. Landshoff, Nuclear Physics B87, 176 (1975).
25) Aachen-CERN-Heidelberg-Munich collaboration: K. Eggert et al., Nuclear Physics B98, 49 and 73 (1975).
26) Stony Brook-Pisa collaboration: the final results of this experiment are described in their recent preprint (R. Kephart et al.).

27) J.D. Bjorken, lectures at 1975 SLAC Summer Institute; S.D. Ellis, M. Jacob and P.V. Landshoff, Nuclear Physics B1o8, 93 (1976); J. Ranft and G. Ranft, Nuclear Physics B11o, 493 (1976); G. Preparata and G. Rossi, Nucl.Phys.B111,111(1976); N. Craigie, Proc.XI Rencontre de Moriond (1976); W. Furmanski and J. Wosiek, Cracow preprint TPJU-7/76; R. Baier, J. Cleymans, K. Kinoshita and B. Petersson, Bielefeld Preprint.
28) S.D. Ellis et al., reference 27; M.Jacob and P.V. Landshoff, Nucl.Phys. B113,395 (1976).
29) J.D. Bjorken, Phys.Rev. D8, 4o98 (1973).
3o) J. Cronin, Proc. VII International Colloquium on Multiparticle Reactions, Tutzing (1976).
31) For a critical discussion of the theoretical basis of such "counting laws", see P.V. Landshoff, ref.23.
32) See the review by P.V. Landshoff, Proc.VII International Colloquium on Multiparticle Reactions, Tutzing (1976) - CERN TH 2182.
33) C. Escobar, Physical Review, to appear.
34) Baier et al., reference 27.
35) See Sivers et al., reference 23.
36) P.V. Landshoff and J.C. Polkinghorne,Phys. Rev.D8, 4157 (1973). The calculations described in detail in this paper assume the production of a pair of single high p_T particles rather than a pair of jets, but they may readily be adapted and the result (12) will be little changed.

LECTURES ON HADRON-NUCLEUS COLLISIONS AT HIGH ENERGIES

A.H. Mueller*†

Dept. of Physics
Columbia University
New York, N.Y. 10027

ABSTRACT

Hadron-nucleus collisions are discussed. It is assumed that the hadron energy is greater than or equal to 200 GeV and that large nuclei like lead or uranium constitute the nuclear targets. It is argued that a relatively large amount of blackness in hadron-nucleus cross sections puts strong constraints on the inclusive spectrum. Reasons are given why the spectrum of fast particles in a hadron-nucleus collision must be very different than such a spectrum in a hadron-hadron collision. A possible reconciliation of an approximate Glauber expansion with the Regge pole expansion in a soft field theory is suggested.

* Alfred P. Sloan Foundation Fellow.

† Also a visitor at the Institute for Advanced Study.

I. INTRODUCTION

In these lectures I would like to discuss some of the gross features of hadron-nucleus collisions at high energy [1-4]. The topics which fall into this category include total and inelastic cross sections along with the fast part of final state spectrum in an inelastic collision. A particle, j, will be considered fast if $p_j/m^2 >> R$ where R is the radius of the nucleus and p_j is the momentum of the particle in the rest system of the nucleus. R will always be assumed to be large, perhaps 6×10^{-13} cm or more. m is some internal mass. Whether $m \lesssim 1/2$ GeV or whether $m \gtrsim 2$ GeV cannot be determined without a fairly detailed model which takes correlations properly into account. Nowhere in our discussion will the detailed properties of the nucleus be important.

In potential theory differential and total cross sections for the scattering of a fast particle off a loosely bound system are governed by the Glauber[5] multiple scattering series. It is possible to formally generalize this series to an arbitrary field theoretic scattering[6] where, say, the external particle impinging on the nucleus is a bound state. The intermediate scatterings involve amplitudes, in general disconnected amplitudes, for n quanta (partons) to go to m quanta. One's intuition for such a process is not terribly well developed. Thus it is difficult to decide, theoretically, whether proton uranium scattering will have a total cross section of $2\pi R^2$ at high energy. What could happen is the following[7]. The fast incoming particle may be representable in terms of a sum of states which scatter off matter in very different ways. Suppose $|\psi>$ the fast particle state is given by

$$| \psi > = \sum_n \alpha_n |\psi_n>$$

where $|\psi_n>$ are, say, states in the Fock space of a field theory. Suppose $|\psi_o>$ does not interact with nuclear matter. We will call such a component an inert component. If it takes a long time for $|\psi_o>$ to change into other $|\psi_n>$ which do interact with nuclear matter then certainly $\sigma_{tot}<2\pi R^2(1-|\alpha_o|^2)$. An example of this is the scattering of K_L^0 on matter at low energies. The K^0 component of K_L^0 is not easily absorbed in matter while the $\bar{K}^0$ component is absorbed. Since it takes a long time for a K^0 to convert into a $\bar{K}^0$ one does not find total absorption of the K_L^0 beam on matter even though the length of the matter is much longer than what one would naively define as the mean free path of the K_L^0. If a particle has components which have very different mean free paths then the relevant mean free path of a particle off matter will have little to do with the particles elementary scatterings. Experimentally, neutron-nucleus scattering appears rather black for large nuclei at 300 GeV. At present we have no good theoretical understanding of this near blackness, although neither is there yet any real contradiction with, say, a Regge pole model of hadron-hadron scattering. Near blackness tells us that a fast proton has very few inert components.

As we shall see in Section IV a very naive view of the multiperipheral-parton model would lead one to say that the spectrum of fast particles in an inelastic hadron-nucleus collision should be very much like that in an inelastic hadron-hadron collision. This is close to the view of Kancheli(8) and Nicolaev(2). The model of Gottfried(9) is similar in its prediction for fast secondaries except that for Gottfried a secondary is fast only if its rapidity is greater than or equal to 1/3 that of the incoming particle in the rest system of the nucleus. I feel that this cannot turn out to be right. If detailed experiments show such a near equality in $\frac{1}{\sigma_{inel}} \frac{d\sigma_{inel}}{dy}$ for the fast secondaries in hadron-hadron and hadron-nucleus collisions then we shall have to completely revise our present understanding of inelastic events in a hadron-hadron collision. For this reason it is absolutely crucial to have more experiments of the type that Busza(10) and collaborators are doing.

The reason for the above feeling can be be stated in the following way. If the inelastic events in a hadron-nucleus collision are similar to those of a hadron-hadron collision, for the fast part of the spectrum, then these inelastic events should be essentially short range correlated. But if the inelastic events in a hadron nucleus collision are short range correlated then the inelastic cross sections will factorize. Thus one would have equations such as

$$\frac{\sigma^{inel}_{\pi\text{-nucleus}}}{\sigma^{inel}_{\text{proton-nucleus}}} = \frac{\sigma^{inel}_{\pi\text{-proton}}}{\sigma^{inel}_{\text{proton-proton}}}$$

Now there is good evidence that neutron nucleus cross sections are fairly black for large nuclei at 300 GeV [(11),(12)]. But blackness and factorization are incompatible. (We are assuming that π-nucleus reactions become reasonably black for large nuclear radii, although these radii may be somewhat larger than that of uranium). Another way of saying this is that if inelastic events in hadron-hadron and hadron-nucleus collisions are really similar, and if hadron-nucleus collisions become black for large nuclear radii, then the inelastic events in hadron-hadron collisions are very non short range correlated.

In Section VI we shall make these statements more precise. There it is shown that a reasonable blackness of hadron-nucleus collisions require a much higher multiplicity of fast secondaries in hadron-nucleus collision than in hadron-hadron collisions.

The evidence for a reasonable blackness in hadron-nucleus collisions leads one to believe that the wave function of the fact incoming hadron is not a short range correlated object, (see the end of Section IV.) but that hadron-hadron inelastic collisions do not see much of the long range correlations present[(13)]. Since the values of the total cross sections and fast secondaries produced in inelastic collisions are properties of the wave function of the fast incoming hadron, nuclear interactions give much detailed information on the hadronic wave function which cannot be obtained easily in hadron-hadron interactions.

Our view, then, is that hadron-hadron collision and hadron-nucleus collisions are quite different. The particles produced in a hadron-nucleus collision will not be predominantly short range correlated and the heights of the central plateau $\frac{1}{\sigma_{inel}} \frac{d\sigma_{inel}}{dy}$ will be much higher than for hadron-hadron collisions. It is somewhat difficult to say exactly where the central region is. Formally, the central region encompasses those rapidities for which $y \gtrsim \ln Rm^2$ and $Y-y >> |$ where Y is the rapidity of the incoming hadron in the rest system of the nucleus. Between 200 and 500 GeV one should be able to tell the difference between hadron-hadron and hadron-nucleus collisions quite clearly even if the central plateau has not emerged

$$\frac{\left.\frac{1}{\sigma_{inel}} \frac{d\sigma_{inel}}{dy}\right|_{\text{had-nucleus}}}{\left.\frac{1}{\sigma_{inel}} \frac{d\sigma_{inel}}{dy}\right|_{\text{had-had.}}} = R_y \text{ should look something like the following}$$

for a hadron-nucleus collision. At $y \approx Y$ Ry should be on the order of one, although perhaps somewhat below one[(2),(14)]. (The value of Ry as $y \to Y$ is a very model dependent quantity.) As one moves away from $y = Y$, Ry should increase smoothly with $Y - y$ and hit a maximum of, say, >4[(1),(15)]. (See Section VI for the exact relation of Ry in the central region as a funciton of the elastic cross section.) This is where we disagree most strongly with Kancheli and Nicolaev. It is our point of view that one must be quite far from the one Regge pole limit, the zeroth order approximation of our Section IV, while Kancheli and Nicolaev minimize the differences between fast secondaries in hadron-hadron and hadron-nucleus collisions. Their view is the more aesthetic, I think, but I do not see how it can be made compatible with near blackness of hadron-nucleus collisions. Finally, as $y \lesssim \ln Rm^2$ we are in the nuclear fragmentation region and I would rather not venture a guess as to exactly what is going on there[(16)].

As far as what is needed experimentally, I would very much like to see both elastic and total cross sections determined from the same experiment at several Fermilab energies to get a precise measure of just how black heavy nuclei are. Also experiments of the Buza type, but somewhat more detailed as far as the single particle and two-particle spectra are concerned. The two particle spectrum is important to see if the inelastic events are short range correlated, although a precise experiment on $\frac{1}{\sigma_{in}} \frac{d\sigma_{in}}{dy}$ would already be suggestive as to correlations.

In Section VII a possible way out of the difficulty of not naturally having a Glauber series is suggested(13). In this situation a Pomeron-nucleus residue would be proportional to $R^2 n_o$ where R is the radius of the nucleus and n_o is a number characterizing the number of independent chains, combs in the language of Kancheli(8), of partons making up the fast hadron. In a scattering each chain has the potentiality of becoming a Pomeron or it may pass through the scatterer and not interact. n_o would be about equal to the height of the central plateau in a hadron-nucleus collision, for a large nucleus, compared to the height in a hadron-hadron collision. n_o, of course, is a property of the fast hadron and not of the nuclear target. The difficulty with such a possibility though, is that the number of independent chains must be equal to R_c. If nuclei are rather black this number may be rather large, $\gtrsim 4$ say, and it is not at all clear how a wave function of a fast hadron manages to have 4 or more noninteracting chains of partons which presumably overlap considerably in space and time.

II NOTATION AND THE GLAUBER EXPANSION

For a scattering of scalar particles a, b into scalar particles a', b' one usually writes the S matrix elements as

$$(a'b'|S|ab) = \delta^3(\vec{p}_a - \vec{p}_{a'})\delta(\vec{p}_b - \vec{p}_{b'}) \frac{-(2\pi)^4 i\ \delta^4(p_a + p_b - p_{a'} - p_{b'})\ T}{\sqrt{(2\pi)^3 2E_a (2\pi)^3 2E_b (2\pi)^3 2E_{a'} (2\pi)^3 2E_{b'}}}$$

where a and a' are of the same particle type as are b and b'. The process is illustrated in Fig. 1. In the above notation T is a scalar amplitude. For our purpose it will prove more convenient to take

$$(a'b'|S-||\ ab) = \frac{-i}{2\pi}\delta^4(p_a + p_b - p_{a'} - p_{b'})\ A$$

where A is not a scalar amplitude. The reason for this notation is that $|A|^2$ has a more direct interpretation as a transition probability than $|T|^2$. The normalization of A is such that

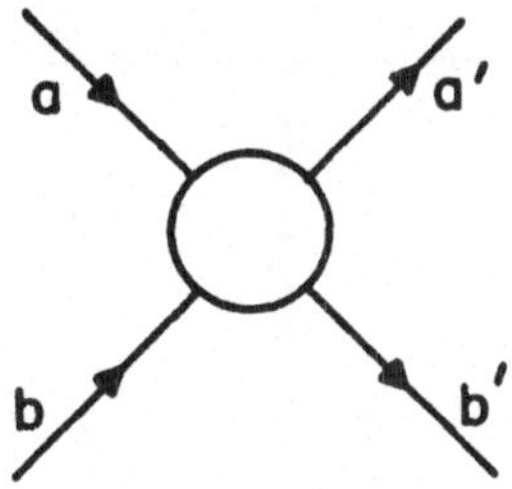

Fig. 1 An elastic scattering.

$$\sigma_{tot} = -2\pi \frac{2E_a 2E_{a'} 2E_b 2E_{b'}}{[s-(m_a+m_b)^2][s-(m_a-m_b)^2]} \text{ im } A(s,o),$$

where $s = (p_a + p_b)^2$ and $t = (p_a - p_{a'})^2$. σ_{tot} is the total cross section for a + b → anything.

We shall mainly be interested in the case where m_b is much larger than all momentum transfers and where $s/_{m_b^2}$ is large. This is the appropriate region for high energy hadron nucleus collisions where b and b' refer to a given nucleus and a and a' refer to a given hadron. b will be taken at rest. Then one can write[5]

$$\sigma_{tot} = 2 \int d^2\underline{x} \text{ Re } \Gamma(x,s)$$

and

$$\sigma_{e\ell} = \int d^2\underline{x} \; |\Gamma(\underline{x},s)|^2$$

where

$$\Gamma(q,s) = i\, A(t,s)$$

is the profile function with $q^2 = -t$ and

$$\Gamma(\underline{x},s) = \frac{1}{2\pi} \int d^2q \; e^{-iq\cdot\underline{x}} \; \Gamma(q,s)$$

In the above q is the momentum transfer variable and $\underline{x}$ is the usual impact parameter variable.

For purposes of illustration it will sometimes be convenient to consider scattering off a static external source at high energies. Such a scattering is shown in Fig. 2. In this case we use the notation

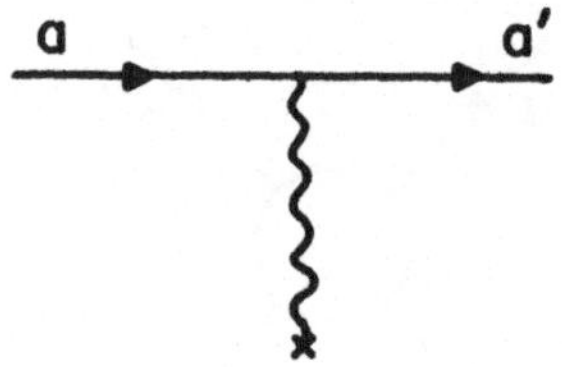

Fig. 2 Scattering of a particle off a potential.

$$(a'|S-\ ||a) = \frac{-i}{2\pi}\,\delta(E_a - E_{a'})\ a$$

$$= \frac{-(2\pi i)\,\delta(E_a - E_{a'})\ t}{\sqrt{(2\pi)^3 2\ E_a (2\pi)^3 2\ E_{a'}}}$$

Then, defining $\gamma = i\ a$,

$$\sigma_{tot} = 2 \int d^2\underline{x}\ \mathrm{Re}\ \gamma(\underline{x},\ p_a)$$

and

$$\sigma_{e\ell} = \int d^2\underline{x}\ |\gamma(\underline{x},p_a)|^2$$

where p_a is the longitudinal component of $\vec{p}_a$.

To see how the Glauber expansion comes about, and for reference later on, consider the scattering of a fast particle off N one dimensional static sources separated one from the next by a distance L. That is, the ℓth source, J_ℓ, is

$$J_\ell(q) = e^{i\ell Lq} J(q)$$

where q is the spatial component of the momentum. e^{iLq} is simply the translation operator, in the momentum representation, for a translation by an amount L. We shall take the potential at high energy to be such that the scattering off a single source is just like an elementary vector meson exchange except for a factor of i. Our scattering model, then, is that of an absorptive potential. We present this model only to illustrate how the Glauber series comes about in potential theory. The model is presumably unrealistic.

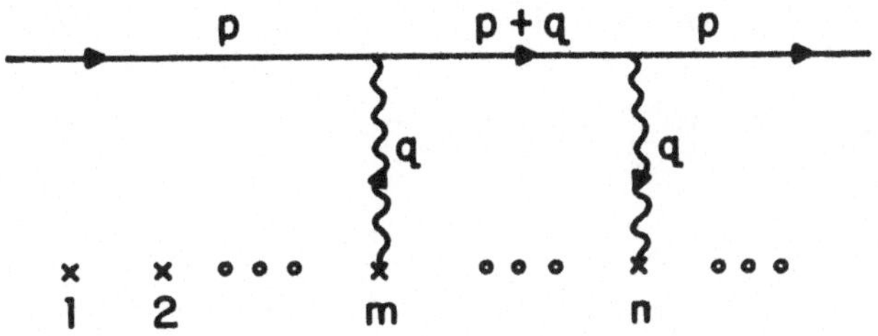

Fig. 3 A double scattering process.

The double scattering term is shown in Fig. 3.

$$T^{(2)}_{nm}(p) = \frac{1}{2\pi}\int dq\ t_n(p,p+q)\ \frac{e^{iqL(n-m)}\ i}{(p+q)^2-m^2+i\varepsilon}\ t_m(p+q,p)$$

where $t_n = t_m = t$ is the scattering matrix off a single source. For on-mass external particles

$$T^{(2)}_{nm}(p) = \frac{1}{2\pi}\int_{-\infty}^{\infty} dq\ t(p,p+q)\ \frac{e^{iqL(n-m)}}{-2pq+i\epsilon}\ t(p+q,p)$$

Suppose, for example, that $J(q) = e^{-q^2R^2}$ with $R << L$.

Then only $q \lesssim \frac{1}{L(n-m)}$ is important in the above integral.

In this region $t(p,p+q) \approx t(p,p) = t(p)$ so

$$T^{(2)}_{nm}(p) \approx t(p)t(p)\ \frac{1}{2\pi}\int_{-\infty}^{\infty} dq\ \frac{e^{iqL(n-m)}\ i}{-2pq+i\epsilon}$$

$$T^{(2)}_{nm}(p) \approx \frac{1}{2p}\ t(p)t(p).$$

In terms of γ

$$\Gamma^{(2)}_{nm}(p) = -\gamma(p)\gamma(p)$$

Defining

$$\Gamma^{(2)}(p) = \sum_{m<n} \Gamma^{(2)}_{nm}$$

one obtains

$$\Gamma^{(2)}(p) = -\ \frac{N(N-1)}{2}\ [\gamma(p)]^2;$$

the procedure for an arbitrary number of scatterings is similar and one obtains

$$\Gamma(p) = N\ \gamma(p)\ \frac{N(N-1)}{2}\ [\gamma(p)]^2\ \ldots$$

or

$$\Gamma(p) = -\prod_{i=1}^{n}(1-\gamma(p)) + 1.$$

We have defined

$$\Gamma(p) = \sum_{n=1}^{\infty} \Gamma^{(n)}(p).$$

The above series for Γ is the Glauber series which can also be written as

$$\mathcal{S}(p) = \prod_{i=1}^{n} s(p)$$

where

$$\mathcal{S}(p) = 1 - \Gamma(p)$$

and

$$s(p) = 1 - \gamma(p).$$

Although we have arrived at the Glauber series in a one-dimensional example, the three dimensional case is not significantly different if one works at a fixed value of the impact parameter.

As is well known different numbers of scatterings are very coherent in the Glauber expansion. Thus, for example,

$$\Gamma(p) = N\ \gamma(p)$$

in the single scattering approximation and this strongly violates unitarity when N is large. The higher numbers of scatterings restore unitarity so that $|\mathcal{S}(p)| < 1$. Not only are unitarity bounds respected in the overall scattering but the Glauber expansion, in potential-theory, preserves unitarity bounds at finite times during the scattering. For example, let the incoming wave be ψe^{ipz-Et}. Then between scatterers m and m + 1 the elastic part of the wave function is $\prod_{1}^{m}(1-\gamma(p))\psi\ e^{ipz-iEt}$

whose norm is less than one. Later, in dealing with Regge pole exchanges, we shall see that any attempt to identify the multiple scattering Glauber series with a series in the number of Regge poles exchanged faces the problem that different numbers of Regge pole exchanges are not necessarily coherent as the hadron crosses an extended set of scatterers. As will be discussed at the end of Section IV it may be possible (only if $\alpha(o) = 1$) for a number of Regge pole exchanges $\leq n_o$ where n_o is independent of R, to be coherent. In such a case one might have an effective Glauber expansion for $\frac{Rm}{n_o}$ not large.

III SPACE-TIME DESCRIPTION OF THE MULTIPERIPHERAL-PARTON MODEL

Before describing hadron-nucleus collisions let us briefly review hadron-hadron scattering in the multiperipheral model(1),(17-19). We shall only be concerned with the gross features of strong interaction amplitudes. Our neglect of spin and the crude treatment of hadronic correlations should thus not be a major defect of this simple ladder graph model of Regge poles. We would like to be able to say that as far as gross features are concerned the multiperipheral model is the most general model of Regge poles. However, this appears not to be the case. As we shall see later the dual model, although identical to the multiperipheral model in its predictions for the gross properties of hadron-hadron collisions, differs significantly in the realm of hadron-nucleus collisions. This serves to emphasize the fact that the properties of very high energy scattering of hadrons off nuclei cannot be completely solved in terms of crude properties of hadron-hadron scattering. On the contrary one gets significant information about hadron dynamics from hadron-nucleus collisions which is difficult to obtain directly from hadron-hadron reactions.

The multiperipheral model amplitude describing the forward elastic scattering of a hadron of momentum p off a hadron of momentum P is illustrated in Fig. 4. We shall choose $P = (m,0,0,0)$ and $p \approx (p + m^2/2p,0,0,p)$ with $p/m >> 1$. Also shown in that figure is an internal line carrying momentum k and vertices a, b and c. a and b are the vertices adjacent to k. An important property of a soft field theory is that longitudinal momenta of adjacent lines in a graph must be of the same order.

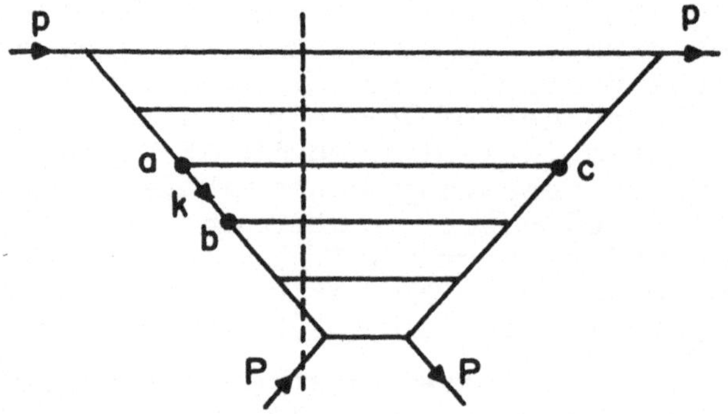

Fig. 4 A multiperipheral graph. The vertical line indicates the partons present just before the interaction with the target.

Thus all the lines adjacent to k have a longitudinal momentum not very different k. Now in its rest system a soft field-theory hadron evolves in time, in the interaction picture, according to a time scale determined by 1/m, where m is a typical internal mass. (This is not the case for a field theory which is not soft in which case arbitrarily rapid interactions, related to the divergences in such a theory, occur.) Now for all vertices close to k the time scales have been dilated and one has $t_c - t_a \approx t_b - t_a \approx k/m^2$. That is, the time scales in a hadron of momentum p vary from p/m^2, for quanta carrying momentum on the order of p, to 1/m for quanta whose momenta are of order m. At each point in a graph there is a local time scale determined by the local longitudinal momentum. The z-positions of the vertices adjacent to k obey $z_c - z_a \approx z_b - z_a \approx k/m^2$ since all quanta have velocities not very different from the velocity of light.

The graph in Fig. 4 is drawn so as to illustrate the time evolution of the scattering. Thus vertices to the left correspond to earlier times than vertices to the right. Only for the bottom few vertices, near the insertion of P, is there not a unique time ordering of vertices corresponding to the Feynman graph. For our purposes such ambiguity in time ordering in these final few vertices can be ignored.

The amplitude, A, of Fig. 4 behaves as

$$A \approx \text{const. } (p.P)^{\alpha-1} = \text{const. } p^{\alpha-1}$$

and this behavior can be interpreted as meaning that

the probability for hadron p to have a soft quantum which can directly interact with P is $p^{\alpha-1}$. In fact the old fashioned perturbation theory interpretation of Fig. 4 gives that part of the wave function of p relevant for the interaction with P. The dashed line in Fig. 4 gives those quanta present at the instant that P interacts with p. These quanta form part of the wave function of p. This can be stated more formally by writing the wave function for p as

$$|\psi_{\underset{\sim}{p}p}> = \sum_{n=1}^{\infty} \frac{1}{\sqrt{n}} \int d^3\vec{p}_1 \ldots d^3\vec{p}_n \delta^3(\vec{p}_1+\vec{p}_2\ldots+\vec{p}_n-\vec{p}) \times$$

$$\times \psi_n(\vec{p}_j\vec{p}_1,\vec{p}_2\ldots\vec{p}_n)|\ \vec{p}_1\ldots\vec{p}_n)$$

where

$$|\psi_{\underset{\sim}{p}p}> = \Omega^{(+)}|\underset{\sim}{p}p>$$

$\Omega^{(+)}$ is the Møller operator and $\vec{p} = (p_x,p_y,p_z) = (\underset{\sim}{p},p_z)$. Our normalization is such that

$$(\psi_{\underset{\sim}{p}'p'}|\psi_{\underset{\sim}{p}p}) = \delta^2(\underset{\sim}{p}'-\underset{\sim}{p})\delta(p'-p).$$

This means

$$1 = \sum_n d^3\vec{p}_1\ldots d^3\vec{p}_n\delta(\vec{p}_1+\vec{p}_2\ldots+\vec{p}_n-\vec{p}) \times$$

$$\times\ |\psi_n(\vec{p}_j\vec{p}_1,\vec{p}_2\ldots\vec{p}_n)|^2$$

The amplitude of Fig. 4 is essentially given by the matrix element

$\mathrm{Im}\, A = \mathrm{const}\ (\psi_{\vec{p}}|\,\mathcal{a}_{\vec{P}} V\delta(E-H_o)V\mathcal{a}_{\vec{P}}\ |\psi_{\vec{p}}$ where V is the ϕ^3 interaction. This means that the two particle scattering state, $|\psi^{(+)}_{\vec{p}\,\vec{P}}>$, is not very different from $\mathcal{a}_{\vec{P}}|\psi_{\vec{p}}>$.

That is, all the structure of the scattering, including the details of the final states produced, are properties of the wave function of a fast isolated hadron and do depend on the properties of the target particle at rest. From this point of view the multipheripheral model is a model for the wave function of a fast hadron, at least for that part of the wave function relevant to a high energy forward scattering process.

When scattering off a large object it is convenient to work in a mixed representation where transverse quantities are expressed in coordinate space and Longitudinal quantities in momentum space. Thus one defines

$$|\psi_{\underline{x}p}> = \frac{1}{(2\pi)} \int d^2\underline{p}\; e^{ip\,\underline{x}} |\psi_{\underline{p}p}>$$

and

$$|\psi_{\underline{x}p}> = \sum_n \frac{1}{\sqrt{n!}} \int dp_1 \ldots dp_n d^2\underline{x}_1 \ldots d^2\underline{x}_n \delta(p_1+p_2 \ldots +p_n-p)$$

$$\psi_n(\underline{x}p; \underline{x}_1, \underline{x}_2 p_2, \ldots \underline{x}_n p_n) |\underline{x}_1 p_1, \underline{x}_2 p_2, \ldots \underline{x}_n p_n>.$$

In a high energy collision $\underline{x}$ is a diagonal coordinate.

IV ZEROTH ORDER PICTURE OF A HADRON-NUCLEUS COLLISION

Consider, again, the hadron-hadron collision shown in Fig. 4. The hadronic matter, at a given time, represented by the quanta shown in Fig. 4 is rather compact in the z direction. (A quantum of longitudinal momentum k can only be $\Delta z \approx \{\max [k,m]\}^{-1}$ from the center of the hadron.) Suppose we decide to replace the hadron P by a nucleus of radius R just before the hadron p reaches the target. Now all those quanta having $k \gtrsim 2Rm^2$ have no time to react as the hadronic matter crosses the nucleus. Thus the spectrum of secondaries of momentum $\gtrsim Rm^2$ should not differ considerably from that of a hadron-hadron collision. Quanta having momenta $\lesssim Rm^2$ may react as the hadron crosses the nucleus so that the spectrum of slow outgoing secondaries is more complicated than in a hadron-hadron collision. We would expect the total inelastic cross section of a fast hadron, p, on a nucleus, R, to be given by three factors: i) $A\pi R^2$ factor reflecting the fact that the reaction takes place at a rather well defined value of the impact parameter. ii) A factor $(p/Rm^2)^{\alpha-1}$ which is the probability of the wave function of p having a quantum of momentum $\lesssim Rm^2$. iii) A factor on the order of 1 which is the probability that a quantum of momentum $\lesssim Rm^2$ interacts with the nucleus. In a moment we shall argue why this last probability is of order 1, but let's accept it for the present. Then, the total hadron nucleus cross section is

$$\sigma = \text{const}\ \pi R^2\ (p/Rm^2)^{\alpha-1}$$

In the case of physical interest $\alpha = 1$ so

$$\sigma = \text{const}\ \pi R^2 \ .$$

Thus the total cross section for a hadron-nucleus collision, in the one Regge pole approximation, is proportional to πR^2, but we have no reason to believe that it will be exactly $2\pi R^2$ for large nuclei. In fact we shall argue in a moment that a one Regge pole exchange is in obvious conflict with blackness of total cross sections. Note that this one Regge pole exchange may reasonably be represented by a set of ladder graphs for all quanta $k \gtrsim Rm^2$ but for $k \lesssim Rm^2$ the interaction is obviously very complicated.

There is another way we can represent a hadron nucleus collision in the one Regge pole approximation. Consider the Feynman graphs shown in Fig. 5, where $k \approx 2Rm^2$. That is, we take all lines whose momentum is $\gtrsim 2Rm^2$ to be part of a single ladder and multiply this amplitude by the scattering of a slightly off shell particle of momentum $k \approx 2mR^2$. (This is a crude decomposition. A more precise decomposition is given in Ref. 1.) Detailed arguments showing that a slightly off shell particle of momentum $k \approx 2Rm^2$ has a cross section proportional to πR^2 are also given in Ref. 1. However, an example may make such detailed, and complicated, arguments unnecessary. Consider the scattering of a ρ meson of, say, 50 GeV/c off uranium. Everyone believes that this cross section will be proportional to the surface area of uranium and not the volume. But the ρ is always off shell by at least 100 MeV. Indeed, the relevant parameter is

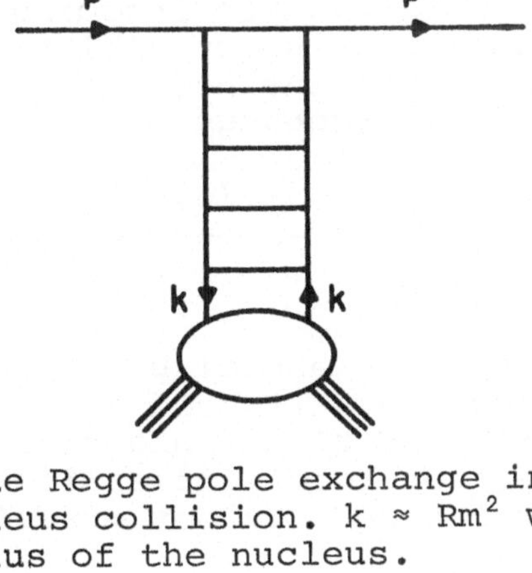

Fig. 5 A one Regge pole exchange in a particle-nucleus collision. $k \approx Rm^2$ where R is the radius of the nucleus.

$(k^2 - m^2)\frac{R}{k}$. So long as the parameter is not large there will be no drastic differences between on and off shell amplitudes. Thus, again, we get a hadron nucleus cross section, in the one Regge pole exchange approximation, proportional to R^2 and not R^3. The detailed calculation which is necessary to show that a R^2 and not R^3 results is very difficult. It would be the same as giving a detailed mechanism showing that an off shell particle has a cross section proportional to R^2 and not R^3 when its longitudinal momentum is $\approx Rm^2$. Such an argument should not depend on a Reggeon calculus or any other ultra high energy model, but rather should use properties of low energy hadronic interactions.

Our understanding of hadron-hadron collisions at high energy in terms of a multiperipheral-parton model has made it seem reasonable that such a model should not be terribly bad for a hadron-nucleus collisions also. We would then say that in a typical inelastic event those secondaries having $p_i \gtrsim Rm^2$ should be distributed as in a hadron-hadron collision while those having $p_i \lesssim Rm^2$ would differ considerably from a hadron-hadron collision. This is the point of view of Kancheli and his followers. Such a picture is certainly not inconsistent with present accelerator data on inelastic collisions. However, the whole theoretical picture is probably not consistent with another piece of experimental information. There are two experiments which measure total neutron-nucleus cross sections at 300 GeV(11),(12). The conclusion is that large nuclei, like lead and uranium, are quite black with respect ot the incoming neutron. This strongly suggests that if we were able to build very large nuclei they would be black as far as all incoming hadrons are concerned. Such a result is incompatible with a one Regge pole dominance of the total inelastic cross section. That is, a one Regge pole exchange model requires

$$\frac{(\sigma_{inel})\ \text{proton-nucleus}}{(\sigma_{inel})\ \text{pion-nucleus}} = \frac{(\sigma_{inel})\ \text{proton-proton}}{(\sigma_{inel})\ \text{pion-proton}}$$

But blackness, relevant for nuclei perhaps somewhat larger than uranium, requires the left hand side of the above equation to be 1 while the right hand side is not 1. Simply said, blackness of total inelastic cross sections and short range correlations are not mutually compatible.

The above difficulty is not unique to a Regge model. Any model which would say that the fast secondaries, over a growing region of rapidity with increasing energy, are identical in hadron-nucleus and hadron-hadron interactions must explain how inelastic events in hadron-hadron collisions appear predominantly short range correlated when such short range correlations are incompatible with blackness of hadron-nucleus cross sections. This difficulty is present, I feel, in both Gottfried's energy flux model and in a planar string model though there the trouble takes the form of equality of all cross sections in hadron-hadron scattering.

Is this difficulty insurmountable or can the Regge model, so succesful in hadron-hadron collisions, be made compatible with large nuclei? We shall argue in detail in Section VI that at present there is no inconsistency, but that the Regge model, including cuts in a soft field theory, puts rather strong constraints on the spectrum of inelastic events. However, I think that our naive parton picture of a hadron-hadron collision cannot be correct if hadron-nucleus cross sections really are black. To illustrate the sort of thing that could be happening consider a high energy deuteron-proton collision in the rest system of the proton. Now if a fast proton has a wave function well represented by the multiperipheral-parton model then the wave function of the deuteron is the product wave function of the neutron and proton wave functions. This is illustrated in Fig. 6. However, the two chains in Fig. 6b are not near each other in impact parameter space and so only one or the other of the chains interacts with the proton at rest. This means that as far as the scattering is concerned we may ignore the other chain completely. The elastic scattering of a deuteron off a proton is well represented by a single Regge pole exchange. (We are neglecting the shadow term in this discussion.) However, if the deuteron collides with a large nucleus both of the chains would interact and, in this simple model, deuteron nucleus scattering would be given by one and two Regge pole exchanges with a well defined relation between the one and two pole terms. Because of the large size of the deuteron we are unable to measure its wave function simply by scattering it off a proton. By scattering the deuteron off a nucleus we do measure the wave function, at least that part of the wave function connected to soft quanta. Attempting to generalize from this example, it may be that the multiperipheral-parton model wave function is a good model for high energy hadron-hadron scattering but that the actual wave function is much more complicated.

a

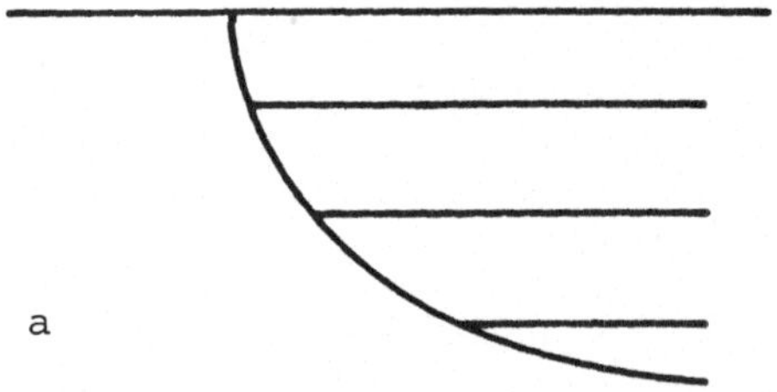

b

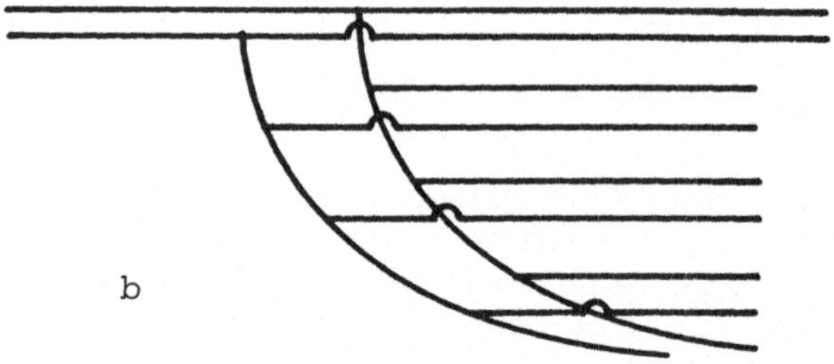

Fig. 6a A one chain, one Reggeon state, in the multiperipheral model.

b A two chain state in the deuteron. One chain of partons is associated with the proton and one chain with the neutron.

It is likely that the actual wave function of a fast hadron is not short range correlated but that in a hadron-hadron collision a part of the wave function is picked out which is short range correlated. To go further we will rely on formalism and say that the scattering off nuclei has taught us that the wave function of a fast hadron is probably significantly more complicated than we had anticipated. The wave function point of view will still prove useful because however complicated it may be, in terms of many Pomeron components, it is a property of the fast hadron and not of the collision. For example, we know immediately that the number of exchanged Pomeron will not grow as $A^{1/3}$ for large A. The wave function point of view is also very useful in determining the size of couplings of a Regge pole to a large nucleus.

V ABRAMOVSKII, KANCHELI, GRIBOV CUTTING RULES (20-23)

In order to proceed further we shall need some relations between different types of events which may occur in a high energy collision. Such relations are furnished by the AKG cutting rules. A derivation of these rules will not be given here, however an example will be given in which the meaning and spirit of the rules can be gleaned. To understand this example consider the scattering of a high energy dueteron off a static potential whose dimension is much larger than the size of the deuteron. We imagine the scattering to take place via an optical potential and only the double scattering term will be discussed. The double scattering term for the forward elastic amplitude is shown in Fig. 7. Fig. 7 shows the scattering taking place in terms of the interaction of the proton and neutron of the deuteron scattering off the potential. The proton and neutron remain free as the deuteron passes the potential since the time scale for the proton and neutron in the deuteron to directly interact with other is slowed by time dilitation. The graph of Fig. 7 is the only nonzero graph, at high energy, for the double scattering term and it correctly gives the forward elastic amplitude for double scattering. The contribution of double scattering to the total deuteron-potential cross section is given by the imaginary part of the graph of Fig. 7. However, the discontinuity of Fig. 7 involves only truly inelastic states (discontinuities through the optical potential) or a pn diffractively produced state. One finds no deuteron in the final state when taking the discontinuity in Fig. 7. Obviously something is wrong. As the deuteron passes the potential the p and n are always in the wave function of the deuteron since the momentum transfer from the potential is much smaller than the inverse radius of the deuteron. Therefore, we would expect to find deuterons, but not a free p-n system, in the final state. The resolution of this

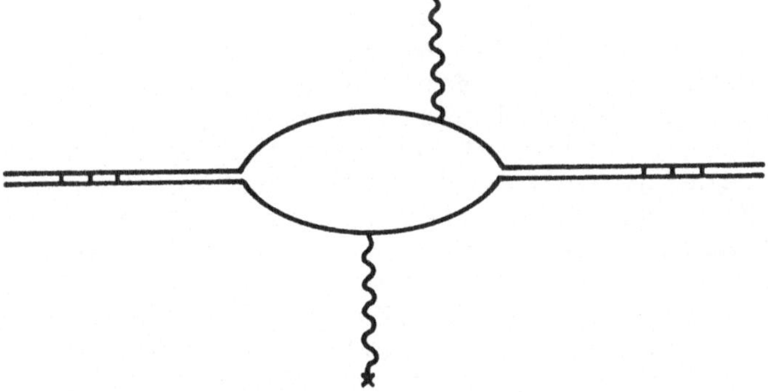

Fig. 7 A double scattering of a deuteron off a potential.

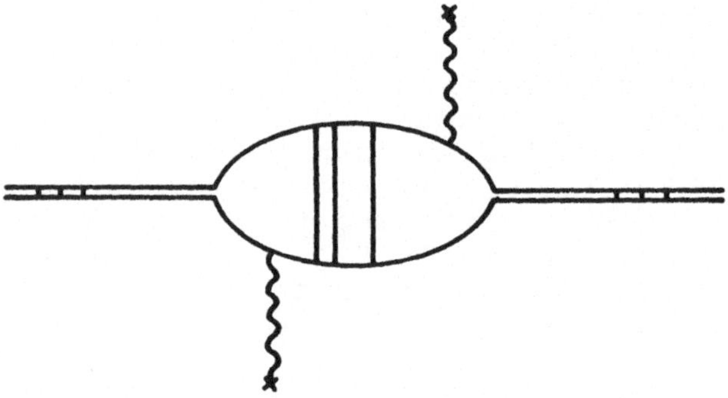

Fig. 8 A double scattering of a deuteron off a potential.

difficulty is provided by the graph of Fig. 8. In Fig. 8 the vertical lines represent the interaction of the p and n as the deuteron passes the potential. As we have said, these graphs are zero. However, there are non-zero discontinuities of Fig. 8. These discontinuities are through a free p-n state and a deuteron. The discontinuities through the p-n state exactly cancel the discontinuity through the p-n state of Fig. 7 leaving the discontinuity through the intermediate deuteron from Fig. 8. Thus though the amplitude of Fig. 8 is zero its various discontinuities are non-trivial and serve to rearrange the probabilities so that one gets the correct final states from the discontinuity of the forward elastic amplitude.

The AKG cutting rules, as derived in Ref. 23, look very much like the deuteron example discussed above. The two Reggeon exchange, the Mandelstam graph, can be written in terms of old fashioned perturbation theory graphs. The old fashioned perturbation theory graphs which survive at high energy only have cuts through two Reggeons, as shown in Fig. 9. These graphs give the correct forward elastic amplitude in the two Reggeon exchange approximation, but in order to get the correct final states one must add in other old fashioned perturbation graphs. There graphs rearrange probabilities so that the correct final states are obtained from the discontinuity of the forward elastic amplitude. The details of the argument are very much like the deuteron case but somewhere more complicated.

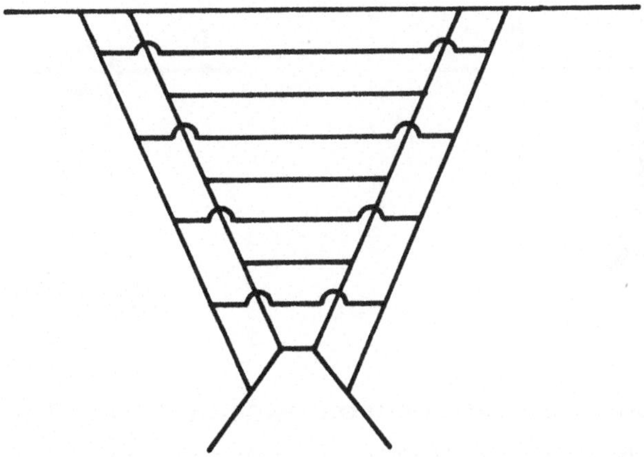

Fig. 9 A two Reggeon exchange graph schematically indicating the time evolution of the partons.

VI SEVERAL POMERON EXCHANGE IN HADRON-NUCLEUS COLLISIONS

We shall consider now multiple Pomeron exchange in hadron-nucleus elastic scattering(1). By using the AKG cutting rules relations between diffractive (including elastic) and truly inelastic final states will be obtained. Let's begin by considering one and two Pomeron exchange as shown in Fig. 10. The graph of Fig. 10a gives a contribution $2\pi R^2 A_1$ where R is the radius, presumed large, of the nucleus and A_1 is an energy independent, R independent number. We are now using the fact that the coupling of a Pomeron to a nucleus at a given impact parameter is not proportional to R(24). This point was discussed in detail in Section V. We shall call the contribution of Fig. 10b $2\pi R^2 A_2$. If we write $\sigma_{tot} = 2\pi R^2 \gamma$, $\sigma_{DD} = 2\pi R^2 \gamma r$, and $\sigma_k = 2\pi R^2 \tilde{\sigma}_k$ then

$$\gamma r = \sum_{\nu=2}^{\infty} (-1)^{\nu} A_{\nu} (2^{\nu-1}-1)$$

and

$$\tilde{\sigma}_k = \sum_{\nu=k}^{\infty} (-1)^{\nu-k} \binom{\nu}{k} 2^{\nu-1} A_{\nu}.$$

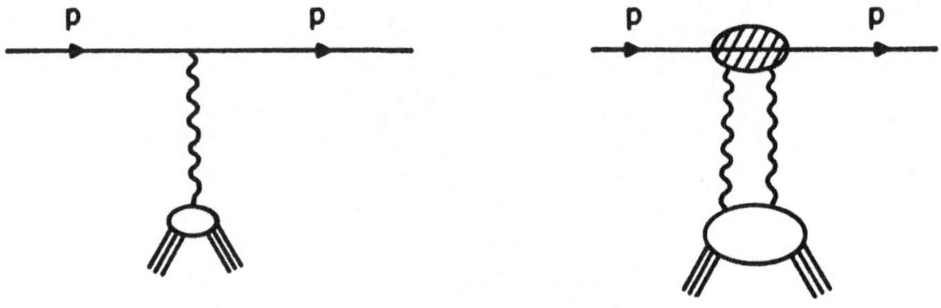

Fig. 10a A one Reggeon exchange amplitude.
b A two Reggeon exchange amplitude.

σ_{DD} is the diffractive, including elastic, cross section. σ_k is the cross section obtained by cutting through exactly k Pomerons and ν is the number of Pomerons exchanged. We are neglecting Reggeon interactions. The above equations are the statement of the AKG cutting rules. For the case of one and two Reggeon exchange

$$\gamma r = A_2$$
$$\tilde{\sigma}_1 = -4A_2 + A_1$$
$$\tilde{\sigma}_2 = 2A_2.$$

If we wish to maximize elastic scattering, for a given total cross section, we should make A_2 as large as possible However, we must keep $A_2 \leq 1/4\ A_1$ so that the maximum amount of diffractive, including elastic, scattering will occur when $A_2 = 1/4\ A$. In that case

$$\gamma r = A_2$$
$$\tilde{\sigma}_2 = 2A_2,$$

so that

$$r = \frac{\sigma_{DD}}{\sigma_{tot}} = \frac{\gamma r}{\gamma_r + \tilde{\sigma}_1 + \tilde{\sigma}_2} = 1/3$$

which is the maximum value which the diffractive part of the cross section can take so long as only one and two Pomerons are allowed. Now define

$$R_c = \left.\left(\frac{\frac{1}{\sigma_{inel}} \frac{d\sigma}{dy}\Big|_A}{\frac{1}{\sigma_{inel}} \frac{d\sigma}{dy}\Big|_{had}} \right)\right|_{\text{cental region}},$$

where had and A refer to hadron and nuclear targets respectively. Then

$$R_c = 2$$

in the above example if we use a single Pomeron approximation for the inelastic part of the hadron-hadron cross section. The necessity of having a large inclusive cross section in the central region when r is near 1/2 is a general feature of a Reggeon model in a soft field theory as long as Reggeon interactions are not allowed. (We expect Reggeon interactions to be small on phenomenological grounds in an intermediate energy region which should extend somewhat above Fermilab energies. A model of Coneschi and Schwimmer[15] involving Reggeon interactions suggests that the bounds stated below may be a little more general than might be expected.) If n is such that

$$\frac{1}{2}\,\frac{1 - 2^{2-n}}{1 - 2^{1-n}} \leq r \leq \frac{1}{2}\,\frac{1 - 2^{1-n}}{1 - 2^{-n}}$$

then

$$R_c \geq n + 1 + \frac{2^{n-1}(2r-1)}{1 - r} ;$$

this means that near blackness can only come about as a result of many Pomeron exchanges and this results in a large inclusive cross section in the central region.

VII DISCUSSION OF POMERON AND GLAUBER SERIES

Consider, again, the one dimensional scattering process discussed in Section II. The double scattering term is shown in Fig. 3. In Section II the discussion was given for a one dimensional model, but there is no essential difference between such a model and a high energy scattering in 3 spatial dimensions at a given value of the impact parameter. Let the state of the hadron as it approaches the first scatterer be ψ_o. We have dropped the center of mass motion. Then between scatterers n and n + 1 the hadron state is

$$s_n s_{n-1} \cdots s_1 \psi_o$$
$$= \left(1 - \sum_{i=1}^{n} \gamma_i + \sum_{1 \leq i \leq j \leq n} \gamma_i \gamma_j \cdots \right) \psi_o$$

(Since we are dealing with a point particle ψ_o is simply a number, say 1.) Thus when the hadron is between scatters n and n + 1 the unscattered part of the wave, ψ_o, the once scattered part of the wave, $- \sum_{i=1}^{n} \gamma_i \psi_o$, the twice twice scattered part of the wave, $- \sum_{1 \leq i \leq j \leq n} \gamma_i \gamma_j$, etc. are all just numbers and so are coherent. That is, there is only one state and different numbers of scatterings just multiply that state by different c-numbers.

Now consider the same arrangement of potentials and compare a single Regge pole exchange term with a double Regge pole term. These exchanges are illustrated in Fig. 11 where the figures are drawn so that the points of emission and absorption of quanta along the z-axis are crudely indicated. Now the incoming hadron is a rather compact object as it approaches the first scatterer and remains so as it traverses the set of scatterers. If we suppose $p/m^2 \gg L.N$ where L is the separation between neighboring scatters, then the early emissions and late absorption of quanta occur before the hadron has reached the first scatterer and after it has passed the last scatterer, respectively. If we look at the time just after the hadron passes the scatterer n we see that the single scattering corresponds to a state of $n \approx \bar{n}(p)$ quanta while the double scattering corresponds to a state of $n \approx 2\bar{n}(p)$ quanta. The fluctuations about these numbers are about $\sqrt{\bar{n}(p)}$, at least if $\alpha(o) < 1$ which we may assume for simplicity. Thus the states which take part in one and two scatterings (one and two Regge pole exchange) are orthogonal. We cannot expect the N independent single

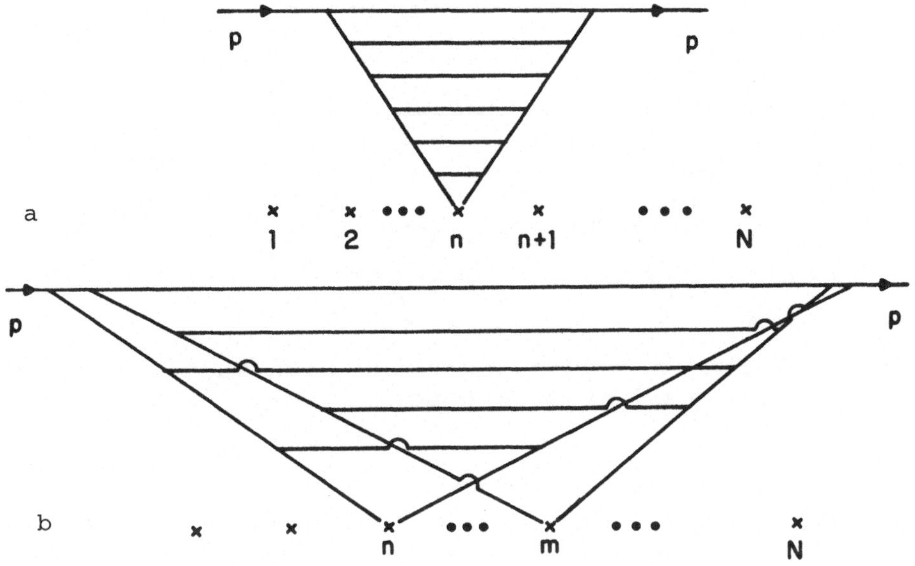

Fig. 11a A single Reggeon exchange.

b A double Reggeon exchange with the time evolution of the quanta schematically indicated.

Regge pole exchanges, the sum of which will be proportional to N and hence not unitary, to be coherent with double Regge pole exchanges. How then is unitarity satisfied? The sort of graphs which must be considered in order to get a unitary result are shown, schematically, in Fig. 12. In these graphs all quanta for which $k/m^2 \gtrsim NL$ are as in a single Regge pole. The interactions which impose unitarity come from quanta having $k/m^2 \lesssim NL$.

I would think that the portions of the graph shown in Fig. 12 which refer to partons having $k/m^2 \leq NL$ must be very similar to those graphs which occur in scattering a particle of momentum $p/m^2 \approx NL$. It is clear that such a scattering can have no factor of N in it and so the one Regge pole exchange, for a particle having $p/m^2 >> NL$, has no factor of N. The emergence of a Glauber-like series is most untransparent in all of this. That is not to say that something like a Glauber series could not be a reasonable approximation in an intermediate energy region. For example, consider the graphs in Fig. 12. This could be part of a Glauber-like series where the partons moving through the nucleus are in the wave function of the incident hadron or one of its law lying excitations. Something this simple cannot be the whole

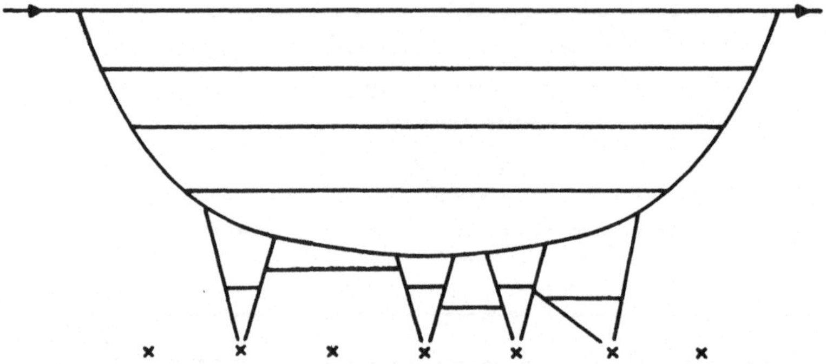

Fig. 12 A single Reggeon exchange with multiple scattering of slow partons.

story, however, since the fast partons in Fig. 12 are uncorrelated and the hadron nucleus cross section would factorize in direct contrast to a Glauber-like expansion.

Going back to the example at the end of Section IV, the wave function of the incident hadron may really not be short range correlated at all, and hadron-hadron scattering simply picks out the short range correlated part. Suppose the incident fast hadron has, say, three independent (noninteracting) chains in its wave function. Lets assume that only one of these chains interacts with the target hadron in a hadron-hadron collision. Now if a fast hadron hits a large nucleus all three of the chains will interact in an inelastic event so $R_c = 3$. In terms of Regge poles there are one, two, and three Regge pole exchanges, but never more than three. The relative weights of the Regge pole exchanges are in accord with the first three terms in a Glauber series with 3 scatterers. (This can be seen by setting $\tilde{\sigma}_k = 0$ for $k \neq 3$ in the equation following Eq. (6.4) of Ref. 1. Such a model is perhaps not an impossibility although we see no reason why 3, no more, chains should sit side by side in a fast hadron and not interact. (Such interactions would spoil short range correlations in inelastic hadron-hadron events.) Such a model is nowhere near as attractive as the simple parton-multiperipheral model having an uncorrelated, one chain, wave function, but experiment, blackness in particular, is forcing us in this direction. If such a model is to be relevant it would be nice to see where the enhancement of the Pomeron nucleus residue, a factor of 3 in our example above, comes from in terms of a process like that shown in Fig. 5. That is, an off shell particle of momentum $p \approx Rm^2$, while it does not get an extra factor of R (for large R) when going off shell, would seem to get an enhancement factor. Overall, I would say that we do not yet understand how the Glauber expansion emerges in a soft-field theory at high energies.

VIII A STRING MODEL EXAMPLE

I would like to discuss an example in the string model, not because it is a realistic model but because it illustrates that there is a possible danger in using the AKG rules for nuclei. In a soft field theory the AKG cutting rules are known to be true and for scattering of a particle off a nucleus one only needs the incident momentum of the particle, p, to be such that $p/m^2 \gg R$ where R is the radius of the nucleus. These cutting rules and the energy necessary for their application follow simply from the softness of the theory and it must be very hard to avoid their consequences. Consider, however, a planar double scattering in the string model as shown in Fig. 13. The p_+'s of the slow strings are chosen to be zero in Fig. 13. Lets group the Regge exchanges into three terms as shown in Fig. 14. Fig. 14a represents the elastic intermediate state, Fig. 14b represents the low mass diffractive states, and Fig. 14c represents the triple Regge coupling. The division between 14b and 14c involves some arbitrariness which need not concern us here. Let $(p_1 - p'_2)^2 = (p_2 - p'_2)^2 = t$. Now the terms shown in 14b and 14c vanish linearly in t. Thus if particles 1,1' and 2,2' are in a nucleus and we are looking

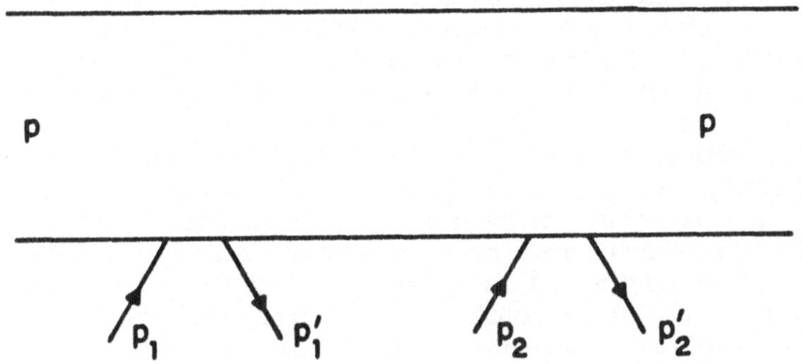

Fig. 13 A double scattering in the string model.

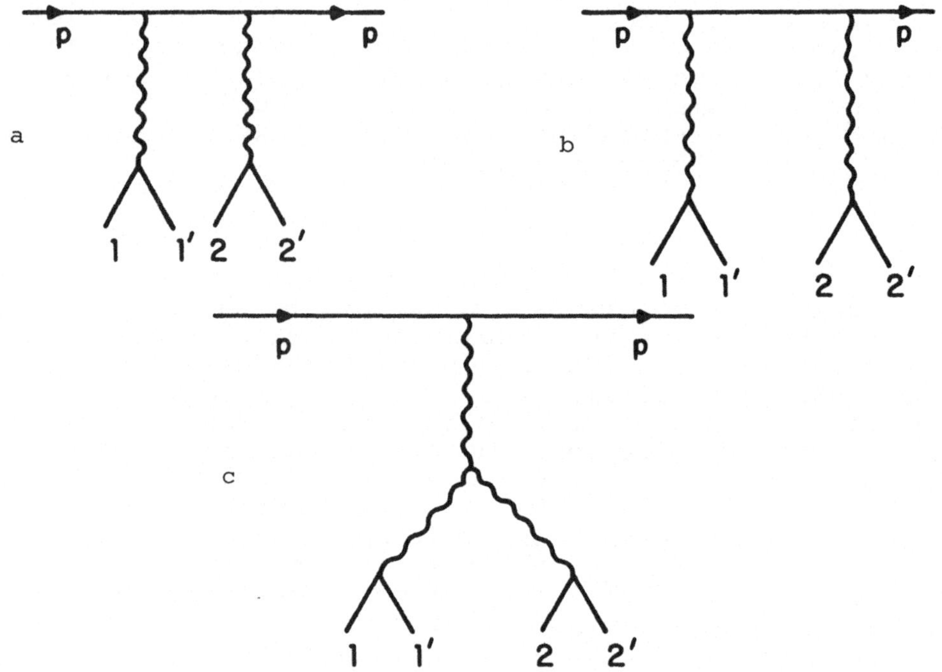

Fig. 14a The elastic intermediate state in the double scattering of a fast string.

b The low mass diffractively excited states in the double scattering of a fast string.

c The triple Reggeon term in the double scattering of a fast string.

at the forward elastic scattering of the hadron off the nucleus, then the term shown in 14b is proportional to $1/R^2 = \langle t \rangle$ while the term shown in 14c is proportional to $\ell np/R^2$. When $\ell np \gg R^2$ the usual AKG relations follow for discontinuities of the diffractive, 1 Reggeon, and two Reggeon types and thus all of Fig. 14 must be simply a renormalization of the Reggeon residue. So long as $\ell ns/R^2 \ll 1$, however, only the elastic term shown in Fig. 14a is relevant and here one has single Reggeon and diffractive discontinuities. The consequences of the AKG rules are thus avoided. In general the dual model, including non planar graphs, will have no simple cutting rules for a fixed number of Reggeon exchanges in the region $\ell ns/R^2 \ll 1$. This is not relevant for hadron-

hadron scattering, but for hadron-nucleus scattering we should be aware that it may not be absolutely impossible to avoid the AKG rules. I see no way of circumventing these cutting rules in a soft field theory, however.

ACKNOWLEDGEMENTS

I have benefited from discussions with many of my colleagues. In particular I am indebted to J. Koplik, M. Baker, J. Weis, J. Bjorken, L. Bertocchi, G. Winbow and A. Krzywicki.

REFERENCES

(1) J. Koplik and A. Mueller, Phys. Rev. D12, 3638 (1975).
(2) N. Nikolaev in lectures presented at the International Topical Meeting on Multiparticle Production at Very High Energies, Trieste (1976).
(3) J. Weis, CERN preprint (1976).
(4) L. Bertocchi in Proceedings of the VIth International Conference on High Energy Physics and Nuclear Structure, Santa Fe and Los Alamos, 1975.
(5) R. Glauber, in Lectures in Theoretical Physics, edited by W.E. Britten and G. Dunham (Interscience, New York, 1959), Vol. 1.
(6) V. Gribov, JETP 30, 709 (1970).
(7) The following example is apparently well known to experts in hadron-nucleus collisions. I thank L. Bertocchi for a discussion.
(8) O. Kancheli, JETP Letters 18, 274 (1973). See also E. Lehman and G. Winbow, Phys. Rev. D10, 2962 (1974).
(9) K. Gottfried, Phys. Rev. Letters 32, 957 (1974).
(10) W. Buza et al., Phys. Rev. Letters 34, 838 (1975).
(11) P. Murthy et al., Nucl. Phys. B92, 269 (1975).
(12) J. Biel et al., Phys. Rev. Letters 36, 1004 (1976).
(13) I have greatly benefited from a number of comments made by J.D. Bjorken concerning such possible wave functions of a fast hadron.
(14) A. Capella and A. Kaidalov, CERN preprint. However, I disagree with the neglect of final state interactions, in the language of Capella and Kaidalov, since it is impossible to have initial and not final state absorptions in a soft field theory. In other words an absorptive model should depress the one Reggeon cut by an amount $\sim e^{-\bar{n}}$ where $\bar{n}$ is the typical number of scatterings at some average impact parameter. I thank A. Krzywicki for discussions on this point.

(15) L. Caneschi and A. Schwimmer, Nucl. Phys. B102, 381 (1975).
(16) L. Caneschi, A. Schwimmer, and R. Jengo. (To be published).
(17) L. Bertocchi, S. Fubini, and M. Tonin, Nuovo Cimento 25, 626 (1962);
D. Amati, A. Stanghellini, and S. Fubini, Nuovo Cimento 26, 6(1962).
(18) R. Feynman, Phys. Rev. Letters 23, 1415 (1969).
(19) J. Kogut and L. Susskind, Phys. Reports 8C, No. 2 (1973).
(20) V. Abramovskii, O. Kancheli, and V. Gribov, Sov. J. Nucl. Phys. 18, 308 (1974).
(21) J. Koplik and A. Mueller, Phys. Letters 58B, 166 (1975).
(22) L. McLerran and J. Weis, Nucl. B100, 329 (1975).
(23) M. Baker and K. Ter-Martirosyan, Phys. Reports (to be published).
(24) For R not too large there may be an R dependence in A_1. However, as $R \to \infty A_1 \to$ constant. If the Glauber expansion is correct, in an approximate sense, as I would expect is the case, then this non asymptotic dependence of A_1 or R is crucial.

HEAVY CLUSTERS AT HIGH ENERGY

M. Anselmino, A. Ballestrero, and E. Predazzi

Istituto di Fisica dell' Università - Torino

Istituto Nazionale di Fisica Nucleare -
Sezione di Torino

Turin, Italy

SUMMARY

It is shown that within a fairly general model of sequential decay, high energy data require the existence of clusters whose mass increases with energy. Arguments are given to suggest that this picture may be more general than the specific model used would lead to conclude.

1) INTRODUCTION

Many features of high energy multiple production are nowadays understood or interpreted in terms of cluster formation and decay.

While the consensus on the origin and nature of clusters is far from universal, the usual folklore demands that they are relatively light objects ($M \simeq 2 \div 3$ GeV/c^2) whose production becomes more and more copious as one goes to higher and higher energies. This picture relies heavily on the short range order hypothesis[1].

In this paper we set to show that the above picture conflicts with the presently accepted general trend of high energy physics once several experimental inputs are combined together within a fairly general probabilistic model of sequential formation and decay of clusters.

It will turn out that the mean value of clusters is asymptotically bounded from above if indeed the ratio of the elastic to the total pp cross section is approaching a non-vanishing constant. Consequently the cluster mass must increase with energy to account for the growth of secondaries. The details can be found in the literature[2).

Arguments will be given that the emerging picture of clusters of increasing mass is very likely an unescapable consequence of factorizable models and is not characteristic of the specific model discussed here.

The only apparent way out of this admittedly rather awkward situation is to assume that the present highest energy data can not yet be considered as asymptotic (in particular the constancy of the ratio of the total to the elastic p-p cross section). Another possibility is, of course, that clusters do not account for all of high energy production and that, in particular,a concomitant diffractive - like process is at work. Whether or not this second component should somehow be subsequently accounted for by some kind of "unitarization" procedure of the cluster component, it seems to us that this attitude would defeat entirely the spirit of a cluster approach.

It must be realized that the use of the term cluster in the present context will ultimately be very different from that of ref. 1.

Our contention, however, is that starting from an unbiased view of clusters, taking at face value the present high energy data and assuming that their gross features must be accounted for by cluster formation, one is naturally led to conclude that their mass must grow with energy.

It is a rewarding fact that cosmic ray data support out findings[3). It is also interesting that fits to accelerator data have been provided along this line [4).

2. BASIC EQUATIONS OF THE SEQUENTIAL MODEL

The main assumptions of the sequential model are the following:

i) particle production goes via a two-step process of cluster formations and decay; these two steps occur independently of one another;

ii) cluster formation takes place after the colliding particles have given rise to a sort of unstable compound which we call "fireball". This is actually an unnecessary assumption but it makes somewhat easier the formal developments without introducing any essential limitation. The fireball keeps boiling off with an independent probability bunches of particles which we call clusters.

iii) these clusters, in turn, emit particles with an independent probability.

iv) Energy-momentum is conserved at each emission and both steps do separately conserve probability.

We call "sequential" the above scheme of both cluster formation and decay in that the emission process may repeat itself an arbitrary number of times (so long as no conservation constraint is violated) until it eventually comes to an end when no further clusters (or particles) are emitted.

The repetitivity of the scheme leads to linear integral equations for the various quantities of physical interest (like the inclusive probabilities) as a consequence of probability conservation and these equations are Volterra-like (in the sense that their series expansion truncates at some point) as a consequence of energy conservation (see ref.2 for details).

Internal quantum numbers have not yet been put in the model although this should not be a major complication. At this stage, a more serious drawback of the model is perhaps the fact that it deals with probabilities and not with amplitudes so that possible quantum mechanical properties of production processes are essentially ignored. Thus, unitarity is beyond the reach of the model as presently formulated. On the other hand, the bonus is that one can make sure that probability is strictly conserved.

The equations of the model are written in terms of two basic, a priori unknown, structure functions which we graphically represent as follows (double, single and broken line denote fireballs, clusters and particles of four momenta P,R and p respectively)

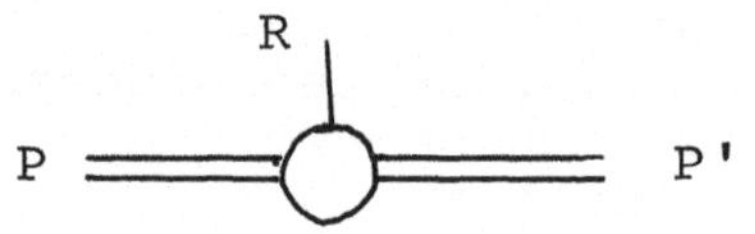

$\equiv V(P,R)$

Fig.1

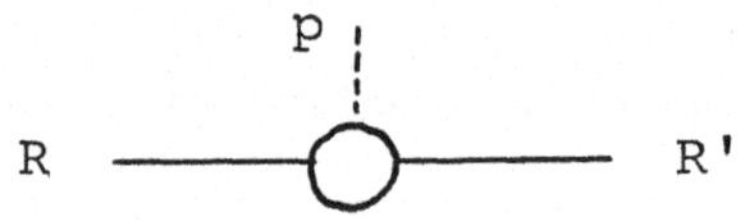

$\equiv g(R,p)$

Fig.2

The use of fireball could be avoided by replacing the graph of fig.1 by the off-shell five point probability

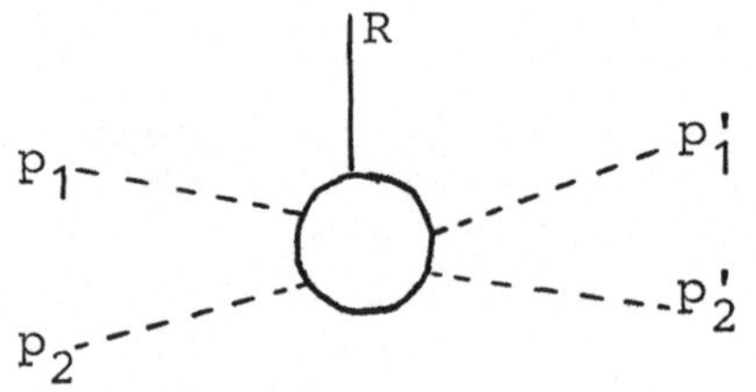

$\equiv V(p_1,p_2;p_1',p_2';R)$.

Fig.3

If we denote by

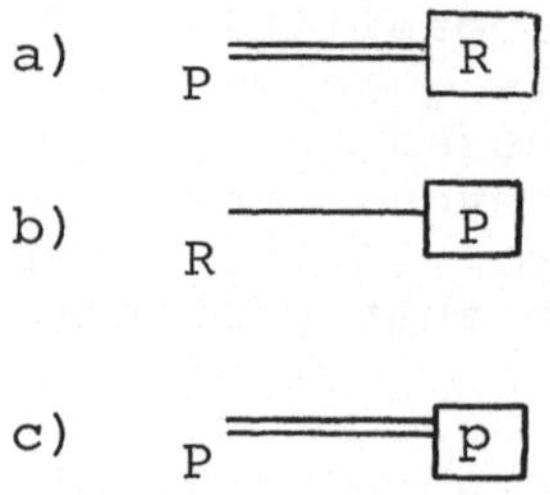

a) $\tau_1^{inc}(P;R)$

b) $\rho_1^{inc}(R;p)$

c) $\sigma_1^{inc}(P;p)$

Fig.4

the single inclusive probabilities for a) the initial state to produce a cluster of four-momentum R,
b) a cluster to emit a particle of four-momentum p,
c) the initial state to emit a particle of four-momentum p, the simplest integral equations of the sequential model are graphically:

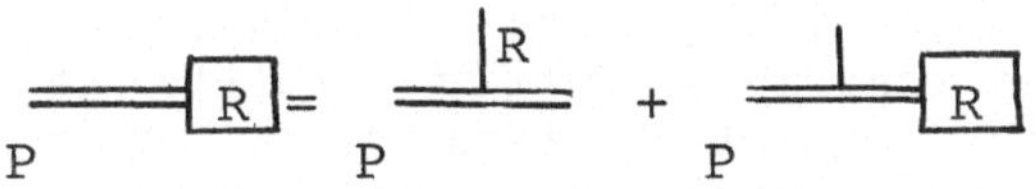

Fig.5

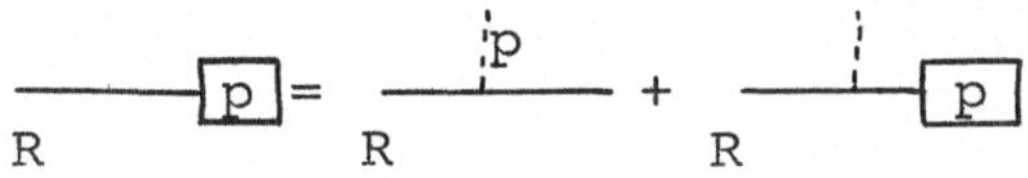

Fig.6

where lines without any label indicate that the integration over the corresponding four-momentum has been performed.

If, given as input the graphs of fig.1 and 2 the equations of fig.5 and 6 are solved to give $\tau_1^{inc}(P,R)$ and $\rho_1^{inc}(R,p)$, the inclusive probability $\sigma_1^{inc}(P,p)$ for the emission of a particle of four-momentum p is obtained by the convolution rule

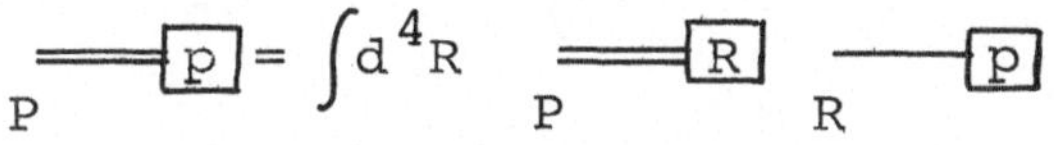

Fig.7

Fig.7 is the simplest of the set of convolution rules[5) that link together the two steps of cluster formation and decay if these two steps occur independently (assumption i).

The analytic ecpressions of the equations represented by the graphs of figs. 5,6,7 are

$$\tau_1^{inc}(P;R)=V(P,R)+\int d^4P'\ V(P,P-P')\,\tau_1^{inc}(P',R) \qquad (1)$$

$$\rho_1^{inc}(R,p)=g(R,p)+\int d^4R'\ g(R;R-R')\,\rho_1^{inc}(R',p) \qquad (2)$$

$$\sigma_1^{inc}(P,p) = \int d^4R \; \tau_1^{inc}(P,R) \; \rho_1^{inc}(R,p) \tag{3}$$

The above equations are all expressed in four-momentum variables and the integrations over the outgoing fireball and cluster four-momenta have already been carried over using the energy-momentum conservation delta function. All other δ and θ functions necessary to ensure mass shell and energy spectrum conditions are still included in the structure functions. Thus, the "actual" inclusive cross section to produce a particle of momentum $\vec{p}$ is the integral over the fourth p_o component of eq.(3) carried out with the help of the mass shell delta function.

For the structure of the higher order equations, we refer to Ref. 2 and 5.

Probability distributions and their moments can be expressed also by means of equations that can be derived within the sequential scheme. Thus, for instance, the probability $P_N(s)$ for producing N clusters and the related moments $\bar{P}_N(s)$ are given by

$$P_N(s) = \int_o^s K(s,x) P_{N-1}(x) dx \qquad (N=1,2,\ldots) \tag{4}$$

$$\bar{P}_N(s) \equiv \langle L(L-1)\ldots(L-N+1)\rangle = N \int_o^s K(s,x)\bar{P}_{N-1}(x)dx + \int_o^s K(s,x)\bar{P}_N(x)dx \tag{5}$$

$$(N=1,2,\ldots) \qquad \bar{P}_o \equiv 1$$

and similar equations express the probability $Q_k(M)$ that a cluster of mass M emits k pions and the related moments $\bar{Q}_k(M)$

$$Q_k(M) = \int_o^M H(M,M')Q_{k-1}(M')dM' \qquad (k=2,3,\ldots) \tag{6}$$

$$\bar{Q}_k(M) \equiv \langle \ell(\ell-1)\ldots(\ell-k+1)\rangle = k\int_0^M H(M,M')\bar{Q}_{k-1}(M')dM' + \int_0^M H(M,M')\bar{Q}_k(M')dM' \quad (k=2,3\ldots) \tag{7}$$

$$\bar{Q}_1(M) = 1+\int_0^M H(M,M')\bar{Q}_1(M')dM' \tag{7'}$$

The kernels K(s,x) and H(M,M') that appear in eqs. (4-7) obtain by integration from the kernels of eqs. (1,2). In particular, probability conservation imposed on eq.(4) gives

$$P_{in}(s) \equiv \sum_{N=1} P_N(s) = \int_0^s K(s,x)dx \equiv \frac{\sigma_{in}(s)}{\sigma_{tot}(s)} \tag{8}$$

where $\sigma_{in}(s)$ and $\sigma_{tot}(s)$ are the actual inelastic and total cross sections. Inserting eq.(5) with N=1 and eq. (7') (they define respectively the mean multiplicity of clusters $\langle N(s)\rangle$ and the mean value of particles emitted by a cluster $\langle k(M)\rangle$) in the convolution rules of Ref. 5 and making the simplifying (and probably incorrect) assumption that the cluster mass distribution is delta like, i.e. that at a given energy clusters of the same mass are created, we get for the actual mean value of particles $\langle n(s)\rangle$

$$\langle n(s)\rangle = \langle N(s)\rangle \; \langle k(M(s))\rangle \tag{9}$$

3. PHENOMENOLOGICAL ANALYSIS OF THE EQUATIONS OF THE SEQUENTIAL DECAY MODEL

The proper way of analyzing the yields of the sequential decay model (of which in Sec. 2 we have given the simplest equations) would be,using the popular Regge-Müller like Ansatz,to characterize the structure functions and solve the various equations nummerically to plot the quantities of physical interest.

At the present stage of investigations, however, we feel that it is more profitable to try to inject phenomenological information in order to extract general predictions. To this aim, all the well established properties of high energy physics can be used such as:

a) growth of σ_{el} and σ_{tot} with energy and asymptotic constancy of the ratio σ_{el}/σ_{tot}
b) growth of the average multiplicity $<n(s)>$
c) growth of the dispersion (Wroblewski's empirical rule[7])
d) leading particle effect
e) cut in the transverse momenta
f) single and double inclusive rapidity distributions.

First of all, we notice that if we use eq.(5) for N=1, we have the cluster multiplicity integral equation

$$<N(s)> = P_{in}(s) + \int_0^s K(s,x) <N(x)> dx \qquad (10)$$

If $<N(s)>$ is assumed to be a non-decreasing function of s as commonly accepted, an immediate majorization of eq. (10) gives

$$<N(s)> \leq \frac{P_{in}(s)}{1-P_{in}(s)} \qquad (11)$$

If we now recall that the ratio $\sigma_{in}/\sigma_{tot} \equiv P_{in}(s)$ is approaching the constant value ~ 4/5 at highest ISR energies[6] (which implies that $P_o(s) \equiv \sigma_{el}(s)/\sigma_{tot}(s)$ is <u>not</u> approaching zero as most models would require) we see that the mean number of clusters is bounded from above by

$$<N(s)> \underset{s\to\infty}{\lesssim} N_o \simeq 4 \qquad (12)$$

contrary to the common belief that the mean number of clusters should grow indefinitely with energy.

Keeping in mind eq. (9) and recalling that the actual mean multiplicity $<n(s)>$ is a (perhaps logarithmically) growing function of s, the first consequence of eq. (12) is that the mean number of particles emitted by a cluster $\langle k(M(s))\rangle$ grows with energy which is possible only if the mass itself of the cluster $M(s)$ grows with s.

It can be shown by arguments similar to the ones used in deriving the upper bound (11) that the previous conclusions applied to the higher order equations of the model imply that for any finite k and $M(s)\to\infty$, $\bar{Q}_k(M)$ as given by eq. (7) may asymptotically become

$$\bar{Q}_k(M) \underset{\substack{M\to\infty \\ k \text{ finite}}}{\sim} k! \frac{[Q_{in}(M)]^{k-1}}{[1-Q_{in}(M)]^k} \tag{13}$$

where

$$Q_{in}(M) \equiv \int_0^M H(M,M')dM' \xrightarrow[M\to\infty]{} 1-O\left(\frac{1}{\langle n(s)\rangle}\right) \tag{14}$$

the last asymptotic equality following from eq. (13) for k=1 together with the previous remark that

$$\langle k(M(s))\rangle \equiv \bar{Q}_1(M(s)) \underset{s\to\infty}{=} O(\langle n(s)\rangle).$$

In particular, from eq. (13) we have for the second moment of cluster decay

$$\langle k(k-1)\rangle \underset{s\to\infty}{\sim} 2\langle k\rangle^2 \tag{15}$$

We now insert the above relation in Wroblewski's empirical rule[7] for the dispersion ($D^2_{(i)} \equiv \langle n_i^2\rangle - \langle n_i\rangle^2$)

$$D_{(i)} = A(\langle n_i\rangle - 1) \tag{16}$$

where the subscript i is a reminder that the particle mean multiplicity is here normalized to σ_{in} rather than to σ_{tot}. For pp scattering the constant A is $\simeq .58$.

Normalizing to σ_{tot}, using the convolution rule[5] for the second moment, after some algebra one finds

$$\langle N\rangle\langle k(k-1)\rangle = \langle k\rangle^2\{\langle N\rangle^2 \frac{\sigma_{tot}}{\sigma_{in}} (1+A^2) - \langle N(N-1)\rangle\}$$

$$-(1+2A^2)\langle N\rangle\langle k\rangle + A^2 \frac{\sigma_{in}}{\sigma_{tot}}. \tag{17}$$

Retaining only higher orders ($<k>^2$) and using eq. (15), we get

$$<N^2> \sim 1.67 <N>^2 - <N> \tag{18}$$

where the numerical values for σ_{in}/σ_{tot} and for A have been used.

It can be shown that no contradiction exists between the present data on double rapidity correlations and the heavy cluster picture within the sequential decay scheme. In particular, using the cut in transverse momenta, one can estimate the single-cluster contribution to the width in rapidity of the double inclusive distribution. This turns out to be of about one rapidity unit as compared to the experimental value of about 1.5 rapidity units at ISR energies. Since the two-cluster contribution is expected to be rather flat, it is reasonable to assume that it must account for the difference. For the detailed calculations which are rather lengthy, we refer the interested reader to the last paper of Ref. 2.

4. CRITICAL REMARKS AND CONCLUSIONS

Several comments are in order in the light of the findings of Sec.3.

First of all, the cosmic ray data of the Brazilian-Japanese collaboration[3)] show evidence of a discrete spectrum of heavier and heavier clusters at 2÷3, 20÷30 and 200÷300 GeV/c^2. Furthermore, as already mentioned, fits to accelerator data have been provided in this spirit[4)].

Secondly, one could argue that the present constancy of the ratio σ_{el}/σ_{tot} is a transient that will disappear at truly asymptotic energies (should this not be the case, incidentally, most popular exchange models would be in severe trouble). In this case all our previous findings would be invalidated and the conclusions should be revised.

A third possibility is that one attributes to a different component or mechanism the responsibility for $\sigma_{el}/\sigma_{tot} \xrightarrow[s\to\infty]{} \text{const.} \neq 0$ postulating that the cluster contribution has instead a vanishing elastic component. In our opinion, however, this would just amount to make rather irrelevant the overall cluster description.

A last possibility could be, of course, that it is just the use of the sequential decay model which leads to this somewhat unexpected prediction of heavy clusters.

This last possibility is ruled out by the observation that factorization alone leads to a conflict between the two simultaneous requirements that

$$\frac{\sigma_{el}}{\sigma_{tot}} \equiv P_o(s) \xrightarrow[s\to\infty]{} P_o \neq o$$

and that $\langle n(s)\rangle$ grows with energy unless some other feature (like the production of heavy clusters) come into play. If $P_n(s)$ is the n-particle production probability, whenever $P_n(s) \leq A(s)\ [P_{in}(s)]^n$, the series $\sum_n P_n(s)$ is uniformly convergent if, asymptotically,

$P_o(\infty) > 0$. In this case, however, also the series

$\sum_n n\, P_n(s) \equiv \langle n(s)\rangle$ is convergent and $\langle n(\infty)\rangle$ is bounded.

It therefore looks as if the conclusion that the elastic cross section must go to zero with respect to the total one if we demand the average multiplicity to grow is an unescapable consequence of factorizable models for direct particle production. This conclusion does not apply any more when the production goes via the intermediate step of cluster formation since in this case it is just the mean cluster value which becomes bounded. As discussed previously, however, in this case the cluster's mass must grow and we are back to the findings of Sec. 3 so that the very foundation of the short range order hypothesis becomes questionable.

A final remark[8)] is that the requirement $P_o(s) \xrightarrow[s\to\infty]{} P_o \neq 0$ is by itself sufficient to essentially rule out the so called uncorrelated cluster models (U.C.M.).

In fact, given a region Y_B-Y_A in rapidity, irrespective of whether or not this region grows with energy, the probability that no clusters are produced in it is, within the assumptions of the UCM

$$P_{AB}(s) = \lim_{\Delta y_i \to 0} \prod_i \left[1-f(s,y_i)\Delta y_i\right] \tag{19}$$

where

$$f(s,y_i)\Delta y_i = \tau_1^{inc}(y,s)\Big|_{y=y_i \in \Delta y_i} \Delta y_i \tag{20}$$

From eq. (19) we have

$$P_{AB}(s) = \exp\left[-\int_{Y_A}^{Y_B} f(s,y)\, dy\right]. \tag{21}$$

Since, by definition,

$$\langle N(s)\rangle_{AB} = \int_{Y_A}^{Y_B} f(s,y)\, dy \tag{22}$$

is the mean number of clusters produced in Y_B-Y_A, we see that $P_{AB}(s)$ must vanish if $\langle N\rangle_{AB}$ grows indefinitely.

On the other hand, if $\sigma_{in}^{(AB)}$ is that portion of the inelastic cross section describing processes for which no cluster is produced in the region Y_B-Y_A, we have

$$P_{AB}(s) = \frac{\sigma_{el}(s)+\sigma_{in}^{(AB)}}{\sigma_{tot}(s)} \geq \frac{\sigma_{el}(s)}{\sigma_{tot}(s)} \tag{23}$$

from which we get

$$\langle N\rangle_{AB} \leq \ln \frac{\sigma_{tot}}{\sigma_{el}} \simeq 1.8 \tag{24}$$

which shows that the UCM is not a viable approach.

This result reinforces our reservations on whether it is at all possible to construct a realistic model with an increasing number of clusters which is not in contradiction with the presently most well established high energy data.

ACKNOWLEDGEMENTS

It is a pleasure to thank Prof. H. Satz, the staff of the Centre for Interdisciplinary Research and the Theoretical Institute of Physics of the University of Bielefeld for support and for their kind hospitality. Two of us (M.A. and A.B.) acknowledge financial help from the Italian National Institute for Nuclear Research. (INFN).

REFERENCES

1) See for instance A. Krzywicki, these proceedings, and references therein.
2) A. Ballestrero, R. Page and E. Predazzi: Nuovo Cimento Letters 15, 57 (1976); see also: M. Anselmino, A. Ballestrero and E. Predazzi: "Evidence for clusters of growing mass" - preprint (1976). In this paper complete references to previous work are given.
3) Brazil-Japan Emulsion chamber Collaboration: Proc. of the 14th Inter. Cosmic Ray Conference, Munich (1975), Vol.7 and references therein.
4) See, for instance: M. Hama, M. Nagasaki and H. Suzuki, Rikkyo Univ. preprint RUP-76-15 (1976)
5) M. Anselmino, A. Ballestrero and E. Predazzi: "Convolution rules for cluster models", University of Bielefeld preprint BI-TP 76/19.
6) U. Amaldi et al. Phys. Lett. 44B, 112 (1973); S.R. Amendolia et al. Phys. Lett. 44B,119 (1973).
7) A. Wroblewski: Proc. III Int. Coll. on Many Body Reactions, Zakopane 1973.
8) M. Anselmino and A. Ballestrero: Nuovo Cimento Letters 15 329 (1976), see also M. Anselmino, A. Ballestrero and E. Predazzi: Nuovo Cimento 35A, 174 (1976)

INCLUSIVE CORRELATIONS IN THE CENTRAL REGION*

Gerald H. Thomas

High Energy Physics Division
Argonne National Laboratory
Argonne, Illinois 60439 U.S.A.

ABSTRACT

The subject of this talk is hadron correlations at very high energies. These correlations are discussed first from the point of view of the cluster picture. The cluster picture attempts to explain why the correlations are present in data by postulating a few simple properties for objects called clusters. Unfortunately, the clusters make no direct contact with known physical mechanisms. To attempt a remedy, the correlations are then discussed from a more fundamental viewpoint. We try to see how people have tried to describe the data as due to resonance production and decay. The dynamics is supposed to be that of Regge exchanges. A simple Mueller-Regge exercise is described and shown to work qualitatively. The last topic considered is that of coherent interferences as a possible source of hadron correlations. It is argued that these interference phenomena are a natural consequence of resonance production. Based on available data, known resonances and their coherent interferences do not account for the observed correlations.

1. INTRODUCTION

To some of us, a fascinating aspect of multihadron production is that the produced hadrons are correlated, both in momenta and in quantum number characteristics. These correlations, observed in Fermilab and ISR data, were unexpected. Most theoretical models, such as the Multiperipheral model, and the Fermi statistical model for simplicity assumed that dynamic correlations were absent, since cosmic ray data did not require such sophistication.

In the last few years, several kinds of correlations have been studied. It has been found that the data can be economically "explained" by the invention of clusters, whose exact nature is not specified, but whose average properties can be used to parametrize the data.[1] The cluster properties one needs are 1) the average number of clusters per unit of rapidity 2) the average number of pions which decay from a cluster and 3) for some applications the average cluster mass. One reason these average properties are sufficient is that the data are presented as 1-and 2-dimensional distributions, and possible structure will be smeared out without a judicious choice of variables.

Our real interest, of course, is to learn why the produced hadrons are correlated. It may be that clusters provide a useful clue, but then we need to know what a cluster is. The most conservative point of view is that the correlations one observes are due to resonances, and that clusters describe the resultant of the process of resonance production and decay.

In this talk we shall discuss to what extent resonances can account for the observed correlations. Included in the generic category of resonances are those coherent interference effects associated with the production of resonances. We also include the dual concept of Regge exchanges, as formulated in the Mueller Regge framework. In order, we shall take up the subject of clusters (Sec.2), resonance production and Regge systematics (Sec.3) and coherent interferences (Sec.4). Though it would be satisfying if these mechanisms were able to totally account for the observed correlations, at present we are not able to come to that conclusion.

2. CLUSTERS

A. Independent Production v.s. Data

When two hadrons collide at very high energies ($E \gtrsim 100$ GeV) a large number of additional hadrons are produced. Early cosmic ray studies revealed that most produced hadrons have a limited transverse momentum, relative to the beam (longitudinal) direction. The final energy is divided about equally between leading particles (those with longitudinal momenta comparable to the beam or target, as viewed in the center of mass) and secondary particles (each of which has a small fraction of the beam or target longitudinal momentum). As the incident energy increases, the average number of secondary particles also increases.

Experiments at Fermilab and the ISR have given us much more detailed information. Though the general picture from cosmic ray data remains, we have quantitative information on the energy dependence of the secondary multiplicities, as well as evidence for correlations between the produced hadrons. Of particular interest has been so called inclusive distributions, the invariant differential distributions of n secondaries, irrespective of how many other secondaries were produced.

Typical inclusive distributions are (n = 0) the inelastic cross section, σ_{inel}; (n = 1) the single particle inclusive distribution $E\frac{d\sigma}{d^3p}$; (n = 2) the two particle inclusive distribution $E_1E_2\frac{d\sigma}{d^3p_1d^3p_2}$; etc. A two point correlation, as opposed to simply a coincidence, is the difference between the observed coincidence and what one might have expected if the events were independent of each other. To be specific, if pions (which form the majority of the secondaries) were produced independent of each other, then the number of pions with momentum $\vec{p}$, per unit of invariant phase space, would be

$$\rho(p) = \frac{1}{\sigma_{inel}} E\frac{d\sigma}{d^3p} \tag{2.1}$$

The number of pions, per unit phase space, expected in coincidence at $\vec{p}_1$ and $\vec{p}_2$ would be the product $\rho(p_1)\rho(p_2)$. Thus if

$$\rho_2(p_1,p_2) = \frac{1}{\sigma_{inel}} E_1E_2\frac{d\sigma}{d^3p_1d^3p_2} \tag{2.2}$$

denotes the observed coincidence rate, then the correlation is

$$C(p_1,p_2) = \rho_2(p_1,p_2) - \rho(p_1)\rho(p_2) \tag{2.3}$$

For independently produced pions, this correlation is zero.

Experimentally the correlation is non zero and positive over some region of the phase space. Indeed there are a number of correlations which have been studied, which are positive. This situation was not expected from the cosmic ray data, nor was it anticipated from theory.

To put our later discussion in context, recall some typical theories. One of the earliest is Fermi's statistical model[2] in which the colliding hadrons are thought to produce a super heated liquid, which subsequently expands until bits of it are cool enough to condense out as pions. The pions are thought to be produced independently, as the simplest of all assumptions. Certain qualitiative details are predicted correctly. The mean multiplicity of pions grows as $E^{\frac{1}{4}}$, a power of the c.m. energy, and the transverse momentum of pions is restricted (though it grows somewhat with energy). It's also not difficult to discuss the leading particle effect in such a model. The difficulty with such statistical approaches is that there is no relativistic version, consistent with quantum mechanics, although Hydrodynamic versions have been reinvestigated recently.[3] Nonetheless, none of those approaches appear sophisticated enough yet to make predictions about correlations.

More consistent with fashion over the last decade or so are S-matrix models. One of the more important ingredients is the idea of Regge pole exchange. The earliest way of making a Regge pole is via ladder Feynman graphs using pion exchanges.[4] The discontinuity of such graphs gives the cross sections for multi-pion production. The kinematics of the ladder graphs causes the pions to be produced with a limited transverse momentum. Moreover the particles at the end of the chain take most of the energy, giving a natural explanation of the leading particle effect. Finally the multiplicity of the secondaries grows with energy ($\approx c\,\ell n\,E$). Thus the qualitiative features of the data are explained.

To be more quantitative, one must go beyond the original ABFST model, and inquire about general multiperipheral models (MPM). The ones which inspire the most confidence are those with the fewest number of parameters; i.e. the simplest versions.

We consider as the simplest class of models those in which pions are produced in a multiperipheral chain via Reggeized vector or tensor exchange. Figure 1 shows a typical chain for $pp \to pp\pi\pi\pi\pi$. The cross section for producing n pions is (aside from irrelevant constant factors)

$$\sigma_n = \frac{(gY)^n}{n!} s^{2\bar{\alpha}-2} \tag{2.4}$$

where $Y = \ell n\, s$, $s = E^2$, and $\bar{\alpha} \simeq 1/2$ is the effective vector-tensor meson intercept. The form of the cross section has a simple interpretation. Because the production is peripheral, the transverse degrees of freedom can be absorbed into the coupling g for producing a pion in the chain. One usually makes a strong ordering assumption in evaluating σ_n. Let y_i be the pion's longitudinal rapidity ($E_i = \kappa_i \cosh y_i$, $\kappa_i^2 = \mu^2 + p_{Ti}^2$, μ is the pion mass, $\vec{p}_{Ti}$ the transverse momentum; the longitudinal momentum p_L is then $\kappa_i \sinh y_i$). For a given event, order the pions' rapidities: $y_o \le y_1 \le y_2 \le \cdots \le y_n \le y_{n+1}$. Then only one Multiperipheral chain is assumed important, that with the same ordering as the pions'.

Each link in the chain contributes a Regge propagator squared factor $(s_{i\,i+1})^{2\bar{\alpha}}$ to σ_n. Since $s_{i\,i+1} = \kappa_i \kappa_{i+1} \exp\,(y_{i+1} - y_i)$

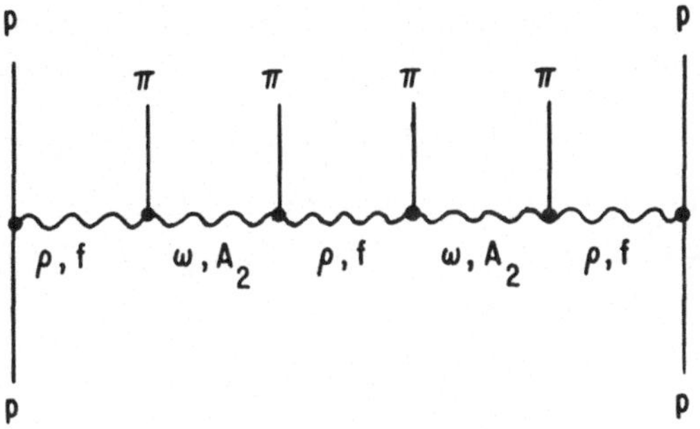

Fig. 1. Multiperipheral graph for $pp \to pp\pi\pi\pi\pi$.

for large $y_{i+1} - y_i$, the product of all such factors is just $\exp(y_{n+1} - y_o)$ times constants. The conservation of energy and longitudinal momentum is approximately satisfied by requiring $y_o \simeq -Y/2$ and $y_{n+1} = Y/2$ (in the c.m.). So, for each $y_1,\ldots,y_n$, the chain contributes

$$(g)^n \; s^{2\bar{\alpha}-2} \qquad\qquad -Y/2 \leq y_1 \leq y_2 \leq \ldots \leq y_n \leq Y/2 \; .$$

Integrating over $y_1,\ldots,y_n$ gives Eq.(2.4) , where s^{-2} comes from flux and kinematic factors.

Though we have certainly oversimplified the MPM, certain consequences follow from Eq.(2.4) which are more general than the arguments used to derive this specific form.

1) Pions have an approximate Poisson multiplicity distribution, with a mean $\langle n\rangle \sim g\ell n\, s$. This is valid as a leading logarithm approximation. A Poisson distribution has no correlation moments.

2) If Eq.(2.4) describes the main source of pions at high energy, then the constraint that $\sigma_{inel} = \sum_n \sigma_n$ is constant with energy (modulo $\ell n\, s$ factors) implies[5]

$$\alpha_P - 1 \equiv 2\bar{\alpha} - 2 + g \qquad (2.5)$$

is zero. Usually, α_P is the Pomeron trajectory intercept, and Eq.(2.5) relates α_P to $\bar{\alpha}$ and g . When $\alpha_P = 1$, then we see we must have $g = 1$ and

$$\langle n\rangle \sim \ell n\, s \quad \text{as} \quad s \to \infty \; . \qquad (2.6)$$

These consequences can be reexpressed in terms of inclusive distributions.

The single particle distribution $\rho(p)$ is expected to be constant, except when y is near $\pm Y/2$. The integral of $\rho(p)$ over $\frac{d^3p}{E} = d^2p_T dy$ gives identically $<n>$. Thus the number of pions per unit of rapidity $\frac{d\sigma}{dy}$ is approximately $<n>/\ell n\, s$, which for the poisson model (2.4) implies $\frac{d\sigma}{dy} \sim g \sim 1$. The Poisson distribution also implies that there is no correlation between the pions. Letting $C(y_1 y_2)$ be $\iint C(p_1 p_2) d^2p_{T1}\, d^2p_{T2}$,

$$C(y_1 y_2) = \frac{1}{\sigma_{inel}} \frac{d\sigma}{dy_1 dy_2} - \frac{1}{\sigma_{inel}} \frac{d\sigma}{dy_1} \frac{1}{\sigma_{inel}} \frac{d\sigma}{dy_2} \tag{2.7}$$

is expected to be zero. For future reference, the fully integrated correlation

$$f_2 = \iint \frac{d^3p_1}{E_1} \frac{d^3p_2}{E_2} C(p_1 p_2)$$

$$= \iint dy_1 dy_2\, C(y_1 y_2) \tag{2.8}$$

is related to the multiplicity moments by

$$f_2 = <n(n-1)> - <n>^2 . \tag{2.9}$$

From Eq. (2.4) it is obvious $f_2 = 0$.

We now compare these expectations with the data. Figure 2 shows a summary of $<n>$ vs $\ell n\, s$ for a variety of very high energy experiments.[6] The curvature of $<n>$ vs $\ell n\, s$ does not by itself indicate that the simplest MPM describes only a small fraction of the data; however the slope of 2 charged particles/unit rapidity

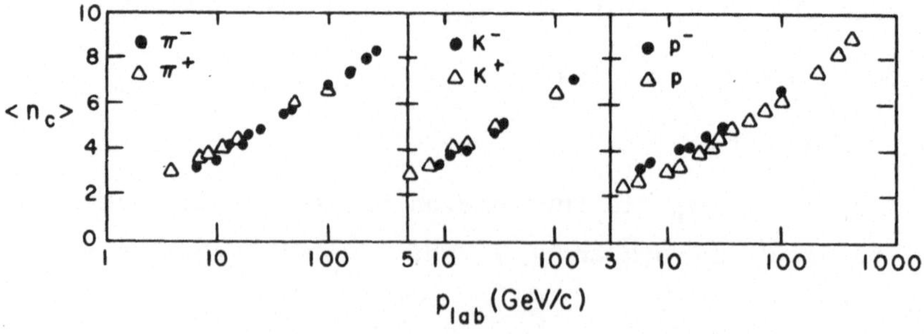

Fig. 2. Average charged multiplicity as a function of the incident beam momentum for $\pi^{\pm}p$, $K^{\pm}p$ and $p^{\pm}p$ interactions [Ref.6].

is a bad sign. This number, assuming 1 neutral for every pair of charged particles, implies $\langle n \rangle / \ell n\, s \sim 3$ asymptotically. The multiplicity distribution is also not Poisson; Figure 3 shows a compilation of f_2 vs $\ell n\, s$.[7] Equally instructive is $C(y_1 y_2)$ (Figure 4) which shows the correlation is a low subenergy effect.

The conclusion is that independent production of pions fails to explain one of the most striking characteristics of the data: the positive correlations.

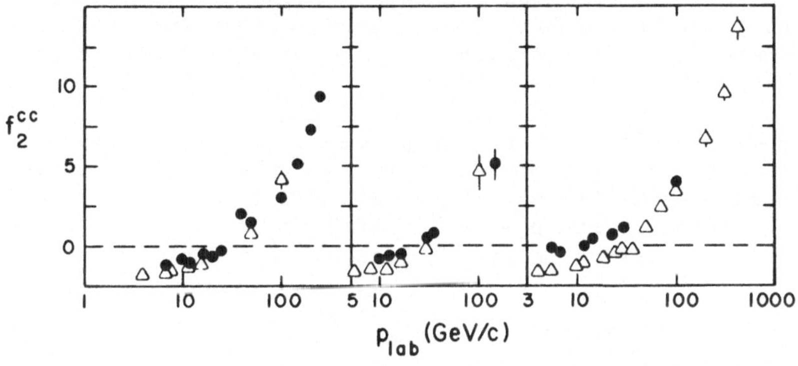

Fig. 3. $f_2^{cc} = \langle n_c (n_c - 1) \rangle - \langle n_c \rangle^2$ as a function of the incident beam momentum for $\pi^{\pm}p$, $K^{\pm}p$ and $p^{\pm}p$ interactions [Ref.7].

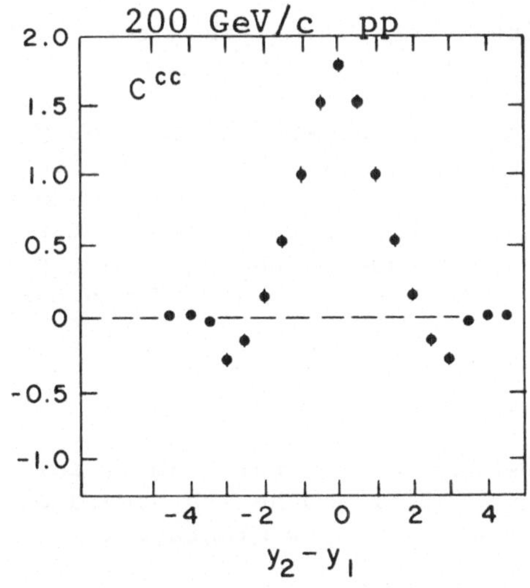

Fig. 4. Typical Fermilab correlation data between charged hadrons [Ref. 8]. Here C is plotted vs. $y_2 - y_1$ when $|y_1| \leq 0.25$.

B. Clusters v.s. Data

Without directly addressing the question about the source of the correlated pions, one can develop a simple formalism in which the above characteristic features of data can be discussed in quantitative fashion. The idea is to keep the notion of independent production, but not the idea that the pions are the objects independently produced. Instead, one imagines the production of a possibly fictitious object called a cluster. This cluster has a certain mass, charge, and pion number distributions, whose average values can be extracted from the data. These average values then become quantitative measures of the correlations observed in the data, assuming the cluster picture can be demonstrated to be compatible with experiments.

Cluster models have been proposed at various levels of sophistication. For our present purposes, the most naive formulation is adequate. We take the simple MPM discussed in Sec.2A as the mechanism for the production of clusters. Thus the production of N clusters follows a Poisson distribution

$$\sigma_N = s^{2\bar{\alpha}-2} \frac{(GY)^N}{N!} \quad . \tag{2.10}$$

Assuming the exchange forces are again $\bar{\alpha} \sim 1/2$ type trajectories, and that the inelastic cross section

$$\sigma = \sum \sigma_N \tag{2.11}$$

is constant (up to logarithms of s), then the cluster coupling G is about unity.

The clusters one produces are collections of pions. One way to express this is to assign a weight G_n for the cluster to decay into n pions. Of course weights can be introduced also to describe the cluster's mass and quantum number distributions. For this talk, we concentrate just on the pion multiplicities. The total coupling G is the sum over each of the partial weights G_n :

$$G = \sum_n G_n \quad . \tag{2.12}$$

Keep in mind our expectation that this sum is about 1 . The average number of pions produced will now be (the average number of clusters) x (the average number of pions per cluster) :

$$\frac{<n>}{Y} = (G) \times \left(\frac{\sum_n nG_n}{\sum_n G_n} \right)$$

$$= \sum_n nG_n \quad . \tag{2.13}$$

So if $G \sim 1$, then $\frac{<n>}{Y} \approx 3$ implies each cluster decays on the average into 3 pions. In similar fashion one argues that the average number of correlated pions is (the average number of clusters) x (the average number of correlated pions per cluster) :

$$\frac{f_2}{Y} = (G) \times \frac{\sum_n n(n-1)G_n}{\sum_n G_n}$$

$$= \sum_n n(n-1)G_n \quad . \tag{2.14}$$

Although the clusters have no correlations between themselves, the pions will necessarily be positively correlated. It is the positivity of the observed correlations which makes cluster models particularly attractive.

We have now constructed a simple version of a multiperipheral cluster model. The main features are illustrated by a numerical exercise. Take $G_3 \approx 1$ and all other $G_n \approx 0$. Then

1) $\sigma_{inel} \approx$ constant (for $\bar{\alpha} \sim 1/2$)
2) $<n>/Y \approx 3$
3) $f_2/<n> \approx 2$.

Of course, our purpose in using the model is to learn the average characteristics of the correlations, so we do not take seriously anything but average properties as deduced from data.

These average properties need not be just single numbers, but can be shapes of distributions. In particular, one of the first successes of cluster models was to predict the shape of the pion rapidity correlation functions $C(y_1,y_2)$.[9] The observed distribution is Gaussian in $(y_1 - y_2)$, with correlation length 2 , and only weakly dependent on $(y_1 + y_2)$. To interpret this in the cluster language, one recalls that rapidities in the central region are related just to the c.m. angle of the produced pion (relative to the beam direction). Shapes in rapidity thus gives information about the angular decay distribution of the cluster. Indeed, an isotropic cluster decay leads to the prediction of a Gaussian correlation in $(y_1 - y_2)$ and a weak $y_1 + y_2$ dependence. The predicted width (correlation length) is also about right.

A number of other experimental distributions have been investigated using the cluster model. Each gives some quantitative constraints on the cluster properties. For example, semi-inclusive correlations ($C(y_1 y_2)$ distributions for fixed charged topologies) constrain the cluster pion multiplicity.[10,11] Correlations between neutral pion multiplicity with the number of produced charged pions tells that (on the average) clusters contain both neutral and charged pions. There are many more distributions. The general situation has been reviewed elsewhere.[11]

There are some simple intuitive tests of the cluster picture one can check without doing a detailed numerical calculation. The idea is to study the distribution of inclusive events as a function of the rapidity gap r, between adjacent hadrons.[12] To be more precise, consider an event with n charged tracks (or n negative tracks; or more generally n hadrons with certain specified characteristics). Order these hadrons by their rapidity. A rapidity gap length is the difference in rapidity between two neighboring hadrons. The inclusive gap distribution is the probability that one finds two hadrons (of certain characteristics) separated by a gap r and no hadrons (with the same characteristic) in between them.

As long as the cluster decay limits the pions to be near the central position (in rapidity) of the cluster, the gap distribution at large gap sizes between hadrons measures directly the gap distribution between clusters. If the clusters are independently made, then the resultant distribution can be shown to be exponential

$$P \propto e^{-\rho r} \tag{2.15}$$

where ρ is the inclusive density of clusters. At small gap sizes, the expectation is that there will be an excess of events over the exponential behavior (2.15) due to the correlations. Figure 5 shows Fermilab data, both for gaps between charged particles, and between negative particles.

The expectations we have outlined are fulfilled. From the graph, one can read off the asymptotic slope ρ to be 1. From Eq. (2.13) this implies the mean number of pions/cluster is 3, to obtain a slope of 3 produced pions/unit of rapidity.

Similar simple arguments have been used[13] to show the clusters must carry non-zero charge some of the time. These arguments suffer from their simplicity, as kinematic effects are totally ignored. However, a more careful analysis leads to the same conclusions.[1]

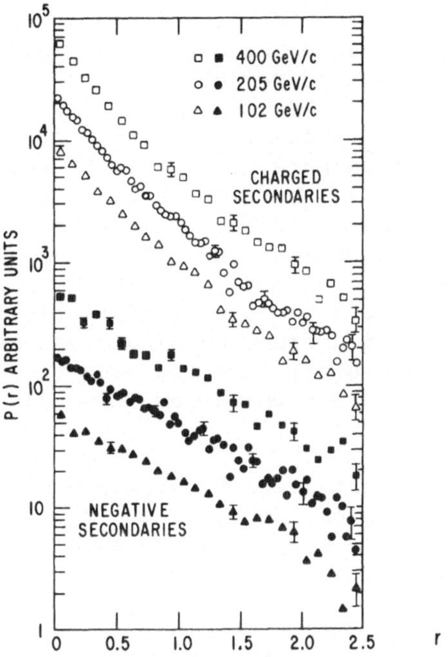

Fig. 5. From Ref. 12, the rapidity gap distribution between produced particles in 102- , 205- , and 405- GeV/c pp collisions. End gaps have been excluded from the data.

3. RESONANCE AND REGGE CHARACTERISTICS

Early measurements of $C(p_1,p_2)$ averaged over the transverse momentum of each particle, leaving only the longitudinal momentum or rapidity dependence. The correlation function one obtains,

$$C(y_1,y_2) = \int d^2p_{T1} \int d^2p_{T2}\, C(p_1,p_2) \quad , \tag{3.1}$$

will in general smear out structure which occurs at a fixed mass $M(M^2 = (p_1 + p_2)^2)$. Thus detailed information about the spectra of the clusters is lost in the study of $C(y_1,y_2)$. In fact, the situation is more severe since one can show that $C(y_1 y_2)$ depends mainly on the angular decay of the cluster and only less importantly on the mass spectra. Thus many cluster models fit the data without making any detailed assumption about the cluster masses.

It is also true that a Mueller-Regge analysis suffers when rapidity variables are used. At large $\pi\pi$ subenergies

$$M^2 = \kappa_1 \kappa_2 \exp|y_1 - y_2| \tag{3.2}$$

and the transverse momentum behavior factors. This is no longer true at small masses. Since the data exist mainly at small masses, rapidity is no longer a good variable.

So if one wants to refine the cluster picture to be the good old physics of known resonance production by known Regge exchanges, then one must look at data where the resonance spectra has not been smeared out. A step in this direction has been made with recent bubble chamber data[14] from Fermilab. One finds that resonance structure will show up more clearly in the two pion correlation function $C(p_1 p_2)$ [Eq. (2.3)] than in the inclusive density $\rho_2(p_1 p_2)$ because the incoherent combinatorial background is removed. Two pions which decay from the same resonance are contained in $C(p_1,p_2)$, whereas pions which come from different resonances are removed.

To obtain sufficient statistics, one would like to make a 1-dimensional projection of $C(p_1,p_2)$. It is argued in Ref.14 that the logical variable to use is the invaraint mass M. The 1-dimensional distribution is:

$$C(M) = \int \frac{d^3 p_1}{E_1} \frac{d^3 p_2}{E_2} \delta\left(M-\sqrt{(p_1+p_2)^2}\right) C(p_1,p_2) \quad . \tag{3.3}$$

All two pion resonances will show up in $C(M)$ as a Breit-Wigner peak; three pion resonances such as $\omega \to \pi^+\pi^-\pi^o$ will give the two pion reflection of the ω Dalitz plot; and so forth.

The correlation (3.3) is the difference of two terms:

$$C_2(M) = \rho_2(M) - \rho_1 \otimes \rho_1(M) \quad . \tag{3.4}$$

In Eq. (3.4), $\rho_2(M)$ is the usual differential cross-section $d\sigma/dM$, divided by σ_{inel}. The second term represents the combinatorial background which we subtract from ρ_2 to obtain $C_2(M)$. This product of single particle distributions at fixed pair mass is

$$\rho_1 \otimes \rho_1(M) = \int \frac{d^3 p_1}{E_1} \frac{d^3 p_2}{E_2} \delta\left(\sqrt{(p_1+p_2)^2} - M\right) \rho_1(p_1) \rho_1(p_2) . \tag{3.5}$$

As described in Ref. 14, this quantity can be directly extracted from the data.

In Fig. 6 the three distributions $\rho_2(M)$, $\rho_1 \otimes \rho_1(M)$, and $C_2(M)$ are presented as functions of the two pion invariant mass. The Figs. 6(a) and 6(b) show the results for $(\pi^+\pi^-)$ and $(\pi^-\pi^-)$, respectively. No selection is made on the location in phase space of the pair.

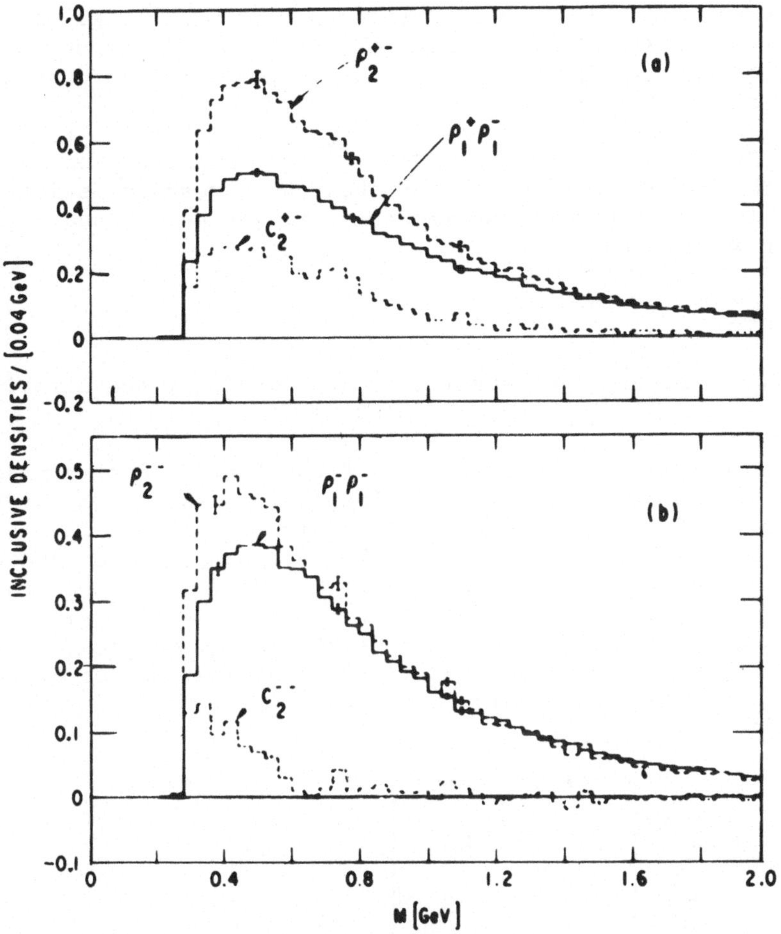

Fig. 6. Displayed as a function of dipion invariant mass are the three distributions ρ_2, $\rho_1 \otimes \rho_1$ and C_2 for (a) $\pi^+\pi^-$ pairs, and (b) $\pi^-\pi^-$ pairs from $pp \to \pi\pi X$ at 205 GeV/c.[14]

It is useful to examine first the (--) data since no resonance-like structure is expected. The distributions $\rho_2(M)$ and $\rho_1 \otimes \rho_1(M)$ are indeed rather smooth. Both show broad enhancements extending from threshold to $\sim$1 GeV. The curves are remarkably similar in shape and magnitude above 0.6 GeV, but differ in the region $M < 0.6$ GeV. The correlation function $C_2^{--}(M)$ shows structure next to the two pion threshold, with width of $\approx$ 200 MeV, and it is nearly zero for $M > 0.6$ GeV. That $C_2^{--}(M)$ is featureless within statistics for all $M \gtrsim 0.6$ GeV conforms to expectations that the exotic $(\pi^-\pi^-)$ system should show no strong correlations. The roughly constant value of $C_2^{--}(M) \approx 0$ above 0.6 GeV also confirms that the definition of the correlation function is reasonable.

The data for (+-) combinations are shown in Fig. 6(a). The distribution $\rho_1^+ \otimes \rho_1^-$ is again smooth, whereas ρ_2^{+-} shows structure in the rho region. The two curves do not coincide, in contrast to (--), until $M \gtrsim 1.6$ GeV.

In Figure 7, the correlations $C^{+-}(M)$ are shown on the same scale. The amount of the ρ contribution is readily estimated as the bump in Fig. 7 above some smoothly falling background curve. The value in Ref. 14 is

$$\langle n_{\rho^0} \rangle = 0.30 \pm 0.04 \; . \tag{3.6}$$

No f is observable, but statistics are possibily the limiting factors.

The striking characteristic of these data is that a major fraction of the correlation lies below the ρ mass. One might hope that $C^{+-}(M)$ is due mainly to the superposition of a ρ and the two pion reflection of the ω (which presumably does account for some of the low mass structure of $C^{+-}(M)$). This expectation predicts that the $\pi^-\pi^-$ correlation, $C^{--}(M)$, is negligible. These expectations are modified somewhat by heavier resonances, such as A_2 and f ; since such resonances are heavier they may be expected to be produced less copiously than ρ and ω .

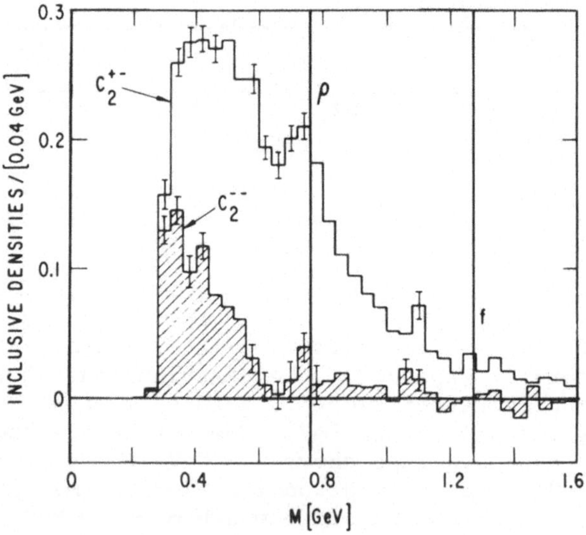

Fig. 7. The correlation functions $C_2(M)$ are displayed versus dipion invariant mass.[14] Shown are results for (+-) and (--) pairs. The nominal locations of the ρ ($M = 0.76$ GeV) and f ($M = 1.27$ GeV) are marked.

To be more quantitative, note that a generous estimate for the ρ production is that it accounts for 10% of all of $C^{+-}(M)$. [The integrated value of C^{+-} is 4.0 ± 0.2; the result follows from Eq. (3.6)]. Allowing an equal amount for ω, though none can be directly observed, the two resonances would account for 20% of $C^{+-}(M)$. One could try model estimates for heavier resonances. For this discussion, let us simply guess that an additional 10% accounts for all heavier (known) resonances. One thus might explain 30% of $C^{+-}(M)$ in terms of known resonances. This then raises the question of what causes the remaining 70%.

Of course maybe the ρ estimate is very wrong. Note that the estimate of 10% is the amount of ρ above background. The estimate depends crucially on the shape of the background below the ρ and possible interferences. If the background is not slowly varying, or if the data were to change, the estimate of the ρ could be substantially different. Notwithstanding this possibility let us accept the data at face value. The question then becomes what other sources of correlation are there besides known two pion resonances and reflections from known m pion resonances, $m \geq 3$. We do not consider the possibility of inventing low mass (Mass $\approx$ 500 MeV) resonances which decay into several coherent pions (three or more).

Whatever these other sources of correlations are, we can enquire whether they have reasonable Mueller-Regge properties.[14] The remainder of this section is taken directly from Ref. 14.

In the Mueller approach to inclusive correlations, the two-particle correlation function $C_2(p_1,p_2)$ is a properly defined discontinuity ("imaginary part") of a four-particle to four-particle forward scattering amplitude.[15] At high energies, the Mueller-Regge approximation for $pp \to \pi\pi X$ is sketched in Fig.8(a). This graph applies to dipion production in the central region of rapidity space. The shaded oval represents a scattering amplitude from which the leading Pomeron exchange is excluded. Note that the shaded oval does not represent the $\pi\pi$ elastic amplitude. Although the Pomerons are each attached to the oval at zero four-vector momentum, the structure of the oval is still that of an amplitude with six external legs. Consequently, there is no general reason to suppose that the dipion mass dependence of Fig.8(a) should closely resemble that of elastic $\pi\pi$ scattering. In particular, in $C_2(M)$ for $pp \to \pi^+\pi^- X$, Fig.8(a) leads us to expect resonance signals at the $\rho, f, g\ldots$ positions, as well as a background. However, the relative contributions of the different resonant partial waves, and the resonance to background ratio need not be related simply to that measured in studies of $\pi^+\pi^-$ elastic scattering. The data in Fig.7 suggest in fact that the ρ resonance to background ratio in the M dependence of $pp \to \pi^+\pi^- X$ is much smaller than in the elastic $\pi^+\pi^-$ amplitude. If true, this is quite striking.

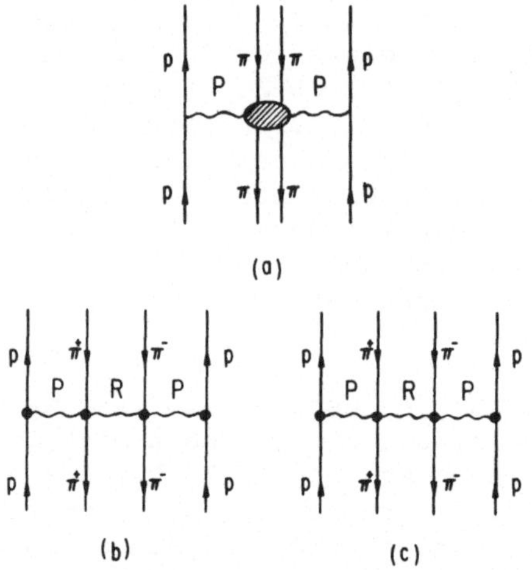

Fig. 8. a) Mueller-Regge diagram for the inclusive reaction $pp \to \pi\pi X$ where the dipion system is produced in the central region of rapidity. The symbol P denotes the Pomeron. b) Mueller-Regge diagram for $pp \to \pi^+\pi^- X$ at large s and large mass M of the $\pi^+\pi^-$ pair. c) Mueller-Regge diagram for $pp \to \pi^-\pi^- X$ at large s and large mass M of the $\pi^-\pi^-$ pair.

As an attempt to obtain a simple parametrization of the M dependence of $C_2(M)$, consider the behavior of Fig.7(a) expected at large M, where a Regge exchange approximation may be used. The resulting graphs are shown in Fig.8(b) and 8(c) for $pp \to \pi^+\pi^- X$ and $pp \to \pi^-\pi^- X$, respectively. For $pp \to \pi^+\pi^- X$, the leading exchanged Reggeon R in Fig.8(b) is the (ρ, f) pair, with intercept $\alpha_R(0) \simeq 0.5$. For $pp \to \pi^-\pi^- X$, somewhat more discussion is necessary.

In two-body phenomenology, duality arguments suggest that the Regge exchanges leading to an exotic system such as $\pi^-\pi^-$ occur in exchange degenerate pairs, and that their contributions cancel in the imaginary part of the scattering amplitude. Applying analogous arguments to $pp \to \pi^-\pi^- X$, we expect that the exchange labeled E in Fig.8(c) is either a low lying trajectory or cut (e.g. $\alpha_E(0) \lesssim 0$) or a normal trajectory (e.g. α_ρ) whose small coupling at $t = 0$ is related to the deviation from exact exchange degeneracy. One admits parenthetically that the concept of an exchange contribution to the discontinuity of the forward amplitude may be altogether wrong in the case of the $\pi^-\pi^-$ channel, but one knows of no other way to proceed.

Assuming that the exchanges in Figs.8(b) and 8(c) are factorizable, the following limiting expressions at large M are obtained in Ref. 14:

$$C_2^{+-}(p_1,p_2) = \beta_{RP}(m_{T1})\ \beta_{RP}(m_{T2})(M^2)^{\alpha_R(0)-1} \quad ; \qquad (3.7)$$

$$\propto M^{-1} \quad , \text{ as } \quad M \to \infty \ .$$

$$C_2^{--}(p_1,p_2) = \beta_{EP}(m_{T1})\ \beta_{EP}(m_{T2})(M^2)^{\alpha_E(0)-1} \quad . \qquad (3.8)$$

The transverse mass m_T is $\sqrt{p_T^2 + \mu^2}$; β_{RP} and β_{EP} are vertex functions whose m_T dependence may be estimated from data on the single pion inclusive yield $\rho_1(p)$. In the central region, the Mueller-Regge approach provides the expansion

$$\rho_1(p) = \beta_{PP}(m_T) + \beta_{RP}(m_T)s^{-1/4} \quad . \qquad (3.9)$$

This equation shows that the vertex functions yield the transverse momentum damping of $\rho_1(p)$.

As is apparent from Fig.7, available data are concentrated at small M where Eqs.(3.7) and (3.8) are not obviously applicable. Inasmuch as one is working with the imaginary parts of amplitudes, one might hope nevertheless that the extrapolation of Eqs.(3.7) and (3.8) to small mass is meaningful in the sense of a dual average. Since the ρ signal is not prominent in $C_2^{+-}(M)$, the average is not difficult to obtain.

Equations (3.7) and (3.8) were compared with the data at low M in Ref. 14 to extract effective values for $\alpha_R(0)$ and $\alpha_E(0)$. The proper comparison should be done at fixed values of all kinematic variables other than M . Present statistics do not permit such a differential comparison. Integrating Eq.(3.7) , one obtains

$$C_2^{+-}(M) = (M^2)^{\alpha_R(0)-1} \iint dp_1 dp_2 \beta_{RP}(m_{T1})\beta_{RP}(m_{T2})\delta(M-\sqrt{(p_1+p_2)^2}) \ . \qquad (3.10)$$

Here dp represents the invariant integration element d^3p/E . The integrations over the p_T and ϕ variables and over the longitudinal position of the dipion pair introduce an added dependence on M through the delta function in Eq.(3.10).

It is a simple matter to show analytically that as $M \to 2\mu$, the integral in Eq.(3.10) provides a threshold suppression factor proportional to $(M^2-4m_\pi^2)^{1/2}$. By contrast, at large M , the transverse momentum dependence of the vertex functions in Eq.(3.10) results in a function which falls rapidly with M . In order to extract the effective Regge power $\alpha_R(0)$ from the data on $C_2^{+-}(M)$, it is first necessary to divide out these important threshold and large M factors. In the absence of a detailed knowledge of $\beta_{RP}(m_T)$, this procedure is necessarily approximative.

Returning to Eq.(3.9), note that at large s, the Mueller-Regge model provides the following prescription for the M dependence of the product $\rho_1^+ \otimes \rho_1^-$:

$$\rho_1^+ \otimes \rho_1^-(M) = \iint dp_1 dp_2 \beta_{PP}(m_{T1})\beta_{PP}(m_{T2})\delta(M-\sqrt{(p_1+p_2)^2}) \quad . \tag{3.11}$$

This integral has exactly the same form as that in Eq.(3.10), except for the replacement of $\beta_{RP}(m_T)$ by $\beta_{PP}(m_T)$. If it is assumed that the m_T dependences of β_{RP} and β_{PP} are identical, one may divide Eq.(3.10) by Eq.(3.11) to obtain

$$R^{+-}(M) \equiv \frac{C^{+-}(M)}{\rho_1^+ \otimes \rho_1^-(M)} \propto (M^2)^{\alpha_R(0)-1} \quad . \tag{3.12}$$

Likewise,

$$R^{--}(M) \propto (M^2)^{\alpha_E(0)-1} \quad . \tag{3.13}$$

Alternatively, assuming that β_{RP} in Eq.(3.7) has the same m_T dependence as β_{EP} in Eq.(3.8), one may work directly with Eq.(3.10) to derive

$$C_2^{--}(M)/C_2^{+-}(M) \propto (M^2)^{\alpha_E(0)-\alpha_R(0)} \quad . \tag{3.14}$$

Motivated by Eq.(3.12), the authors of Ref.14 displayed in Fig.9 $R^{+-}(M)$ on a logarithmic scale. Beyond $M \simeq 2.5$ GeV, the errors are large, and therefore the data are not shown. The form of Eq.(3.12) requires that the data points in Fig. 9 should fall on one straight line. Quite obviously they do not. An excursion is visible in the ρ region, as expected. For M values below the ρ position, a rough fit[14] to the data suggests an effective trajectory with rather high intercept $\alpha(0) \simeq 0.6$. For $M \gtrsim 1$ GeV, the intercept is much lower, $\alpha(0) \simeq 0.35$. In view of the approximations made in Ref.14, it is hazardous to propose strong conclusions. However, attention was called to the fact that the Mueller-Regge expectation of $\alpha(0) \simeq 0.5$ lies well within the range of values of $\alpha(0)$ which provide an acceptable average fit to the data in Fig. 9.

In Fig.10, the ratio C_2^{--}/C_2^{+-} is displayed, as suggested by Eq.(3.14). Here the statistics preclude an examination of the M

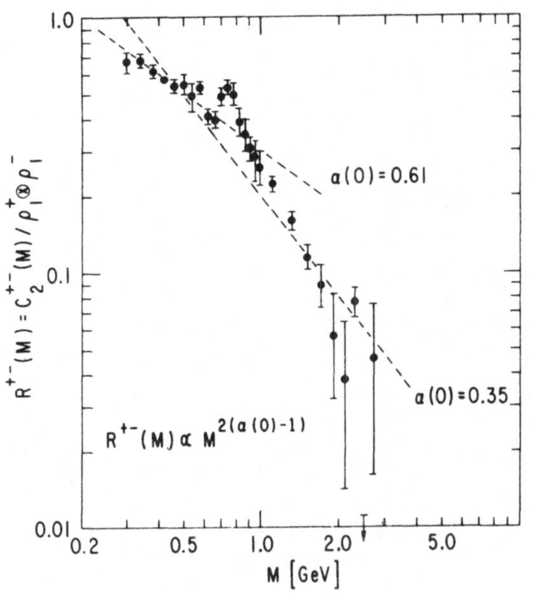

Fig. 9. From Ref. 14, the mass dependence of the ratio $R^{+-}(M) = C^{+-}(M)/\rho_1^+ \otimes \rho_1^-(M)$ for $pp \to \pi^+\pi^- X$ at 205 GeV/c. Two straight lines are drawn. They correspond to two choices of the parameter $\alpha(0)$ in the expression $R^{+-}(M) = g M^{2(\alpha(0)-1)}$ suggested by Mueller-Regge analysis.

dependence for $M \gtrsim 1$ GeV. Shown for comparison, is the M^{-2} dependence expected if $\alpha_R(0)-\alpha_E(0) = 1$ in Eq.(3.14). It was observed that this form is consistent with the fall of C_2^{--}/C_2^{+-}. In the Mueller-Regge framework this result suggests that the process $pp \to \pi^-\pi^- X$ is mediated by a rather low lying exchange, with intercept $\alpha_E(0) \simeq -0.5$. The duality expectation is borne out, in that the exchange degenerate pair of ρ and f with intercept $\alpha(0) \simeq 0.5$ do not contribute to the M dependence of $pp \to \pi^-\pi^- X$. This result[14] is an important verification that standard duality notions are applicable in inclusive processes. It stands in contrast to the list of embarassing failures associated with early scaling criteria.[15]

In terms of rapidity, the Mueller-Regge expectation for the correlation function C^{--} is

$$C^{--}(\Delta y) \propto \exp[(\alpha_E(0) - \alpha_P(0))\,|\Delta y|]. \tag{3.15}$$

The low intercept $\alpha_E(0) \simeq -0.5$ provides a correlation length $\lambda^{--} \equiv (\alpha_E - \alpha_P)^{-1} \simeq 2/3$, much shorter than the (+-) value $\lambda^{+-} \simeq 2$.

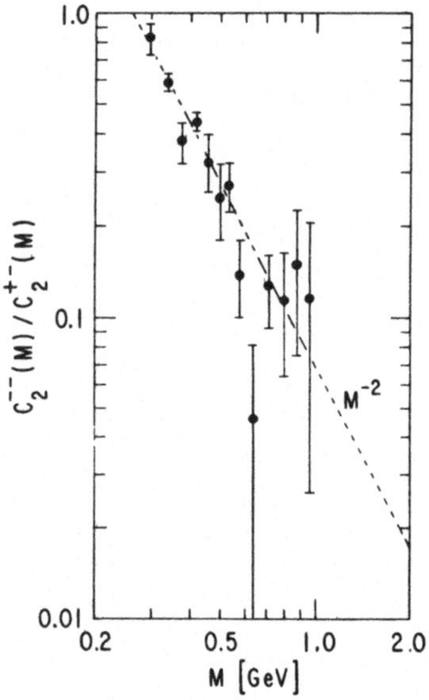

Fig. 10. From Ref.14, the mass dependence of the ratio $C_2^{--}(M)/C_2^{+-}(M)$. The dashed line is drawn to show the M dependence expected if the ratio is proportional to M^{-2}. It is not a fit to the data.

The fact that $\lambda^{--} \simeq \frac{1}{3}\lambda^{+-}$ is to be contrasted with expectations of cluster models in which typically $\lambda^{--} \simeq \lambda^{+-}$. Previously a small value of λ^{--} was suggested, based upon the observation[16] of a sharp peak near $\Delta y = 0$ in the rapidity variation of the two dimensional distribution $C^{--}(\Delta y, \phi)$ <u>for values of ϕ near zero</u>. The extraction[14] of a small value of λ^{--} from the one-dimensional $C(M)$ is perhaps more direct. However, the two observations are surely related since small M and small $(\Delta y, \phi)$ are correlated kinematically.

An examination of the absolute magnitudes of $C_2^{+-}(M)$ and $C_2^{--}(M)$ was also discussed in Ref. 14. The structure of Eq.(3.10) shows that

$$\frac{C_2^{--}(M)}{C_2^{+-}(M)} = \left(\frac{\bar{\beta}_{EP}}{\bar{\beta}_{RP}}\right)^2 M^{-2} \quad . \tag{3.16}$$

In obtaining Eq.(3.16) the vertex function $\bar{\beta}$ has been averaged over p_T, and $\alpha_R(0) - \alpha_E(0) = 1$. The data in Fig.10 provide the estimate

$$\left(\frac{\bar{\beta}_{EP}}{\bar{\beta}_{RP}}\right)^2 \simeq \frac{1}{15} \quad , \quad \text{or} \quad \left|\frac{\bar{\beta}_{EP}}{\bar{\beta}_{RP}}\right| \simeq \frac{1}{4} \quad .$$

It was noted that the exchange denoted E couples relatively weakly, at least in the imaginary part of the inclusive amplitude for $pp \to \pi^-\pi^-X$. Whether E is a factorizable singularity (or pair of singularities) which plays a role in the $\pi^-\pi^-$ elastic amplitude is open to question. However, presumably α_E does contribute to $pp \to \pi^+\pi^-X$ in much the same fashion as in $pp \to \pi^-\pi^-X$. Therefore, a consistent phenomenological study of $C_2^{+-}(M)$ would require a reanalysis of its M dependence in terms of both $\alpha_R(0)$ and $\alpha_E(0)$. The data was not judged to warrant such a detailed treatment.

Turning to $pp \to \pi^+\pi^-X$, and using Eqs.(3.7) and (3.9)-(3.12), one may express $R^{+-}(M)$ as

$$R^{+-}(M) = \left(\frac{\bar{\beta}_{RP}}{\bar{\beta}_{PP}}\right)^2 M^{2(\alpha_R(0)-1)} \quad . \tag{3.17}$$

The data in Fig.9 provide the estimate $(\bar{\beta}_{RP}/\bar{\beta}_{PP})^2 \simeq 0.25 \pm 0.05$. Recognizing that the trajectory R represents the (ρ,f) exchange degenerate pair, whose couplings are presumed to be equal in magnitude, it was deduced that[14]

$$\left|\frac{\bar{\beta}_{fP}}{\bar{\beta}_{PP}}\right| \simeq \left|\frac{\bar{\beta}_{\rho P}}{\bar{\beta}_{PP}}\right| \simeq 0.35 \pm 0.04 \quad . \tag{3.18}$$

Fits to single particle spectra[17] provide values of these ratios in the range 0.4 to 0.7 . While these different estimates are in fairly close agreement, they suggest that the size of $C_2^{+-}(M)$ is somewhat smaller ($\sim 50\%$) than expected in the Mueller-Regge framework based on factorization and the properties of single particle inclusive spectra.

As a final remark from Ref.14, a comment was made on the energy dependence of $C_2(M)$. In all short-range order models, $C_2(p_1,p_2)$ is expected to approach an energy independent constant value as $s \to \infty$. This remains true if one integrates over all variables except M and $y_{\pi\pi}$. If the dipion system is distributed uniformly in rapidity, the integral over $y_{\pi\pi}$ is proportional to $\ell n\, s$. Therefore, one may expect that $C_2^{\pm-}(M)/\ell n\, s$ will become independent of s as $s \to \infty$. It would be useful to check this expectation with data at other energies.

4. COHERENT INTERFERENCE PHENOMENA

Our study so far has not been able to locate the source of the pion correlations (or equivalently, we can't yet define what a cluster is). One mechanism we have not looked into is that of interference effects.

As argued in Ref.14, the use of the mass variable is also advantageous if one wishes to investigate correlations due to identical particle symmetry (Bose-Einstein symmetry for pions). When two identical pions have nearly equal momenta p_1 and p_2, a quantum mechanical confusion can occur. A measure of when the momenta are close is given by $(p_1 - p_2)^2$, the invariant magnitude of the difference. Note that

$$(p_1 - p_2)^2 = 4\mu^2 - M^2 \tag{4.1}$$

so that the region of confusion $p_1 \approx p_2$ is when $M \approx 2\mu$. Thus some of the low mass enhancements in Fig.7 could in principle be due to this phenomenon.

The purpose of the present section is to explore the possibility that some of the low mass structure is due to quantum mechanical interferences, and to give an estimate by means of a simple model, of the size and range one might reasonably expect for such effects.[18] To anticipate our conclusion, we do not find a sufficiently large contribution to allow for an understanding of the $C(M)$ data. There still remains a large fraction of $C(M)$ unexplained. The amount we do find is largest for $C^{--}(M)$, (though there is some effect in $C^{+-}(M)$) and may be observable, though present data give no unambiguous evidence for its existence. It seems evident that more data are needed to clarify the mechanisms which are responsible for the observed correlations. Particularly data from πp reactions as well as pp data at higher energies might prove useful.

One strong motivation for studying interference phenomena as an important source for correlations is the large literature on the subject, and the conclusion by some that the effect might account for all the $\pi^-\pi^-$ correlations. To our knowledge, there are no analytic calculations in the literature of the interference effect for inclusive distributions which attempt to estimate how big the effect might be. This has suggested our doing an explicit calculation. Before turning to the calculation, we shall first try to describe the interference phenomena we are trying to calculate, and how the calculation relates to similar ideas in the literature.

The standard example of what we will calculate is the Hanbury-Brown and Twiss[19] effect in astronomy. There, photons (Bose particles) detected in coincidence, which have been emitted from a single source (a sun) will show correlations which depend on the size of the emitter. These correlations are sometimes called second order interference phenomena. A good discussion of the effect for pions is given by Goldhaber, Goldhaber, Lee and Pais.[20]

Two like pions ($\pi^-\pi^-$ say) produced in a localized region V of space, radius R must be described by a symmetric wave function $\psi_{p_1p_2}(x_1x_2)$ where p_1 and p_2 are the momenta, x_1 and x_2 the coordinates and ψ the superposition of plane waves describing the pions inside the region V. The probability $P(p_1p_2)$ of observing the pions with momenta p_1 and p_2 is $\int_V dx_1dx_2|\psi|^2$ integrated over the region V. The result is, up to constants;

$$P(p_1,p_2) \propto 1 + e^{-\lambda|(p_1-p_2)^2|} \qquad (4.2)$$

The parameter λ turns out to be of order R^2. In the discussion, we have not been careful about whether p is a 3-vector or 4-vector: for 3-vectors the discussion is simply carried out; for applications, Eq.(4.2) is the correct expression with p_i being 4-vectors and $(p_1 - p_2)^2$ given by Eq.(4.1).

The result of saying that identical particles are produced from a local region V is that when $(p_1 - p_2)^2$ is small compared to $1/R^2$ the probability $P(p_1p_2)$ is enhanced. The second term in Eq.(4.2) is thus precisely a measure of the correlation in the same sense as Eq.(2.3). The argument we have followed can be called a space-time argument. It does not reliably estimate the size of the effect, but tells us the range of the effect is specified by R. A reasonable guess for R in hadron collisions is 1 fermi, or to the same accuracy, $1/\mu$, the pion compton wave length. Thus in the variable M, one expects the interference phenomena to extend from threshold $(M = 2\mu)$ to (about) one pion mass above threshold $(M \simeq 3\mu)$. A large literature has accumulated on the space-time viewpoint of the interferences.[21] As with the applications in astronomy, the idea is to measure the properties of the volume in which the pions are made. By studying both transverse and longitudinal momenta $\pi\pi$ correlations, one hopes to measure both the transverse and longitudinal hadron interaction volume radii.

It is possible to take a completely different point of view from the space time approach.[22] That is what one could call the S-matrix point of view. In this point of view, one works entirely in momentum space. Pions are correlated in production by specific dynamics. For example, production of ρ mesons makes correlated pions. Indeed, if one just considered resonance production, but allowed the amplitudes for different resonance to interfere, it is possible for two identical pions to display the same second order interference phenomena discussed in the space-time language. This ideas is treated in detail in Refs.21. To create a resonance with width Γ necessarily implies the produced pions "live together" a certain time $1/\Gamma$. If two resonances have similar momenta, so will the decay products: all the products will have come from a localized volume of space, and coherent interference effects can result. Here Γ sets the scale for the range of the effect. We will discuss this idea in more detail below, as it forms the basis of our calculation.

As a final comment on the current literature, we note that though interference effects can in fact exist for both like and unlike charged pion pairs, one usually expects the effect to be more dramatic for like pions since Bose-Einstein symmetry requires the interferences to be coherent. Thus one often calls the interference a Bose, or Bose-Einstein effect. Claims have been made for the existence of the Bose effect and for phenomenological support (Eq. Ranft, Ref. 22) . Our point of view is that in those cases one has used data plotted in rapidity, so one could not separate out that part of the correlation whose range was greater than (the reasonable expectation of) 1 pion mass above threshold. In this sense the use of the $\pi\pi$ mass as one of the variables makes the analysis cleaner. Similarly calculations of $C(p_1 p_2)$ which do not study the resultant mass dependence may be in serious disagreement with data.

Our calculation of the size of coherent interference effects depends on knowing the ρ experimental inclusive cross section. There is presumably also ω production, but its explicit addition would make the present calculation unmanageable. One might however expect a similar ratio for $\pi\pi$ interferences due to ω production, as is due to ρ . Indeed some accounting is made for π's which do not result from ρ decay. We assume that some of the π's one produces are made independently. With these reservations, we assert that if there is ρ production, then the resultant $\pi\pi$ correlations will show not only a ρ resonance signal, but also a threshold enhancement due to coherent interferences of the produced ρ's and π's .

The way in which we will generate the interference effect can be seen already in the $\pi_1^-\pi_2^+\pi_3^-$ final state.[23] We assume a ρ resonance in each $\pi^+\pi^-$ combination. The amplitude for three pions is

$$A_3(p_1,p_2,p_3) = G_2^{-+}(p_1,p_2)G_1^-(p_3) + G_2^{+-}(p_2,p_3)G_1^-(p_1) \ . \quad (4.3)$$

The square of A_3 integrated over p_2 is proportional to the rate of observing two negative pions in coincidence. The rate has two types of terms: 1) the absolute square of each term in Eq.(4.3) in which p_1 and p_3 are not correlated and 2) the interference term which correlates p_1 and p_3. We wish to calculate the mass dependence of the interference term, which we call $C(M_{13})$ since it is, up to a constant, the contribution from 3π production to the inclusive correlation. The interference term $C(M_{13})$ is given explicitly as:

$$C(M_{13}) = \int \frac{d^3p_1}{E_1}\frac{d^3p_2}{E_2}\frac{d^3p_3}{E_3}\, 2M_{13}\,\delta\left((p_1+p_3)^2-M_{13}^2\right)$$
$$\times\, 2\,\mathrm{Re}\left\{G_2^{-+}(p_1,p_2)G_1^{-}(p_3)G_2^{+-}(p_2,p_3)^*G_1(p_1)^*\right\}. \qquad (4.4)$$

Our conclusions will be biased because we follow tradition here and ignore local energy momentum conservation for centrally produced pions. Since the momenta of the pions are meant to be small (in the overall c.m. frame) this is not a strong bias.

To get a feeling for the mass dependence of $C(M_{13})$, we imagine that $G_1^{\pm}(p)$ is real and a function only of the transverse momentum p_T, and that $G_2^{-+}(p_1,p_2) = G_2^{+-}(p_2,p_1)$ is a function only of the dipion mass M_{12}. The expression for $C(M_{13})$ then factors:

$$C(M_{13}) = 2M_{13}H_o(M_{13})P_o(M_{13}) \qquad (4.5)$$

where

$$H_o(M_{13}) = 2\mathrm{Re}\left\{\int \frac{d^3p_2}{E_2}\, G_2^{+-}(M_{12})G_2^{-+}(M_{12})^*\right\}, \qquad (4.6)$$

and

$$P_o(M_{13}) = \int \frac{d^3p_1}{E_1}\frac{d^3p_3}{E_3}\, G_1(p_1)G_1(p_3)\,\delta((p_1+p_3)^2 - M_{13}^2)\,. \qquad (4.7)$$

Each of these factors has a comprehensible behavior which follows from general characteristics of G_1 and G_2. It can be shown that

$$H_o(M_{13}) = \frac{\pi}{M_{13}\sqrt{M_{13}^2-4\mu^2}}\iint_A dM_{12}^2\,dM_{23}^2\; \mathrm{Re}(G_2(M_{12})G_2(M_{23})^*) \qquad (4.8)$$

where μ is the pion mass. The allowed area A is expressed simply

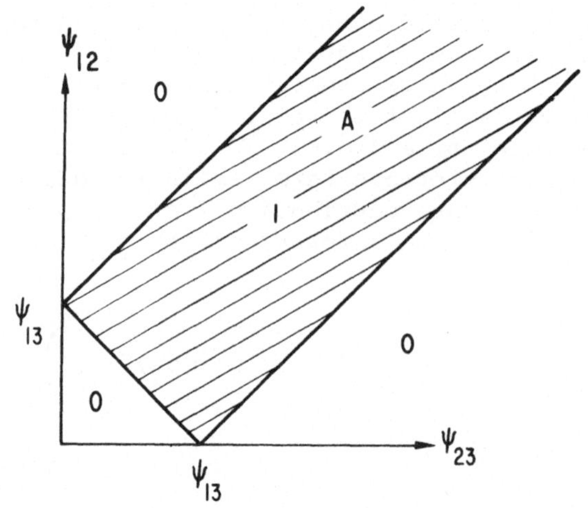

Fig. 11. The allowed region for M_{13} in the M_{12}-M_{23} plane for a 3π final state is presented in terms of the boost angles ψ_{ij} [Eq.4.9].

in terms of the Lorentz boost angles ψ_{ij} defined by

$$M_{ij}^2 = 2\mu^2 + 2\mu^2 \cosh\psi_{ij} \; . \tag{4.9}$$

The region A has

$$\Theta(\psi_{12}+\psi_{23}-\psi_{13}) - \Theta(\psi_{12}-\psi_{23}-\psi_{13}) - \Theta(\psi_{23}-\psi_{12}-\psi_{13})$$

non-zero. This function is symmetric in ψ_{12}, ψ_{23} and ψ_{13} . For a given ψ_{13} , the allowed region is displayed in Fig. 11. When M_{13} approaches threshold, the allowed region A approaches the line $\psi_{12} = \psi_{23}$ with vanishing area. Thus $M_{13}\sqrt{M_{13}^2-4\mu^2}\, H_o(M_{13})$ must necessarily vanish, assuming G_2 is sufficiently damped at large mass. Away from threshold H_o suddenly gets a large contribution since the allowed band A cuts the region where the resonances overlap. As M_{13} continues to increase, $M_{13}\sqrt{M_{13}^2-4\mu^2}\, H_o$ as given by (4.8) would tend to a constant until the region A ceased to intersect the resonance overlap region, in which case H_o vanishes. Thus H_o will necessarily have a peak near threshold whose position is a function of the resonance width.

We note that when G_2 is given by a Breit-Wigner form, the real and imaginary parts will behave somewhat differently. The imaginary part of G_2 contributes as described above. The real

part of G_2 gives only a small contribution to $H_o(M_{13})M_{13}\sqrt{M_{13}^2-4\mu^2}$ as soon as M_{13} is a couple of resonance widths above threshold. This is because the integral in (4.8) becomes approximately the product $|\int dM^2 G_2(M^2)|^2$. Since $\mathrm{Re}G_2$ is antisymmetric about the resonance peak, it gives no contribution to this region of M_{13}^2 .

The second factor P_o of (4.5) describes the behavior of the phase space. If there were no p_T damping, $[G_1 \equiv 1]$ then $P_o(M_{13}) \propto M_{13}\sqrt{M_{13}^2-4\mu^2}$. With p_T damping, $P_o(M)$ shows a more gentle behavior, decreasing to zero for sufficiently large M .

As a specific example, consider $G_1 = e^{-R^2 p_T^2}$. One can then show that $P_o(M_{13})$ has the approximate form

$$P_o(M_{13}) \propto Y\Theta(-\gamma - \ln\frac{\mu^2 R^2}{4} - \psi_{13}) \tag{4.10}$$

$$\gamma = 0.57721...$$

$$Y = \ln s \ ,$$

except very near $\psi_{13} = 0$, in which case $P_o(M_{13}) \propto \psi_{13}$.

The resultant expectation for $C(M_{13})$ is for a threshold enhancement proportional to $H_o(M_{13})M_{13}$. Thus we see that the mechanism of resonance production and transverse momentum damping of pions together can cause a threshold enhancement. The mechanism is the same as that causing the second order interference effects discussed in refs. 19-22. The region where the effect is strongest is characterized by the resonance width.

So far we have considered only the $\pi^+\pi^-\pi^-$ contribution to the inclusive correlation function. In Ref. 18, the contributions from any number of pions are summed analytically to obtain $C(p_1 p_2)$. The mass projections are presented in Figs. 12 and 13. The relevant result is the ratio of the coherent enhancements to ρ^o production. On the one hand three body resonances have been left out, so the model is not expected to fit $C(p_1 p_2)$ directly. On the other hand the ratios may be expected to be more reliable.

The area under C_2^{+-} for $0.28 \le M < 0.42$ GeV is 0.09; for the same mass range the area under C_2^{--} is 0.15. These numbers are to be compared with the mean number of produced ρ^o's in the model of 0.32. This is the area under the ρ^o peak in C_2^{+-} , taking the mass interval 0.7 - 0.9 GeV and a background estimate of 0.15. The total f_2^{+-} of 0.9 is the sum of the ρ , background under the ρ , threshold effect, and a contribution of 0.34 which lies outside the ρ^o region and outside the threshold again ($M \lesssim 0.42$ GeV). Our

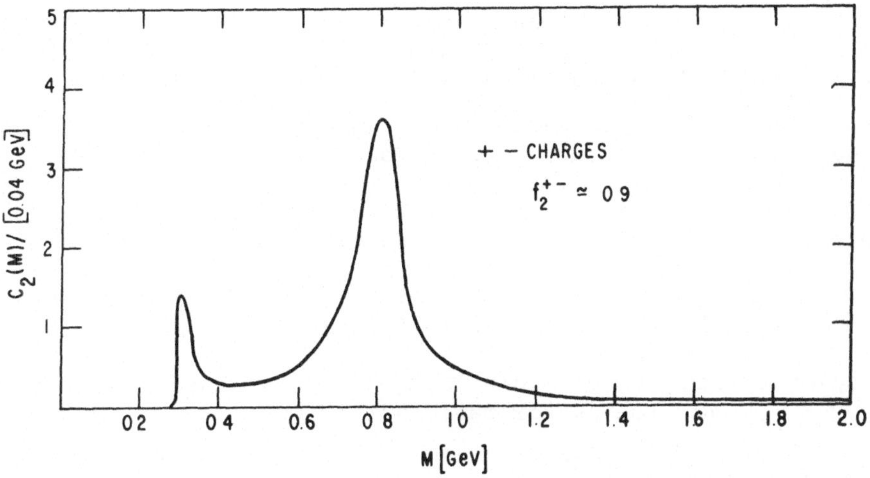

Fig. 12. The model calculation of $C_2^{+-}(M)$ (the integral of $C_2(p_1,p_2)$ for fixed invariant mass in bins of 40 MeV) as described in the text, (Cf. Ref. 18).

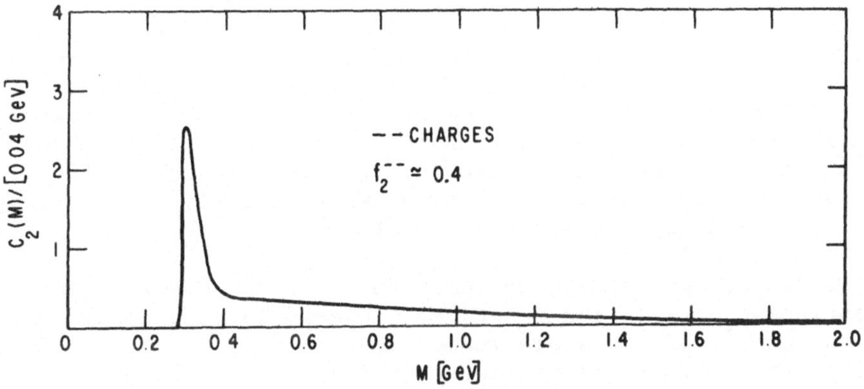

Fig. 13. The model calculation of $C_2^{--}(M)$ as described in the text, (Cf. Ref. 18).

estimate is that the fraction of threshold pions per produced ρ^o is $0.09/0.32 = 0.28$ for +– and $0.15/0.32 = 0.47$ for –– . There are almost twice as many threshold pairs in –– compared to +– .

If we take as an experimental value 0.25 for the number of ρ's per inelastic event seen in 205 GeV/c pp collisions, we estimate a 0.07 contribution to the experimental f_2^{+-} and a 0.12

contribution to f_2^{--} from the threshold region. The effect is respectively 2% and 20% of the experimental totals $f_2^{+-} = 4.0$ and $f_2^{--} = 0.8$. Thus the interference effects are not expected to be large. This is also consistent with the shapes of $C_2^{\pm -}(M)$ in the data (Fig.7). The experimental distributions are broader than expected if they were due solely to interferences of resonances.

5. CONCLUSIONS

In the course of these discussions, various aspects of high energy hadron correlations have been discussed. We have seen that the cluster picture can provide an average description of data, but fails on the details. The most striking failure is for mass distributions, where no straight forward accounting can be made for why a large fraction of the $\pi\pi$ correlation lies below the ρ mass. Clusters don't tell us why the pions are correlated since we don't know what the clusters are.

We made an estimate of the known resonance contributions to C^{+-} in an effort to determine whether they might account for the correlations. We could count up 30% of C^{+-} but the remaining 70% was still a mystery. Nevertheless, certain general Mueller-Regge expectations of the correlations are borne out. The $\pi^+\pi^-$ correlation, like $\pi^+\pi^-$ elastic scattering has a strong Regge exchange component (ρ-f exchanges). As would be expected by exchange degeneracy arguments, this Regge component is absent in $\pi^-\pi^-$ inclusive correlations, again analogous to $\pi^-\pi^-$ elastic scattering.

Finally, we described an estimate of Bose effects. More generally we found one would naturally expect coherent interference effects whenever resonances are produced, and whenever the dynamics are such that one adds amplitudes coherently to make different resonances. We found that the ratio of the number of correlated $\pi^+\pi^-$ pairs (within one pion mass of threshold) to ρ^o's is about 1/4 ; the number of $\pi^-\pi^-$ pairs to ρ^o's is about 1/2. Given only a small ρ^o signal, we would not expect a large number of coherent pion pairs.

At present therefore, we are unable to understand the pion correlations as manifestations of known physics. Of the many possibilities, perhaps our data on ρ production is too crude and there really are many more ρ resonances than 10%. Alternatively, perhaps we have ignored some very important backgrounds, which might go away at higher energies. Finally the possibility does exist that we are seeing something new. Of the possibilities listed, I would guess the last is the least likely. Certainly more data would help, as well as more thinking about the problem.

ACKNOWLEDGMENTS

In writing this talk I have greatly benefitted from numerous conversations with C. Sorensen. I also would like to thank my collaborators and colleagues, especially E. Berger, for encouraging me to think about these problems, and for a variety of good ideas.

REFERENCES

1. There is a large literature on the subject of clusters. One of the more recent and most detailed fits to data using the cluster model is by A. Arneodo and G. Plaut, Nucl. Phys. B107, 262 (1976).
2. E. Fermi, Progr. Theoret. Phys. 5, 570 (1950); Phys. Rev. 92, 452 (1953); 93, 1434 (1954).
3. P. Carruthers and F. Zachariasen, Theories and Experiments in High-Energy Physics, eds. A. Perlmutter and S.M. Widmayer (Plenum Press, N.Y. 1975), p. 349.
4. ABFST: D. Amati, S. Fubini, and A. Stanghellini, Nuovo Cimento 26, 896 (1962); L. Bertocchi, S. Fubini, and M. Tonin, Nuovo Cimento 25, 626 (1962).
5. G.F. Chew and A. Pignotti, Phys. Rev. 276, 2112 (1968).
6. J. Whitmore, Phys. Reports 27C, 187 (1976), Fig. 8a .
7. Ibid., Fig. 8c .
8. Ibid., Fig. 73 .
9. E.g. E.L. Berger and A. Krzywicki, Phys. Letters 36B, 380 (1971).
10. E.L. Berger and C.C. Fox, Phys. Letters 47B, 162 (1973).
11. A clear account of clusters and their properties can be found in E.L. Berger, Nucl. Phys. B85, 61 (1975).
12. C. Quigg, P. Pirila, and G.H. Thomas, Phys. Rev. Letters 34, 290 (1975).
13. C. Quigg, P. Pirila, and G.H. Thomas, Phys. Rev. D12, 92 (1975).
14. E.L. Berger, R. Singer, G.H. Thomas and T. Kafka, Phys. Rev. 15, 206 (1977).
15. Various review articles may be consulted for introductory material: E.g. E.L. Berger in Proc. II Int. Colloquium on Multiparticle Dynamics, Helsinki, 1971, Edited by E. Byckling, K. Kajantie, H. Satz and J. Tuominiemi (Univ. of Helsinki, 1971); Chan Hong-Mo in Proc. Int. Conf. on High Energy Collisions, Oxford, 1972, Ed. J.R. Smith (RHEL, Chilton, 1972).
16. Aachen-CERN-Heidelberg-Munich ISR collaboration, K. Eggert et al., Nucl. Phys. B86, 201 (1975); Michigan State-Argonne-Fermilab-Iowa State-Maryland collaboration, B.Y. Oh et al., Phys. Letters 56B, 400 (1975); G. A. Smith, paper presented at the Third International Winter Meeting on Fundamental Physics, Parador de Sierra Nevada, Spain, 1975.
17. R.C. Brower, R.N. Cahn, and J. Ellis, Phys. Rev. D7, 1080 (1973); J.R. Freeman and C. Quigg, Phys. Letters, 47B, 39 (1973); S. Pinsky and G. Thomas, Phys. Rev. D9, 1350 (1974).

18. G. Thomas, ANL-HEP-PR-76-33, Phys. Rev. to be published.
19. R. Hanbury Brown and R.Q. Twiss, Proc. R. Soc. A248, 300 (1957).
20. G. Goldhaber, S. Goldhaber, W. Lee and A. Pais, Phys. Rev. 120, 300 (1960).
21. V.G. Grishin, G.I. Kopylov and M.I. Podgoretskii, Yad. Fiz. 13, 1116 (1971) [English Trans.: Sov. J. Nucl. Phys. 13, 638 (1971)];
V.G. Grishin, G.I. Kopylov and M.I. Podgoretskii, Yad. Fiz. 14, 600 (1971) [English Trans.: Sov. J. Nucl. Phys. 14, 335 (1972)];
G.I. Kopylov and M.I. Podgoretskii, Yad. Fiz. 14, 1081 (1971) [English Trans.: Sov. J. Nucl. Phys. 14, 604 (1972)].
G.I. Kopylov, Yad. Fiz. 15, 178 (1972) [English Trans.: Sov. J. Nucl. Phys. 15, 103 (1972)];
E.V. Shuryak, Phys. Lett. 44B, 387 (1973);
G. Cocconi, Phys. Lett. 49B, 459 (1974);
G.I. Kopylov, Phys. Lett. 50B, 472 (1974);
G.I. Kopylov and M.I. Podgoretskii, Yad. Fiz. 19, 434 (1974) [English Trans.: Sov. J. Nucl. Phys. 19, 215 (1974)];
William J. Knox, Phys. Rev. D10, 65 (1974).
22. R.F. Amann and P.M. Shah, Phys. Lett. 42B, 353 (1972);
J. Steinhoff, Nucl. Phys. B55, 132 (1973);
M. Biyajima and O. Miyamura, Phys. Lett. 53B, 181 (1974); 57B, 376 (1975);
A. Arneodo and G. Plaut, Nucl. Phys. B97,51 (1975);
J. Ranft and G. Ranft, Phys. Lett. 57B, 373 (1975) and references therein.
23. The argument given here is basically that used by Podgoretskii and co-workers[21], except we must take into account the transverse momentum damping of produced mesons.

*Work performed under the auspices of the United States Energy Research and Development Administration.

RECENT DEVELOPMENTS IN DUAL STRING MODELS

P. Di Vecchia

NORDITA

Copenhagen, Denmark

INTRODUCTION

Dual string models have been quite successful in explaining many important features of hadron physics (1). They present, however, also some shortcomings as non-physical values of the space-time dimension and of the intercept of the Regge trajectory which imply the existence of tachyonic states. Many attempts have been made to construct more realistic models, but all of them have been unsuccessful. The existing models were in fact so tight that it seemed very difficult to modify or generalize them without losing some of their good features.

The situation has somewhat changed in the course of the last year when new dual string models have been constructed and a general procedure for classifying all the possible dual string models has been proposed.

The first approach (2) is based on an extension of the gauge algebra of the Neveu-Schwarz-Ramond (NSR) Model obtained by the addition of some internal degrees of freedom. Using this new set of non trivial gauge algebras one has been able to construct a family of new dual models with different values of critical dimension and Regge intercept. The introduction of new degrees of freedom gives, however, a value for the critical dimension that is lower than $D = 4$. Consequently, the only physically reasonable new dual model is the one with an $O(2)$ internal symmetry (3), while any model with a nonabelian symmetry has a negative value for the critical dimension (4),(5). Although these models are no better than the conventional Veneziano model and the NSR model, they seem to me rather important in showing that new consistent dual models can be constructed. This gives some hope that more physical models can be found.

A very important step along this direction has been made by W. Nahm,(6) who has given a classification of dual models in terms of the subgroups of the modular group. His starting point is the fundamental assumption (which is satisfied by the known dual models) that the partition function of the most general dual model is a modular function with respect to some subgroup of the modular group. In this way Nahm has been able to construct the spectrum of dual models with critical dimension $D = 8, 6$ and also 4. In particular there is a model, that Nahm calls the "correct model" which has critical dimension $D = 4$ and no tachyon. Although this model has very nice features as a zero mass pion and trajectories spaced by units of $1/4$, it gives however a spectrum of hadrons which is not quite the physical spectrum. Nevertheless it would be very interesting to construct the scattering amplitude and the gauge algebra of such a model with $D = 4$. In particular its gauge algebra could presumably provide an extension of the conformal algebra other than supersymmetry and this would be by itself very interesting.

The paper is organized as follows:

Sections (2) and (4) are devoted to a review of the conventional string model and the NSR model. In Section (3) we discuss the problem of the longitudinal oscillations of a string and their connection with a two dimensional Yang-Mills theory.

In Section (5) we describe in some detail the $O(2)$ symmetric model and its relation with the nonlinear σ model. Section (6) is devoted to the $O(N)$ symmetric string, while in Section (7) we discuss the calculation of a zero point energy of a string. Finally in Sect. (8) we describe the Nahm approach.

2) THE CONVENTIONAL STRING MODEL

The motion of a relativistic string is described by the Nambu-Goto action

$$S = - \frac{1}{2\pi\alpha'\hbar c^2} \int_{\tau_i}^{\tau_f} d\tau \int_{\sigma_i}^{\sigma_f} d\sigma \; L \tag{2.1}$$

where the Lagrangian

$$L = \sqrt{-g} = \sqrt{-\det(\partial_\mu X \cdot \partial_\nu X)} \tag{2.2}$$

represents the area of the world sheet spanned by the string and $X^A(\xi^\mu)$ is the coordinate of the string. A is a Minkowski vector index and $\xi^\mu = (\xi^0, \xi^1) \equiv (\tau, \sigma)$ are the two variables which describe the world sheet of the string.

By construction action (2.1) is invariant under any reparametrization of the variables ξ^μ :

$$\xi^\mu \rightarrow f^\mu (\xi^\mu) \tag{2.3}$$

Consequently the two-dimensional energy-momentum tensor

$$\theta_{\mu\nu} = L\, g_{\mu\nu} - \frac{\partial L}{\partial(\partial^\mu X)} \cdot \partial_\nu X = 0 \tag{2.4}$$

is identically vanishing. There is in fact no energy-momentum density in the (τ,σ) space since the physics does not depend on the way we choose those variables. The two-dimensional energy-momentum tensor $\theta_{\mu\nu}$ must not be confused with the Minkowski energy momentum tensor θ_{AB} which is not vanishing and determines the momentum and the angular momentum of the string.

Because of the gauge invariance of (2.1) under (2.3) one can choose a special gauge where the equations of motion become linear and so they can be easily solved. The most convenient gauge is the orthonormal gauge which is characterized by

$$(\partial_\pm X)^2 = 0 \tag{2.5}$$

where

$$\partial_\pm = \frac{\partial}{\partial \xi^\pm} ; \qquad \xi^\pm = \xi^0 \pm \xi^1 \tag{2.6}$$

In this gauge the equation of motion becomes just the massless Klein-Gordon equation in two dimensions

$$\partial_\mu \partial^\mu X = 0 \tag{2.7}$$

The boundary conditions are given by

$$\frac{dX}{d\sigma} (\tau; \sigma = \sigma_i, \sigma_f) = 0 \tag{2.8}$$

while the energy-momentum tensor (2.4) becomes

$$\theta_{\mu\nu} = \partial_\mu X \cdot \partial_\nu X - \frac{1}{2} g_{\mu\nu}\, \partial_\rho X \cdot \partial^\rho X = 0 \tag{2.9}$$

Eqs. (2.7) and (2.8) can be derived from the following linear Lagrangian:

$$L = - \frac{1}{4\pi\alpha'\hbar c^2} \partial_\mu X \cdot \partial_\mu X \tag{2.10}$$

It is easy to see that the orthonormal gauge does not specify uniquely the gauge. One can still make gauge transformations which preserve the orthonormal conditions (2.5). They are conformal transformations which are characterized by gauge functions f^μ satisfying the conditions:

$$\partial_\mu f_\nu + \partial_\nu f_\mu - \frac{1}{2} g_{\mu\nu} \partial_\rho f^\rho = 0 \tag{2.11}$$

In terms of the light cone variables the conformal transformations are given by

$$\delta\xi^+ = f^+(\xi^+) \; ; \quad \delta\xi^- = f^-(\xi^-) \tag{2.12}$$

being $f^\pm$ two arbitrary functions of $\xi^\pm$ respectively. The group of the conformal transformations is of course the group of invariance of the linearized Lagrangian (2.10).

It is convenient to restrict ourselves to those conformal transformations that leave invariant the parametrization of the ends of the string which are described by $\sigma = 0, \pi$ respectively. One must then impose the constraints

$$\delta\sigma \; (\tau; \sigma = 0,\pi) = 0 \tag{2.13}$$

which imply

$$f^+(\xi) = f^-(\xi) = f(\xi)$$

$$f(\xi + 2\pi) = f(\xi) \tag{2.14}$$

The generators of the conformal transformations which leave invariant the ends of the string are given by

$$L_n = \frac{\pi}{2} \int_0^\pi d\sigma \left\{ \theta_{++} \, e^{in(\tau+\sigma)} + \theta_{--} \, e^{-in(\tau-\sigma)} \right\} \tag{2.15}$$

and they are identically vanishing because of (2.9) for any integer n .

We want now to construct the quantum theory for the string and, as is the case for any gauge theory, either of two procedures can be followed (7). The first one is the covariant quantization procedure which consists in preserving all the dynamical variables, including those that possibly could be eliminated by a gauge transformation, and then in quantizing all of these. That means that the coordinate $X_A(\tau,\sigma)$ becomes an operator which satisfies canonical commutation relations:

$$\left[X^A(\tau,\sigma)\ ,\ \Pi^B(\tau,\sigma')\right] = i\ \delta(\sigma-\sigma')\ g^{AB}$$

$$\left[X^A(\tau,\sigma)\ ,\ X^B(\tau,\sigma')\right] = \left[\Pi^A(\tau,\sigma)\ ,\ \Pi^B(\tau,\sigma')\right] = 0 \qquad (2.16)$$

where $\Pi^A(\tau,\sigma)$ is the conjugate momentum that can be obtained from Lagrangian (2.10). Since the most general solution of eqs. (2.7) and (2.8) is given by:

$$X(\tau,\sigma) = q_0 + 2p_0\tau + i\sum_{n=1}^{\infty}\frac{\cos n\sigma}{\sqrt{n}}\left[a_n\, e^{-in\tau} - a_n^+\, e^{in\tau}\right] \qquad (2.17)$$

and

$$\Pi^A = \frac{1}{2\pi\alpha'}\ \frac{dX^A}{d\tau} \qquad (2.18)$$

one gets the following commutation relations for the oscillators appearing in (2.17)

$$\left[a_{n,A}\ ,\ a^+_{m,B}\right] = \delta_{n,m}\ g_{AB}$$

$$\left[q_{0A}\ ,\ p_{0B}\right] = i\ g_{AB} \qquad (2.19)$$

Also the gauge generators L_n become operators and, since they can be expressed as bilinear expressions in terms of the harmonic oscillators a_n , it is necessary to choose the ordering of the oscillators in such a way that L_n are free of infinities. Actually the only operator which has a re-ordering problem is L_0 which we shall define in the quantum theory with the normal ordering. Therefore the quantum expression of L_0 is given by

$$L_0 = - p_0^2 - \sum_{n=1}^{\infty} n\, a_n^+ a_n \tag{2.20}$$

As a consequence of the normal ordering procedure the commutation relations of the L_n operators get an additional C number with respect to the corresponding classical Poisson brackets:

$$\left[L_n , L_m\right] = (n-m)\, L_{n+m} + \frac{D}{12} n(n^2-1)\, \delta_{n,-m} \tag{2.21}$$

where D is the dimension of the Minkowski space-time. It can also be seen that this additional C-number is a consequence of the Schwinger terms in the energy-momentum tensor (2.9) commutators (8). The physical states in the quantum theory are defined by:

$$L_n \,|\, \text{Phys}\rangle = 0 \qquad n > 0$$

$$(L_0 + \alpha_0) \,|\, \text{Phys}\rangle = 0 \tag{2.22}$$

where the additional constant α_0 is allowed in the quantum theory because of the normal ordering of L_0 .

The Virasoro algebra and the conditions (2.22) are extremely important for the elimination of negative norm states from the physical spectrum, which are induced by the time Minkowski component. We will see that any consistent dual model has a gauge algebra and this allows one to construct a theory which is simultaneously in agreement with the principles of a quantum theory (no negative probability) and those of a relativistic theory (relativistic covariance).

The second procedure of quantization is the non covariant one that consists in eliminating first all the superfluous degrees of freedom and then quantizing only the physical variables. The advantage of this procedure is that one eliminates immediately the negative norm states. On the other hand, one loses the manifest Lorentz covariance, which one then must check at the end of the quantization, even though the classical action already was Lorentz invariant.

In the case of a string the non covariant quantization can be done in the transverse gauge, where the variables τ and σ are specified by (7)

$$X_+ = 2P_+ \alpha' \tau$$

$$2P_+ \sigma = \int_0^{\sigma} d\sigma' \dot{X}_+ \tag{2.23}$$

and P is the total momentum of the string. With this choice of variables the only independent degrees of freedom are $X^i(\tau,\sigma)$, $\pi^i(\sigma,\tau)$ [i runs only over the transverse directions], q_{0+} and p_{0-}. They satisfy the following canonical commutation relations:

$$\left[X^i(\sigma,\tau)\ ,\ \pi^j(\sigma',\tau)\right] = i\ \delta^{ij}\ \delta(\sigma-\sigma') \qquad i,j = 1\ ,\ \cdots\ D-2$$

$$\left[q_{0+}\ ,\ p_{0-}\right] = i \tag{2.24}$$

The other commutators are vanishing. The hamiltonian in the transverse gauge is given by

$$H = 2\alpha'\ P^+P^- = \alpha'\ P_\perp^2 + \frac{1}{2}\sum_n (a^+_{n,\perp}\ a_{n,\perp} + a_{n\perp}\ a^+_{n\perp}) \tag{2.25}$$

Because of the infinite number of harmonic oscillators H is divergent when the oscillators become operators satisfying the commutation relations:

$$\left[a_{n,i}\ ,\ a^+_{m,j}\right] = \delta_{ij}\ \delta_{n,m} \tag{2.26}$$

This divergence can be eliminated if we normal order H allowing at most the appearance of a finite C-number α_0. The quantum hamiltonian becomes then

$$H = \alpha'\ p_\perp^2 + \sum n\ a^+_{n,\perp}\ a_{n,\perp} - \alpha_0 \tag{2.27}$$

This normal ordering procedure must also be followed for defining the other physical operators of the theory, as for example the Lorentz generators. The quantum theory will be then Lorentz invariant if the generators of the Lorentz group satisfy the algebra of the Lorentz group. In the transverse gauge this has been possible only when the dimension of the space-time D and the Regge intercept α_0 take the following values:

$$\alpha_0 = 1\ ;\quad D = 26 \tag{2.28}$$

In this case the physical spectrum is given by the following partition function:

$$A(w) = \sum_m d(m)\, e^{2i\pi\alpha' m^2 w} = \frac{1}{q} \prod_{n=1}^{\infty} \left(\frac{1}{1-q^n}\right)^{24} \tag{2.29}$$

where $q = e^{2i\pi w}$ and $d(m)$ is the degeneracy of states with mass m.

Finally, the scattering of strings is described by the dual N-point amplitude

$$A_N = \int dV(z) \prod_{i>j} (z_i - z_j)^{2\alpha' p_i p_j} \tag{2.30}$$

In particular for $N = 4$ one gets the well known B function:

$$A_4 = \frac{\Gamma(-\alpha_s)\ \Gamma(-\alpha_t)}{\Gamma(-\alpha_s - \alpha_t)} \quad , \quad \alpha_s = 1 + \alpha' s \tag{2.31}$$

3) LONGITUDINAL STRING OSCILLATIONS

It was pointed out by A. Patrascioiu (9) that the choice of the transverse gauge eliminates possible longitudinal motions of the string. Therefore when one quantizes the string in this singular gauge one gets the restrictions (2.28) on the value of D and α_0. However, those restrictions may disappear when one includes also the possible longitudinal motions.

This problem has been recently considered again by several authors (10),(11). In particular, it seems to me particularly interesting the approach of the authors of Ref. (10), who start from a string with massive ends, the action of which is given by

$$S = \int_{\tau_i}^{\tau_f} d\tau \left\{ -\mu \sqrt{\dot{X}_0^2} - \mu \sqrt{\dot{X}_\pi^2} - \gamma \int_0^\pi d\sigma\, L_{N.G} \right\} \tag{3.1}$$

where $L_{N.G}$ is the Nambu-Goto Lagrangian (2.2), $\gamma = -\frac{1}{2\pi\alpha'\hbar c^2}$ and $\dot{X}_{0,\pi} = \frac{dX_{0,\pi}}{d\tau}$. They show that the limit $\mu \to 0$ is rather tricky and the result, that one gets varies according to whether one takes this limit directly on (3.1) or after having set up the hamiltonian formalism. It is important to note that this feature is not peculiar to a string but it is also present in the case of the massive pointlike particle described by the following Lagrangian

$$S = -\mu \int \sqrt{\dot{X}^2}\, d\tau \qquad (3.2)$$

This Lagrangian describes in fact also a zero mass particle if we take the limit $\mu \to 0$ not directly on the Lagrangian (3.2), but on the Hamiltonian.

In the time-like gauge $[X^0 \sim \tau]$ the hamiltonian corresponding to (3.1) is given by:

$$H = \left[\vec{P}_0^{\,2} + \mu^2\right]^{\frac{1}{2}} + \left[\vec{P}_\pi^{\,2} + \mu^2\right]^{\frac{1}{2}} + \gamma \int_0^\pi d\sigma \sqrt{\vec{K}^2 + \vec{X}'^2} \qquad (3.3)$$

where

$$K^\mu = \frac{\partial L}{\partial \dot{X}_\mu} \quad ; \quad P_0^{\ \mu} = \frac{\mu\, \dot{X}_0^{\ \mu}}{\sqrt{-\dot{X}_0^{\ 2}}} \quad ; \quad P_\pi^{\ \mu} = \frac{\mu\, X_\pi^{\ \mu}}{\sqrt{-\dot{X}_\pi^{\ 2}}} \qquad (3.4)$$

and $\dot{X}_\mu = \frac{\partial X_\mu}{\partial \tau}$, $X'_\mu = \frac{\partial X_\mu}{\partial \sigma}$.

It is especially simple to consider the two dimensional case $[D = 2]$ where $\vec{K} = 0$ and the hamiltonian (3.3) becomes:

$$H = \left[\vec{P}_0^{\,2} + \mu^2\right]^{\frac{1}{2}} + \left[\vec{P}_\pi^{\,2} + \mu^2\right]^{\frac{1}{2}} + \gamma |X(\sigma=\pi) - X(\sigma=0)| \qquad (3.5)$$

The main point is that, when we take the limit $\mu \to 0$, the kinetic term corresponding to the two ends of the string does not vanish and we get:

$$H = |\vec{P}_0| + |\vec{P}_\pi| + \gamma |X(\pi) - X(0)| \qquad (3.6)$$

where the potential part of H is given by the Coulomb potential in 1 space dimension. We note that the kinetic piece in (3.6) would have been absent if we had started from (3.1) directly with $\mu = 0$.

The authors of Ref. (10) show then that the hamiltonian (3.6) reproduces the simplest longitudinal modes proposed by A. Patrascioiu.

It is particularly interesting to analyze the same system in the light-cone gauge $[X^+ \sim \tau]$ where the hamiltonian is given by:

$$H = P^- = \frac{\mu^2}{2P_0^+} + \frac{\mu^2}{2P_\pi^+} + \gamma |X_\pi^- - X_0^-| \tag{3.7}$$

and

$$P^\pm = P_0^\pm \pm P_\pi^\pm \tag{3.8}$$

After the canonical change of variables

$$K = \frac{P_\pi^+ - P_0^+}{2P^+} \quad ; \quad \rho = P^+ (X_\pi^- - X_0^-) \tag{3.9}$$

the mass square operator is given by

$$M^2 = 2P^+ H = \mu^2 \left[\frac{1}{\frac{1}{2}-K} + \frac{1}{\frac{1}{2}+K}\right] + \gamma|\rho| \tag{3.10}$$

The quantum spectrum of (3.10) is obtained by writing the Schroedinger equation:

$$\gamma \int dK' \; G(K,K') \; \psi(K') + \left[\frac{\mu_0^2}{\frac{1}{2}-K} + \frac{\mu_\pi^2}{\frac{1}{2}+K} - M^2\right] \psi(K) = 0 \tag{3.11}$$

where in general one can have different masses at the ends of the string and

$$G(K,K') = P\left[\frac{1}{\pi(K-K')^2}\right] = \lim_{\varepsilon\to 0} \frac{1}{\pi}\int_0^\infty \rho \cos \rho(K-K') \; e^{-\varepsilon\rho} d\rho \tag{3.12}$$

Eq. (3.11) was also derived by 't Hooft (12) in a completely different framework. He considered a two dimensional Yang-Mills theory and in the gauge $A^+=0$ for $N\to\infty$ [N is the number of colours] he has been able to prove that the quarks are confined and the meson $(q\bar{q})$ spectrum is given exactly by eq. (3.11). 't Hooft has also solved eq. (3.11) getting asymptotically linear rising trajectories and no tachyon. All this suggests a very strong connection between the string model and the Yang-Mills theory in two dimensions. This analogy has been carried on further by I. Bars (13) who computed form factors, three point functions, etc. finding a complete agreement between the two theories.

One can, therefore, draw the following conclusions on the problems discussed in this section:

i) The string seems also to have longitudinal motions. Among these are oscillations with no folds and oscillations with any number of folds. The point, where the string folds, is characterized by moving with the speed of light.

ii) However, it must still be proven that one gets a consistent quantum theory with no constraint on α_0 and D when one includes these additional motions.

iii) The motions with no fold give a quantum spectrum, which is the same as the one obtained by 't Hooft in the two dimensional Yang-Mills theory. But what about the folded modes? Do they also appear in the Yang-Mills theory?

iv) Is it possible to construct dual models with the contribution of these longitudinal modes? Can one incorporate conformal invariance without having $\alpha_0 = 1$?

It is very important to give an answer to these questions to see whether one can construct a quantum relativistic theory of free and interacting strings without any constraint on α_0 and D .

4) SPINNING STRING

In the previous section we have seen that the quantization of a conventional string gives quite unphysical values for D and α_0 . We want now to generalize the conventional string in order to construct a realistic dual model.

A possible generalization of the string can be obtained by adding another variable to σ and τ ; one gets then a membrane described by $X^A(\tau,\sigma_1,\sigma_2)$. However this system is very difficult to solve because one cannot find a gauge where the eqs. of motion become linear (14). On the other hand introducing an additional variable σ one loses the beauty and the simplicity of a one dimensional structure.

Another way of generalizing the conventional string is given by

$$X^A(\tau,\sigma) \quad \rightarrow \quad X^A(\tau,\sigma;\, \theta_a) \tag{4.1}$$

where θ_a is a two-dimensional spinor, which is an odd element of a Grassmann algebra, i.e. it satisfies the following anticommutation relations:

$$\theta_a\, \theta_b + \theta_b\, \theta_a = 0 \tag{4.2}$$

One can expand $X^A(\tau,\sigma;\, \theta_a)$ in power series in θ and because of (4.2) the series contain only three terms:

$$X^A(\tau,\sigma;\ \theta_a) = \phi^A(\tau,\sigma) + i\bar{\theta}\ \psi^A(\tau,\sigma) + \frac{i}{2}\,\bar{\theta}\theta\ F^A(\theta,\sigma) \qquad (4.3)$$

where $\phi^A(\tau,\sigma)$ is the coordinate of the string while ψ^A and F^A describe the spin distribution along the string.

Zumino (15) has constructed a non linear Lagrangian for the spinning string. In order to make more natural the approach followed by Zumino let me reformulate first the Nambu-Goto action (2.1) in a different but equivalent way.

In the orthonormal gauge the motion of a spinless string is described by the massless Klein-Gordon equation and by the vanishing of the two-dimensional energy-momentum tensor. Only the eq. of motion (2.7) can be obtained from the linearized Lagrangian (2.10), while the condition $\theta_{\mu\nu} = 0$ must be added by hand. An extension of the Lagrangian (2.10) which gives both the eq. of motion and the constraint (2.9), can be obtained by writing L in a curved space. One gets:

$$L\ (g^{\mu\nu},\ X) = -\frac{1}{2}\sqrt{-g}\ \ g^{\mu\nu}\ \partial_\mu X\ \partial_\nu X \qquad (4.4)$$

Then by varying (4.4) with respect to X one gets the eq. of motion:

$$\partial_\mu \left[\sqrt{-g}\ g^{\mu\nu}\ \partial_\nu X\right] = 0 \qquad (4.5)$$

while the variation with respect to $g^{\mu\nu}$ gives the vanishing of the energy momentum tensor:

$$\partial_\mu X\ \partial_\nu X - g_{\mu\nu}\,\frac{1}{2}\,g^{\rho\sigma}\ \partial_\rho X\ \partial_\sigma X = 0 \qquad (4.6)$$

The action corresponding to (4.4) is invariant under general coordinate transformations (2.3). Note that (4.4) is just an equivalent way of rewriting the Nambu-Goto Lagrangian. This can be easily seen from the following identity:

$$\sqrt{\det\,(\partial_\mu X\ \partial_\nu X)} = +\frac{1}{2}\sqrt{-g}\ g^{\mu\nu}\ \partial_\mu X\ \partial_\nu X \qquad (4.7)$$

which can be obtained taking the determinant of (4.6). In particular in the orthonormal gauge which corresponds in this other language to a conformally flat space, the metric tensor is given by

$$g_{\mu\nu} = F(X)\ \ \eta_{\mu\nu} \qquad (4.8)$$

where $\eta_{\mu\nu}$ is the flat metric.

Using this particular $g_{\mu\nu}$ in (4.5) and (4.7) one gets the eqs. of motion (2.7) and the constraints (2.9) of the linearized theory.

In the case of a spinning string we have, in addition to τ and σ, also the Grassmann variables θ_a and if we proceed in analogy with the spinless string we must write down a Lagrangian which is invariant under any reparametrization in the superspace defined by the variables $z^M \equiv (\xi^\mu, \theta^m)$. This has been done by Zumino (15) who has written the following Lagrangian:

$$L = (-\frac{1}{2} E_a{}^M \partial_M X E^{aN} \partial_N X - if(-1)^m \gamma_\alpha^{bc} E_c{}^M E_b{}^N \partial_N E_M{}^\alpha)D \qquad (4.9)$$

where $E^A{}_M$ is the vierbein in superspace which satisfies the following relation:

$$E_M{}^A E^N{}_A = \delta^N{}_M \qquad (4.10)$$

and X is the supercoordinate of the string. X has also a Minkowski vector index that we omit for the sake of simplicity. M , N [A,B] are curved [flat indices] and $M \equiv (\mu,m)$ $[A \equiv (\alpha,a)]$ for bosonic and femionic indices respectively. D is the generalized determinant

$$D = \det E_M{}^A \qquad (4.11)$$

Finally f is an arbitrary constant. The first term in (4.9) is the one that one expects as the generalization of (4.4), while the second term is quite unexpected, but it is necessary to reproduce the constraints of the NSR model.

The action corresponding to (4.9) can then be obtained integrating (4.9) over d^4z :

$$S = \int d^4z \quad L \qquad (4.12)$$

where the integration over the Grassmann variables is defined by the following rules

$$\int d\,\theta = 0 \qquad \int \theta\, d\,\theta = 1 \qquad (4.13)$$

The action (4.12) is invariant under general coordinate transformations in superspace

$$z^M \rightarrow f(z^M) \qquad (4.14)$$

and under local rotations

$$\delta\, E_M{}^A = E_M{}^B\, X_B{}^A \tag{4.15}$$

where

$$X_B{}^A = \begin{pmatrix} -\ell \varepsilon^\alpha_\beta & (\gamma_\beta \phi)^a \\ 0 & \frac{1}{2}\,\ell(\gamma_5)^a_b \end{pmatrix} \tag{4.16}$$

being ℓ and φ arbitrary superfields.

As in (4.8) one can go to the orthonormal gauge which corresponds to choose a conformally flat superspace where the vierbein is given by

$$E_M{}^\alpha = \Lambda(z)\,\bar{E}_M{}^\alpha \quad ; \quad E_M{}^a = \Lambda^{\frac{1}{2}}\bar{E}_M{}^a \quad ; \quad D = \Lambda \tag{4.17}$$

and $\bar{E}_M{}^A$ is the flat vierbein:

$$\bar{E}_\mu{}^\alpha = \delta_\mu{}^\alpha \qquad \bar{E}_\mu{}^a = 0$$

$$\bar{E}_m{}^\alpha = i(\gamma^\alpha)_{mn}\,\theta^n \qquad \bar{E}_m{}^a = \delta_m{}^a \tag{4.18}$$

If one inserts the expressions (4.17) in the eq. of motion derived from (4.12) varying the action with respect to $X(z)$ and in the constraints which are obtained by varying with respect to the supervierbein $E_M{}^A$, one gets the eq. of motion and the constraints for the spinning string in the orthonormal gauge. They are respectively:

$$\bar{D}DX = 0 \tag{4.19}$$

and

$$J_a{}^\mu = \gamma^\nu\,\gamma^\mu\,\partial_\nu X \cdot DX = 0 \tag{4.20}$$

where D is the flat covariant derivative:

$$D_a = \frac{\partial}{\partial\bar{\theta}_a} + i(\gamma^\mu\theta)_a\,\partial_\mu \tag{4.21}$$

In the orthonormal gauge the action (4.12) becomes:

$$S = \frac{1}{8\pi\alpha'}\int d^2\xi \int d^2\theta\; \bar{D}X \cdot DX \tag{4.22}$$

and after the integration over the Grassmann variables one gets:

$$S = \frac{1}{2\pi\alpha'} \int d^2\xi \left\{ -\frac{1}{2} \partial_\mu \phi \, \partial^\mu \phi - \frac{i}{2} \bar{\psi} \gamma^\mu \partial_\mu \psi + \frac{1}{2} F^2 \right\} \tag{4.23}$$

In terms of the component fields the constraint (4.20) becomes:

$$\theta_{\mu\nu} = 0 \tag{4.24}$$

$$j_a{}^\mu = \gamma^\nu \gamma^\mu \partial_\nu \phi \psi = 0 \tag{4.25}$$

where $\theta_{\mu\nu}$ is the energy-momentum corresponding to the Lagrangian (4.23) while $j_a{}^\mu$ is the spinor current which corresponds to the invariance under supersymmetry of (4.23). From (4.19) one gets also the following eqs. of motion:

$$\partial_\mu \partial^\mu \phi = \gamma^\mu \partial_\mu \psi = F = 0 \tag{4.26}$$

Finally one can also derive the following boundary conditions from Lagrangian (4.23):

$$\begin{aligned} &\frac{\partial\phi}{\partial\sigma} (\sigma=0,\pi;\tau) = 0 \\ &\psi_1(\xi) = \psi_2(\xi) = \psi(\xi) \\ &\psi(\xi+2\pi) = \pm\, \psi(\xi) \end{aligned} \tag{4.27}$$

where the $+(-)$ sign refers to the fermion (boson) sector. The linearized action (4.23) is now not only invariant under conformal transformations which act on ξ and θ as follows:

$$\begin{aligned} &\delta\, \xi^\pm = f^\pm (\xi^\pm) \\ &\delta\, \theta^\pm = \dot{f}^\pm (\xi^\pm) \frac{\theta^\pm}{2} \end{aligned} \tag{4.28}$$

but also under supergauge transformations which are given by:

$$\begin{aligned} &\delta\, \xi^\pm = i\, \alpha^\pm(\xi^\pm)\, \theta^\pm \\ &\delta\, \theta^\pm = \alpha^\pm(\xi^\pm) \end{aligned} \tag{4.29}$$

where

$$\theta^\pm = \frac{1}{2}(1 \pm \gamma^5)\, \theta \tag{4.30}$$

As in the case of the conventional string we can restrict ourselves to those transformations (4.28) and (4.29) which leave unchanged the parametrization of the ends of the string. One gets the following constraints on the functions f and α :

$$f^{+}(\xi) = f^{-}(\xi) = f(\xi)$$

$$f(\xi) = f(\xi+2\pi) \tag{4.31}$$

and

$$\alpha^{+}(\xi) = \alpha^{-}(\xi) = \alpha(\xi)$$

$$\alpha(\xi+2\pi) = \pm\, \alpha(\xi) \tag{4.32}$$

where the sign $+$ is for the fermion sector and the sign $-$ for the boson sector.

The generators of those transformations can be constructed via the Noether theorem from (4.24) and (4.25) which give the Noether currents corresponding to the symmetries (4.28) and (4.29). One gets the well-known operators L_n , G_n $[F_n]$ of the Neveu Schwarz [Ramond] model which, according to (4.24) and (4.25), must be identically vanishing

$$L_n = G_n = 0 \tag{4.33}$$

In the quantum theory we have again reordering ambiguity only with the operator L_0 which can be again defined with the normal ordering procedure. The physical states are characterized by the physical conditions:

$$L_n|\text{Phys}\rangle = G_n|\text{Phys}\rangle = 0 \qquad n > 0 \tag{4.34}$$

$$(L_0 + \alpha_0)|\text{Phys}\rangle = 0$$

In the meson sector the gauge algebra is given by

$$\left[L_n , L_m\right] = (n-m)\, L_{n+m} + \frac{D}{8}\, n(n^2-1)\, \delta_{n,-m}$$

$$\left\{G_n , G_m\right\} = 2\, L_{n+m} + \frac{D}{8}\, (4n^2-1)\, \delta_{n,-m}$$

$$\left[L_n , G_m\right] = (\tfrac{1}{2}n - m)\, G_{n+m} \tag{4.35}$$

where again the C-numbers are a consequence of the normal ordering procedure or equivalently of the Schwinger terms in the commutation relations of the energy-momentum tensor (4.24) and of the spinor current (4.25).

In the transverse gauge the quantum theory is consistent with the theory of relativity if:

$$\begin{array}{lll} \text{Meson sector :} & \alpha_0 = \tfrac{1}{2} & D = 10 \\ \text{Fermion sector :} & \alpha_0 = 0 & D = 10 \end{array} \qquad (4.36)$$

If $D = 10$ the spectrum of the boson sector is given by the following partition function:

$$A^{NS}(w) = \frac{1}{q^{\frac{1}{2}}} \prod_{n=1}^{\infty} \left[\frac{1}{1-q^n} (1+q^{n-\frac{1}{2}}) \right]^8 \qquad (4.37)$$

while the spectrum of the fermion sector is given by

$$A^R(w) = 32 \prod_{n=1}^{\infty} \left[\frac{1}{1-q^n} (1+q^n) \right]^8 \qquad (4.38)$$

Finally, the scattering of strings in the meson sector is given by the Neveu Schwarz N point amplitude and it can be written in a manifestly supersymmetric way as follows (16):

$$B_N = \int \frac{\prod^n dz_i \prod^n d\theta_i}{d\, V_{abc}} \prod_{i>j} (z_i - z_j + \tilde{\theta}_i \tilde{\theta}_j)^{2\alpha' p_i p_j} \qquad (4.39)$$

where

$$d\, V_{abc} = \frac{dz_a\, dz_b\, dz_c}{\left[z_a - z_b + \tilde{\theta}_a \tilde{\theta}_b\right]\left[z_b - z_c + \tilde{\theta}_b \tilde{\theta}_c\right]\left[z_c - z_a + \tilde{\theta}_c \tilde{\theta}_a\right]} \qquad (4.40)$$

with $\tilde{\theta}_a^{\,i} = \sqrt{z_a}\, \theta_a^{\,i}$.

In particular one gets the following 4-point amplitude:

$$B = \frac{\Gamma(1-\alpha_s)\, \Gamma(1-\alpha_t)}{\Gamma(1-\alpha_s-\alpha_t)} \quad , \quad \alpha_s = 1 + \alpha' s \qquad (4.41)$$

We conclude this section mentioning two recent papers by L. Brink and J. O. Winnberg (17), where the operator formalism of the NSR model is reformulated in a manifest supersymmetric way, and by P. Goddard and R. Horsley (18), where they study in some detail the formal properties of the fermion transition vertex.

5) O(2) SYMMETRIC STRING

The string models that we have described in the previous sections have the serious shortcomings of having tachyons and critical dimensions $D = 26$ and 10 . Furthermore, they do not incorporate any internal symmetry degrees of freedom, apart from trivial Chan-Paton factors.

One way out of these problems has been suggested by E. Cremmer and J. Scherk (19); it consists in compactifying some of the space-time dimensions and to interpret them as due to an internal symmetry. These authors show that the dual models obtained with this procedure are self-consistent. For example, in the case of the NSR model one gets an $O(6) \sim SU(4)$ internal symmetry.

Another way to construct more realistic models would be to generalize the procedure described in (4.1) by incorporating internal degrees of freedom. This can be done by considering a string described by the supercoordinate (2), (3), (4)

$$X^A\ (\tau,\sigma,\ \theta_a{}^i) \tag{5.1}$$

where the Grassmann variables $\theta_a{}^i$ transform according to a certain representation of a compact group $G = [O(N)\ ,\ SU(N)\ \text{etc.}]$.

The simplest case is given by $G = O(2) \equiv U(1)$ with a couple of spinors $\theta_a{}^i$ which transform according to the adjoint representation of $O(2)$.

Expanding X^A in power series of θ's one gets

$$X^A(\tau,\sigma,\theta a^i) = \phi^A(\tau,\sigma) + i\ \bar\theta^i\ \psi^{A;i}(\tau,\sigma) + i\ \bar\theta^i\theta^j\ (F^{ij})^A +$$

$$+ \frac{i}{2}\ \varepsilon^{ij}\ \bar\theta^i\ \gamma^5\theta^j\ D^A(\tau,\sigma) + \frac{1}{2}\ \varepsilon^{ij}\ \bar\theta^i\ \gamma^\mu\ \theta^j\ B_\mu{}^A(\tau,\sigma) +$$

$$+ i\ \bar\theta^i\theta^i\ \bar\theta^j(\chi^j(\tau,\sigma))^A + (\bar\theta^i\theta^i)^2G^A(\tau,\sigma) \tag{5.2}$$

By following the same procedure used by Zumino for constructing a non linear Lagrangian for the NSR model, one could construct a non linear Lagrangian also for the $O(2)$ symmetric model. However, we have seen in the previous cases that one does not lose any physical feature working in the orthonormal gauge where the eqs. of motion become linear. Therefore, in the case of the $O(2)$ symmetric model we can just as well write the action directly in the orthonormal gauge:

$$S \simeq \int_{\tau_i}^{\tau_f} d\tau \int_{\sigma_i}^{\sigma_f} d\sigma \int d^4\theta \; X^2 \tag{5.3}$$

where X is given by:

$$X^A = \left[1 - \frac{1}{8}(\overline{\theta}^i\theta^i)^2 \, \Box\right] \phi^A(\tau,\sigma) + i\overline{\theta}^i \left[1 + \frac{i}{2}\,\overline{\theta}^j\theta^j \, \not\partial\right] \psi^{i;A}(\tau,\sigma) +$$

$$+ \frac{i}{2}\,\overline{\theta}^i\theta^i \, F^A(\tau,\sigma) + \frac{i}{2}\,\varepsilon^{ij}\,\overline{\theta}^i\,\gamma^5 \left[D(\tau,\sigma) - \not\partial C(\tau,\sigma)\right]\theta^j \tag{5.4}$$

which is obtained from (5.2) imposing the supercovariant constraints:

$$(\overline{D}^i D^j - \frac{1}{2}\,\overline{D}^K D^K \, \delta^{ij}) \, X = 0 \tag{5.5}$$

with

$$D^i_a = \frac{\partial}{\partial \overline{\theta}^i_a} + i\,(\gamma^\mu\,\theta^i)_a\,\partial_\mu \tag{5.6}$$

The constraint (5.5) can be imposed because it does not spoil the supersymmetry property of the action (5.3). Performing the integration over the Grassmann variables θ one gets the following action:

$$S \simeq \int_{\tau_i}^{\tau_f} d\tau \int_{\sigma_i}^{\sigma_f} d\sigma \left\{-\frac{1}{2}\,\partial_\mu\,\phi\,\partial^\mu\,\phi - \frac{1}{2}\,\partial_\mu\,C\,\partial^\mu\,C - \frac{i}{2}\,\overline{\psi}^i\,\gamma^\mu\,\partial_\mu\,\psi^i + \right.$$

$$\left. + \frac{1}{2}\,(F^2 + D^2)\right\} \tag{5.7}$$

which gives the following eqs. of motion:

$$\Box\phi = \Box C = F = D = 0$$

$$\not\partial\psi^i = 0 \tag{5.8}$$

According to the procedure described in Ref. (3) one gets also two kinds of boundary conditions as in the case of the NSR model. They are respectively:

Meson sector
$$\begin{cases} \dfrac{d}{d\sigma}\,\phi(\sigma=0,\pi;\tau) = \dfrac{d}{d\sigma}\,C(\sigma=0,\pi;\tau) = 0 \\ \psi^i_1(\sigma=0;\tau) = +\,\psi^i_2(\sigma=0;\tau) \;\; ; \;\; \psi^i_1(\sigma=\pi;\tau) = -\,\psi^i_2(\sigma=\pi,\tau) \end{cases}$$

$$\text{Fermion sector}\begin{cases}\frac{d}{d\sigma}\,\phi(\sigma=0,\pi;\tau) = \frac{d}{d\sigma}\,C(\sigma=0;\tau) = \frac{d}{d\tau}\,C(\sigma=\pi;\tau) = 0 \\ \psi_1^i(\sigma=0;\tau) = \psi_2^i(\sigma=0;\tau)\;;\quad \psi_1^1(\sigma=\pi;\tau) = -\,\psi_2^1(\sigma=\pi;\tau) \\ \qquad\qquad\qquad\qquad\qquad\qquad\quad \psi_1^2(\sigma=\pi;\tau) = \psi_2^2(\sigma=\pi;\tau)\end{cases} \tag{5.9}$$

As pointed out in Ref. (3) they can be rewritten in a more general way, but it is not very instructive here to discuss these technical details.

The action (5.7) is invariant under the following kinds of transformations:

1) Supergauge transformations

$$\delta\xi^{\pm} = i\,\alpha^{i\pm}(\xi^{\pm})\,\theta^{i\pm}$$
$$\delta\theta^{\pm i} = \alpha^{i\pm}(\xi^{\pm}) + i\,\dot{\alpha}^{j\pm}(\xi^{\pm})\,\theta^{j}\,\theta^{i}$$

2) Conformal transformations

$$\delta\xi^{\pm} = f^{\pm}(\xi^{\pm})$$
$$\delta\theta^{\pm i} = \dot{f}^{\pm}(\xi^{\pm})\,\frac{\theta^{i\pm}}{2}$$

3) Conformal O(2) transformations

$$\delta\xi^{\pm} = 0$$
$$\delta\theta^{\pm i} = t^{\pm}(\xi^{\pm})\,\varepsilon^{ij}\,\theta^{j\pm} \tag{5.10}$$

where f, α and t are arbitrary functions.

Imposing to those transformations to leave invariant the parametrization of the ends of the string one gets in the meson sector

$$
\begin{aligned}
& f^{+}(\xi) = f^{-}(\xi) = f(\xi) \\
& f(\xi + 2\pi) = f(\xi) \\
& \alpha^{i+}(\xi) = \alpha^{i-}(\xi) = \alpha(\xi) \\
& \alpha^{i}(\xi + 2\pi) = -\alpha^{i}(\xi) \\
& t^{+}(\xi) = t^{-}(\xi) \\
& t(\xi + 2\pi) = t(\xi)
\end{aligned}
\tag{5.11}
$$

Finally, the constraints are given by the vanishing of the supercurrent as in the case of the NSR model:

$$J^{\mu} = \varepsilon^{ij}\, \overline{D}^{i} X\, \gamma^{\mu} D^{j} X = 0 \tag{5.12}$$

In terms of the component fields (5.12) implies:

$$\theta_{\mu\nu} = 0 \quad ; \quad j_a^{\mu;i} = 0 \quad ; \quad t^{\mu} = 0 \tag{5.13}$$

where $\theta_{\mu\nu}$ is the energy-momentum tensor corresponding to the action (5.7) and

$$
\begin{aligned}
& j_a^{\mu;i} = (\delta^{ij}\, \partial_{\nu}\phi - \varepsilon^{ij}\, \partial_{\nu} C)\, \gamma^{\nu} \gamma^{\mu} \psi^{j} \\
& t^{\mu} = \varepsilon^{ij}\, \overline{\psi}^{i} \gamma^{\mu} \psi^{j}
\end{aligned}
\tag{5.14}
$$

In terms of the currents (5.13) one can construct via the Noether theorem the generators of the transformations (5.10). One gets the L_n's from the energy-momentum tensor, the G_n^i from the spinor current and finally from t_μ a new kind of gauge operators T_n which generate O(2) conformal transformations.

Because of the vanishing of the supercurrent (5.12) classically one gets the following relations:

$$L_n = T_n = G_n^i = 0 \tag{5.15}$$

The quantum theory can be constructed following the same procedure as in the case of the other string models. One must again normal order the operator L_0 and the conditions on the physical states are:

$$L_n|\text{Phys}\rangle = G_n^i|\text{Phys}\rangle = T_n|\text{Phys}\rangle = 0 \qquad n > 0 \tag{5.16}$$

$$(L_0 + \alpha_0)|\text{Phys}\rangle = T_0|\text{Phys}\rangle = 0 \tag{5.17}$$

The gauge algebra is given by:

$$\left[L_n , L_m\right] = (n-m)\; L_{n+m} + \frac{D}{4}\, n(n^3-1)\; \delta_{n,-m}$$
$$\left[L_n , G^i_m\right] = \left(\frac{n}{2} - m\right)\; G^i_{n+m}$$
$$\left\{G^i_n , G^j_m\right\} = 2\, \delta^{ij}\; L_{n+m} + i\, \varepsilon^{ij}(n-m)\; T_{n+m} + \frac{D}{4}(4n^2-1)\; \delta_{n,-m}$$
$$\left[T_n , T_m\right] = \frac{D}{4}\, n\; \delta_{n,-m}$$
$$\left[T_n , G^i_m\right] = i\, \varepsilon^{ij}\; G^j_{n+m}$$
$$\left[T_n , L_m\right] = m\; T_{n+m} \qquad (5.18)$$

where again the C-numbers are a consequence of the normal ordering of L_0 .

With respect to the other models we have, however, the new feature of the existence of an internal symmetry. This symmetry is not realized as a flavour symmetry because the particles of the physical spectrum are not classified according to multiplets of $O(2)$. It is instead realized as a kind of "colour" symmetry because the condition (5.17) tells us that the physical states must be $O(2)$ singlets. This is of course a consequence of the requirement that the physical quantities be invariant under any reparametrization of the variables τ , σ , θ^i_a . In the following we summarize the main physical features of the $O(2)$ symmetric model:

i) The intercept of the leading Regge trajectory is $\alpha_0 = 1$, but the lowest state in both sectors is massless. So no tachyon is present in the spectrum.

ii) The scattering amplitude is given by the following N-point dual amplitude

$$B_N = \lim_{M_0 \to 0} \frac{1}{\alpha' M_0^2} \int dV \prod_{i>j} \left(z_i - z_j + \sum_{k=1}^{2} \tilde{\theta}^{(k)}_i\, \tilde{\theta}^{(k)}_j\right)^{2\alpha' p_i p_j} \qquad (5.19)$$

where

$$dV = \frac{\prod_{i=1}^{n} dz_i \prod_{i=1}^{n} \prod_{k=1}^{2} d\,\tilde{\theta}^{(k)}_i \left[z_a - z_b + \tilde{\theta}_a \tilde{\theta}_b\right]\left[z_b - z_c + \tilde{\theta}_b \tilde{\theta}_c\right]\left[z_c - z_a + \tilde{\theta}_c \tilde{\theta}_a\right]}{dz_a\, dz_b\, dz_c} \qquad (5.20)$$

with $\tilde{\theta}^{(k)}_i = \sqrt{z_i}\; \theta^{(k)}_i$

iii) B_N is perfectly well defined for any value D of the space-time dimension. However, the quantum theory of the $O(2)$ symmetric string is Lorentz invariant if

$$D = 2 \tag{5.21}$$

iv) The 4-point amplitude is given by:

$$B_4 = \frac{\Gamma(2-\alpha_s)\,\Gamma(2-\alpha_t)}{\Gamma(2-\alpha_s-\alpha_t)} \quad ; \quad \alpha_s = 1 + \alpha' s \tag{5.22}$$

If $D = 2$ one has only forward and backward scattering and from (5.22) one gets:

Forward scattering: $B_4 \sim - \alpha' s$

Backward scattering: $B_4 \sim 0$ (5.23)

One gets therefore a dual model with no pole in the 4-point amplitude!! In addition one can also check that only the ground state pole is present in the N-point amplitude. This can be easily understood in terms of the string. In fact in the transverse gauge as we have seen in Sect. (1) the only independent degrees of freedom of a string are the transverse ones. But if $D = 2$ there is no transverse direction. Therefore the string cannot oscillate and classically it must collapse to a point; so one gets only one particle in the physical spectrum.

v) Although the spectrum contains only one particle the scattering amplitude for any number of ground states is not trivial if one introduces Chan-Paton factors to implement isospin. The scattering amplitude that one obtains in this way has the following properties:
1) Cyclic symmetry and factorization
2) The only coupled particle is the ground state
3) Asymptotic Regge behavior with $\alpha_0 = 1$
4) Adler zeroes.

It has been shown (20) that those properties imply that the underlying theory is the non linear σ-model which is described by the following Lagrangian

$$L = - \frac{1}{2} \partial_\mu \vec{n}\, \partial^\mu \vec{n} \tag{5.24}$$

and by the constraint

$$\vec{n}^2 = 1 \tag{5.25}$$

In conclusion the boson sector of the O(2) symmetric string model is equivalent to the non linear σ model, while the fermion sector has not yet been investigated.

6) O(N) SYMMETRIC STRING

In this section we want to discuss the problems that arise if we try to extend the procedure of Sect. 5 to an O(N) symmetric string. The O(2) algebra (5.18) can be easily generalized to an O(N) gauge algebra. One gets (2):

$$\left[G_m^{i_1\cdots i_R}, G_n^{j_1\cdots j_S}\right]_{(-1)^{RS+1}} = i^{-RS}\left\{\left[m(2-S) - n(2-R)\right]G_{n+m}^{i_1\cdots i_R j_1\cdots j_S} - i\sum_{h=1}^{R}\sum_{k=1}^{S}(-1)^{h+k+S}\,\delta^{i_h,j_k}\,G_{n+m}^{i_1\cdots\hat{i}_h\cdots i_R j_1\cdots\hat{j}_k j_S}\right\} \tag{6.1}$$

where one has a commutator (anticommutator) if RS is even (odd). $G_n^{i_1\cdots i_R}$ is a completely antisymmetric tensor of the global O(N) group and it is a bosonic (fermionic) quantity if R is even (odd). The hat on the indices i_h and j_k means that these indices are omitted. The algebra (6.1) contains just 2^N independent generators of which $2^N/2$ are bosons and $2^N/2$ are fermions.

There are, however, difficulties in constructing Lagrangian models for O(N) with $N > 2$. The action corresponding to an O(N) symmetric model must be invariant under the following transformations

$$\delta\xi^{\pm} = i\,(2-n)\,\alpha^{\pm}_{i_1\cdots i_n}(\xi^{\pm})\,\theta^{\pm i_1}\cdots\theta^{\pm i_n}$$

$$\delta\theta^{\pm i} = n\,\alpha^{\pm}_{i_1\cdots i_{n-1}}{}^{i}(\xi^{\pm})\,\theta^{i_1}\cdots\theta^{i_{n-1}} + i\,\dot{\alpha}^{\pm}_{i_1\cdots i_n}(\xi^{\pm})\theta^{i_1}\cdots\theta^{i_n}\,\theta^{i} \tag{6.2}$$

whose generators satisfy the commutation relations (6.1). In particular the algebra (6.1) contains the conformal algebra as a subalgebra; therefore the action of the O(N) symmetric string must be invariant under the conformal group and in particular the action must be a dimensionless quantity. This happens for the conventional string action (2.10) because $d^2\xi$ has dimension 2 while the dimension of the Lagrangian is -2. This is also true for the NSR model and the O(2) symmetric string model where the dimension of the integration volume is 1 and 0 respectively, which is exactly matched by the dimension of the Lagrangians (4.23) and (5.7) respectively. In the case of an O(3) symmetric model, however, the dimension of the volume of integration is -1. Therefore, in order to cancel this negative dimension we need fields with non canonical dimensions, which are not allowed in the case of a free field theory. A complete classification of the supersymmetric infinite graded algebras containing a Virasoro subalgebra has been given by P. Ramond and J. H. Schwarz and they too arrived at the conclusion that only few of these algebras [Neveu-Schwarz, O(2) and SU(2)] can be realized by means of commuting or anticommuting oscillators (21).

Although one cannot construct Lagrangian models for the O(N) symmetric string with $N > 2$, one can, however, construct the N-point amplitude. It can be easily obtained as a natural generalization of (4.39) and (5.19) and they are given by (4),(5):

$$B_N = \int dV \prod_{i<j} \left(z_i - z_j + \sum_{\ell=1}^{N} \tilde{\theta}_i^{\ell} \tilde{\theta}_j^{\ell}\right)^{2\alpha' p_i p_j} \tag{6.3}$$

where

$$dV = \frac{\prod_{i=1}^{n} dz_i \prod_{k=1}^{N} \prod_{i=1}^{n} d\tilde{\theta}_i^{(k)}}{dz_a\, dz_b\, dz_c} \left[z_a - z_b + \tilde{\theta}_a^k \theta_b^k\right]\left[z_b - z_c + \tilde{\theta}_b^k \tilde{\theta}_c^k\right]\cdot\left[z_c - z_a + \tilde{\theta}_c^k \tilde{\theta}_a^k\right] \tag{6.4}$$

In particular for $N = 4$ one gets:

$$B_4 = \frac{\Gamma(N-\alpha_s)\,\Gamma(N-\alpha_t)}{\Gamma(N-\alpha_s-\alpha_t)} \tag{6.5}$$

The invariance under the projective group which is a subgroup of the conformal group corresponding to the operators L_0, L_1 and L_{-1}, imposes that the mass of the ground state be

$$\alpha' M_0^2 = \frac{N}{2} - 1 \tag{6.6}$$

Therefore, the O(N) symmetric model does not contain any tachyon; however, except for $N = 1$, 2, negative norm states are present in the physical spectrum for any positive value of the space-time dimension. This means that we have now models all having too many internal degrees of freedom and consequently the value of the critical dimension is negative.

We have seen that if $N > 2$ we are unable to construct a Lagrangian model for the O(N) symmetric string. There is, however, an exception in the case of a subalgebra of (6.1) if $N = 4$ (4).

When $N = 4$ one has the following gauge generators L_n, G_n^i, T_n^{ij}, Γ_n^i and Δ_n with $i = 1,2,3,4$. One can then define the following linear combinations:

$$\mathcal{L}_n = L_n + \frac{n^2}{2}\Delta_n$$

$$\mathcal{G}^i_n = G^i_n + in\,\Gamma^i_n$$

$$\mathcal{T}^{ij}_n = \frac{1}{2}\left[T^{ij}_n + \frac{1}{2}\varepsilon^{ijhk}\,T^{hk}_n\right] \tag{6.7}$$

They satisfy a subalgebra of (6.1) for $N = 4$, which is the same algebra that is obtained assuming that the θ^i_a's transform as SU(2) doublets. This can be easily seen if we define the following SU(2) doublets in terms of the fourvector $\mathcal{G}^i_n$:

$$G^a \equiv \left(\frac{\mathcal{G}^4 - i\mathcal{G}^3}{\sqrt{2}}, \frac{\mathcal{G}^1 - i\mathcal{G}^2}{\sqrt{2}}\right)$$

$$\overline{G}^a = \left(\frac{\mathcal{G}^4 + i\mathcal{G}^3}{\sqrt{2}}, \frac{\mathcal{G}^1 + i\mathcal{G}^2}{\sqrt{2}}\right) \tag{6.8}$$

Using G^a and $\overline{G}^a$ one gets the following gauge algebra for the SU(2) symmetric string:

$$\left[L_n, L_m\right] = (n-m)\,L_{n+m} + \frac{D}{2}n\,(n^2-1)\,\delta_{n,-m}$$

$$\left\{G^a_n, G^b_m\right\} = \left\{\overline{G}^a_n, \overline{G}^b_m\right\} = 0$$

$$\left\{G^a_n, \overline{G}^b_m\right\} = 2\,\delta^{ab}\left[L_{n+m} + \frac{D}{2}(4n^2-1)\,\delta_{n,-m}\right] + 2(n-m)\,\sigma^{ab}_i\,T^i_{n+m}$$

$$\left[L_n, T^i_m\right] = -\,m\,T^i_{n+m}$$

$$\left[L_n, G^a_m\right] = \left(\frac{n}{2} - m\right)G^a_{n+m}$$

$$\left[T^i_n, G^a_m\right] = \frac{1}{2}(\sigma^i)^{ab}\,G^b_{n+m}$$

$$\left[T^i_n, T^j_m\right] = i\,\varepsilon^{ijk}\,T^k_{n+m} + \frac{D}{2}\,n\,\delta_{n,-m} \tag{6.9}$$

where

$$T^i_n = \varepsilon^{ijk}\,\mathcal{T}^{jk}_n \tag{6.10}$$

and $(\sigma^i)^{ab}$ are the Pauli matrices.

In the orthonormal gauge the Lagrangian of the SU(2) symmetric string is given by

$$S \simeq \int_{\tau_i}^{\tau_f} d\tau \int_{\sigma_i}^{\sigma_f} d\sigma \left\{ -\tfrac{1}{2}\, \partial_\mu \phi^i \partial^\mu \phi^i - \frac{i}{2} \bar{\psi}^i \gamma_\mu \partial^\mu \psi^i + \tfrac{1}{2} F^i F^i \right\} \tag{6.11}$$

where the index i runs from 1 to 4 and all the fields in (6.11) transform as $O(4)$ fourvectors.

The physical states are characterized by the following conditions:

$$\mathcal{G}_n^i |\text{Phys}\rangle = \mathcal{L}_n |\text{Phys}\rangle = \mathcal{T}_n^{ij} |\text{Phys}\rangle = 0 \tag{6.12}$$

$$(\mathcal{L}_0 + \alpha_0) |\text{Phys}\rangle = \mathcal{T}_0^{ij} |\text{Phys}\rangle = 0 \tag{6.13}$$

One of the two equations in (6.13) implies that the physical states transform as singlets of the SU(2) group. Therefore, the SU(2) symmetry of the Lagrangian (6.11) is again realized as a "colour" symmetry.

We conclude this section showing that in addition to a SU(2) "colour" symmetry we have also a SU(2) "flavour" symmetry. The Lagrangian (6.11) is in fact manifestly invariant under $O(4) = SU(2) \otimes SU(2)$ transformation. Now the generators of one of the two SU(2) groups are the colour generators $\mathcal{T}^{ij}$, while those corresponding to the other SU(2) group, which we call I^i leave the action invariant and reproduce the gauge operators in the sense that:

$$\left[I_i \, , Q \right] = \Sigma\, Q \tag{6.14}$$

where Q is any gauge operator.

This implies that the physical spectrum of the SU(2) symmetric model is classified according to multiplets of this SU(2) group. Therefore, we have also a flavour symmetry.

However, the beautiful feature of having automatically a flavour symmetry without tachyons in the physical spectrum, is somewhat obscured by the fact that the model has negative norm states for any positive value of the space-time dimension.

7) ZERO POINT FLUCTUATION

In the previous sections we have seen that the quantization procedure imposes for a string model a constraint on the value of the Regge intercept α_0 and of the dimension of the space-time D , if one wants to have a Lorentz invariant theory. The values of the mass of the ground state M_0 and of the dimension of the space-time D are summarized in the following table for all the models considered in the previous sections:

Model	$\alpha' M_0^2$	D
Veneziano	-1	26
Neveu-Schwarz	$-\frac{1}{2}$	10
Ramoud	0	10
O(2) boson sector	0	2
O(2) fermion sector	0	2
O(N)	N/2-1	<0
SU(2)	1	<0

(7.1)

These values are always in agreement with those obtained by Brink and Nielsen (22) using zero point energy considerations.

In the most general dual model the mass square is given by:

$$\alpha' M^2 = \sum_{n=1}^{\infty} \sum_{i,j} \tfrac{1}{2}(n+\alpha_i) \left[a_{n,j}^{+(i)} a_{n,j}^{(i)} + (-1)^{F_i} a_{n,j}^{(i)} a_{n,j}^{+(i)} \right] \qquad (7.2)$$

where $\sum_i$ is a sum over different kinds of oscillators characterized by the number α_i . $\sum_j$ is, on the other hand a sum over the effective components of each kind of oscillator. Finally

$$F_i \begin{cases} = 0 & \text{for bosonic oscillators} \\ = +1 & \text{for fermionic oscillators} \end{cases} \qquad (7.3)$$

In the quantum theory a and a^+ become operators satisfying canonical [anti]commutation relations:

$$\left[a_{n,j}^{(i)}, a_{m,k}^{+(h)}\right]_{\pm} = \delta_{n,m}\, \delta^{ih}\, \delta_{jk} \tag{7.4}$$

Using these commutation relations (7.2) becomes:

$$\alpha' M^2 = \sum_{j,i} \sum_{n=0}^{\infty} (n+\alpha_i)\, a_{n,j}^{+(i)}\, a_{n,j}^{(i)} + \alpha' M_0^2 \tag{7.5}$$

where

$$\alpha' M_0^2 = \tfrac{1}{2} \sum_{j,i} (-1)^{F_i} \sum_{n=0}^{\infty} (n+\alpha_i) \tag{7.6}$$

is the mass square of the ground state.

Due to the infinite number of degrees of freedom of a string the series (7.6) is divergent; so if we want to compute the mass of the ground state we must give a meaning to the infinite series (7.6). Note that the appearance of an infinite series is not peculiar to the string model. We have a divergent series for any system with an infinite number of degrees of freedom. The same problem is for example present in computing the zero point energy in the Casimir effect (23), which generates an attraction between two conducting planes in the vacuum. The calculation is usually done introducing a cut-off for the high frequencies and picking up the finite term which does not depend on the cut-off. The same procedure has been applied by Brink and Nielsen for computing the zero point energy in the case of a string. An equivalent regularization procedure has been suggested by Gliozzi (24) and it is a kind of "dimensional" regularization. He suggested to replace the sum appearing in (7.6) with the Riemann ζ function

$$\sum_n (n+\alpha_i) \rightarrow \sum_n (n+\alpha_i)^d = \zeta(d,\alpha_i) \tag{7.7}$$

which is an analytic function in the whole complex plane except $d = 1$. However, we are interested only at the value of ζ for $d = -1$ where ζ is perfectly defined. One gets

$$\zeta(-1, \alpha_i) = \tfrac{1}{2} \left(- \frac{1}{6} - \alpha_i^2 + \alpha_i\right) \tag{7.8}$$

Inserting this expression in (7.6) one gets:

$$\alpha' M_0^2 = \frac{1}{4} \sum_i (-1)^F D_{\text{eff}}^i \left(- \frac{1}{6} - \alpha_i^2 + \alpha_i\right) \tag{7.9}$$

One can easily check that the previous relation is satisfied for all the existing dual models and it is presumably a relation that any dual model must satisfy.

8) NAHM APPROACH

One of the most important ingredients of a dual model is the gauge algebra which is responsible for the decoupling of the negative norm states from the physical spectrum. Therefore, a possible way to construct more realistic models is to search for new gauge algebras. This approach has led to the construction of the supersymmetric models and to the discovery of supersymmetry itself, which was discovered in the NSR model and extended then by Wess and Zumino (25) to a fourdimensional field theory. However, this approach seems at the moment quite difficult to pursue because it would presumably mean finding an extension of the conformal algebra other than supersymmetry.

Nahm (6) has recently proposed a different method of constructing more realistic models based on studying the properties of the partition function which describes the physical spectrum. The partition function has been defined in (2.29) and it is given by

$$A(w) = \sum_m d(m)\, e^{2i\pi\alpha' m^2 w} \tag{8.1}$$

In the statistical bootstrap approach it is more convenient to use the following function:

$$\tau(E) = \sum_m \int d(m)\, \delta(E - \sqrt{k^2+m^2})\, d^R K \tag{8.2}$$

where R is the number of space dimensions. It turns out that the experimental data are very well reproduced by the following function

$$\tau(E) = A\, e^{\beta E} \tag{8.3}$$

where

$$A = \frac{1}{2\alpha'} \; ; \quad \beta = 2\pi\sqrt{\alpha'} \; ; \quad \sqrt{\alpha'} = 0.95 \text{ GeV}^{-2} \tag{8.4}$$

Is it possible to understand these properties ? What kind of partition function should one have in order to reproduce (8.3) and (8.4) ?

The fundamental assumption that has been made by Nahm is the following:

1) A(w) is a modular function for a subgroup G of finite index ν_G of the modular group Γ

Let me start with defining the modular group. The modular group Γ is the set of the following transformations acting on a complex variable w :

$$w \rightarrow w' = \frac{aw+b}{cw+d} = \varphi w \tag{8.5}$$

where a, b, c, d are integers and satisfy the condition $ad-bc = 1$. More precisely, the previous group is $SL(2,Z)$ and the modular group is obtained from $SL(2,Z)$ dividing by ± 1:

$$\Gamma = \frac{SL(z,Z)}{(\pm 1)} \tag{8.6}$$

taking into account the fact that a, b, c, d and $-a$, $-b$, $-c$, $-d$ generate the same transformation on w. It is easy to see that Γ transforms the complex upper half plane H into itself. The index ν_G of a subgroup G of the modular group is given by the number of cosets of G.

Finally $A(w)$ is a weakly modular function with respect to a subgroup G of Γ if it transforms as follows under any transformation φ of G:

$$A\left(\frac{aw+b}{cw+d}\right) = (cw+d)^{-k/2} A(w) \tag{8.7}$$

where k is called the weight of $A(w)$.
A weakly modular function, which is also meromorphic in H, is called a modular function.

Since we do not want to have a physical spectrum increasing more than exponentially in the mass we can limit ourselves to a partition function which is a holomorphic function in H (see Eq. (8.1)).

Each subgroup G of Γ is characterized by two integers r and s, which appear in the two basic transformations of G:

$$w \rightarrow w + r \tag{8.8}$$

$$\frac{1}{w} \rightarrow \frac{1}{w} + s \tag{8.9}$$

The most general transformation of G can then be obtained as a product of a finite number of transformations (8.8) and (8.9). For example, the modular group itself is characterized by $r = s = 1$, and any other transformation of Γ can be generated by a finite product of (8.8) and (8.9) with $r = s = 1$. From assumption 1) and from the definition of a partition function (8.1) we can derive a number of physical consequences.

Because of (8.7) the partition function $A(w)$ is left invariant under (8.8):

$$A(w + r) = A(w) \tag{8.10}$$

That means that the masses of the physical particles are quantized in units of $1/r$:

$$\alpha' m^2 = \frac{n}{r} \tag{8.11}$$

where n is any integer. Therefore the choice of a subgroup G of Γ implies a well-defined quantization relation for the masses of hadrons.

The transformation (8.9) has also a physical meaning because it implies [see Ref. (6)] that the function $\tau(E)$ defined in (8.2) is given by:

$$\tau(E) = (2\alpha')^{-k/2} \sum_{n=1}^{N_0} d_0(n) \exp 4\pi \left[\sqrt{\frac{\alpha' n}{s}}\, E\right] \tag{8.12}$$

where $k = D-2$ is the number of transverse dimensions. Therefore if one assumes that $A(w)$ is a modular function with respect to G the two integers r and s have an immediate physical meaning.

Given the subgroup G , assumption 1) is not strong enough to fix uniquely the partition function $A(w)$. In order to eliminate this arbitrariness, Nahm has made the following additional assumption:

2) A(w) has no zeroes and only one simple pole in the fundamental domain H/G

The fundamental domain for G is an open set D which does not contain any pair of distinct equivalent points. Two points z_1 and $z_2 \in H$ are equivalent if a transformation $\varphi \in G$ exists such that $\varphi z_1 = z_2$. This second assumption that, as we shall see does not stand on as firm a ground as the first one, fixes completely $A(w)$ once the subgroup G is given. Using assumption 2) , one can prove that

$$k = D - 2 = \frac{24}{\nu_G} \tag{8.13}$$

where ν_G is the index of G and

$$\tau(E) = (2\alpha')^{-\frac{D-2}{2}} \exp[\beta E] \tag{8.14}$$

with

$$\beta = 4\pi \left[\frac{\alpha'}{s}\right]^{\frac{1}{2}} \tag{8.15}$$

All the consistent dual models that we have described in the previous sections satisfy both assumptions 1) and 2) . In particular in the case of the conventional string model the relevant group is the modular group itself, and $A(w)$ in (2.29) is the only modular function satisfying assumption 2) . Because of (8.13) and (8.15) the conventional string model has:

$$D = 26 \qquad \text{and} \qquad \beta = 4\pi\sqrt{\alpha'} \tag{8.16}$$

In the case of the spinning string the relevant subgroups are $G_0[r=2\ ;\ s=1]$ and $\Gamma_0(2)\ [r=1\ ;\ s=2]$ for the Neveu-Schwarz and the Ramond model respectively and the partition functions in (4.37) and in (4.38) are characterized by assumption 2) . Since both the previous subgroups have index $\nu_G = 3$ then the dimension of the space-time is $D = 10$ and

$$\beta = 2\pi\sqrt{2\alpha'} \tag{8.17}$$

Following assumptions 1) and 2) one can construct partition functions corresponding to possible dual models with critical dimensions $D = 8,6$ as well as 4 . In particular for $D = 4$ one can construct several possible partition functions corresponding to different subgroups of Γ . The most promising among these subgroups is that which corresponds to a model that Nahm calls the "correct model". It corresponds to a subgroup of Γ which is generated by $\Gamma(4)$ plus the additional transformation

$$w' = \frac{2w + 3}{w + 2w} \tag{8.18}$$

where $\Gamma(4)$ is the congruence subgroup of order 4 . This subgroup G is characterized by $r = 4$ and $s = 1$ and has index $\nu_G = 12$, that implies:

$$D = 4 \qquad \text{and} \qquad \beta = 2\pi\sqrt{\alpha'} \tag{8.19}$$

which agree with (8.4).
The partition function of the "correct model" is given by

$$A(w) = \prod_{\substack{n=1,2\,\cdots \\ r=1/2,3/2\cdots}} (1-q^n)^{-2}\,(1+q^r)^3\,(1+q^{r/2})^2\,(1+q^{2n})^2 = \tag{8.20}$$
$$= 1 + 2q^{1/4} + 4q^{1/2} + 8q^{3/4} + 12q + 20q^{5/4} + \cdots$$

In addition to the orbital modes, which give the factor $\left(\frac{1}{1-q^n}\right)^2$ and to the Neveu-Schwarz-like modes, which give the factor $(1-q^r)^3$, there are also two other infinite sets of new oscillators that the rotational invariance forces us to interpret as being due to some kind of internal symmetry such as strangeness and charm. The low energy spectrum which is given in fig. (1) does not contain any tachyon. It presents the following serious problems:

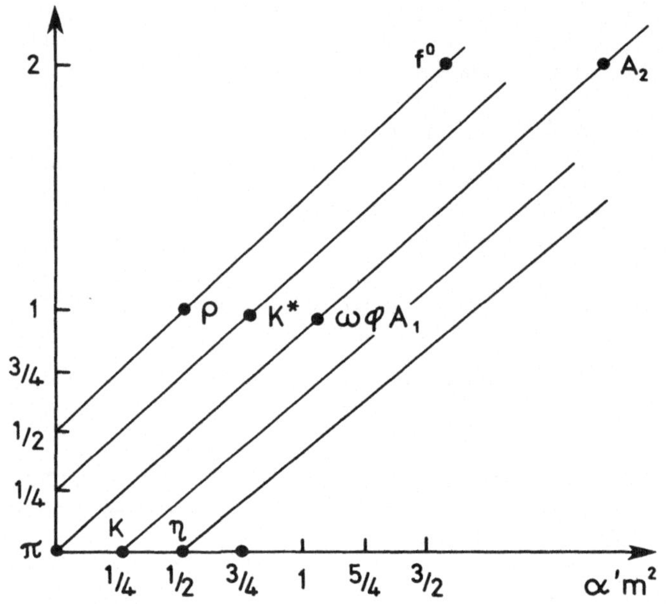

fig. 1

i) No isotopic spin
ii) ω and A_2 are shifted $\frac{1}{2}$ unit too high
iii) There is a "strange" scalar meson at the same mass as the K^* and there are exotic states with strangeness larger than 1 with mass $\alpha' m^2 = 1$
iv) The lowest charmed mesons have a mass $\alpha' m^2 = 4$, which is too low with respect to the experimental value. They are, however, stable.

This shows that the "correct model" is not so physical after all. It has the correct value of the space-time dimension but the spectrum of particles does not quite coincide with the spectrum that comes out from experiments. It does not contain any tachyon, but it is not clear whether a tachyon could be present in some other sector of the model, as in the case of the Ramond model which shows a tachyon in the meson sector (Neveu-Schwarz model). Nevertheless, I think that it would be very interesting to work out the model explicitly and find the gauge algebra, which could also have interesting applications outside the domain of dual models.

We conclude this section with a discussion of the meaning of assumptions 1) and 2). One of the reasons for using these assumptions is due to the fact that all the existing dual models satisfy them. However, we must not forget that they also have some non-physical features; therefore it is not very healthy to imitate them too much. In particular assumption 2) does not seem to agree with a phenomenological estimate done by B. Mortensen, B. S. Nielsen, and H. B. Nielsen (26), who constructed the partition function inserting directly the experimental data listed in the Rosenfeld tables. Although one must be careful in drawing too many conclusions from data, which are not so precisely known, the experimental partition function seems to show at least one zero in a region sufficiently far away from the real axis that we can be confident that it is not a spurious zero introduced by the errors in the experimental data. We can, therefore, hope that a more physical partition function may be found if one changes slightly assumption 2) .

On the other hand, assumption 1) seems to me more reliable and it can presumably be proved using the path integral formulation of a string. Another way of understanding this assumption is by using the requirements of duality and unitarity that any dual model must satisfy (28).

At the tree level, for instance, in the case of a four point amplitude, duality states that the scattering amplitude corresponding to the duality diagram in fig. (2) must contain both the poles in the s and t channels. This statement can be generalized in the case of a one-loop diagram which is given in fig. (3a)

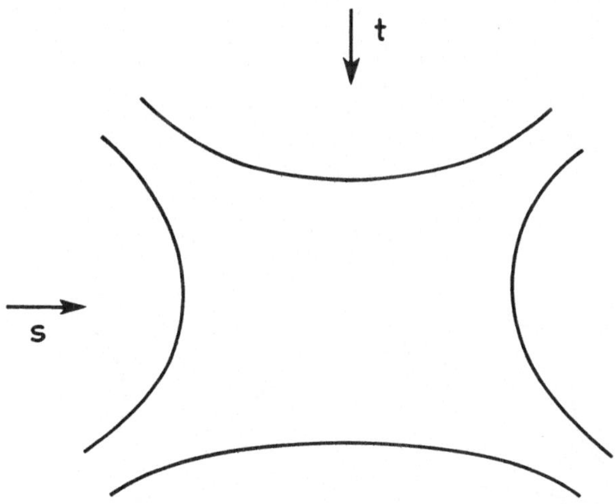

fig. 2

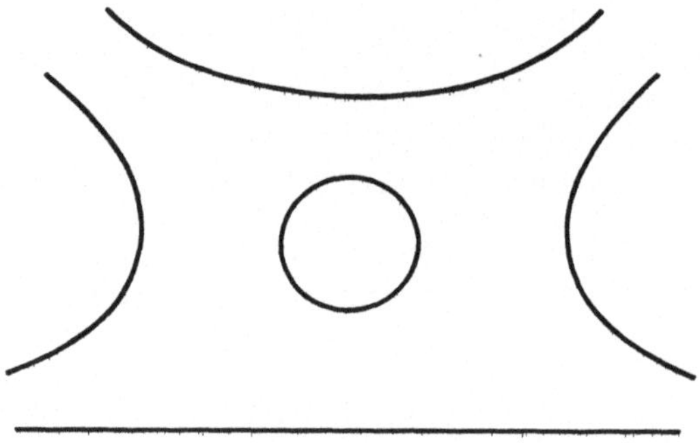

fig. 3a

and which is topologically equivalent to the one in fig. (3b). Because of this equivalence one must be able to write the corresponding scattering amplitude either in a way where the unitarity cut is explicitly displayed [fig. (3a)] or in a way where we see explicitly a closed string (pomeron) going into the vacuum [fig. (3b)]. In the first case the N-point amplitude is given by:

$$B_N\,(p_1 \cdot\cdot p_N) = \int \frac{d\omega}{\omega} \prod_{i=1}^{N-1} \frac{dx_i}{x_i}\, F_N\,(\nu_i\;,\; w\;;\; p_i) \tag{8.21}$$

where the integration is performed in the domain defined by

$$0 < \omega < x_1 < \cdots < x_{N-} < x_{N-1} < x_N \equiv 1 \tag{8.22}$$

and

$$F_N = \mathrm{Tr}\left\{\omega^{-L_0+\alpha' m^2}\, PV\,(p_1\,,x_1) \qquad V(p_N,x_N)\right\} \tag{8.23}$$

P is the projection operator on the physical states and m is the mass of the external particles. We have the following

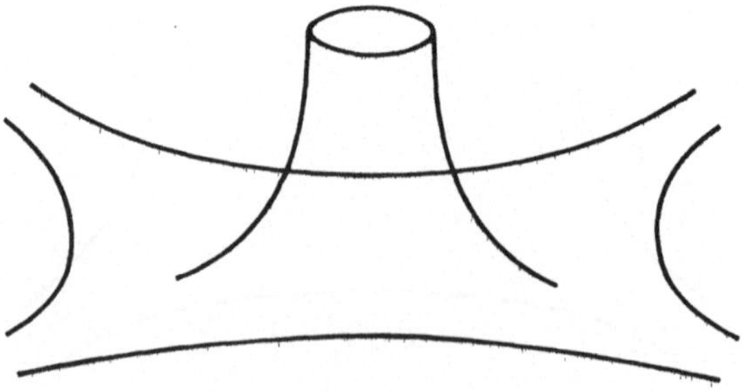

fig. 3b

relations among the various variables:

$$\omega = e^{2i\pi w} \quad ; \quad x_i = e^{2i\pi\nu_i} \tag{8.24}$$

The amplitude corresponding to the diagram in fig. (3b) is given by:

$$B_N = \int_0^1 \frac{dr}{r} \int \prod_{i=1}^{N-1} \frac{d\rho_i}{\rho_i} \; G_N\,(\nu_i' \;, w' \;; p_i) \tag{8.25}$$

with

$$G_N(\nu_i' \;, w' \;; p_i) = {}_0\langle 0|\; \overline{\mathrm{Tr}} \left\{ (r^2)^{-L_0+\alpha' m^2} P \quad V(p_1,\rho_1) \cdots V(p_N,\rho_N) \right\} |0\rangle_0 \tag{8.26}$$

where $|0\rangle_0$ is the vacuum state of the zero mode and $\overline{\mathrm{Tr}}$ is the trace taken over only the non zero modes. The integral in (8.25) is performed on the unit circle with the following restrictions

$$\theta_N = 0 < \theta_{N-1} < \theta_{N-2} < \cdots < \theta_1 < 2\pi \tag{8.27}$$

and

$$r^2 = e^{2i\pi w'} \quad ; \quad \rho_i = e^{2i\pi\nu'} \tag{8.28}$$

Because of the topological equivalence between the two diagrams in figs. (3a) and (3b) there must exist a transformation from the variables (ω, x_i) to the variables (r^2,ρ_i) that shows explicitly the equivalence between (8.21) and (8.25). This is the Jacobi transformation which is given by

$$\nu_i' = \frac{\nu_i}{w} \quad ; \quad w' = -\frac{1}{w} \tag{8.29}$$

Using it in (8.21) one can easily prove that (8.21) and (8.25) are identical, when the function G_N transforms in the following way:

$$G_N(\nu_i', w'; p_i) = e^{-\frac{2i\pi}{w}\left(\sum_{j=1}^{N} \sqrt{\alpha'}\, p_i\, \nu_i\right)^2} w^{-D/2+N+1}\, G_N(\nu_i, w; p_i) \tag{8.30}$$

where D is the dimension of the space-time which emerges from the trace taken over the zero mode in (8.21). Eq. (8.30) is satisfied in the case of the conventional string model and of the Neveu Schwarz model, but it is not satisfied in the Ramond model. This is due to the fact that the Jacobi transformation is contained in Γ and G_θ which correspond respectively to the conventional string model and to the Neveu-Schwarz model, but it is not contained in $\Gamma_0(2)$ which corresponds to the Ramond model. One can check, however, that eq. (8.30) becomes in the case of the Ramond model the following equation:

$$G_N(\nu_i', w'; p_i) = e^{-\frac{2i\pi}{cw+d}\left(\sum_{i=1}^{N} \sqrt{\alpha'}\, p_i\, \nu_i\right)^2} (cw+d)^{-D/2+N+1} G_N(\nu_i, w, p_i) \tag{8.31}$$

where

$$\nu_i' = \frac{\nu i}{cw+d} \quad ; \quad w' = \varphi w = \frac{aw+b}{cw+d} \tag{8.32}$$

with φ belonging to $\Gamma_0(2)$.

We can conclude, therefore, that, as a consequence of the requirements of duality and unitarity, the integrand of the loop diagram in (8.26) must satisfy the eq. (8.31) where φ belongs to a subgroup G of the modular group.

In particular the integrand of a planar loop with no external legs $N = 0$ is the partition function within an inessential proportionality factor. In this case (8.31) reduces to

$$G_0(w') = (cw+d)^{-\frac{D-2}{2}} G_0(w) \tag{8.33}$$

which reproduces the assumption 1) of Nahm. Eq. (8.31) is more general than (8.33) and may be a good starting point for the explicit construction of new dual models.

Assumption 1) is, therefore, very important being the consequence of the postulates of duality and unitarity.

REFERENCES

1. For a general review of dual models see for instance:
   a) Dual theory, ed. M. Jacob (North-Holland, Amsterdam, 1974)
   b) P. Frampton, Dual resonance models (Benjamin, Reading, Mass., 1974)
   c) J. Scherk, Rev. Mod. Phys. 47, 123 (1975)
2. M. Ademollo, L. Brink, A. D'Adda, R. D'Auria, E. Napolitano, S. Sciuto, E. Del Giudice, P. Di Vecchia, S. Ferrara, F. Gliozzi, R. Musto and R. Pettorino, Phys. Letters 62B, 105 (1976)
3. M. Ademollo, L. Brink, A. D'Adda, R. D'Auria, E. Napolitano, S. Sciuto, E. Del Giudice, P. Di Vecchia, S. Ferrara, F. Gliozzi, R. Musto, R. Pettorino and J. H. Schwarz, Nucl. Phys. B111, 77 (1976)

4. M. Ademollo, L. Brink, A. D'Adda, R. D'Auria, E. Napolitano, S. Sciuto, E. Del Giudice, P. Di Vecchia, S. Ferrara, F. Gliozzi, R. Musto and R. Pettorino, Nucl. Physics B114, 297 (1976)
5. D. J. Bruce, D. B. Fairlie and R. G. Yates, Nucl. Physics B108, 310 (1976)
6. W. Nahm, Nucl. Physics, B114, 174 (1976)
7. P. Goddard, J. Goldstone, C. Rebbi and C. Thorn, Nucl. Phys. B56, 109 (1973)
8. S. Fubini, A. Hanson and R. Jackiw, Phys. Rev. D7, 1732 (1973)
9. A. Patrascioiu, Nucl. Phys. B81, 525 (1974)
10. W. Bardeen, I. Bars, A. Hanson and R. Peccei, Phys. Rev. 13D, 2364 (1976)
11. F. Rohrlich, Phys. Rev. Letters 34, 842 (1975); Nucl. Physics B112, 177 (1976)
12. G. 't Hooft, Nucl. Phys. B75, 461 (1975)
13. I. Bars, Phys. Rev. Letters 26, 1521 (1976)
14. P. Collins and R. Tucker, Nucl. Physics B112, 150 (1976)
15. B. Zumino, Proceeding of the Boston Conference on Gauge Theories, MIT press
16. D. B. Fairlie and D. Martin, Nuovo Cimento 18A, 373 (1973)
C. Montonen, Nuovo Cimento 19A, 69 (1974)
17. L. Brink and J. O. Winnberg, Nucl. Physics B103, 445 (1976)
18. P. Goddard and R. Horsley, Nucl. Physics B111, 272 (1976)
19. E. Cremmer and J. Scherk, Nucl. Phys. B103, 399 (1976)
20. H. Osborn, Nuovo Cimento Letters 2, 717 (1969) and 3, 135 (1970)
J. Ellis, Nucl. Phys. B21, 217 (1971)
J. Schwarz and D. J. Wallace, Phys. Rev. D6, 723 (1972)
21. P. Ramond and J. H. Schwarz, Phys. Letters 64B, 75 (1976)
E. Corrigan and D. Olive, private communication
22. L. Brink and H. B. Nielsen, Phys. Lett. 45B, 332 (1973)
23. See for instance T. H. Boyer, Annals of Physics 56, 474 (1970)
24. F. Gliozzi, private communication
25. J. Wess and B. Zumino, Nucl. Phys. B70, 39 (1974)
26. H. B. Nielsen, Lectures given at the Scottish Summer School (1976)
27. W. Nahm, private communication
28. L. Brink, A. D'Adda and P. Di Vecchia, to be published.

WHO NEEDS CLUSTERS IN THE DUAL BOOTSTRAP?*

Y. Zarmi

Department of Theoretical Physics
The University of Bielefeld
Bielefeld, Germany

and

Department of Nuclear Physics
The Weizmann Institute
Rehovot, Israel

ABSTRACT

We review recent progress in the planar Dual Bootstrap program. The delicate interplay between the analytic properties of Reggeon amplitudes and the proper counting of the contribution of all intermediate states to the planar unitarity summation is exposed. This interplay guarantees a pure Regge-pole-type solution to the bootstrap program. The naive expectation that planar unitarity will enforce the existence of Regge cuts does not materialize at the planar level. The notion of clusters

* This work is dedicated to the memory of Wing Sum Lam, a dear friend and colleague.

Material covered is essentially based on the work of Refs. 9 and 12 of this paper.

Research supported in part by the U.S.-Israel Bi-national Science Foundation.

as narrow resonances is not used. Rather, they are viewed as bins of arbitrary size into which all available phase space in the intermediate state is divided.

TOPOLOGICAL EXPANSION

The Topological Expansion (T.E.) or Dual Unitarization approach[1] to strong interaction amplitudes is an intermediate step in the search for a fundamental description of hadron physics. It emerges as a method of studying Quantum chromodynamics[2] (the theory of coloured quarks and gluons (-QCD) when the non-perturbative $\frac{1}{N}$ expansions are employed.[3] In such expansions terms with complicated topologies are damped by powers of $\frac{1}{N}$.

The TE enables us to relate several aspects of hadron physics, and introduces a small dimensionless parameter $(\frac{1}{N})$ into strong interaction dynamics. It has yielded many interesting results in the physics of low energy[4], present accelarator energies[5] and super asymptotic energies.[6]

PLANAR UNITARITY

The first step of the TE is the calculation of the zero order term - the planar S-matrix. At the present state of art this is impossible. However, planar scattering amplitudes are expected to obey a relation called planar unitarity.[1] For a process $x \to y$ one has

$$\operatorname{Im} A^{(\text{planar})}_{x \to y} = \sum_n A^{(\text{planar})}_{x \to n} \, A^{(\text{planar})\,*}_{y \to n} \tag{1}$$

where the unitarity summation is over all possible intermediate states, and the two planar amplitudes are "sewn" together in a planar manner. Planar unitarity becomes a bootstrap scheme when one assumes that all planar amplitudes are dominated at high energy by simple Regge poles. In view of the simple properties of these amplitudes in energy variables (only right hand cuts, no third double spectral functions) one expects that they should not have Regge cuts. The problem is now to see whether these two conjectures are indeed possible and consistent with planar unitarity.

MULTIPERIPHERAL CLUSTER PRODUCTION

In order to convert the set of nonlinear equations represented by eq. (1) into a powerful calculational tool, one imposes on the production amplitudes some multiperipheral assumptions. Intending to use the information provided by duality, one replaces the assumption of multiperipheral production of stable particles[7,8] by a weaker one: that the unitarity sum in eq. (1) can be saturated by intermediate states composed of some relatively long lived resonant states (clusters), which are produced multiperiherally and then decay into stable particles. Since resonances are narrow, the interference amongst particles originating from different resonances may be neglected. The phenomenological validity of this approach has been studied extensively.[4,5] It has some theoretical justification[3] within the framework of QCD[3]. The approach is appealing both theoretically and phenomenologically, as it relates properties of low energy resonances (spectrum, couplings) to those of high energy amplitudes. In particular, the resulting bootstrap constraints depend explicitly on the cluster $(\text{mass})^2$ cut-off $\bar{s}$ (or, in rapidity language, the cluster length cut-off $\bar{L}$). A study of various models adopting this approach indicates that[9] they can lead to a leading Regge-pole to leading Regge-pole bootstrap. However, in general they end up with output amplitudes that have in addition to the imput Regge-poles, some Regge-cuts as well. This happens even when duality constraints in the form of Finite Energy (Mass) Sum Rules (FESR, FMSR) are taken into account. Thus, complete consistency of the scheme is not achieved (pure-pole input amplitudes lead to poles and cuts in the output). Moreover, the starting conjecture[1] that planar amplitudes do not have cuts in the J-plane is not established.

A common feature of all these models is that their classification of final state events in terms of clusters (instead of the observed stable particles) leads to double - or under - counting of events. We mention here two typical examples both in energy and rapidity variables (see Fig. 1). For every two adjacent clusters:

1. Maximal Cluster Size, Minimal Gap Size

$$s_{th} \le s_1, s_2 \le \bar{s} \qquad\qquad o \le L_1, L_2 \le \bar{L} \tag{2}$$

$$\bar{G}/s_{th} \le s_{12}/s_1 s_2 \qquad\qquad \bar{g} \le g$$

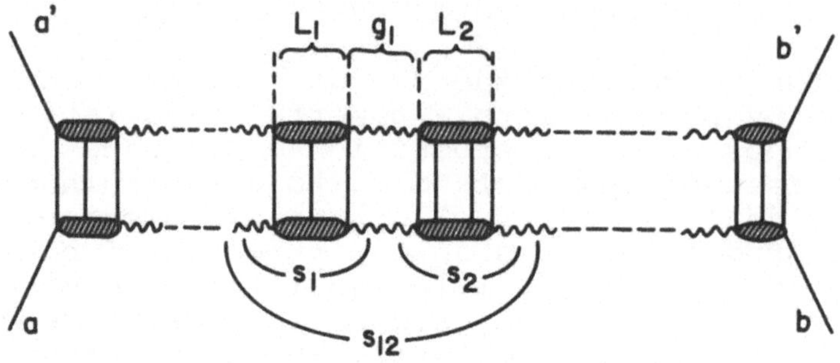

Fig. 1 Kinematics in a multiperipheral cluster chain.

s_{th} is a typical threshold for hadron amplitudes.
$\bar{s} \tilde{=} s_{th} e^{\bar{L}}$ is the cluster size cut-off. The dynamical minimal-gap constraint guarantees that a leading Reggeon exchange across the gap is a good approximation. However, this condition neglects events with gaps which are smaller than $\bar{G}(\bar{g})$; hence it under-counts.

2. Symmetric No Double Counting Condition (NDC)

$$s_{th} \leq s_1, s_2 \leq \bar{s} \qquad o \leq L_1, L_2 \leq \bar{L}$$
$$\bar{s} \leq s_{12} \qquad \bar{L} \leq L_1 + g_1 + L_2 \tag{3}$$

This NDC is typical to models[4,5,10,11] trying to avoid double counting à la duality. However, if we view each cluster as populated by stable particles, we see that double-counting is committed. This is easiest seen in rapidity. At the equality point, $L_1 + g_1 + L_2 = \bar{L}$, two clusters may be confused with a single one of size $\bar{L}$. Moreover, even for $L_1 + g_1 + L_2 > \bar{L}$ there are different inequivalent divisions of the same rapidity interval into clusters.

PROPER COUNTING - BINS INSTEAD OF CLUSTERS

The simplest way to count events uniquely is to abandon the concept of dynamical clusters, simply divide phase space into populated bins with $(\text{mass})^2$ up to $\bar{s}$ (rapidity length up to $\bar{L}$) and impose the constraint[9,11,12] (see Fig. 1):

$$s_{th} \leq s_1 \leq \bar{s} \qquad o \leq L_1 \leq \bar{L} \tag{4}$$
$$\bar{s}/s_{th} \leq s_{12}/s_2 \qquad \bar{L} \leq L_1 + g_1$$

This asymmetric NDC should be applied all along the multiperipheral chain. Each event is counted then once and only once. Hence a bin is defined in correlation with the gap lying between it and the next bin. Bins have nothing to do with narrow resonances. Since the cut-off $\bar{s}(\bar{L})$ is chosen arbitrarily, results should not depend on it.

Let us stress at the outset that this approach has its own weak points. First of all, Reggeon exchanges between bins are assumed even when the gaps are small (however, this corresponds to keeping only leading terms). Moreover, if particle production does proceed via an intermediate stage of narrow resonances, then the brutal division of phase space into bins will sometimes assign particles coming from one resonance to two different (adjacent) bins. As a result, the contribution of a single direct channel resonance will be approximated by that of a single crossed channel Regge-Pole - a bad approximation. The model is therefore expected to be good only when resonances have large widths so that overlap is substantial.

DEFINITION OF AMPLITUDES, FMSR

The planar particle scattering amplitude (see Fig. 2a) is normalized so that

$$I_m A_{ab \to a'b'} \xrightarrow[s\to\infty]{} \frac{\gamma_{aa'}(t)\gamma_{bb'}(t)}{\Gamma(\alpha(t)+1)} (s/s_o)^{\alpha(t)} \tag{5}$$

The scale factor will be (as in the dual model) $s_o = 1/\alpha' \cong 1$. Studying planar amplitudes with external Reggeon legs one finds that certain factors that introduce kinematical singularities have to be extracted.[12,13,14] Having done so, our Reggeon-particle (Rp) amplitude (see Fig. 2b) is extracted from the six-point function such that

$$\frac{1}{2i} D_{isc\ M^2} A_{R_1(t_1^-)a \to R_1'(t_1^+)a'} \xrightarrow[M^2\to\infty]{} \frac{\gamma_{aa'}(t)g(t,t_1^{\pm})}{\Gamma(\alpha(t)+1)} (M^2/s_o)^{\alpha(t)-\alpha(t_1^-)-\alpha(t_1^+)} \tag{6}$$

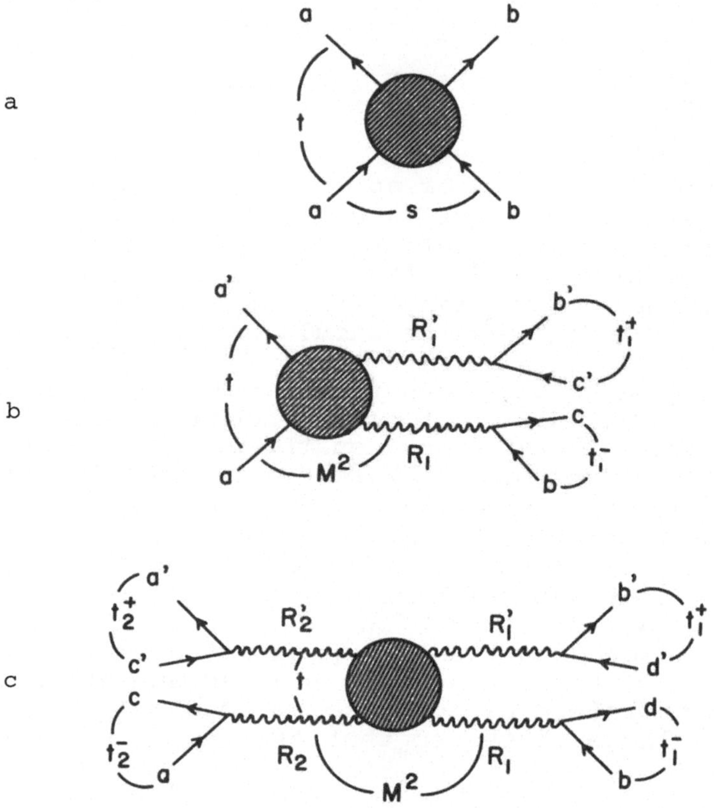

Fig. 2 Kinematics of particle amplitude (a), Reggeon-particle amplitude (b) and Reggeon-Reggeon amplitude (c).

The Reggeon-Reggeon (RR) amplitude is extracted from an eight-point function (see Fig. 2c) such that

$$\frac{1}{2i}\, D\mathrm{isc}_{M^2}\, A_{R_1(t_1^-)R_2(t_2^-)\;\to\;R_1'(t_1^+)R_2'(t_2^+)} \xrightarrow[M^2\to\infty]{} \frac{g(t,t_1)\,g(t,t_2^{\pm})}{\Gamma(\alpha(t)+1)} \left(M^2/s_o \right)^{\alpha(t)-\alpha(t_1^-)-\alpha(t_2^-)-\alpha(t_1^+)-\alpha(t_2^+)} \qquad (7)$$

Here γ- is the particle-Reggeon coupling and g is the triple-Regge coupling. For simplicity, we shall deal with the case of only <u>one</u> pair of exchange degenerate trajectories (the important extension to more than one pair does not change our conclusions). Particle scattering amplitudes satisfy the standard FESR

$$\int_{s_{th}}^{\bar{s}} s^n I_m A_{ab\to a'b'} ds \cong \frac{\gamma_{aa'}(t)\gamma_{bb'}(t)}{\Gamma(\alpha(t)+1)} \frac{(\bar{s}/s_o)^{\alpha(t)+n+1}}{\alpha(t)+n+1} \quad , \quad n = o,1,2,... \tag{8}$$

Abstracting from the Dual Model, Rp amplitudes satisfy

$$\int_{s_{th}}^{\bar{s}} (M^2)^n \frac{1}{2i} \mathrm{Disc}_{M^2} A_{Ra\to R'a'} dM^2 = \frac{\gamma_{aa'}(t) g(t,t_1^{\pm})}{\Gamma(\alpha(t)+1)} \frac{(s/s_o)^{\alpha(t)-\alpha_{c,1}+n}}{\alpha(t)-\alpha_{c,1}+n} \tag{9}$$

Similarly, guided by the Dual Model, we expect RR amplitudes to obey[12)]

$$\int_{s_{th}}^{\bar{s}} (M^2)^n (M_\perp^2)^{\alpha_{c,2}+1} \frac{1}{2i} \mathrm{Disc}_{M^2} A_{R_1R_2\to R_1'R_2'} (M_\perp^2 P_\pm^2, t, t_1^\pm, t_2^\pm) \cong \frac{g(t,t_1^\pm) g(t,t_2^\pm)}{\Gamma(\alpha(t)+1)} \frac{(\bar{s})^{\alpha(t)-\alpha_{c,1}+1}}{\alpha(t)-\alpha_{c,1}+n} \tag{1o}$$

where the interchange $1 \leftrightarrow 2$ is obviously allowed as well. In eqs. (9,1o)

$$\alpha_{c,i} = \alpha(t_i^-) + \alpha(t_i^+) - 1 \quad , \quad i=1,2 \tag{11}$$

Only leading pole contributions were kept in eqs. (9-11). Notice the peculiar dependence of eq. (1o) on

$$M_\perp^2 = M^2 + P_\perp^2 \tag{12}$$

It reflects the more complicated nature of RR amplitudes. A_{RR} depends on $P_\perp^2$ (transverse momentum of the M^2-intermediate state with respect to the line of motion of the incoming particles, a,b, in the c.m. frame). The particular forms of FMSR (9,1o) will turn out to be crucial in attaining a cut-free internally consistent bootstrap scheme.

They will also affect the resulting J-plane integral equations. The r.h.s. of eqs. (8-1o) has the form of FMSR in the Dual Model (no contribution from lower limit of integration - this has to do with the low energy singularity structure of amplitudes, e.g., single particle pole). Clearly, models in which the s^{α} (or $(M^2)^{\alpha}$) behaviour is extrapolated down to threshold do not satisfy these FMSR precisely.

PLANAR BOOTSTRAP IN ENERGY PLANE-TWO BINS

In the first application of planar unitarity to planar amplitudes[15)] phase space was simply divided into two bins with a Regid division line (see Fig. 3). For simplicity, the division line was put exactly in the middle:

$$s_{th} \leq s_1, s_2 \leq \sqrt{s_{th}\ s} \qquad o \leq L_1, L_2 \leq Y/2 \qquad (13)$$

Notice that here NDC (4) is obeyed automatically. For particle scattering the equation implied by Fig. 3 is (using the dual example)

$$A_{ab\to a'b'} = \pi \int d\phi_1 \eta_1 \ (s/s_o)^{\alpha_{c,1}}$$
$$\int_{s_{th}}^{\bar{s}} ds_1 A_{aR\to a'R'}(s_1, t, t_1^{\pm}) \int_{s_{th}}^{\bar{s}} ds_2 A_{Rb\to R'b'}(s_2, t, t_1^{\pm}) . \qquad (14)$$

Here $d\phi_1$ is the standard loop integration over $t_1^{\pm}$, and

$$\eta_1 = \Gamma(-\alpha(t_1^-))\Gamma(-\alpha(t_1^+))\cos \pi(\alpha(t_1^-) - \alpha(t_1^+)) . \qquad (15)$$

The cosine factor is due to the interference of two phase factors, $e^{-i\pi\alpha(t_1^-)} \cdot e^{+i\pi\alpha(t_1^+)}$. Notice that on both sides of eq. (14) A stands for the imaginary part of an amplitude. Both s_1 and s_2 integrals in eq. (14) are precisely of the form eq. (9). Hence, we have

$$A_{ab\to a'b'} = \frac{\gamma_{aa'}(t)\gamma_{bb'}(t)}{\Gamma(\alpha(t)+1)} \ (s/s_o)^{\alpha(t)} \ I_{RV}(\alpha, t) . \qquad (16)$$

The bootstrap is therefore completed if we have

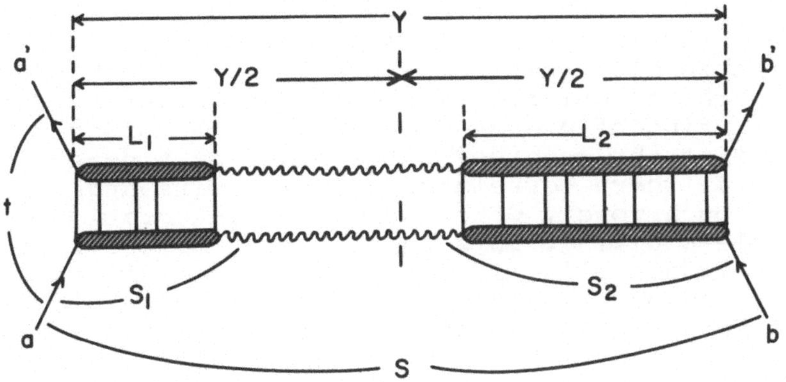

Fig. 3 Planar bootstrap with two (equal) bins.

$$I_{RV}(\alpha,t) \equiv \frac{\pi}{\Gamma(\alpha(t)+1)} \int d\phi_1 \eta_1 \frac{g^2(t,t_1^{\pm})}{(\alpha(t)-\alpha_{c,1})} = 1 \quad (17)$$

where, for simplicity $s_{th} \cong s_o$ ($= 1/\alpha_o \cong 1$) is assumed. This constaint, originally derived in Ref. 15), has been studied numerically[16,17], and yields reasonable values for the Reggeon intercept and slope and for the triple-Reggeon coupling; it is also consistent with the usual additive quark model splitting between the ρ, K^* and ϕ trajectories.[18] Notice that, as expected, the bootstrap condition (17) does not depend on the bin-size cut-off. In fact, the same conclusions hold if two unequal bins are chosen ($s_i \le \bar{s}_i$ i=1,2, or $L_i \le \bar{L}_i$, so that $\frac{\bar{s}_1 \bar{s}_2}{s_{th}} = s$, or $\bar{L}_1 + \bar{L}_2 = Y$). Similar conclusions may be easily reached for Rp and RR amplitudes. Thus, the correct FMSR and proper counting of events guarantee a pure pole-type solution to the bootstrap equation. Nothing essentially new is learned by studying amplitudes with external Reggeon legs (Rather than particle legs).

DUAL BOOTSTRAP IN ENERGY PLANE - MORE THAN TWO BINS

Unlike with two bins, for three or more bins one has to impose NDC (4) explicitly in the unitarity integral, so that adjacent bins do not sweep through each other (double-counting!). There is, of course, the question of what order of integration over bins and gaps should be chosen. For simple smooth amplitudes (e.g. $(M^2)^{\alpha-\alpha_c}$ down to threshold) this is no problem. But when FMSR are used, the singularity structure at low enery and proper counting of events dictate that the order of integrations is the reverse of the order of bin-assignment (see Fig. 4, where correlated adjacent bins and gaps are connected by arrows). Other orders of integration end up with double counting. This complication arises from the fact that middle bins (i.e. with four Reggeon legs) may be empty, and one should not confuse an empty bin with a single particle intermediate state (such a bin has $s = s_{th}$, or $L = o$). This cannot happen with end bins that have at least two particle legs (out of the four). They are always populated. With the correct order of integrations the bootstrap is again solved by a pure pole-type amplitude. Let us stress that for all bin assignment procedures, NDC (4) guarantees that integrals over all bins are of either forms of eqs. (9) or (1o). Again, the only outcome of the bootstrap is the $\bar{s}(\bar{L})$ independent constraint (17) (in all three cases of particle, Rp and RR scattering).

Since we want later on to study integral equations in the J-plane, let us mention here the results (for details, see Ref. 12) for bins with a cut-off $\bar{s}$ ($\bar{L}$ in rapidity) which is energy independent. At a given $s > \bar{s}$ ($Y > \bar{L}$) the full unitarity sum for (e.g.) particle scattering is

$$A_{ab \to a'b'}(s > \bar{s}, t) = \sum_{n=2}^{n_{max}} A^{(n)}_{ab \to a'b'} \qquad (18)$$

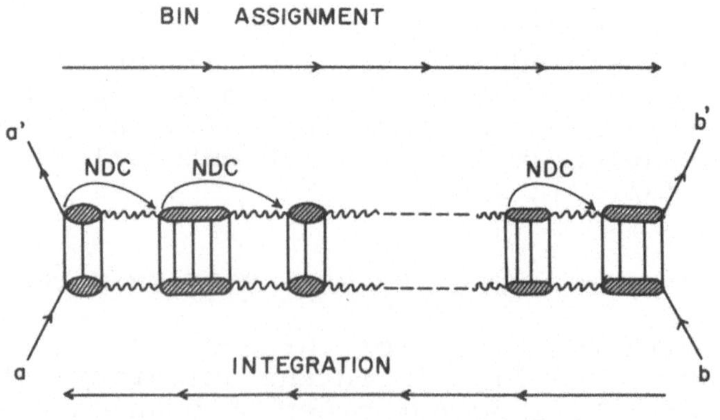

Fig. 4 Planar bootstrap with more than two bins.

Here $A^{(n)}$ is the contribution of those events with precisely n populated bins, and

$$n_{max} = \left[\frac{\ln s/s_{th}}{\ln \bar{s}/s_{th}}\right] + 1 = [Y/\bar{L}] + 1 \qquad (19)$$

A tedious calculation shows that when NDC (4) and FMSR (9, 1o) are employed, then

$$A^{(n+1)}_{ab\to a'b'} = \frac{\gamma_{aa'}(t)\gamma_{bb'}(t)}{\Gamma(\alpha(t)+1)} \{\theta(\bar{s}^{n+1}-s)\theta(s-\bar{s}^{n})s^{\alpha(t)}$$

$$(I_{RV}(\alpha,t))^{n} - \theta(s-\bar{s}^{n})C_n(s,t)I_{RV}(\alpha,t) + \theta(s-\bar{s}^{n+1})C_{n+1}(s,t)\} \qquad (2o)$$

C_n are known multiple integrals over triple-Reggeon couplings. Their s dependence exhibits Regge-cuts in the J-plane. Clearly, if the bootstrap condition (17) holds (it is independent of $\bar{s}$!) then in the sum (18) all C_n cancel. Only the $s^{\alpha(t)}$ piece of $A^{(n_{max})}$ survives to give

$$A_{ab\to a'b'}(s > \bar{s},t) = \frac{\gamma_{aa'}(t)\gamma_{bb'}(t)}{\Gamma(\alpha(t)+1)} s^{\alpha(t)} \qquad (21)$$

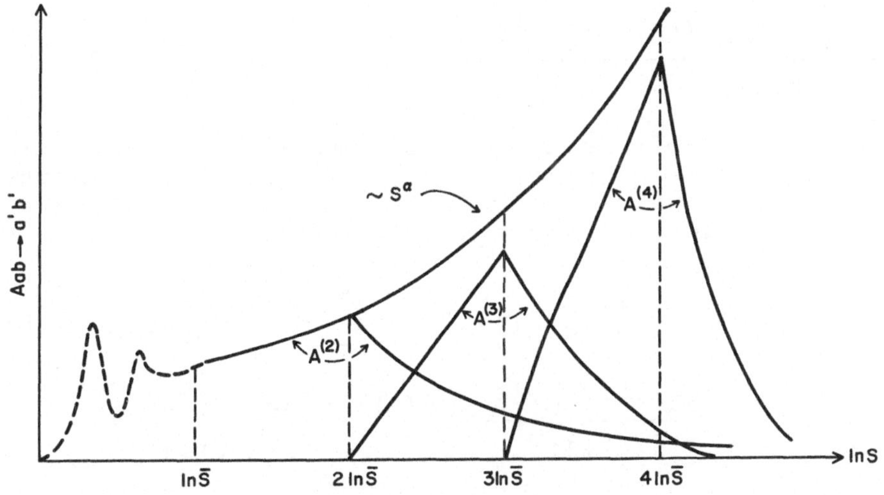

Fig. 5 Decomposition of solution to bootstrap (for $s>\bar{s}$) into n-bin contributions (see eq. (18) versus logs - qualitative only.

Similar results can be shown to hold for Rp and RR amplitudes. The way in which all n-bin terms add up to a single s^{α} form is depicted in Fig. 5. Each term depends on $\bar{s}$, but the sum does not (this is only a qualitative representation).

DUAL BOOTSTRAP IN THE J-PLANE

First, let us study the Mellin transform of $A^{(n+1)}$ of eq. (2o). Its form is [12] (eq. (17) being used).

$$\tilde{A}^{(n+1)}_{ab\to a'b'}(J,t) = \frac{\gamma_{aa'}(t)\gamma_{bb'}(t)}{\Gamma(\alpha(t)+1)} \frac{(\bar{s})^{n(\alpha(t)-J)}}{J-\alpha(t)} (I_{RV}(J,t))^n (1-(\bar{s})^{\alpha(t)-J} I_{RV}(J,t)) \tag{22}$$

with

$$I_{RV}(J,t) = \frac{\pi}{\Gamma(\alpha(t)+1} \int d\phi_1 \eta_1 \frac{g^2(t,t_1^{\pm})}{(J-\alpha_{c,1})(\alpha(t)-\alpha_{c,1})} \tag{23}$$

Summing up all contributions we easily find

$$\tilde{A}_{ab\to a'b'}(J,t) - \tilde{A}^{(1)}_{ab\to a'b'}(J,t) = \sum_{n=2}^{\infty} A^{(n)}_{ab\to a'b'}(J,t) = \frac{\gamma_{aa'}(t)\gamma_{bb'}(t)}{\Gamma(\alpha(t)+1)} \frac{(\bar{s})^{\alpha(t)-J}}{J-\alpha(t)} \tag{24}$$

$\tilde{A}^{(1)}$ is the cut-off Mellin transform, with s integrated only up to $\bar{s}$. This is precisely the correct form expected for an amplitude which is dominated at high s($>\bar{s}$) by a single pole! Indeed, using eq. (5) we have

$$\tilde{A}_{ab\to a'b'}(J,t) - A^{(1)}_{ab\to a'b'}(J,t) = \int_{\bar{s}}^{\infty} s^{-J-1} A_{ab\to a'b'}(J,t)$$

$$= \int_{\bar{s}}^{\infty} s^{-J-1} \frac{\gamma_{aa'}(t)\gamma_{bb'}(t)}{\Gamma(\alpha(t)+1)} s^{\alpha(t)} = \frac{\gamma_{aa'}(t)\gamma_{bb'}(t)}{\Gamma(\alpha(t)+1)} \frac{(\bar{s})^{\alpha(t)-J}}{J-\alpha(t)} . \tag{25}$$

Let us now turn to the task of constructing integral equations. Fig. 6 depicts all integral relations which can be written down for our amplitudes. Of these only Figs. 6c,d constitute integral equations. Figs. 6a,b are only consistency checks. Let us state that all four relations end up with the same bin cut-off independent bootstrap condition (17) and with a pure input pole leading to a pure output pole amplitude. We study as an example Fig. 6c (yielding an integral equation for the Rp amplitude). The Mellin transform

$$\tilde{A}_{R_1 a}(J,t,t_1^{\pm}) \equiv \int_{s_{th}}^{\infty} dM^2 (M^2)^{-J-1} A_{R_1 a\to R_1' a'}(M_1^2 t, t_1^{\pm}) \tag{26}$$

is expected to have a single pole at $J = \alpha(t)-\alpha_{c,1}-1)$ (see high M^2 behaviour - eq. (6)). In the naive approach (smooth power behaviour down to threshold, no bin-gap correlation via e.g., NDC (4)) one has a completely factorized J-diagonal integral equation:

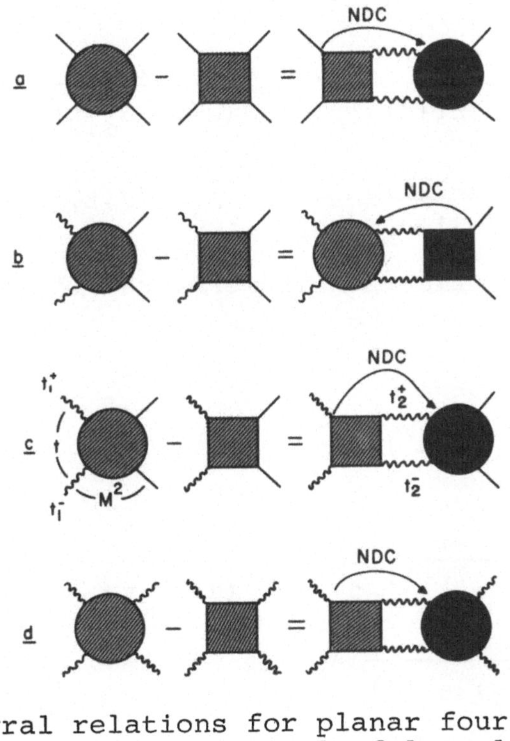

Fig. 6 Integral relations for planar four-point functions. Square-single bin; circle-ful amplitude.

$$\tilde{A}_{R_1a}(J_1t,t_1^{\pm}) - \tilde{A}^{(1)}_{R_1a}(J,t,t_1^{\pm}) = \int d\phi_2\eta_2\tilde{A}^{(1)}_{R_1R_2}(J,t,t_1^{\pm},t_2^{\pm}) \qquad (27?)$$

$$\frac{1}{J-\alpha_{c,2}}\tilde{A}_{R_2a}(J,t,t_2^{\pm})$$

Here $\tilde{A}^{(1)}_{R_1a}$ plays the role of an inhomogeneous term. Eq. (27?) is of the standard CGL type[8]. This, however, is not our case. First, due to NDC (4), the cut-off Mellin transform on the r.h.s. of eq. (27) is converted into a FMSR precisely of the form eq. (1o). Moreover, again, owing to NDC (4), the full Mellin transform $\tilde{A}_{R_2a}(J,t,t_2^{\pm})$ does not appear immediately on the r.h.s. of eq. (27). The lower limit of energy integration over this amplitude is not fixed (as it is in the definition eq. (26)). Rather, it is affected by the NDC. Consequently, it is impossible to obtain eq. (27?). Instead, an integral equation not diagonal in J can be derived:

$$\tilde{A}_{R,a}(J,t,t_1^{\pm}) - \tilde{A}^{(1)}_{R_1a}(J,t,t_1^{\pm}) = (\bar{s})^{\alpha(t)-J-\alpha_{c,1}-1} \cdot \qquad (27)$$

$$\frac{g(t,t_1^{\pm})}{\Gamma(\alpha(t)+1} \; \frac{\pi}{\Gamma(\alpha(t)+1)} \int d\phi_2\eta_2 \frac{g(t,t_2^{\pm})}{(\alpha(t)-\alpha_{c,2})}$$

$$\int \frac{dJ'}{2\pi i(J'+1)(J-J'+\alpha_{c,1}-\alpha_{c,2})}\tilde{A}_{R_2a}(J,t,t_2^{\pm}) \; .$$

For a slight simplification in form, use the shifts

$$J = j-\alpha_{c,1}-1 \quad , \quad J' = j'-\alpha_{c,2}-1 \qquad (28)$$

to find

$$\tilde{A}_{R_1a}(j-\alpha_{c,1}-1,t,t_1^{\pm}) - \tilde{A}^{(1)}_{R_1a}(j-\alpha_{c,1}-1,t,t_1^{\pm}) = (\bar{s})^{\alpha(t)-j}$$

$$\frac{g(t,t_1^{\pm})}{\Gamma(\alpha(t)+1} \; \frac{\pi}{\Gamma(\alpha(t)+1)} \int d\phi_2\eta_2 \frac{g(t,t_2^{\pm})}{(\alpha(t)-\alpha_{c,2})} \qquad (29)$$

$$\int \frac{dj'}{2\pi i(j'-\alpha_{c,2})(j-j')}\tilde{A}_{R_2a}(j'-\alpha_{c,2}-1,t,t_2^{\pm})$$

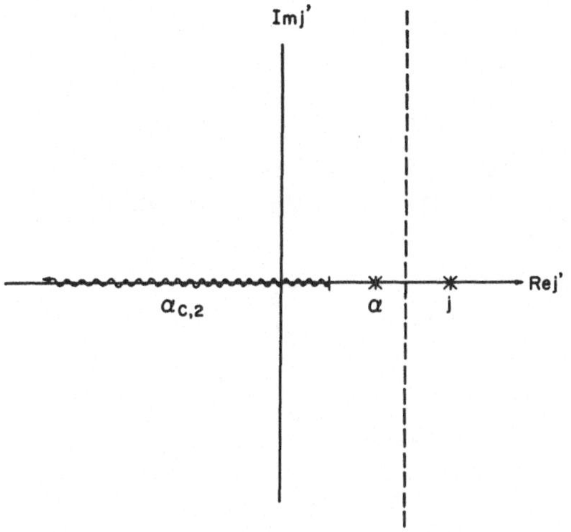

Fig. 7 Integration contour (dashed line) and singularities in j' plane.

The integration path over j' is shown in Fig. 7. After the shift in J,J', $\tilde{A}_{R_i a}$ (i = 1,2) are expected to have a Regge-pole at j,j' = α(t) (corresponding to J = α(t) - $\alpha_{c,1}$ - 1, J' = α(t) - $\alpha_{c,2}$ - 1).

Closing the integration contain of Fig. 7 to the left we pick up the singularities of $\tilde{A}_{R_2 a}(j'-\alpha_{c,2}-1,t,t_2^{\pm})$ (in particular, its pole at j' = α(t)) and, potentially, the pole at j' = $\alpha_{c,2}$. The latter leads to cuts in the resulting amplitude. Luckily enough, this does not happen. We are looking for a solution whose high energy behaviour is given solely by a simple Regge-pole (no cuts!) and which obeys FMSR (9). In J-plane language FMSR (9) means that the Rp amplitudes have a zero at J(J') = -1 (j = $\alpha_{c,1}$, j' = $_{c,2}$) - so that the n=o FMSR can be satisfied. The general form of the expected solution is thus

$$\tilde{A}_{R_2 a}(j'-\alpha_{c,2}-1,t,t_2^{\pm}) = \frac{\gamma_{aa'}(t)\,g(t,t_2^{\pm})}{\Gamma(\alpha(t)+1)}\;\frac{h(j',t,t_2^{\pm})}{j'-\alpha(t)}\;\frac{(j'-\alpha_{c,2})}{(\alpha(t)-\alpha_{c,2})} \qquad (3o)$$

Here h is a smooth regular function of j' which falls sufficiently fast for $j' \to -\infty$ and obeys $h(j'=\alpha(t), t,t_2) \equiv 1$. Thanks to zero at $j' = \alpha_{c,2}$ in eq. (3o), the j' integration in eq. (29) (the contour being closed to the left) only picks up the residue of the pole at $j' = \alpha(t)$, yielding

$$\tilde{A}_{R_1 a}(j-\alpha_{c,1}-1,t,t_1^{\pm}) - \tilde{A}^{(1)}_{R_1 a}(j-\alpha_{c,1}-1,t,t_1^{\pm}) = \frac{\gamma_{aa'}(t)g(t,t_1^{\pm})}{\Gamma(\alpha(t)+1)} \frac{(\bar{s})^{\alpha(t)-j}}{j-\alpha(t)} I_{RV}(\alpha,t) \tag{31}$$

With the by now well recognized bootstrap condition (17) we thus find the right solution.

Thus, without even going through the painstaking job of solving the integral equation we see that the planar bootstrap program is solved by planar amplitudes whose high energy behaviour is dominated by a single Regge pole (no cuts), and whose low energy behaviour is constrained by FMSR (8-1o). In fact, the only constraints of the bootstrap are eq. (17) and the relation

$$\tilde{A}^{(1)}_{R_1 a}(j-\alpha_{c,1}-1,t,t_1^{\pm}) = \frac{\gamma_{aa'}(t)g(t,t_1^{\mp})}{\Gamma(\alpha(t)+1)} \frac{1}{j-\alpha(t)} \left\{ h(j,t,t_1^{\pm}) \frac{j-\alpha_{c,1}}{\alpha(t)-\alpha_{c,1}} - (\bar{s})^{\alpha(t)-j} \right\} \tag{32}$$

which ensures that FMSR (9) is satisfied for n=o.

If the j' contour is closed to the right (in general, this is not allowed) a solution of a more restricted form (with $h(j,t,t_1^{\pm}) \equiv 1$ for all j) results. This solution was already discussed in the literature (for $\bar{s} = 1$)[19].

Eq. (29) can be converted into an equation diagonal in j at the expense of having to separate out the two-bin term[9])

$$[\tilde{A}_{R_1a}(j-\alpha_{c,1}-1,t,t_1^{\pm}) - \tilde{A}^{(1)}_{R_1A}(j-\alpha_{c,1}-1,t,t_1^{\pm})] =$$

$$\tilde{A}^{(2)}_{R_1a}(j-\alpha_{c,1}-1,t,t_1^{\pm}) + g(t,t_1^{\pm})(\bar{s})^{2(\alpha(t)-j)} \cdot \tag{33}$$

$$\frac{\pi}{\Gamma(\alpha(t)+1)} \int d\phi_2 \eta_2 \frac{g(t_1 t_2^{\pm})}{(\alpha(t)-\alpha_{c,2})} \cdot \frac{1}{j-\alpha_{c,2}} \cdot$$

$$[\tilde{A}_{R_2a}(j-\alpha_{c,2}-1,t,t_2^{\pm}) - \tilde{A}^{(1)}_{R_2a}(j-\alpha_{c,2}-1,t,t_2^{\pm})] .$$

This is an integral equation with a factorized kernel (though not of the CGL type[8]) for $\tilde{A} - \tilde{A}^{(1)}$ rather than for $\tilde{A}$. The inhomogeneous term is now $\tilde{A}^{(2)}$.

It is only a matter of further work to show that eqs. (29) and (33) both can be solved by a series solution precisely analogous to eqs. (22 - 24) (with the appropriate alterations due to particle legs being replaced by Reggeon legs). The same holds for RR scattering (Fig. 6d) as well.

CONCLUDING REMARKS

1. The delicate interplay between analyticity and proper counting of events in planar unitarity is essential for internal consistency of the bootstrap. Abandoning either element destroys full consistency (cuts arise), although at the pole-consistency is possible.

2. Instead of "dynamical" clusters (narrow resonances), we have used "mathematical" clusters - bins of phase space (sets of particles) over which averaging à la duality was performed.

3. Proper counting of events via NDC (4) converts various cut-off Mellin transforms into "good" FMSR, and yields J-plane integral equations not of the usual CGL type[8].

4. Cut cancellation is of a "promotion" type. A sum of positive definite terms ($A^{(n)}$) all of which have J-

plane cuts add up to a pure Regge pole. This is unlike the cancellation of the AFS cut[7] which comes about due to the existence of different discontinuity terms with opposite signs[20]. Here, at a given energy, the only remaining contribution is an s^{α} term arising solely from the contribution of $A^{(n_{max})}$ coming from the configuration with maximum allowed number of populated bins. This configuration includes the events of uniform particle distribution. In the ordinary multiperipheral model[7,8] these are the most probable events, and to leading order in log s they give the leading s^{α} behaviour and the multiplicity (~log s). Here, by properly including all possible events (not only the more probable ones) we are able to get rich of non-leading cuts.

5. Multi-Regge kinematics are essential to the approach. Regge exchanges across small gaps had to be assumed (this corresponds to keeping leading terms only-which is what we do anyway). t_{min} effects were neglected, as they only affect singularities with $j \leq \alpha_c - 1$, which are lower than anything we keep track of. Adding more trajectories does not change our conclusions[9,18].

OPEN QUESTIONS

1. Extension to models with dynamical clusters is of great interest. The combination of this approach with ours will necessarily deepen our understanding of the nature of resonances, the dynamics of their production and decay. Moreover, the phenomenological analysis of high energy production data certainly indicates that clustering effects exist[21].

2. The low energy behaviour of planar amplitudes (constrained by FMSR-analyticity) plays a delicate role in achieving a self consistent bootstrap scheme. The integral equations we find are not of the standard CGL type[8]. In the discussion of higher topologies in the TE one usually assumes standard integral equations (calculation of bare Pomeron-cylinder[4,17,22] and Reggeon Field theory[23]). Off-hand, this seems to be justified, as one is only interested in the high energy behaviour of amplitudes. However, as the discussion (here and elsewhere[9] indicates this is not obvious. The iteration of approximate forms for input amplitudes (ignoring, say, low energy constraints) may result in the accumulation of significant quantitative errors in the output amplitude.

3. The inclusion of sister trajectories[24)] in the bootstrap scheme is still an open problem (see discussion in ref. 12).

ACKNOWLEDGEMENT

This review would have not been possible without the extensive collaboration with J.R. Freeman and G. Veneziano, which led to the new results described here. For this they are deeply thanked. Many thanks are due to M. Bishari for lengthy and helpful discussions.

REFERENCES

1. G. Veneziano, Nucl. Phys. B74 (1974) 365.
2. G. 't Hooft, Nucl. Phys. B75 (1974) 461;
C.G. Callan, N. Coote and D.J. Gross, Phys. Rev. D13 (1976) 1649.
3. G. 't Hooft, Nucl. Phys. B72 (1974) 461;
G. Veneziano, Phys. Letters, 52B (1974) 22o, and CERN preprint TH-22oo (1976).
4. G.F. Chew and C. Rosenzweig, Nucl. Phys. B1o4 (1976) 29o;
Chan Hong-Mo, J. Kwiecinski and R.G. Roberts, Phys. Letters 6oB (1976) 367;
Chan Hong-Mo, K.I. Konishi, J. Kwieinski and R.G. Roberts, Phys. Letters 6oB (1976) 469;
C. Schmid, D.M. Webber and C. Sorensen, ETH preprint (1976);
Chan Hong-Mo, Review Talk in this conference, and references therein.
5. P. Aurenche, Ng Sing Wai, Tsun Sheung Tsun, Chan Hong-Mo, A. Gula, T. Inami and R.G. Roberts, Rutherford preprint RL-75-o84 (1975);
N. Papadopoulos, C. Schmid, C. Sorensen and D.M. Webber, Nucl. Phys. B1o1 (1975) 189;
G.F. Chew and C. Rosenzweig, Phys. Letters 58B (1975) 93;
B.R. Webber, Cavendish (Cambridge) preprint HEP 76/5 (1976);
Chan Hong-Mo, Ken-Ichi Konishi, J. Kwiecinski and G.R. Roberts, Rutherford preprint, RL-76-o56 (1976).
6. M. Ciafaloni and G. Veneziano, Phys. Letters 56B (1975) 271;
M. Ciafaloni, G. Marchesini and G. Veneziano, Nucl. Phys. B98 (1975) 472, 493;
G. Marchesini, CERN preprint TH. 2o69 (1975).

7. D. Amati, S. Fubini and A. Stanghelini, Nuov. Cim. 26 (1962) 86.
8. G.F. Chew, M.L. Goldberger and F.E. Low, Phys. Rev. Letters 22 (1969) 2o8;
M.L. Goldberger in Proceedings of the International School of Physics "Enrico Fermi". Course LIV, ed. R. Gatto (Academic Press, (1972)).
9. J.R. Freeman and Y. Zarmi, Nucl. Phys. B112 (1976) 3o3.
1o. J. Kwiecinski and N. Sakai, Nucl. Phys. B1o1 (1976)44.
11. J. Finkelstein and J. Koplik, Columbia preprint co-2271-72 (1975).
12. J.R. Freeman, G. Veneziano and Y. Zarmi, CERN preprint TH-2211 (1976).
13. J. Kwiecinski, Nuovo Cim. Letters 3 (1972) 19;
M.B. Einhorn, J. Ellis and J. Finkelstein, Phys. Rev. D5 (1972) 2o63;
M. Bishari, G.F. Chew and J. Koplik, Nucl. Phys. B72 (1974) 61.
14. J. Kwiecinski, Nucl. Phys. B97 (1975) 475.
15. C. Rosenzweig and G. Veneziano, Phys. Letters 52B (1974) 365.
16. M. Schaap and G. Veneziano, Nuovo Cim. Letters 12 (1975) 2o4.
17. M. Bishari, Phys. Letters 59B (1975) 461.
18. K.I. Konishi, Rutherford preprint RL-76-o36 (1976).
19. M. Bishari and G. Veneziano, Phys. Letters 58B (1975) 445.
2o. S. Mandelstam, Nuovo Cim. 3o (1963) 1127, 1148;
J.C. Polkingharne, Phys. Letters 4 (1963) 24.
21. A. Krzywicki, Review Talk, this conference, and references therein;
G. Thomas, Review Talk, this conference, and references therein;
E. Predazzi, Review Talk, this conference, and references therein.
22. G.F. Chew and C. Rosenzweig, Phys. Rev. D12 (1975) 39o7;
N. Sakai, Nucl. Phys. B99 (1975) 167;
L.A.P. Baĺazs, Phys. Rev. D11 (1975) 1o71.
23. For a review see H.D.I. Abarbanel, J.D. Bronzan, R.L. Sugar and A.R. White, Phys. Reports 21C (1975) 119.
24. P. Hoyer, N.A. Törnquist and B.R. Webber, Nucl. Phys. B115 (1976) 429.

LIST OF PARTICIPANTS

A. Actor, Theoretische Physik der Universität Dortmund
S. Albeverio, Institute of Mathematics, University of Oslo
M.F. G. Dias d'Almeida, Porto, Portugal
I. Andric, Institut "R. Boskovic", Zagreb
C.M. Vassalo Serra Alves, Porto, Portugal
M.C. Oliveira Amorim, Porto, Portugal
M. Bace, Inst. für Theor. Physik, Universität Heidelberg
R. Baier, Fakultät für Physik, Universität Bielefeld
M.L. Machado Cerqueira Bastos, Vila Nova de Gaia, Portugal
K. Baumann, Fakultät für Physik, Universität Bielefeld
H. Behncke, Inst. für Mathematik, Universität Osnabrück
M. Le Bellac, CERN, Theory Division, Genève
J. Bellissard, CNRS, Marseille
M. Benayoun, Collège de France, Paris
N. Bilic, Inst. "Ruder Boskovic", Zagreb
P. Blanchard, Fakultät für Physik, Universität Bielefeld
F. Bopp, Gesamthochschule Siegen
S.K. Bose, Universität Kaiserslautern
G.C. Branco, The City College of New York
O. Bratteli, CNRS, Marseille
E. Brüning, Fakultät für Physik, Universität Bielefeld
D. Buchholz, II Inst. für Theor. Physik, Hamburg
F.J. Lage Campelo Calheiros, Porto, Portugal
V. Canuto, NASA, New York
P. Caraveo, Università degli Studi di Milano
T. Celik, Hacettepe University, Ankara, Turkey
Chan Hong Mo, Rutherford Lab., Berkshire
P.S. Collecott, Cambridge University, Cambridge
J. Cleymans, Fakultät für Physik, Universität Bielefeld
M. Daniel, University of Athens
S. Dimopoulos, Enrico Fermi Inst., University of Chicago
A.M. Din, Inst. of Theor. Physics, Göteborg University
J.A. Dixon, CNRS, Marseille

R. Dobbertin, Laboratoire de Physique Théorique, Université Paris VII, Paris
G. Dorfmeister, Fakultät für Physik, Universität Bielefeld
F. Dustmann, Fakultät für Physik, Universität Bielefeld
W. Driessler, Fakultät für Physik, Universität Bielefeld
J.-P. Eckmann, Dépt. de Physique Théor., Université de Genève
G. Eilam, Israel Institute of Technology, Haifa
J. Engels, Fakultät für Physik, Universität Bielefeld
V. Enss, Fakultät für Physik, Universität Bielefeld
W. Ernst, Fakultät für Physik, Universität Bielefeld
E. Etim, Department of Physics, University of Ibadan, Nigeria
D. de Falco, Istituto di Fisica, Università de Salerno
M.M. Coelho Ribeiro de Faria, Porto, Portugal
F. Fleischer, Fakultät für Physik, Universität Bielefeld
P. Fré, Istituto di Fisica Teorica, Università di Torino
K. Fredenhagen, DESY, Hamburg
S. Fredriksson, The Royal Institute of Technology, Stockholm
A. Frigerio, Università degli Studi di Milano
H. Fritzsch, CERN, Genève
J. Fröhlich, Princeton University, Princeton
H. Galic, Inst. "Ruder Boskovic", Zagreb
A. Gandolfi, Università degli Studi di Parma, Parma
P. Garbaczewski, Inst. of Theor. Physics, University of Wroclaw
W.D. Garber, Inst. für Theor. Physik, Universität Göttingen
L. Garrido, Universidad de Barcelona, Barcelona
H.R. Gerhold, Österreichische Akademie der Wissenschaften, Wien
B. Gidas, University of Washington, Seattle
M.T. Rodriques dos Santos Goncalves, Porto, Portugal
E.H. de Groot, CERN, Genève
H. Grosse, Inst. für Theor. Physik, Universität Wien
F. Guerra, Istituto di Fisica, Università di Salerno
H. Hahn, Inst. für Theor. Physik A, TU Braunschweig
C.J. Hamer, Department of Applied Mathematics and Physics, University of Cambridge
B. Hasslacher, California Institute of Technology, Pasadena, USA
G. Hegerfeldt, Inst. für Theor. Physik, Universität Göttingen
P. Hertel, Fachbereich Naturwissenschaften, Universität Osnabrück
A.C. Hirshfeld, Theoretische Physik III, Universität Dortmund
L. Hudson, Dept. of Mathematics, University of Nottingham
H. Inagaki, Comitato Nazionale per l'Energia Nucleare, Frascati, Roma

P.D.F. Ion, Institut für Mathematik, Universität Heidelberg
F. Jegerlehner, Fakultät für Physik, Universität Bielefeld
J. Jersak, Inst. für Theor. Physik der RWTH Aachen
K. Johnson, Massachusetts Inst. of Technology, Cambridge
G. Jona-Lasinio, Università degli Studi, Roma
K. Kajantie, Dept. of Physics, University of Helsinki
S.S. Kanval, I.I.T. Kanpur, Kanpur, India
M. Karowski, FU Berlin Fb. Physik, Berlin
R.K. Kaul, Department of Physics, Dehli, India
M. Kiera, Insti. für Theor. Physik der RWTH Aachen
K. Kinoshita, Dept. of Physics, Kyushu University, Fukuoka
S. Kitakado, Inst. of Physics, University of Tokyo
A. Knoth, Fakultät für Physik, Universität Bielefeld
I. Koch, Dept. of Mathematics, Bedford College, London
J. Kogut, Cornell University, Ithaca, N.Y., USA
K. Koller, DESY, Hamburg
K. Konishi, Rutherford Lab., Chilton, Didcot, Oxford
A. Krzywicki, Université de Paris-Sud, Orsay
F. Kuypers, Fakultät für Physik, Universität Freiburg
P. Landshoff, CERN, Genève
R. Lima, Faculty of Science, University of Porto
P. Leyland, CNRS, Marseille
K. Litwin, Niels Bohr Institute, København
H. Lotsch, Inst. für Theor. Physik, Universität Wien
Y. Loubatières, Dépt. de Physique Mathematique, Université du Languedoc, Montpellier
J. Lukierski, Inst. of Theoretical Physics, University of Wroclaw
L.E. Lundberg, Matematisk Institut, Universitet København
M. Magg, Inst. für Theor. Physik, RWTH Aachen
J. Magnen, Ecole Polytechnique, Paris
S. Mallik, Inst. für Theor. Physik, Universität Bern
W.J. Marciano, Rockefeller University, New York
P. Martin, Dept. of Physics, ETH Lausanne
M. Mebkhout, CNRS, Marseille
D. Miller, Freie Universität Berlin
M. Mizouchi, Okayama College of Science, Okayama-shi, Japan
M.C. de Oliveira Gomes Moreira, Porto, Portugal
O.M. Vaz Moreira, Porto Portugal
A. Mueller, Institute for Advanced Studies, Princeton
P. Mulders, Inst. for Theor. Physics, Toernooiveld, Nijmegen
H. Narnhofer, Inst. für Theor. Physik, Universität Wien
K. Napiorkowski, Wydzial Fizyki, University of Warsaw
A.T. Ogielski, Inst. of Theor. Physics, University of Wroclaw
M.A. Marques de Oliveira, Porto, Portugal

A. Ostebee, Institute for Theoretical Physics, State University of New York, Stony Brook
R. Page, Centro Atomico Bariloche, Buenos Aires
L.S. Panta, Dept. Math. Physics, University of Birmingham
W. Pesch, Inst. für Theor. Physik, TU Hannover
B. Petersson, Fakultät für Physik, Universität Bielefeld
Ch. Pfister, Theoretische Physik, ETH Zürich
O. Piguet, Max-Planck-Institut für Physik, München
F.M. Pires, Porto, Portugal
H. Pohlmeyer, II Inst. für Theor. Physik, Universität Hamburg
E. Predazzi, Istituto di Fisica Teorica, University of Turin
P.J. Provost, Physique Théorique, Université de Nice
W. Pusz, Faculty of Physics, University of Warsaw
L. O'Raifeartaigh, Institut des Hautes Etudes Scientifiques Bures-sur-Yvette, France
M. do Ceu Fernandes de sa Ramalho, Porto, Portugal
M. Ramon-Medrano, Universidad de Madrid, Madrid
J. Randa, Dept. of Physics, University of Manchester
M. Reed, Dept. of Mathematics, Duke University, Durham
H. Reeh, Inst. für Theor. Physik, Universität Göttingen
J.E. Roberts, CNRS, Marseille
D.W. Robinson, CNRS, Marseille
M. Romerio, Institut de Physique, Neuchâtel
L. Rytel, Inst. of Theor. Physics, University of Wroclaw
C. Sachrajda, Stanford University SLAC, Stanford, USA
H. Satz, Fakultät für Physik, Universität Bielefeld
P. Scanzano, Fakultät für Physik, Universität Bielefeld
K. Schilling, Gesamthochschule Wuppertal, Wuppertal
U.E. Schröder, Inst. für Theor. Physik, Universität Frankfurt am Main
W. Schröder, Fakultät für Physik, Universität Bielefeld
F. Schumacher, Fakultät für Physik, Universität Bielefeld
L. Sertorio, Istituto di Fisica, University of Torino
Q. Shafi, Fakultät für Physik, Universität Freiburg
K. Sibold, Max-Planck-Institut für Physik, München
M. da Silva, Porto, Portugal
B.S. Skagerstam, University of St. Andrews, St. Andrews, Scotland
A.W. Smith, Dept. of Applied Mathematics, University of Cambridge
G. Sommer, Fakultät für Physik, Universität Bielefeld
P.P. Srivastava, Centro Brasileiro de Pesquisas Fisicas, Rio de Janeiro
I.O. Stamatescu, Inst. für Theor. Physik, Universität Heidelberg
J.P. Steinhardt, Coral Gables, USA

O. Steinmann, Fakultät für Physik, Universität Bielefeld
A. Stern, Chicago, Ill., USA
P. Stichel, Fakultät für Physik, Universität Bielefeld
A. Stoffel, Inst. für Theor. Physik, TH Aachen
R.F. Streater, Dept. of Mathematics, Bedford College, London
L. Streit, Fakultät für Physik, Universität Bielefeld
R.L. Stuller, Imperial College of Science and Technology, Dept. of Physics, London
E. Suhonen, Dept. of Theor. Physics, University of Oulu
I. Szczyrba, Faculty of Physics, University of Warsaw
J. Tarski, Fakultät für Physik, Universität Bielefeld
P. Tataru-Mihai, Fakultät für Physik, Gesamthochschule Wuppertal
R. Tegen, Universität Hamburg, Hamburg
G. Thomas, Argonne National Laboratory, Argonne
H.-J. Thun, Fachbereich Physik, FU Berlin
A. Tounsi, Physique Théorique, Université de Paris
P. Tsilimigras, Nuclear Research Center, Attiki, Athens
L. Turki, Inst. für Theor. Physik, Universität Karlsruhe
A. Ungkitchanukit, Royal Holloway College, London
H. Uschersohn, TFT, Helsinki
P. di Veccia, Nordita, Copenhagen
G. Velo, Università degli Studi, Bologna
I. Ventura, Universidade de São Paulo, São Paulo
H. Watanabe, Dept. of Applied Sciences, Kyushu University, Fukuoka
F. Widder, Inst. für Theor. Physik, Universität Graz
G. Wilk, Institute of Nuclear Research, Warsaw
J. Willrodt, FB Mathematik, Gesamthochschule Siegen
M. Winnink, Institut voor Theoretische Natuurkunde, Universiteitskomplex Paddepoel, Groningen
E. de Wolf, Dept. Naturkunde, Universität Antwerpen
S. Woronowicz, Dept. of Mathematics, University of Warszawa
W. Wreszinski, Instituto de Fisica, Universidade de Sao Paulo, São Paulo
D. Wright, Bedford College, London
B.C. Yunn, Fakultät für Physik, Universität Kaiserslautern
Y. Zarmi, Physics Dept., Weizmann Institute, Rehovot, Israel

INDEX